TOYOTA
TUNDRA/SEQUOIA
2000-06 REPAIR MANUAL
Deleted

CHILTON'S

Covers all U.S. and Canadian models of
2WD and 4WD Toyota Tundra 2000-2006
and Sequoia 2001-2006

by Mike Stubblefield

CHILTON Automotive Books

PUBLISHED BY **HAYNES NORTH AMERICA, Inc.**

AUTOMOTIVE
PARTS &
ACCESSORIES
ASSOCIATION MEMBER

Manufactured in USA
© 2002, 2009 Haynes North America, Inc.
ISBN-13: 978-1-56392-761-4
ISBN-10: 1-56392-761-6
Library of Congress Control No. 2009923701

Haynes Publishing Group
Sparkford Nr Yeovil
Somerset BA22 7JJ England

Haynes North America, Inc
861 Lawrence Drive
Newbury Park
California 91320 USA

ABCDE
FGHIJ
KLMNO
PQ

3M2

Contents

Mechanic, author and photographer with 2000 Toyota Tundra

ACKNOWLEDGMENTS

We are grateful for the help and cooperation of the Toyota Motor Company for their assistance with technical information and certain illustrations. Technical writers who contributed to this project include Bob Henderson, Rob Maddox, Larry Warren and John Wegmann. Technical consultants who contributed to this project include Jamie Sarté and Brad Conn.

About this manual

ITS PURPOSE

The purpose of this manual is to help you get the best value from your vehicle. It can do so in several ways. It can help you decide what work must be done, even if you choose to have it done by a dealer service department or a repair shop; it provides information and procedures for routine maintenance and servicing; and it offers diagnostic and repair procedures to follow when trouble occurs.

We hope you use the manual to tackle the work yourself. For many simpler jobs, doing it yourself may be quicker than arranging an appointment to get the vehicle into a shop and making the trips to leave it and pick it up. More importantly, a lot of money can be saved by avoiding the expense the shop must pass on to you to cover its labor and overhead costs. An added benefit is the sense of satisfaction and accomplishment that you feel after doing the job yourself.

USING THE MANUAL

The manual is divided into Chapters. Each Chapter is divided into numbered Sections. Each Section consists of consecutively numbered paragraphs.

At the beginning of each numbered Section you will be referred to any illustrations which apply to the procedures in that Section. The reference numbers used in illustration captions pinpoint the pertinent Section and the Step within that Section. That is, illustration 3.2 means the illustration refers to Section 3 and Step (or paragraph) 2 within that Section.

Procedures, once described in the text, are not normally repeated. When it's necessary to refer to another Chapter, the reference will be given as Chapter and Section number. Cross references given without use of the word "Chapter" apply to Sections and/or paragraphs in the same Chapter. For example, "see Section 8" means in the same Chapter.

References to the left or right side of the vehicle assume you are sitting in the driver's seat, facing forward.

Even though we have prepared this manual with extreme care, neither the publisher nor the author can accept responsibility for any errors in, or omissions from, the information given.

➡NOTE

A *Note* provides information necessary to properly complete a procedure or information which will make the procedure easier to understand.

✳ CAUTION

A *Caution* provides a special procedure or special steps which must be taken while completing the procedure where the Caution is found. Not heeding a Caution can result in damage to the assembly being worked on.

✳ WARNING

A *Warning* provides a special procedure or special steps which must be taken while completing the procedure where the Warning is found. Not heeding a Warning can result in personal injury.

Introduction to the Toyota Tundra and Sequoia

The Tundra was manufactured beginning in 2000 and is equipped with either a 3.4L V6 engine, a 4.0L V6 engine or a 4.7 liter V8 engine. The Sequoia was manufactured beginning in 2001 and is only available with the 4.7 liter V8 engine.

All engines are equipped with the Electronic Fuel Injection (EFI) system.

The engine drives the rear wheels through either a manual or automatic transmission via a driveshaft and solid rear axle. A transfer case and driveshaft are used to drive the front axle on 4WD models. All models are available with either 2WD or 4WD.

The front suspension is fully independent: It consists of upper and lower control arms, a stabilizer bar, and integral coil spring/shock absorber assemblies. A solid axle at the rear is suspended by leaf springs and shock absorbers (Tundra models) or coil springs and shock absorbers (Sequoia models). Sequoia models have four rear suspension arms (two upper and two lower arms) and a lateral control rod.

The steering gear is a rack-and-pinion type and is connected to the steering knuckles by tie-rods.

The front brakes are disc type on all models. The rear brakes on Tundra models are drum type; Sequoia models have rear disc brakes. Power assist is standard on all models. Some models are equipped with anti-lock brakes.

Vehicle identification numbers

Modifications are a continuing and unpublicized process in vehicle manufacturing. Since spare parts manuals and lists are compiled on a numerical basis, the individual vehicle numbers are essential to correctly identify the component required.

VEHICLE IDENTIFICATION NUMBER (VIN)

This very important identification number is stamped on a plate attached to the left side of the dashboard just inside the windshield on the driver's side of the vehicle (see illustration). The VIN also appears on the Vehicle Certificate of Title and Registration. It contains information such as where and when the vehicle was manufactured, the model year and the body style.

Counting from the left, the engine code is the eighth digit and the model year code is the tenth digit.

Model year codes:

Y	2000
1	2001
2	2002
3	2003
4	2004
5	2005
6	2006

ENGINE SERIAL NUMBER

On V6 engines, the engine serial number is located on the left side of the engine block, near the oil filter (see illustration). On V8 engines it's located on the top front of the engine block, between the cylinder heads (see illustration).

MANUFACTURER CERTIFICATION LABEL

The Manufacturer Certification label is affixed to the front door pillar. The plate contains the name of the manufacturer, the month and year of production, the Gross Vehicle Weight Rating (GVWR) and the certification statement (see illustration).

VEHICLE EMISSIONS CONTROL INFORMATION (VECI) LABEL

The emissions control information label is found under the hood. This label contains information on the emissions control equipment installed on the vehicle, as well as tune-up specifications (see illustration).

TRANSFER CASE AND TRANSMISSION IDENTIFICATION NUMBER

The transfer case and manual transmission identification number is stamped into the case of the component. Automatic transmission numbers are stamped onto ID plates (see illustration).

The Vehicle Identification Number (VIN) is visible through the driver's side of the windshield

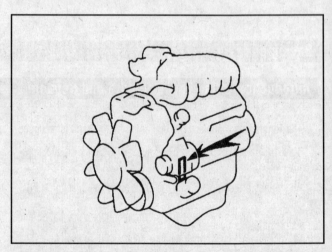

The engine serial number on the V6 engine is stamped onto a pad on the left side of the engine block, above the oil filter cartridge

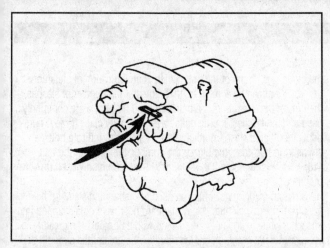

The engine serial number on the V8 engine is stamped onto a pad at the top front of the engine block, just below the front of the intake manifold

The manufacturer's certification label is affixed to the driver's side door jamb

The Vehicle Emissions Control Label (VECI) is located on the underside of the hood

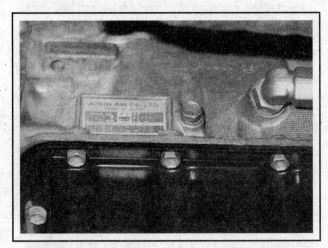

The identification number on automatic transmissions is stamped into a plate located on the side of the transmission

Buying parts

Replacement parts are available from many sources, which generally fall into one of two categories - authorized dealer parts departments and independent retail auto parts stores. Our advice concerning these parts is as follows:

Retail auto parts stores: Good auto parts stores will stock frequently needed components which wear out relatively fast, such as clutch components, exhaust systems, brake parts, tune-up parts, etc. These stores often supply new or reconditioned parts on an exchange basis, which can save a considerable amount of money. Discount auto parts stores are often very good places to buy materials and parts needed for general vehicle maintenance such as oil, grease, filters, spark plugs, belts, touch-up paint, bulbs, etc. They also usually sell tools and general accessories, have convenient hours, charge lower prices and can often be found not far from home.

Authorized dealer parts department: This is the best source for parts which are unique to the vehicle and not generally available elsewhere (such as major engine parts, transmission parts, trim pieces, etc.).

Warranty information: If the vehicle is still covered under warranty, be sure that any replacement parts purchased - regardless of the source - do not invalidate the warranty!

To be sure of obtaining the correct parts, have engine and chassis numbers available and, if possible, take the old parts along for positive identification.

MAINTENANCE TECHNIQUES

There are a number of techniques involved in maintenance and repair that will be referred to throughout this manual. Application of these techniques will enable the home mechanic to be more efficient, better organized and capable of performing the various tasks properly, which will ensure that the repair job is thorough and complete.

Fasteners

Fasteners are nuts, bolts, studs and screws used to hold two or more parts together. There are a few things to keep in mind when working with fasteners. Almost all of them use a locking device of some type, either a lockwasher, locknut, locking tab or thread adhesive. All threaded fasteners should be clean and straight, with undamaged threads and undamaged corners on the hex head where the wrench fits. Develop the habit of replacing all damaged nuts and bolts with new ones. Special locknuts with nylon or fiber inserts can only be used once. If they are removed, they lose their locking ability and must be replaced with new ones.

Rusted nuts and bolts should be treated with a penetrating fluid to ease removal and prevent breakage. Some mechanics use turpentine in a spout-type oil can, which works quite well. After applying the rust penetrant, let it work for a few minutes before trying to loosen the nut or bolt. Badly rusted fasteners may have to be chiseled or sawed off or removed with a special nut breaker, available at tool stores.

If a bolt or stud breaks off in an assembly, it can be drilled and removed with a special tool commonly available for this purpose. Most automotive machine shops can perform this task, as well as other repair procedures, such as the repair of threaded holes that have been stripped out.

Flat washers and lockwashers, when removed from an assembly, should always be replaced exactly as removed. Replace any damaged washers with new ones. Never use a lockwasher on any soft metal surface (such as aluminum), thin sheet metal or plastic.

Fastener sizes

For a number of reasons, automobile manufacturers are making wider and wider use of metric fasteners. Therefore, it is important to be able to tell the difference between standard (sometimes called U.S. or SAE) and metric hardware, since they cannot be interchanged.

All bolts, whether standard or metric, are sized according to diameter, thread pitch and length. For example, a standard 1/2 - 13 x 1 bolt is 1/2 inch in diameter, has 13 threads per inch and is 1 inch long. An M12 - 1.75 x 25 metric bolt is 12 mm in diameter, has a thread pitch of 1.75 mm (the distance between threads) and is 25 mm long. The two bolts are nearly identical, and easily confused, but they are not interchangeable.

In addition to the differences in diameter, thread pitch and length, metric and standard bolts can also be distinguished by examining the bolt heads. To begin with, the distance across the flats on a standard bolt head is measured in inches, while the same dimension on a metric bolt is sized in millimeters (the same is true for nuts). As a result, a standard wrench should not be used on a metric bolt and a metric wrench should not be used on a standard bolt. Also, most standard bolts have slashes radiating out from the center of the head to denote the grade or strength of the bolt, which is an indication of the amount of torque that can be applied to it. The greater the number of slashes, the greater the strength of the bolt. Grades 0 through 5 are commonly used on automobiles. Metric bolts have a property class (grade) number, rather than a slash, molded into their heads to indicate bolt strength. In this case, the higher the number, the stronger the bolt. Property class numbers 8.8, 9.8 and 10.9 are commonly used on automobiles.

Strength markings can also be used to distinguish standard hex nuts from metric hex nuts. Many standard nuts have dots stamped into one side, while metric nuts are marked with a number. The greater the number of dots, or the higher the number, the greater the strength of the nut.

Metric studs are also marked on their ends according to property class (grade). Larger studs are numbered (the same as metric bolts), while smaller studs carry a geometric code to denote grade.

It should be noted that many fasteners, especially Grades 0 through 2, have no distinguishing marks on them. When such is the case, the only way to determine whether it is standard or metric is to measure the thread pitch or compare it to a known fastener of the same size.

Standard fasteners are often referred to as SAE, as opposed to metric. However, it should be noted that SAE technically refers to a non-metric fine thread fastener only. Coarse thread non-metric fasteners are referred to as USS sizes.

Since fasteners of the same size (both standard and metric) may have different strength ratings, be sure to reinstall any bolts, studs or nuts removed from your vehicle in their original locations. Also, when replacing a fastener with a new one, make sure that the new one has a strength rating equal to or greater than the original.

Tightening sequences and procedures

Most threaded fasteners should be tightened to a specific torque value (torque is the twisting force applied to a threaded component such as a nut or bolt). Overtightening the fastener can weaken it and cause it to break, while undertightening can cause it to eventually come loose. Bolts, screws and studs, depending on the material they are made of and their thread diameters, have specific torque values, many of which are noted in the Specifications at the end of each Chapter. Be sure to follow the torque recommendations closely. For fasteners not assigned a specific torque, a general torque value chart is presented here as a guide. These torque values are for dry (unlubricated) fasteners threaded into steel or cast iron (not aluminum). As was previously mentioned, the size and grade of a fastener determine the amount of torque that can safely be applied to it. The figures listed here are approximate for Grade 2 and Grade 3 fasteners. Higher grades can tolerate higher torque values.

Fasteners laid out in a pattern, such as cylinder head bolts, oil pan bolts, differential cover bolts, etc., must be loosened or tightened in sequence to avoid warping the component. This sequence will normally be shown in the appropriate Chapter. If a specific pattern is not given, the following procedures can be used to prevent warping.

Initially, the bolts or nuts should be assembled finger-tight only. Next, they should be tightened one full turn each, in a criss-cross or diagonal pattern. After each one has been tightened one full turn, return to the first one and tighten them all one-half turn, following the same

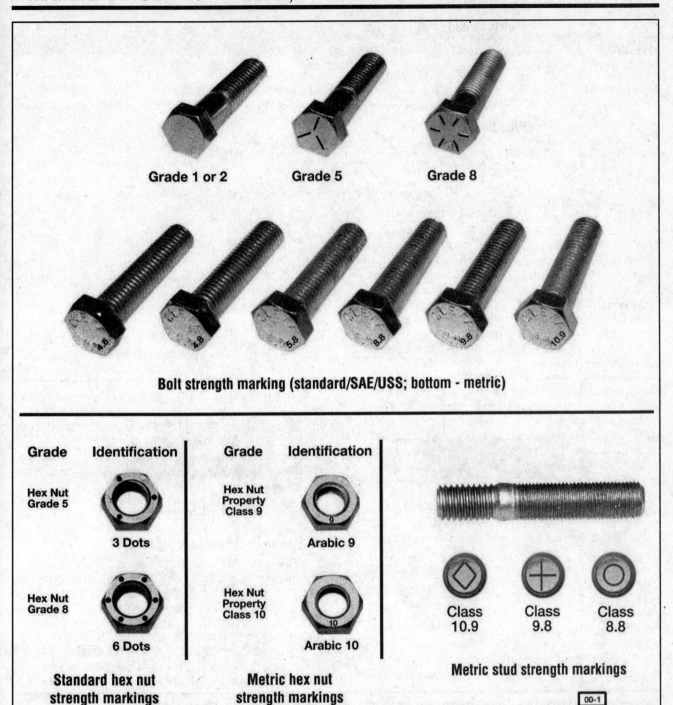

Grade 1 or 2   Grade 5   Grade 8

Bolt strength marking (standard/SAE/USS; bottom - metric)

Grade	Identification
Hex Nut Grade 5	3 Dots
Hex Nut Grade 8	6 Dots

Standard hex nut strength markings

Grade	Identification
Hex Nut Property Class 9	Arabic 9
Hex Nut Property Class 10	Arabic 10

Metric hex nut strength markings

Class 10.9   Class 9.8   Class 8.8

Metric stud strength markings

00-1

pattern. Finally, tighten each of them one-quarter turn at a time until each fastener has been tightened to the proper torque. To loosen and remove the fasteners, the procedure would be reversed.

Component disassembly

Component disassembly should be done with care and purpose to help ensure that the parts go back together properly. Always keep track of the sequence in which parts are removed. Make note of special characteristics or marks on parts that can be installed more than one way, such as a grooved thrust washer on a shaft. It is a good idea to lay the disassembled parts out on a clean surface in the order that they were removed. It may also be helpful to make sketches or take instant photos of components before removal.

When removing fasteners from a component, keep track of their locations. Sometimes threading a bolt back in a part, or putting the washers and nut back on a stud, can prevent mix-ups later. If nuts and bolts cannot be returned to their original locations, they should be kept in a compartmented box or a series of small boxes. A cupcake or muffin tin is ideal for this purpose, since each cavity can hold the bolts and nuts from a particular area (i.e. oil pan bolts, valve cover bolts, engine

Metric thread sizes

	Ft-lbs	Nm
M-6	6 to 9	9 to 12
M-8	14 to 21	19 to 28
M-10	28 to 40	38 to 54
M-12	50 to 71	68 to 96
M-14	80 to 140	109 to 154

Pipe thread sizes

1/8	5 to 8	7 to 10
1/4	12 to 18	17 to 24
3/8	22 to 33	30 to 44
1/2	25 to 35	34 to 47

U.S. thread sizes

1/4 - 20	6 to 9	9 to 12
5/16 - 18	12 to 18	17 to 24
5/16 - 24	14 to 20	19 to 27
3/8 - 16	22 to 32	30 to 43
3/8 - 24	27 to 38	37 to 51
7/16 - 14	40 to 55	55 to 74
7/16 - 20	40 to 60	55 to 81
1/2 - 13	55 to 80	75 to 108

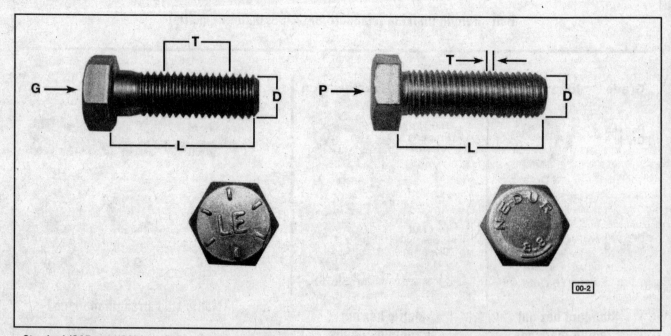

00-2

Standard (SAE and USS) bolt dimensions/grade marks

G   Grade marks (bolt strength)
L   Length (in inches)
T   Thread pitch (number of threads per inch)
D   Nominal diameter (in inches)

Metric bolt dimensions/grade marks

P   Property class (bolt strength)
L   Length (in millimeters)
T   Thread pitch (distance between threads in millimeters)
D   Diameter

mount bolts, etc.). A pan of this type is especially helpful when working on assemblies with very small parts, such as the carburetor, alternator, valve train or interior dash and trim pieces. The cavities can be marked with paint or tape to identify the contents.

Whenever wiring looms, harnesses or connectors are separated, it is a good idea to identify the two halves with numbered pieces of masking tape so they can be easily reconnected.

Gasket sealing surfaces

Throughout any vehicle, gaskets are used to seal the mating surfaces between two parts and keep lubricants, fluids, vacuum or pressure contained in an assembly.

Many times these gaskets are coated with a liquid or paste-type gasket sealing compound before assembly. Age, heat and pressure can sometimes cause the two parts to stick together so tightly that they are

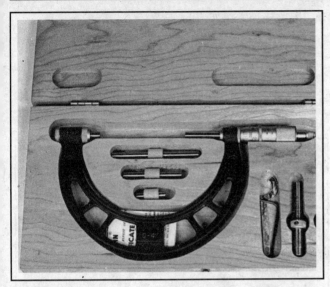

Micrometer set

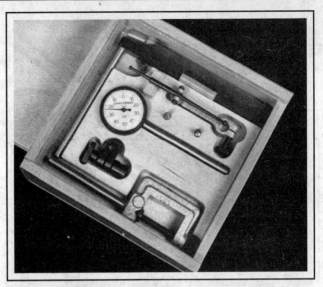

Dial indicator set

very difficult to separate. Often, the assembly can be loosened by striking it with a soft-face hammer near the mating surfaces. A regular hammer can be used if a block of wood is placed between the hammer and the part. Do not hammer on cast parts or parts that could be easily damaged. With any particularly stubborn part, always recheck to make sure that every fastener has been removed.

Avoid using a screwdriver or bar to pry apart an assembly, as they can easily mar the gasket sealing surfaces of the parts, which must remain smooth. If prying is absolutely necessary, use an old broom handle, but keep in mind that extra clean up will be necessary if the wood splinters.

After the parts are separated, the old gasket must be carefully scraped off and the gasket surfaces cleaned. Stubborn gasket material can be soaked with rust penetrant or treated with a special chemical to soften it so it can be easily scraped off. A scraper can be fashioned from a piece of copper tubing by flattening and sharpening one end. Copper is recommended because it is usually softer than the surfaces to be scraped, which reduces the chance of gouging the part. Some gaskets can be removed with a wire brush, but regardless of the method used, the mating surfaces must be left clean and smooth. If for some reason the gasket surface is gouged, then a gasket sealer thick enough to fill scratches will have to be used during reassembly of the components. For most applications, a non-drying (or semi-drying) gasket sealer should be used.

Hose removal tips

❋❋ WARNING:

If the vehicle is equipped with air conditioning, do not disconnect any of the A/C hoses without first having the system depressurized by a dealer service department or a service station.

Hose removal precautions closely parallel gasket removal precautions. Avoid scratching or gouging the surface that the hose mates against or the connection may leak. This is especially true for radiator hoses. Because of various chemical reactions, the rubber in hoses can bond itself to the metal spigot that the hose fits over. To remove a hose, first loosen the hose clamps that secure it to the spigot. Then, with slip-joint pliers, grab the hose at the clamp and rotate it around the spigot. Work it back and forth until it is completely free, then pull it off. Silicone or other lubricants will ease removal if they can be applied between the hose and the outside of the spigot. Apply the same lubricant to the inside of the hose and the outside of the spigot to simplify installation.

As a last resort (and if the hose is to be replaced with a new one anyway), the rubber can be slit with a knife and the hose peeled from the spigot. If this must be done, be careful that the metal connection is not damaged.

If a hose clamp is broken or damaged, do not reuse it. Wire-type clamps usually weaken with age, so it is a good idea to replace them with screw-type clamps whenever a hose is removed.

TOOLS

A selection of good tools is a basic requirement for anyone who plans to maintain and repair his or her own vehicle. For the owner who has few tools, the initial investment might seem high, but when compared to the spiraling costs of professional auto maintenance and repair, it is a wise one.

To help the owner decide which tools are needed to perform the tasks detailed in this manual, the following tool lists are offered: *Maintenance and minor repair, Repair/overhaul* and *Special*.

The newcomer to practical mechanics should start off with the *maintenance and minor repair* tool kit, which is adequate for the simpler jobs performed on a vehicle. Then, as confidence and experience grow, the owner can tackle more difficult tasks, buying additional tools as they are needed. Eventually the basic kit will be expanded into the *repair and overhaul* tool set. Over a period of time, the experienced do-it-yourselfer will assemble a tool set complete enough for most repair and overhaul procedures and will add tools from the special category when it is felt that the expense is justified by the frequency of use.

Maintenance and minor repair tool kit

The tools in this list should be considered the minimum required for performance of routine maintenance, servicing and minor repair work. We recommend the purchase of combination wrenches (box-end and open-end combined in one wrench). While more expensive than open end wrenches, they offer the advantages of both types of wrench.

Combination wrench set (1/4-inch to 1 inch or 6 mm to 19 mm)

Adjustable wrench, 8 inch
Spark plug wrench with rubber insert
Spark plug gap adjusting tool
Feeler gauge set
Brake bleeder wrench
Standard screwdriver (5/16-inch x 6 inch)
Phillips screwdriver (No. 2 x 6 inch)
Combination pliers - 6 inch

Hacksaw and assortment of blades
Tire pressure gauge
Grease gun
Oil can
Fine emery cloth
Wire brush
Battery post and cable cleaning tool
Oil filter wrench

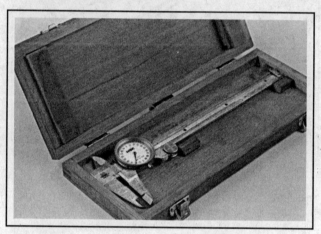

Dial caliper

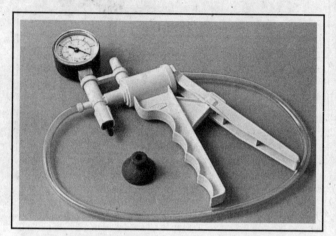

Hand-operated vacuum pump

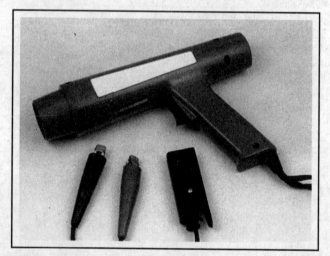

Timing light

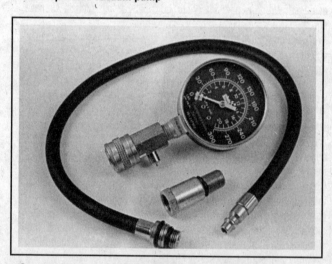

Compression gauge with spark plug hole adapter

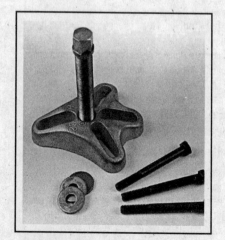

Damper/steering wheel puller

General purpose puller

Hydraulic lifter removal tool

Funnel (medium size)
Safety goggles
Jackstands (2)
Drain pan

➡**Note: If basic tune-ups are going to be part of routine maintenance, it will be necessary to purchase a good quality stroboscopic timing light and combination tachometer/dwell meter. Although they are included in the list of special tools, it is mentioned here because they are absolutely necessary for tuning most vehicles properly.**

Repair and overhaul tool set

These tools are essential for anyone who plans to perform major repairs and are in addition to those in the maintenance and minor repair tool kit. Included is a comprehensive set of sockets which, though expensive, are invaluable because of their versatility, especially when various extensions and drives are available. We recommend the 1/2-inch drive over the 3/8-inch drive. Although the larger drive is bulky and more expensive, it has the capacity of accepting a very wide range of large sockets. Ideally, however, the mechanic should have a 3/8-inch drive set and a 1/2-inch drive set.

Valve spring compressor

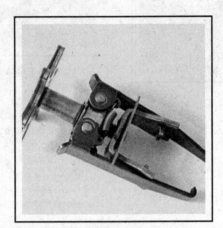

Valve spring compressor

Ridge reamer

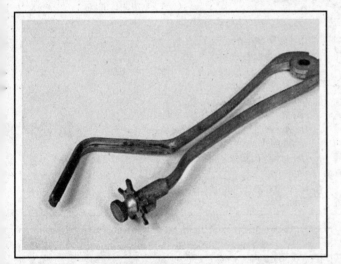

Piston ring groove cleaning tool

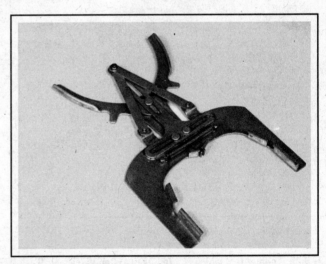

Ring removal/installation tool

Ring compressor

Cylinder hone

Brake hold-down spring tool

Torque angle gauge

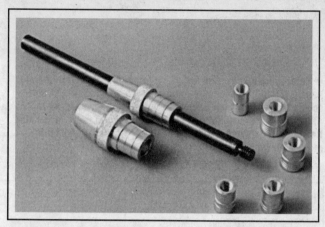

Clutch plate alignment tool

Socket set(s)
Reversible ratchet
Extension - 10 inch
Universal joint
Torque wrench (same size drive as sockets)
Ball peen hammer - 8 ounce
Soft-face hammer (plastic/rubber)
Standard screwdriver (1/4-inch x 6 inch)
Standard screwdriver (stubby - 5/16-inch)
Phillips screwdriver (No. 3 x 8 inch)
Phillips screwdriver (stubby - No. 2)
Pliers - vise grip
Pliers - lineman's
Pliers - needle nose
Pliers - snap-ring (internal and external)
Cold chisel - 1/2-inch
Scribe
Scraper (made from flattened copper tubing)
Centerpunch
Pin punches (1/16, 1/8, 3/16-inch)
Steel rule/straightedge - 12 inch
Allen wrench set (1/8 to 3/8-inch or 4 mm to 10 mm)
A selection of files
Wire brush (large)
Jackstands (second set)
Jack (scissor or hydraulic type)

➡**Note: Another tool which is often useful is an electric drill with a chuck capacity of 3/8-inch and a set of good quality drill bits.**

Special tools

The tools in this list include those which are not used regularly, are expensive to buy, or which need to be used in accordance with their manufacturer's instructions. Unless these tools will be used frequently, it is not very economical to purchase many of them. A consideration would be to split the cost and use between yourself and a friend or friends. In addition, most of these tools can be obtained from a tool rental shop on a temporary basis.

This list primarily contains only those tools and instruments widely available to the public, and not those special tools produced by the vehicle manufacturer for distribution to dealer service departments. Occasionally, references to the manufacturer's special tools are included in the text of this manual. Generally, an alternative method of doing the

job without the special tool is offered. However, sometimes there is no alternative to their use. Where this is the case, and the tool cannot be purchased or borrowed, the work should be turned over to the dealer service department or an automotive repair shop.

Valve spring compressor
Piston ring groove cleaning tool
Piston ring compressor
Piston ring installation tool
Cylinder compression gauge
Cylinder ridge reamer
Cylinder surfacing hone
Cylinder bore gauge
Micrometers and/or dial calipers
Hydraulic lifter removal tool
Balljoint separator
Universal-type puller
Impact screwdriver
Dial indicator set
Stroboscopic timing light (inductive pick-up)
Hand operated vacuum/pressure pump
Tachometer/dwell meter
Universal electrical multimeter
Cable hoist
Brake spring removal and installation tools
Floor jack

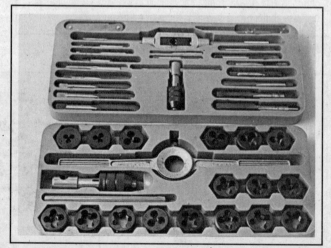

Tap and die set

Buying tools

For the do-it-yourselfer who is just starting to get involved in vehicle maintenance and repair, there are a number of options available when purchasing tools. If maintenance and minor repair is the extent of the work to be done, the purchase of individual tools is satisfactory. If, on the other hand, extensive work is planned, it would be a good idea to purchase a modest tool set from one of the large retail chain stores. A set can usually be bought at a substantial savings over the individual tool prices, and they often come with a tool box. As additional tools are needed, add-on sets, individual tools and a larger tool box can be purchased to expand the tool selection. Building a tool set gradually allows the cost of the tools to be spread over a longer period of time and gives the mechanic the freedom to choose only those tools that will actually be used.

Tool stores will often be the only source of some of the special tools that are needed, but regardless of where tools are bought, try to avoid cheap ones, especially when buying screwdrivers and sockets, because they won't last very long. The expense involved in replacing cheap tools will eventually be greater than the initial cost of quality tools.

Care and maintenance of tools

Good tools are expensive, so it makes sense to treat them with respect. Keep them clean and in usable condition and store them properly when not in use. Always wipe off any dirt, grease or metal chips before putting them away. Never leave tools lying around in the work area. Upon completion of a job, always check closely under the hood for tools that may have been left there so they won't get lost during a test drive.

Some tools, such as screwdrivers, pliers, wrenches and sockets, can be hung on a panel mounted on the garage or workshop wall, while others should be kept in a tool box or tray. Measuring instruments, gauges, meters, etc. must be carefully stored where they cannot be damaged by weather or impact from other tools.

When tools are used with care and stored properly, they will last a very long time. Even with the best of care, though, tools will wear out if used frequently. When a tool is damaged or worn out, replace it. Subsequent jobs will be safer and more enjoyable if you do.

HOW TO REPAIR DAMAGED THREADS

Sometimes, the internal threads of a nut or bolt hole can become stripped, usually from overtightening. Stripping threads is an all-too-common occurrence, especially when working with aluminum parts, because aluminum is so soft that it easily strips out.

Usually, external or internal threads are only partially stripped. After they've been cleaned up with a tap or die, they'll still work. Sometimes, however, threads are badly damaged. When this happens, you've got three choices:

1) Drill and tap the hole to the next suitable oversize and install a larger diameter bolt, screw or stud.

2) Drill and tap the hole to accept a threaded plug, then drill and tap the plug to the original screw size. You can also buy a plug already threaded to the original size. Then you simply drill a hole to the specified size, then run the threaded plug into the hole with a bolt and jam nut. Once the plug is fully seated, remove the jam nut and bolt.

3) The third method uses a patented thread repair kit like Heli-Coil or Slimsert. These easy-to-use kits are designed to repair damaged threads in straight-through holes and blind holes. Both are available as kits which can handle a variety of sizes and thread patterns. Drill the hole, then tap it with the special included tap. Install the Heli-Coil and the hole is back to its original diameter and thread pitch.

Regardless of which method you use, be sure to proceed calmly and carefully. A little impatience or carelessness during one of these relatively simple procedures can ruin your whole day's work and cost you a bundle if you wreck an expensive part.

WORKING FACILITIES

Not to be overlooked when discussing tools is the workshop. If anything more than routine maintenance is to be carried out, some sort of suitable work area is essential.

It is understood, and appreciated, that many home mechanics do not have a good workshop or garage available, and end up removing an engine or doing major repairs outside. It is recommended, however, that the overhaul or repair be completed under the cover of a roof.

A clean, flat workbench or table of comfortable working height is an absolute necessity. The workbench should be equipped with a vise that has a jaw opening of at least four inches.

As mentioned previously, some clean, dry storage space is also required for tools, as well as the lubricants, fluids, cleaning solvents, etc. which soon become necessary.

Sometimes waste oil and fluids, drained from the engine or cooling system during normal maintenance or repairs, present a disposal problem. To avoid pouring them on the ground or into a sewage system, pour the used fluids into large containers, seal them with caps and take them to an authorized disposal site or recycling center. Plastic jugs, such as old antifreeze containers, are ideal for this purpose.

Always keep a supply of old newspapers and clean rags available. Old towels are excellent for mopping up spills. Many mechanics use rolls of paper towels for most work because they are readily available and disposable. To help keep the area under the vehicle clean, a large cardboard box can be cut open and flattened to protect the garage or shop floor.

Whenever working over a painted surface, such as when leaning over a fender to service something under the hood, always cover it with an old blanket or bedspread to protect the finish. Vinyl covered pads, made especially for this purpose, are available at auto parts stores.

Jacking and towing

JACKING

✽✽ WARNING:

The jack supplied with the vehicle should only be used for changing a tire or placing jackstands under the frame. Never work under the vehicle or start the engine while this jack is being used as the only means of support.

The vehicle should be on level ground with the hazard flashers on, the wheels blocked, the parking brake applied and the transmission in Park (automatic) or Reverse (manual). If a tire is being changed, loosen the lug nuts one-half turn and leave them in place until the wheel is raised off the ground.

Place the jack under the vehicle (see illustrations):

Front: Under the frame rail, where the crossmember is attached.

Rear: Under the rear axle housing.

Operate the jack with a slow, smooth motion until the wheel is raised off the ground. Remove the lug nuts, pull off the wheel, install the spare and thread the lug nuts back on with the beveled sides facing in. Tighten them snugly, but wait until the vehicle is lowered to tighten them completely.

Lower the vehicle, remove the jack and tighten the nuts (if loosened or removed) in a criss-cross pattern.

TOWING

As a general rule, the vehicle should be towed with professional towing equipment. If towed from the front, the rear wheels should be placed on a towing dolly. If a 4WD model is towed from the rear, the front wheels should be placed on a towing dolly; if a dolly is not available, turn the ignition key to the ACC position, place the transfer case in the 2H position and the transmission in Neutral.

When a vehicle is towed with the rear wheels raised, the steering wheel must be clamped in the straight ahead position with a special device designed for use during towing. The ignition key must not be in the LOCK position, since the steering lock mechanism isn't strong enough to hold the front wheels straight while towing.

Equipment specifically designed for towing should be used. It should be attached to the main structural members of the vehicle, not the bumpers or brackets. Safety is a major consideration when towing and all applicable state and local laws must be obeyed. A safety chain system must be used at all times. Remember that power steering and power brakes will not work with the engine off.

Front jacking point

Rear jacking point

Booster battery (jump) starting

Observe these precautions when using a booster battery to start a vehicle:

a) Before connecting the booster battery, make sure the ignition switch is in the Off position.

b) Turn off the lights, heater and other electrical loads.

c) Your eyes should be shielded. Safety goggles are a good idea.

d) Make sure the booster battery is the same voltage as the dead one in the vehicle.

e) The two vehicles MUST NOT TOUCH each other!

f) Make sure the transaxle is in Neutral (manual) or Park (automatic).

g) If the booster battery is not a maintenance-free type, remove the vent caps and lay a cloth over the vent holes.

Connect the red jumper cable to the positive (+) terminals of each battery (see illustration).

Connect one end of the black jumper cable to the negative (-) terminal of the booster battery. The other end of this cable should be connected to a good ground on the vehicle to be started, such as a bolt or bracket on the body.

Start the engine using the booster battery, then, with the engine running at idle speed, disconnect the jumper cables in the reverse order of connection.

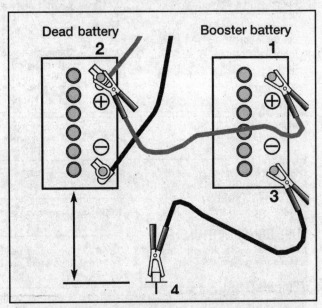

Make the booster battery cable connections in the numerical order shown (note that the negative cable of the booster battery is NOT attached to the negative terminal of the dead battery)

CONVERSION FACTORS

LENGTH (distance)

Inches (in)	X	25.4	= Millimeters (mm)	X 0.0394	= Inches (in)
Feet (ft)	X	0.305	= Meters (m)	X 3.281	= Feet (ft)
Miles	X	1.609	= Kilometers (km)	X 0.621	= Miles

VOLUME (capacity)

Cubic inches (cu in; in^3)	X	16.387	= Cubic centimeters (cc; cm^3)	X 0.061	= Cubic inches (cu in; in^3)
Imperial pints (Imp pt)	X	0.568	= Liters (l)	X 1.76	= Imperial pints (Imp pt)
Imperial quarts (Imp qt)	X	1.137	= Liters (l)	X 0.88	= Imperial quarts (Imp qt)
Imperial quarts (Imp qt)	X	1.201	= US quarts (US qt)	X 0.833	= Imperial quarts (Imp qt)
US quarts (US qt)	X	0.946	= Liters (l)	X 1.057	= US quarts (US qt)
Imperial gallons (Imp gal)	X	4.546	= Liters (l)	X 0.22	= Imperial gallons (Imp gal)
Imperial gallons (Imp gal)	X	1.201	= US gallons (US gal)	X 0.833	= Imperial gallons (Imp gal)
US gallons (US gal)	X	3.785	= Liters (l)	X 0.264	= US gallons (US gal)

MASS (weight)

Ounces (oz)	X	28.35	= Grams (g)	X 0.035	= Ounces (oz)
Pounds (lb)	X	0.454	= Kilograms (kg)	X 2.205	= Pounds (lb)

FORCE

Ounces-force (ozf; oz)	X	0.278	= Newtons (N)	X 3.6	= Ounces-force (ozf; oz)
Pounds-force (lbf; lb)	X	4.448	= Newtons (N)	X 0.225	= Pounds-force (lbf; lb)
Newtons (N)	X	0.1	= Kilograms-force (kgf; kg)	X 9.81	= Newtons (N)

PRESSURE

Pounds-force per square inch (psi; lbf/in^2; lb/in^2)	X	0.070	= Kilograms-force per square centimeter (kgf/cm^2; kg/cm^2)	X 14.223	= Pounds-force per square inch (psi; lbf/in^2; lb/in^2)
Pounds-force per square inch (psi; lbf/in^2; lb/in^2)	X	0.068	= Atmospheres (atm)	X 14.696	= Pounds-force per square inch (psi; lbf/in^2; lb/in^2)
Pounds-force per square inch (psi; lbf/in^2; lb/in^2)	X	0.069	= Bars	X 14.5	= Pounds-force per square inch (psi; lbf/in^2; lb/in^2)
Pounds-force per square inch (psi; lbf/in^2; lb/in^2)	X	6.895	= Kilopascals (kPa)	X 0.145	= Pounds-force per square inch (psi; lbf/in^2; lb/in^2)
Kilopascals (kPa)	X	0.01	= Kilograms-force per square centimeter (kgf/cm^2; kg/cm^2)	X 98.1	= Kilopascals (kPa)

TORQUE (moment of force)

Pounds-force inches (lbf in; lb in)	X	1.152	= Kilograms-force centimeter (kgf cm; kg cm)	X 0.868	= Pounds-force inches (lbf in; lb in)
Pounds-force inches (lbf in; lb in)	X	0.113	= Newton meters (Nm)	X 8.85	= Pounds-force inches (lbf in; lb in)
Pounds-force inches (lbf in; lb in)	X	0.083	= Pounds-force feet (lbf ft; lb ft)	X 12	= Pounds-force inches (lbf in; lb in)
Pounds-force feet (lbf ft; lb ft)	X	0.138	= Kilograms-force meters (kgf m; kg m)	X 7.233	= Pounds-force feet (lbf ft; lb ft)
Pounds-force feet (lbf ft; lb ft)	X	1.356	= Newton meters (Nm)	X 0.738	= Pounds-force feet (lbf ft; lb ft)
Newton meters (Nm)	X	0.102	= Kilograms-force meters (kgf m; kg m)	X 9.804	= Newton meters (Nm)

VACUUM

Inches mercury (in. Hg)	X	3.377	= Kilopascals (kPa)	X 0.2961	= Inches mercury
Inches mercury (in. Hg)	X	25.4	= Millimeters mercury (mm Hg)	X 0.0394	= Inches mercury

POWER

Horsepower (hp)	X	745.7	= Watts (W)	X 0.0013	= Horsepower (hp)

VELOCITY (speed)

Miles per hour (miles/hr; mph)	X	1.609	= Kilometers per hour (km/hr; kph)	X 0.621	= Miles per hour (miles/hr; mph)

FUEL CONSUMPTION *

Miles per gallon, Imperial (mpg)	X	0.354	= Kilometers per liter (km/l)	X 2.825	= Miles per gallon, Imperial (mpg)
Miles per gallon, US (mpg)	X	0.425	= Kilometers per liter (km/l)	X 2.352	= Miles per gallon, US (mpg)

TEMPERATURE

Degrees Fahrenheit = (°C x 1.8) + 32   Degrees Celsius (Degrees Centigrade; °C) = (°F - 32) x 0.56

*It is common practice to convert from miles per gallon (mpg) to liters/100 kilometers (l/100km), where mpg (Imperial) x l/100 km = 282 and mpg (US) x l/100 km = 235

FRACTION/DECIMAL/MILLIMETER EQUIVALENTS

DECIMALS TO MILLIMETERS

Decimal	mm	Decimal	mm
0.001	0.0254	0.500	12.7000
0.002	0.0508	0.510	12.9540
0.003	0.0762	0.520	13.2080
0.004	0.1016	0.530	13.4620
0.005	0.1270	0.540	13.7160
0.006	0.1524	0.550	13.9700
0.007	0.1778	0.560	14.2240
0.008	0.2032	0.570	14.4780
0.009	0.2286	0.580	14.7320
		0.590	14.9860
0.010	0.2540		
0.020	0.5080		
0.030	0.7620		
0.040	1.0160	0.600	15.2400
0.050	1.2700	0.610	15.4940
0.060	1.5240	0.620	15.7480
0.070	1.7780	0.630	16.0020
0.080	2.0320	0.640	16.2560
0.090	2.2860	0.650	16.5100
		0.660	16.7640
0.100	2.5400	0.670	17.0180
0.110	2.7940	0.680	17.2720
0.120	3.0480	0.690	17.5260
0.130	3.3020		
0.140	3.5560		
0.150	3.8100	0.700	17.7800
0.160	4.0640	0.710	18.0340
0.170	4.3180	0.720	18.2880
0.180	4.5720	0.730	18.5420
0.190	4.8260	0.740	18.7960
		0.750	19.0500
0.200	5.0800	0.760	19.3040
0.210	5.3340	0.770	19.5580
0.220	5.5880	0.780	19.8120
0.230	5.8420	0.790	20.0660
0.240	6.0960		
0.250	6.3500		
0.260	6.6040	0.800	20.3200
0.270	6.8580	0.810	20.5740
0.280	7.1120	0.820	21.8280
0.290	7.3660	0.830	21.0820
0.300	7.6200	0.840	21.3360
0.310	7.8740	0.850	21.5900
0.320	8.1280	0.860	21.8440
0.330	8.3820	0.870	22.0980
0.340	8.6360	0.880	22.3520
0.350	8.8900	0.890	22.6060
0.360	9.1440		
0.370	9.3980		
0.380	9.6520		
0.390	9.9060	0.900	22.8600
0.400	10.1600	0.910	23.1140
0.410	10.4140	0.920	23.3680
0.420	10.6680	0.930	23.6220
0.430	10.9220	0.940	23.8760
0.440	11.1760	0.950	24.1300
0.450	11.4300	0.960	24.3840
0.460	11.6840	0.970	24.6380
0.470	11.9380	0.980	24.8920
0.480	12.1920	0.990	25.1460
0.490	12.4460	1.000	25.4000

FRACTIONS TO DECIMALS TO MILLIMETERS

Fraction	Decimal	mm	Fraction	Decimal	mm
1/64	0.0156	0.3969	33/64	0.5156	13.0969
1/32	0.0312	0.7938	17/32	0.5312	13.4938
3/64	0.0469	1.1906	35/64	0.5469	13.8906
1/16	0.0625	1.5875	9/16	0.5625	14.2875
5/64	0.0781	1.9844	37/64	0.5781	14.6844
3/32	0.0938	2.3812	19/32	0.5938	15.0812
7/64	0.1094	2.7781	39/64	0.6094	15.4781
1/8	0.1250	3.1750	5/8	0.6250	15.8750
9/64	0.1406	3.5719	41/64	0.6406	16.2719
5/32	0.1562	3.9688	21/32	0.6562	16.6688
11/64	0.1719	4.3656	43/64	0.6719	17.0656
3/16	0.1875	4.7625	11/16	0.6875	17.4625
13/64	0.2031	5.1594	45/64	0.7031	17.8594
7/32	0.2188	5.5562	23/32	0.7188	18.2562
15/64	0.2344	5.9531	47/64	0.7344	18.6531
1/4	0.2500	6.3500	3/4	0.7500	19.0500
17/64	0.2656	6.7469	49/64	0.7656	19.4469
9/32	0.2812	7.1438	25/32	0.7812	19.8438
19/64	0.2969	7.5406	51/64	0.7969	20.2406
5/16	0.3125	7.9375	13/16	0.8125	20.6375
21/64	0.3281	8.3344	53/64	0.8281	21.0344
11/32	0.3438	8.7312	27/32	0.8438	21.4312
23/64	0.3594	9.1281	55/64	0.8594	21.8281
3/8	0.3750	9.5250	7/8	0.8750	22.2250
25/64	0.3906	9.9219	57/64	0.8906	22.6219
13/32	0.4062	10.3188	29/32	0.9062	23.0188
27/64	0.4219	10.7156	59/64	0.9219	23.4156
7/16	0.4375	11.1125	15/16	0.9375	23.8125
29/64	0.4531	11.5094	61/64	0.9531	24.2094
15/32	0.4688	11.9062	31/32	0.9688	24.6062
31/64	0.4844	12.3031	63/64	0.9844	25.0031
1/2	0.5000	12.7000	1	1.0000	25.4000

Automotive chemicals and lubricants

A number of automotive chemicals and lubricants are available for use during vehicle maintenance and repair. They include a wide variety of products ranging from cleaning solvents and degreasers to lubricants and protective sprays for rubber, plastic and vinyl.

CLEANERS

Carburetor cleaner and choke cleaner is a strong solvent for gum, varnish and carbon. Most carburetor cleaners leave a dry-type lubricant film which will not harden or gum up. Because of this film it is not recommended for use on electrical components.

Brake system cleaner is used to remove brake dust, grease and brake fluid from the brake system, where clean surfaces are absolutely necessary. It leaves no residue and often eliminates brake squeal caused by contaminants.

Electrical cleaner removes oxidation, corrosion and carbon deposits from electrical contacts, restoring full current flow. It can also be used to clean spark plugs, carburetor jets, voltage regulators and other parts where an oil-free surface is desired.

Demoisturants remove water and moisture from electrical components such as alternators, voltage regulators, electrical connectors and fuse blocks. They are non-conductive and non-corrosive.

Degreasers are heavy-duty solvents used to remove grease from the outside of the engine and from chassis components. They can be sprayed or brushed on and, depending on the type, are rinsed off either with water or solvent.

LUBRICANTS

Motor oil is the lubricant formulated for use in engines. It normally contains a wide variety of additives to prevent corrosion and reduce foaming and wear. Motor oil comes in various weights (viscosity ratings) from 0 to 50. The recommended weight of the oil depends on the season, temperature and the demands on the engine. Light oil is used in cold climates and under light load conditions. Heavy oil is used in hot climates and where high loads are encountered. Multi-viscosity oils are designed to have characteristics of both light and heavy oils and are available in a number of weights from 0W-20 to 20W-50.

Gear oil is designed to be used in differentials, manual transmissions and other areas where high-temperature lubrication is required.

Chassis and wheel bearing grease is a heavy grease used where increased loads and friction are encountered, such as for wheel bearings, balljoints, tie-rod ends and universal joints.

High-temperature wheel bearing grease is designed to withstand the extreme temperatures encountered by wheel bearings in disc brake equipped vehicles. It usually contains molybdenum disulfide (moly), which is a dry-type lubricant.

White grease is a heavy grease for metal-to-metal applications where water is a problem. White grease stays soft under both low and high temperatures (usually from -100 to +190-degrees F), and will not wash off or dilute in the presence of water.

Assembly lube is a special extreme pressure lubricant, usually containing moly, used to lubricate high-load parts (such as main and rod bearings and cam lobes) for initial start-up of a new engine. The assembly lube lubricates the parts without being squeezed out or washed away until the engine oiling system begins to function.

Silicone lubricants are used to protect rubber, plastic, vinyl and nylon parts.

Graphite lubricants are used where oils cannot be used due to contamination problems, such as in locks. The dry graphite will lubricate metal parts while remaining uncontaminated by dirt, water, oil or acids. It is electrically conductive and will not foul electrical contacts in locks such as the ignition switch.

Moly penetrants loosen and lubricate frozen, rusted and corroded fasteners and prevent future rusting or freezing.

Heat-sink grease is a special electrically non-conductive grease that is used for mounting electronic ignition modules where it is essential that heat is transferred away from the module.

SEALANTS

RTV sealant is one of the most widely used gasket compounds. Made from silicone, RTV is air curing, it seals, bonds, waterproofs, fills surface irregularities, remains flexible, doesn't shrink, is relatively easy to remove, and is used as a supplementary sealer with almost all low and medium temperature gaskets.

Anaerobic sealant is much like RTV in that it can be used either to seal gaskets or to form gaskets by itself. It remains flexible, is solvent resistant and fills surface imperfections. The difference between an anaerobic sealant and an RTV-type sealant is in the curing. RTV cures when exposed to air, while an anaerobic sealant cures only in the absence of air. This means that an anaerobic sealant cures only after the assembly of parts, sealing them together.

Thread and pipe sealant is used for sealing hydraulic and pneumatic fittings and vacuum lines. It is usually made from a Teflon compound, and comes in a spray, a paint-on liquid and as a wrap-around tape.

CHEMICALS

Anti-seize compound prevents seizing, galling, cold welding, rust and corrosion in fasteners. High-temperature anti-seize, usually made with copper and graphite lubricants, is used for exhaust system and exhaust manifold bolts.

Anaerobic locking compounds are used to keep fasteners from vibrating or working loose and cure only after installation, in the absence of air. Medium strength locking compound is used for small nuts, bolts and screws that may be removed later. High-strength locking compound is for large nuts, bolts and studs which aren't removed on a regular basis.

Oil additives range from viscosity index improvers to chemical treatments that claim to reduce internal engine friction. It should be noted that most oil manufacturers caution against using additives with their oils.

Gas additives perform several functions, depending on their chemical makeup. They usually contain solvents that help dissolve gum and varnish that build up on carburetor, fuel injection and intake parts. They also serve to break down carbon deposits that form on the inside surfaces of the combustion chambers. Some additives contain upper cylinder lubricants for valves and piston rings, and others contain chemicals to remove condensation from the gas tank.

MISCELLANEOUS

Brake fluid is specially formulated hydraulic fluid that can withstand the heat and pressure encountered in brake systems. Care must be taken so this fluid does not come in contact with painted surfaces or plastics. An opened container should always be resealed to prevent contamination by water or dirt.

Weatherstrip adhesive is used to bond weatherstripping around doors, windows and trunk lids. It is sometimes used to attach trim pieces.

Undercoating is a petroleum-based, tar-like substance that is designed to protect metal surfaces on the underside of the vehicle from corrosion. It also acts as a sound-deadening agent by insulating the bottom of the vehicle.

Waxes and polishes are used to help protect painted and plated surfaces from the weather. Different types of paint may require the use of different types of wax and polish. Some polishes utilize a chemical or abrasive cleaner to help remove the top layer of oxidized (dull) paint on older vehicles. In recent years many non-wax polishes that contain a wide variety of chemicals such as polymers and silicones have been introduced. These non-wax polishes are usually easier to apply and last longer than conventional waxes and polishes.

Safety first!

Regardless of how enthusiastic you may be about getting on with the job at hand, take the time to ensure that your safety is not jeopardized. A moment's lack of attention can result in an accident, as can failure to observe certain simple safety precautions. The possibility of an accident will always exist, and the following points should not be considered a comprehensive list of all dangers. Rather, they are intended to make you aware of the risks and to encourage a safety conscious approach to all work you carry out on your vehicle.

ESSENTIAL DOS AND DON'TS

DON'T rely on a jack when working under the vehicle. Always use approved jackstands to support the weight of the vehicle and place them under the recommended lift or support points.

DON'T attempt to loosen extremely tight fasteners (i.e. wheel lug nuts) while the vehicle is on a jack - it may fall.

DON'T start the engine without first making sure that the transmission is in Neutral (or Park where applicable) and the parking brake is set.

DON'T remove the radiator cap from a hot cooling system - let it cool or cover it with a cloth and release the pressure gradually.

DON'T attempt to drain the engine oil until you are sure it has cooled to the point that it will not burn you.

DON'T touch any part of the engine or exhaust system until it has cooled sufficiently to avoid burns.

DON'T siphon toxic liquids such as gasoline, antifreeze and brake fluid by mouth, or allow them to remain on your skin.

DON'T inhale brake lining dust - it is potentially hazardous (see Asbestos below).

DON'T allow spilled oil or grease to remain on the floor - wipe it up before someone slips on it.

DON'T use loose fitting wrenches or other tools which may slip and cause injury.

DON'T push on wrenches when loosening or tightening nuts or bolts. Always try to pull the wrench toward you. If the situation calls for pushing the wrench away, push with an open hand to avoid scraped knuckles if the wrench should slip.

DON'T attempt to lift a heavy component alone - get someone to help you.

DON'T rush or take unsafe shortcuts to finish a job.

DON'T allow children or animals in or around the vehicle while you are working on it.

DO wear eye protection when using power tools such as a drill, sander, bench grinder, etc. and when working under a vehicle.

DO keep loose clothing and long hair well out of the way of moving parts.

DO make sure that any hoist used has a safe working load rating adequate for the job.

DO get someone to check on you periodically when working alone on a vehicle.

DO carry out work in a logical sequence and make sure that everything is correctly assembled and tightened.

DO keep chemicals and fluids tightly capped and out of the reach of children and pets.

DO remember that your vehicle's safety affects that of yourself and others. If in doubt on any point, get professional advice.

STEERING, SUSPENSION AND BRAKES

These systems are essential to driving safety, so make sure you have a qualified shop or individual check your work. Also, compressed suspension springs can cause injury if released suddenly - be sure to use a spring compressor.

AIRBAGS

Airbags are explosive devices that can CAUSE injury if they deploy while you're working on the vehicle. Follow the manufacturer's instructions to disable the airbag whenever you're working in the vicinity of airbag components.

ASBESTOS

Certain friction, insulating, sealing, and other products - such as brake linings, brake bands, clutch linings, torque converters, gaskets, etc. - may contain asbestos or other hazardous friction material. Extreme care must be taken to avoid inhalation of dust from such products, since it is hazardous to health. If in doubt, assume that they do contain asbestos.

FIRE

Remember at all times that gasoline is highly flammable. Never smoke or have any kind of open flame around when working on a vehicle. But the risk does not end there. A spark caused by an electrical short circuit, by two metal surfaces contacting each other, or even by static electricity built up in your body under certain conditions, can ignite gasoline vapors, which in a confined space are highly explosive. Do not, under any circumstances, use gasoline for cleaning parts. Use an approved safety solvent.

Always disconnect the battery ground (-) cable at the battery before working on any part of the fuel system or electrical system. Never risk spilling fuel on a hot engine or exhaust component. It is strongly recommended that a fire extinguisher suitable for use on fuel and electrical fires be kept handy in the garage or workshop at all times. Never try to extinguish a fuel or electrical fire with water.

FUMES

Certain fumes are highly toxic and can quickly cause unconsciousness and even death if inhaled to any extent. Gasoline vapor falls into this category, as do the vapors from some cleaning solvents. Any draining or pouring of such volatile fluids should be done in a well ventilated area.

When using cleaning fluids and solvents, read the instructions on the container carefully. Never use materials from unmarked containers.

Never run the engine in an enclosed space, such as a garage. Exhaust fumes contain carbon monoxide, which is extremely poisonous. If you need to run the engine, always do so in the open air, or at least have the rear of the vehicle outside the work area.

THE BATTERY

Never create a spark or allow a bare light bulb near a battery. They normally give off a certain amount of hydrogen gas, which is highly explosive.

Always disconnect the battery ground (-) cable at the battery before working on the fuel or electrical systems.

If possible, loosen the filler caps or cover when charging the battery from an external source (this does not apply to sealed or maintenance-free batteries). Do not charge at an excessive rate or the battery may burst.

Take care when adding water to a non maintenance-free battery and when carrying a battery. The electrolyte, even when diluted, is very corrosive and should not be allowed to contact clothing or skin.

Always wear eye protection when cleaning the battery to prevent the caustic deposits from entering your eyes.

HOUSEHOLD CURRENT

When using an electric power tool, inspection light, etc., which operates on household current, always make sure that the tool is correctly connected to its plug and that, where necessary, it is properly grounded. Do not use such items in damp conditions and, again, do not create a spark or apply excessive heat in the vicinity of fuel or fuel vapor.

SECONDARY IGNITION SYSTEM VOLTAGE

A severe electric shock can result from touching certain parts of the ignition system (such as the spark plug wires) when the engine is running or being cranked, particularly if components are damp or the insulation is defective. In the case of an electronic ignition system, the secondary system voltage is much higher and could prove fatal.

HYDROFLUORIC ACID

This extremely corrosive acid is formed when certain types of synthetic rubber, found in some O-rings, oil seals, fuel hoses, etc. are exposed to temperatures above 750-degrees F (400-degrees C). The rubber changes into a charred or sticky substance containing the acid. *Once formed, the acid remains dangerous for years. If it gets onto the skin, it may be necessary to amputate the limb concerned.*

When dealing with a vehicle which has suffered a fire, or with components salvaged from such a vehicle, wear protective gloves and discard them after use.

Troubleshooting

CONTENTS

This section provides an easy reference guide to the more common problems which may occur during the operation of your vehicle. These problems and their possible causes are grouped under headings denoting various components or systems, such as Engine, Cooling system, etc. They also refer you to the chapter and/or section which deals with the problem.

Remember that successful troubleshooting is not a mysterious "black art" practiced only by professional mechanics. It is simply the result of the right knowledge combined with an intelligent, systematic approach to the problem. Always work by process of elimination, starting with the simplest solution and working through to the most complex - and never overlook the obvious. Anyone can run the gas tank dry or leave the lights on overnight, so don't assume that you are exempt from such oversights.

Finally, always establish a clear idea of why a problem has occurred and take steps to ensure that it doesn't happen again. If the electrical system fails because of a poor connection, check the other connections in the system to make sure that they don't fail as well. If a particular fuse continues to blow, find out why - don't just replace one fuse after another. Remember, failure of a small component can often be indicative of potential failure or incorrect functioning of a more important component or system.

ENGINE AND PERFORMANCE

1   Engine will not rotate when attempting to start

1   Battery terminal connections loose or corroded. Check the cable terminals at the battery; tighten cable clamp and/or clean off corrosion as necessary (see Chapter 1).

2   Battery discharged or faulty. If the cable ends are clean and tight on the battery posts, turn the key to the On position and switch on the headlights or windshield wipers. If they won't run, the battery is discharged.

3   Automatic transmission not engaged in park (P) or Neutral (N).

4   Broken, loose or disconnected wires in the starting circuit. Inspect all wires and connectors at the battery, starter solenoid and ignition switch (on steering column).

5   Starter motor pinion jammed in flywheel ring gear. If manual transmission, place transmission in gear and rock the vehicle to manually turn the engine. Remove starter (Chapter 5) and inspect pinion and flywheel (Chapter 2) at earliest convenience.

6   Starter solenoid faulty (Chapter 5).

7   Starter motor faulty (Chapter 5).

8   Ignition switch faulty (Chapter 12).

9   Engine seized. Try to turn the crankshaft with a large socket and breaker bar on the pulley bolt.

2   Engine rotates but will not start

1   Fuel tank empty.

2   Battery discharged (engine rotates slowly). Check the operation of electrical components as described in previous Section.

3   Battery terminal connections loose or corroded. See previous Section.

4   Fuel not reaching the fuel injectors. Check for clogged fuel filter or lines and defective fuel pump. Also make sure the tank vent lines aren't clogged (Chapter 4).

5   Low cylinder compression. Check as described in Chapter 2.

6   Valve clearances not properly adjusted (Chapter 1).

7   Water in fuel. Drain tank and fill with new fuel.

8   Dirty fuel injector(s) (Chapter 4).

9   Wet or damaged ignition components (Chapters 1 and 5).

10   Worn, faulty or incorrectly gapped spark plugs (Chapter 1).

11   Timing belt broken (Chapter 2B).

3   Starter motor operates without turning engine

1   Starter pinion sticking. Remove the starter (Chapter 5) and inspect.

2   Starter pinion or flywheel/driveplate teeth worn or broken. Remove the inspection cover and inspect.

4   Engine hard to start when cold

1   Battery discharged or low. Check as described in Chapter 1.

2   Fuel not reaching the fuel injectors. Check the fuel filter, lines and fuel pump (Chapters 1 and 4).

3   Defective spark plugs (Chapter 1).

4   Fault with the fuel injection or engine management system (Chapter 4 or 6).

5   Engine hard to start when hot

1   Air filter dirty (Chapter 1).

2   Fuel not reaching fuel injectors (see Chapter 4). Check for a vapor lock situation, brought about by clogged fuel tank vent lines.

3   Bad engine ground connection.

4   Fault with the fuel injection or engine management system (Chapter 4 or 6).

6   Starter motor noisy or engages roughly

1   Pinion or flywheel/driveplate teeth worn or broken. Remove the inspection cover on the left side of the engine and inspect.

2   Starter motor mounting bolts loose or missing.

7   Engine starts but stops immediately

1   Loose or damaged wire harness connections in the ignition system or at the alternator.

2   Intake manifold vacuum leaks. Make sure all mounting bolts/nuts are tight and all vacuum hoses connected to the manifold are attached properly and in good condition (Chapters 2A, 2B and 4).

3   Insufficient fuel flow to fuel injectors (Chapter 4).

8   Engine 'lopes' while idling or idles erratically

1   Vacuum leaks. Check mounting bolts at the intake manifold for tightness. Make sure that all vacuum hoses are connected and in good condition. Use a stethoscope or a length of fuel hose held against your ear to listen for vacuum leaks while the engine is running. A hissing sound will be heard. A soapy water solution will also detect leaks. Check the intake manifold gasket surfaces.

2   Leaking EGR valve or plugged PCV valve (see Chapters 1 and 6).

3   Air filter clogged (Chapter 1).

4   Fuel pump not delivering sufficient fuel (Chapter 4).

5   Leaking head gasket. Perform a cylinder compression check (Chapter 2).

6   Timing chain worn (Chapter 2).

7   Camshaft lobes worn (Chapter 2).

8   Valve clearance out of adjustment (Chapter 1).

9   Valves burned or otherwise leaking (Chapter 2).

10   Ignition system not operating properly (Chapters 1 and 5).

11   Dirty or clogged injector(s). (Chapter 4).

9   Engine misses at idle speed

1   Spark plugs faulty or not gapped properly (Chapter 1).

2   Faulty spark plug wires (Chapter 1).

3   Wet or damaged distributor components (Chapter 1).
4   Short circuits in ignition, coil or spark plug wires.
5   Sticking or faulty emissions systems (see Chapter 6).
6   Clogged fuel filter and/or foreign matter in fuel. Remove the fuel filter (Chapter 1) and inspect.
7   Vacuum leaks at intake manifold or hose connections. Check as described in Section 8.
8   Low or uneven cylinder compression. Check as described in Chapter 2.
9   Clogged or dirty fuel injectors (Chapter 4).

10   Excessively high idle speed

1   Sticking throttle linkage (Chapter 4).
2   Intake air leak (Chapters 2 and 4).
3   Malfunction in the engine management system (Chapter 6).

11   Battery will not hold a charge

1   Alternator drivebelt defective or not adjusted properly (Chapter 1).
2   Battery cables loose or corroded (Chapter 1).
3   Alternator not charging properly (Chapter 5).
4   Loose, broken or faulty wires in the charging circuit (Chapter 5).
5   Short circuit causing a continuous drain on the battery.
6   Battery defective internally.

12   Alternator light stays on

1   Fault in alternator or charging circuit (Chapter 5).
2   Alternator drivebelt defective or not properly adjusted (Chapter 1).

13   Alternator light fails to come on when key is turned on

1   Faulty bulb (Chapter 12).
2   Defective alternator (Chapter 5).
3   Fault in the printed circuit, dash wiring or bulb holder (Chapter 12).

14   Engine misses throughout driving speed range

1   Fuel filter clogged and/or impurities in the fuel system. Check fuel filter (Chapter 1) or clean system (Chapter 4).
2   Faulty or incorrectly gapped spark plugs (Chapter 1).
3   Defective spark plug wires (Chapter 1).
4   Emissions system components faulty (Chapter 6).
5   Low or uneven cylinder compression pressures. Check as described in Chapter 2.
6   Weak or faulty ignition coil(s) (Chapter 5).
7   Weak or faulty ignition system (Chapter 5).
8   Vacuum leaks at intake manifold or vacuum hoses (see Section 8).
9   Dirty or clogged fuel injector (Chapter 4).
10 Leaky EGR valve (Chapter 6).
11 Idle speed out of adjustment (Chapter 1).

15   Hesitation or stumble during acceleration

1   Ignition system not operating properly (Chapter 5).
2   Dirty or clogged fuel injector(s) (Chapter 4).
3   Low fuel pressure. Check for proper operation of the fuel pump and for restrictions in the fuel filter and lines (Chapter 4).
4   Fault with the fuel injection or engine management system (Chapter 4 or 6).

16   Engine stalls

1   Fuel filter clogged and/or water and impurities in the fuel system (Chapter 1).

2   Emissions system components faulty (Chapter 6).
3   Faulty or incorrectly gapped spark plugs (Chapter 1). Also check the spark plug wires (on models so equipped) (Chapter 1).
4   Vacuum leak at the throttle body, the intake manifold or vacuum hoses. Check as described in Section 8.
5   Valve clearances incorrect (Chapter 1).
6   Fault with the fuel injection or engine management system (Chapter 4 or 6).

17   Engine lacks power

1   Faulty or incorrectly gapped spark plugs (Chapter 1).
2   Air filter dirty (Chapter 1).
3   Faulty ignition coil(s) (Chapter 5).
4   Automatic transmission fluid level incorrect, causing slippage (Chapter 1).
5   Clutch slipping (Chapter 8).
6   Fuel filter clogged and/or impurities in the fuel system (Chapters 1 and 4).
7   EGR system not functioning properly (Chapter 6).
8   Use of sub-standard fuel. Fill tank with proper octane fuel.
9   Low or uneven cylinder compression pressures. Check as described in Chapter 2C.
10 Air leak at intake manifold (check as described in Section 8).
11 Fault with the fuel injection or engine management system (Chapter 4 or 6).
12 Brakes binding (Chapters 1 and 10).

18   Engine backfires

1   EGR system not functioning properly (Chapter 6).
2   Thermostatic air cleaner system not operating properly (Chapter 6).
3   Vacuum leak (refer to Section 8).
4   Valve clearances incorrect (Chapter 1).
5   Damaged valve springs or sticking valves (Chapter 2).
6   Intake air leak (see Section 8).

19   Engine surges while holding accelerator steady

1   Intake air leak (see Section 8).
2   Fuel pump not working properly (Chapter 4).
3   Fault with the fuel injection or engine management system (Chapter 4 or 6).

20   Pinging or knocking engine sounds when engine is under load

1   Incorrect grade of fuel. Fill tank with fuel of the proper octane rating.
2   Carbon build-up in combustion chambers. Remove cylinder heads and clean combustion chambers (Chapter 2).
3   Incorrect spark plugs (Chapter 1).

21   Engine diesels (continues to run) after being turned off

1   Incorrect spark plug heat range (Chapter 1).
2   Intake air leak (see Section 8).
3   Carbon build-up in combustion chambers. Remove the cylinder heads and clean the combustion chambers (Chapter 2).
4   Valves sticking (Chapter 2).
5   Valve clearances incorrect (Chapter 1).
6   EGR system not operating properly (Chapter 6).
7   Check for causes of overheating (Section 27).

22 Low oil pressure

1 Improper grade of oil.
2 Oil pump worn or damaged (Chapter 2).
3 Engine overheating (refer to Section 27).
4 Clogged oil filter (Chapter 1).
5 Clogged oil strainer (Chapter 2).
6 Oil pressure gauge not working properly (Chapter 2C).

23 Excessive oil consumption

1 Loose oil drain plug.
2 Loose bolts or damaged oil pan gasket (Chapter 2).
3 Loose bolts or damaged front cover gasket (Chapter 2).
4 Front or rear crankshaft oil seal leaking (Chapter 2).
5 Loose bolts or damaged rocker arm cover gasket (Chapter 2).
6 Loose oil filter (Chapter 1).
7 Loose or damaged oil pressure switch (Chapter 2).
8 Pistons and cylinders excessively worn (Chapter 2).
9 Piston rings not installed correctly on pistons (Chapter 2).
10 Worn or damaged piston rings (Chapter 2).
11 Intake and/or exhaust valve oil seals worn or damaged (Chapter 2).
12 Worn valve stems.
13 Worn or damaged valves/guides (Chapter 2).

24 Excessive fuel consumption

1 Dirty or clogged air filter element (Chapter 1).
2 Incorrect ignition timing (Chapter 1).
3 Incorrect idle speed (Chapter 1).
4 Low tire pressure or incorrect tire size (Chapter 11).
5 Fuel leakage. Check all connections, lines and components in the fuel system (Chapter 4).
6 Dirty or clogged fuel injectors (Chapter 4).
7 Fault with the fuel injection or engine management system (Chapter 4 or 6).

25 Fuel odor

1 Fuel leakage. Check all connections, lines and components in the fuel system (Chapter 4).
2 Fuel tank overfilled. Fill only to automatic shut-off.
3 Charcoal canister filter in Evaporative Emissions Control system clogged (Chapter 1).
4 Vapor leaks from Evaporative Emissions Control system lines (Chapter 6).

26 Miscellaneous engine noises

1 A strong dull noise that becomes more rapid as the engine accelerates indicates worn or damaged crankshaft bearings or an unevenly worn crankshaft. To pinpoint the trouble spot, remove the spark plug wire (or ignition coil electrical connector on coil-over-plug systems) from one plug at a time and crank the engine over. If the noise stops, the cylinder with the removed plug wire indicates the problem area. Replace the bearing and/or service or replace the crankshaft (Chapter 2).

2 A similar (yet slightly higher pitched) noise to the crankshaft knocking described in the previous paragraph, that becomes more rapid as the engine accelerates, indicates worn or damaged connecting rod bearings (Chapter 2). The procedure for locating the problem cylinder is the same as described in Paragraph 1.

3 An overlapping metallic noise that increases in intensity as the engine speed increases, yet diminishes as the engine warms up indi-

cates abnormal piston and cylinder wear (Chapter 2). To locate the problem cylinder, use the procedure described in Paragraph 1.

4 A rapid clicking noise that becomes faster as the engine accelerates indicates a worn piston pin or piston pin hole. This sound will happen each time the piston hits the highest and lowest points in the stroke (Chapter 2). The procedure for locating the problem piston is described in Paragraph 1.

5 A metallic clicking noise coming from the water pump indicates worn or damaged water pump bearings or pump. Replace the water pump with a new one (Chapter 3).

6 A rapid tapping sound or clicking sound that becomes faster as the engine speed increases indicates "valve tapping" or improperly adjusted valve clearances. This can be identified by holding one end of a section of hose to your ear and placing the other end at different spots along the valve cover. The point where the sound is loudest indicates the problem valve. Adjust the valve clearance (Chapter 1). If the problem persists, you likely have a collapsed valve lifter or other damaged valve train component. Changing the engine oil and adding a high viscosity oil treatment will sometimes cure a stuck lifter problem. If the problem still persists, the lifters, pushrods and rocker arms must be removed for inspection (see Chapter 2).

7 A steady metallic rattling or rapping sound coming from the area of the timing chain cover indicates a worn, damaged or out-of-adjustment timing chain. Service or replace the chain and related components (Chapter 2).

COOLING SYSTEM

27 Overheating

1 Insufficient coolant in system (Chapter 1).
2 Drivebelt defective or not adjusted properly (Chapter 1).
3 Radiator core blocked or radiator grille dirty or restricted (Chapter 3).
4 Thermostat faulty (Chapter 3).
5 Fan not functioning properly (Chapter 3).
6 Radiator cap not maintaining proper pressure. Have cap pressure tested by gas station or repair shop.
7 Defective water pump (Chapter 3).
8 Improper grade of engine oil.
9 Inaccurate temperature gauge (Chapter 12).

28 Overcooling

1 Thermostat faulty (Chapter 3).
2 Inaccurate temperature gauge (Chapter 12).

29 External coolant leakage

1 Deteriorated or damaged hoses. Loose clamps at hose connections (Chapter 1).
2 Water pump seals defective. If this is the case, water will drip from the weep hole in the water pump body (Chapter 3).
3 Leakage from radiator core or header tank. This will require the radiator to be professionally repaired (see Chapter 3 for removal procedures).
4 Engine drain plugs or water jacket freeze plugs leaking (see Chapters 1 and 2).
5 Leak from coolant temperature switch (Chapter 3).
6 Leak from damaged gaskets or small cracks (Chapter 2).
7 Damaged head gasket. This can be verified by checking the condition of the engine oil as noted in Section 30.

30 Internal coolant leakage

➡Note: Internal coolant leaks can usually be detected by examining the oil. Check the dipstick and inside the valve cover for water deposits and an oil consistency like that of a milkshake.

1 Leaking cylinder head gasket. Have the system pressure tested or remove the cylinder head (Chapter 2) and inspect.
2 Cracked cylinder bore or cylinder head. Dismantle engine and inspect (Chapter 2).
3 Loose cylinder head bolts (tighten as described in Chapter 2).

31 Abnormal coolant loss

1 Overfilling system (Chapter 1).
2 Coolant boiling away due to overheating (see causes in Section 27).
3 Internal or external leakage (see Sections 29 and 30).
4 Faulty radiator cap. Have the cap pressure tested.
5 Cooling system being pressurized by engine compression. This could be due to a cracked head or block or leaking head gaskets.

32 Poor coolant circulation

1 Inoperative water pump. A quick test is to pinch the top radiator hose closed with your hand while the engine is idling, then release it. You should feel a surge of coolant if the pump is working properly (Chapter 3).
2 Restriction in cooling system. Drain, flush and refill the system (Chapter 1). If necessary, remove the radiator (Chapter 3) and have it reverse flushed or professionally cleaned.
3 Loose drivebelt (Chapter 1).
4 Thermostat sticking (Chapter 3).
5 Insufficient coolant (Chapter 1).

33 Corrosion

1 Excessive impurities in the water. Soft, clean water is recommended. Distilled or rainwater is satisfactory.
2 Insufficient antifreeze solution (refer to Chapter 1 for the proper ratio of water to antifreeze).
3 Infrequent flushing and draining of system. Regular flushing of the cooling system should be carried out at the specified intervals as described in (Chapter 1).

CLUTCH

➡Note: All clutch related service information is located in Chapter 8, unless otherwise noted.

34 Fails to release (pedal pressed to the floor - shift lever does not move freely in and out of Reverse)

1 Freeplay incorrectly adjusted.
2 Clutch contaminated with oil. Remove clutch plate and inspect.
3 Clutch plate warped, distorted or otherwise damaged.
4 Diaphragm spring fatigued. Remove clutch cover/pressure plate assembly and inspect.
5 Leakage of fluid from clutch hydraulic system. Inspect master cylinder, operating cylinder and connecting lines.
6 Air in clutch hydraulic system. Bleed the system.
7 Insufficient pedal stroke. Check and adjust as necessary.
8 Piston seal in master or release cylinder deformed or damaged.
9 Lack of grease on pilot bearing.

35 Clutch slips (engine speed increases with no increase in vehicle speed)

1 Worn or oil-soaked clutch plate.
2 Clutch plate not broken in. It may take 30 or 40 normal starts for a new clutch to seat.
3 Diaphragm spring weak or damaged. Remove clutch cover/pressure plate assembly and inspect.
4 Debris in master cylinder preventing the piston from returning to its normal position.
5 Clutch hydraulic line damaged internally (not allowing fluid to return to the clutch master cylinder).
6 Binding in the release mechanism.

36 Grabbing (chattering) as clutch is engaged

1 Oil on clutch plate. Remove and inspect. Repair any leaks.
2 Worn or loose engine or transmission mounts. They may move slightly when clutch is released. Inspect mounts and bolts.
3 Worn splines on transmission input shaft. Remove clutch components and inspect.
4 Warped pressure plate or flywheel. Remove clutch components and inspect.
5 Diaphragm spring fatigued. Remove clutch cover/pressure plate assembly and inspect.
6 Clutch linings hardened or warped.
7 Clutch lining rivets loose.

37 Squeal or rumble with clutch engaged (pedal released)

1 Improper pedal adjustment. Adjust pedal freeplay.
2 Release bearing binding on transmission shaft. Remove clutch components and check bearing. Remove any burrs or nicks, clean and relubricate before reinstallation.
4 Clutch rivets loose.
5 Clutch plate cracked.
6 Fatigued clutch plate torsion springs. Replace clutch plate.

38 Squeal or rumble with clutch disengaged (pedal depressed)

1 Worn or damaged release bearing.
2 Worn or broken pressure plate diaphragm fingers.
3 Pilot bearing worn or damaged.

39 Clutch pedal stays on floor when disengaged

Binding release bearing.

MANUAL TRANSMISSION

➡Note: All manual transmission service information is located in Chapter 7A, unless otherwise noted.

40 Noisy in Neutral with engine running

1 Input shaft bearing worn.
2 Damaged main drive gear bearing.
3 Insufficient transmission oil (Chapter 1).
4 Transmission oil in poor condition. Drain and fill with proper

grade oil. Check old oil for water and debris (Chapter 1).

5   Noise can be caused by variations in engine torque. Change the idle speed and see if noise disappears.

41   Noisy in all gears

1   Any of the above causes, and/or:
2   Worn or damaged output gear bearings or shaft.

42   Noisy in one particular gear

1   Worn, damaged or chipped gear teeth.
2   Worn or damaged synchronizer.

43   Slips out of gear

1   Stiff shift lever seal.
2   Shift linkage binding.
3   Broken or loose input gear bearing retainer.
4   Dirt between clutch lever and engine housing.
5   Worn linkage.
6   Damaged or worn check balls, fork rod ball grooves or check springs.
7   Worn mainshaft or countershaft bearings.
8   Loose engine mounts (Chapter 2).
9   Excessive gear endplay.
10 Worn synchronizers.

44   Oil leaks

1   Excessive amount of lubricant in transmission (see Chapter 1 for correct checking procedures). Drain lubricant as required.
2   Rear oil seal or speedometer oil seal damaged.
3   To pinpoint a leak, first remove all built-up dirt and grime from the transmission. Degreasing agents and/or steam cleaning will achieve this. With the underside clean, drive the vehicle at low speeds so the air flow will not blow the leak far from its source. Raise the vehicle and determine where the leak is located.

45   Difficulty engaging gears

1   Clutch not releasing completely.
2   Loose or damaged shift linkage. Make a thorough inspection, replacing parts as necessary.
3   Insufficient transmission oil (Chapter 1).
4   Transmission oil in poor condition. Drain and fill with proper grade oil. Check oil for water and debris (Chapter 1).
5   Worn or damaged striking rod.
6   Sticking or jamming gears.

46   Noise occurs while shifting gears

1   Check for proper operation of the clutch (Chapter 8).
2   Faulty synchronizer assemblies.

AUTOMATIC TRANSMISSION

➡**Note: Due to the complexity of the automatic transmission, it's difficult for the home mechanic to properly diagnose and service. For problems other than the following, the vehicle should be taken to a reputable mechanic.**

47   Fluid leakage

1   Automatic transmission fluid is a deep red color, and fluid leaks should not be confused with engine oil which can easily be blown by air flow to the transmission.
2   To pinpoint a leak, first remove all built-up dirt and grime from the transmission. Degreasing agents and/or steam cleaning will achieve this. With the underside clean, drive the vehicle at low speeds so the air flow will not blow the leak far from its source. Raise the vehicle and determine where the leak is located. Common areas of leakage are:
   a) *Fluid pan: tighten mounting bolts and/or replace pan gasket as necessary (Chapter 1).*
   b) *Rear extension: tighten bolts and/or replace oil seal as necessary.*
   c) *Filler pipe: replace the rubber oil seal where pipe enters transmission case.*
   d) *Transmission oil lines: tighten fittings where lines enter transmission case and/or replace lines.*
   e) *Vent pipe: transmission overfilled and/or water in fluid (see checking procedures, Chapter 1).*
   f) *Speedometer connector: replace the O-ring where speedometer cable enters transmission case.*

48   General shift mechanism problems

Chapter 7B deals with checking and adjusting the shift cable on automatic transmissions. Common problems which may be caused by out-of-adjustment linkage are:
   a) *Engine starting in gears other than P (park) or N (Neutral).*
   b) *Indicator pointing to a gear other than the one actually engaged.*
   c) *Vehicle moves with transmission in P (Park) position.*

49   Transmission will not downshift with the accelerator pedal pressed to the floor

Chapter 7B deals with adjusting the throttle valve cable to enable the transmission to downshift properly (V6 models). On V8 models this could be due to a malfunction in the transmission's electronic control system.

50   Engine will start in gears other than Park or Neutral

Chapter 7B deals with adjusting the Park/Neutral position switch installed on automatic transmissions.

51   Transmission slips, shifts rough, is noisy or has no drive in forward or Reverse gears

1   There are many probable causes for the above problems, but the home mechanic should concern himself only with one possibility: fluid level.
2   Before taking the vehicle to a shop, check the fluid level and condition as described in Chapter 1. Add fluid, if necessary, or change the fluid and filter if needed. If problems persist, have a professional diagnose the transmission.

DRIVESHAFT

➡**Note: Refer to Chapter 8, unless otherwise specified, for service information.**

52   Leaks at front of driveshaft

Defective transmission rear seal. See Chapter 7 for replacement procedure. As this is done, check the splined yoke for burrs or roughness that could damage the new seal. Remove burrs with a fine file or whetstone.

53 Knock or clunk when transmission is under initial load (just after transmission is put into gear)

1   Loose or disconnected rear suspension components. Check all mounting bolts and bushings (Chapters 7 and 10).
2   Loose driveshaft bolts. Inspect all bolts and nuts and tighten them securely.
3   Worn or damaged universal joint bearings. Inspect the universal joints (Chapter 8).
4   Worn sleeve yoke and mainshaft spline.

54 Metallic grating sound consistent with vehicle speed

Pronounced wear in the universal joint bearings. Replace U-joints or driveshafts, as necessary.

55 Vibration

➡Note: Before blaming the driveshaft, make sure the tires are perfectly balanced and perform the following test.

1   Install a tachometer inside the vehicle to monitor engine speed as the vehicle is driven. Drive the vehicle and note the engine speed at which the vibration (roughness) is most pronounced. Now shift the transmission to a different gear and bring the engine speed to the same point.
2   If the vibration occurs at the same engine speed (rpm) regardless of which gear the transmission is in, the driveshaft is NOT at fault since the driveshaft speed varies.
3   If the vibration decreases or is eliminated when the transmission is in a different gear at the same engine speed, refer to the following probable causes.
4   Bent or dented driveshaft. Inspect and replace as necessary.
5   Undercoating or built-up dirt, etc. on the driveshaft. Clean the shaft thoroughly.
6   Worn universal joint bearings. Replace the U-joints or driveshaft as necessary.
7   Driveshaft and/or companion flange out of balance. Check for missing weights on the shaft. Remove driveshaft and reinstall 180-degrees from original position, then recheck. Have the driveshaft balanced if problem persists.
8   Loose driveshaft mounting bolts/nuts.
9   Defective center bearing, if so equipped.
10 Worn transmission rear bushing (Chapter 7).

56 Scraping noise

Make sure the dust cover on the sleeve yoke isn't rubbing on the transmission extension housing.

57 Whining or whistling noise

Defective center bearing.

AXLES AND DIFFERENTIAL

➡Note: For differential servicing information, refer to Chapter 8, unless otherwise specified.

58 Noise - same when in drive as when vehicle is coasting

1   Road noise. No corrective action available.
2   Tire noise. Inspect tires and check tire pressures (Chapter 1).

3   Front wheel bearings loose, worn or damaged (Chapters 1 and 10).
4   Insufficient differential oil (Chapter 1).
5   Defective differential.

59 Knocking sound when starting or shifting gears

Defective or incorrectly adjusted differential.

60 Noise when turning

Defective differential.

61 Vibration

1   See probable causes under Driveshaft. Proceed under the guidelines listed for the driveshaft. If the problem persists, check the rear wheel bearings by raising the rear of the vehicle and spinning the wheels by hand. Listen for evidence of rough (noisy) bearings. Remove and inspect (Chapter 8).
2   Worn driveaxle CV joint (Chapter 8).

62 Oil leaks

1   Pinion oil seal damaged (Chapter 8).
2   Axleshaft oil seals damaged (Chapter 8).
3   Differential cover leaking. Tighten mounting bolts or replace the gasket as required.
4   Loose filler or drain plug on differential (Chapter 1).
5   Clogged or damaged breather on differential.

TRANSFER CASE

➡Note: Unless otherwise specified, refer to Chapter 7C for service and repair information.

63 Gear jumping out of mesh

1   Interference between the control lever and the console.
2   Internal wear or incorrect adjustments.

64 Difficult shifting

1   Lack of oil.
2   Internal wear, damage or incorrect adjustment.

65 Noise

1   Lack of oil in transfer case.
2   Noise in 4H and 4L, but not in 2H indicates cause is in the front differential or front axle.
3   Noise in 2H, 4H and 4L indicates cause is in rear differential or rear axle.
4   Noise in 2H and 4H but not in 4L, or in 4L only, indicates internal wear or damage in transfer case.

BRAKES

➡Note: Before assuming a brake problem exists, make sure the tires are in good condition and inflated properly, the front end alignment is correct and the vehicle is not loaded with weight in an unequal manner. All service procedures for the brakes are included in Chapter 9, unless otherwise noted.

66   Vehicle pulls to one side during braking

1   Defective, damaged or oil contaminated brake pad on one side. Inspect as described in Chapter 1. Refer to Chapter 9 if replacement is required.

2   Excessive wear of brake pad material or disc on one side. Inspect and repair as necessary (Chapter 9).

3   Loose or disconnected front suspension components. Inspect and tighten all bolts securely (Chapters 1 and 11).

4   Defective caliper assembly. Remove caliper and inspect for stuck piston or damage.

5   Scored or out-of-round rotor (Chapter 9).

6   Loose caliper mounting bolts (Chapter 9).

7   Incorrect wheel bearing adjustment (Chapter 1).

67   Noise (high-pitched squeal)

1   Brake pads worn out. Replace pads with new ones immediately!

2   Glazed or contaminated pads.

3   Dirty or scored disc.

4   Bent support plate.

68   Excessive brake pedal travel

1   Partial brake system failure. Inspect entire system (Chapter 1) and correct as required.

2   Insufficient fluid in master cylinder. Check (Chapter 1) and add fluid - bleed system if necessary.

3   Air in system. Bleed system.

4   Drum brakes out of adjustment. Check the operation of the automatic adjusters.

5   Defective proportioning valve. Replace valve and bleed system.

69   Brake pedal feels spongy when depressed

1   Air in brake lines. Bleed the brake system.

2   Deteriorated rubber brake hoses. Inspect all system hoses and lines. Replace parts as necessary.

3   Master cylinder mounting nuts loose. Inspect master cylinder bolts (nuts) and tighten them securely.

4   Master cylinder faulty.

5   Incorrect shoe or pad clearance.

6   Deformed rubber brake lines.

7   Soft or swollen caliper seals.

8   Poor quality brake fluid. Bleed entire system and fill with new approved fluid.

70   Excessive effort required to stop vehicle

1   Power brake booster not operating properly.

2   Excessively worn linings or pads. Check and replace if necessary.

3   One or more caliper pistons seized or sticking. Inspect and replace caliper as required.

4   Brake pads or linings contaminated with oil or grease. Inspect and replace as required.

5   New pads or linings installed and not yet seated. It'll take a while for the new material to seat against the disc or drum.

6   Worn or damaged master cylinder or caliper assemblies. Check particularly for frozen pistons.

7   Also see causes listed under Section 69.

71   Pedal travels to the floor with little resistance

Little or no fluid in the master cylinder reservoir caused by leaking caliper piston(s) or loose, damaged or disconnected brake lines. Inspect entire system and repair as necessary.

72   Brake pedal pulsates during brake application

1   Wheel bearings damaged, worn or out of adjustment (Chapter 1).

2   Disc not within specifications. Remove the disc and check for excessive lateral runout and parallelism. Have the discs resurfaced or replace them with new ones. Also make sure that all discs are the same thickness.

3   Out-of-round rear brake drums. Remove the drums and have them resurfaced or replace them with new ones.

73   Brakes drag (indicated by sluggish engine performance or wheels being very hot after driving)

1   Output rod adjustment incorrect at the brake pedal.

2   Obstructed master cylinder fill port.

3   Master cylinder piston seized in bore.

4   Caliper sticking.

5   Piston cups in master cylinder or caliper assembly deformed.

6   Parking brake assembly will not release.

7   Clogged brake lines.

8   Brake pedal height improperly adjusted.

9   Wheel cylinder sticking.

10   Improper shoe-to-drum clearance. Adjust as necessary.

74   Rear brakes lock up under light brake application

1   Tire pressures too high.

2   Tires excessively worn (Chapter 1).

75   Rear brakes lock up under heavy brake application

1   Tire pressures too high.

2   Tires excessively worn (Chapter 1).

3   Front brake pads contaminated with oil, mud or water. Clean or replace the pads.

4   Front brake pads excessively worn.

5   Defective master cylinder or caliper assembly.

SUSPENSION AND STEERING

➡ Note: All service procedures for the suspension and steering systems are included in Chapter 10, unless otherwise noted.

76   Vehicle pulls to one side

1   Tire pressures uneven (Chapter 1).

2   Defective tire (Chapter 1).

3   Excessive wear in suspension or steering components (Chapter 1).

4   Wheel alignment incorrect.

5   Front brakes dragging. Inspect as described in Section 73.

6   Wheel bearings improperly adjusted (Chapter 1 or 8).

7   Wheel lug nuts loose.

77   Shimmy, shake or vibration

1   Tire or wheel out-of-balance or out-of-round. Have them balanced on the vehicle.

2   Worn wheel bearings (Chapter 1 or 8).

3   Shock absorbers and/or suspension components worn or damaged.

Check for worn bushings in the upper and lower links.

4 Wheel lug nuts loose.
5 Incorrect tire pressures.
6 Excessively worn or damaged tire.
7 Loosely mounted steering gear housing.
8 Steering gear improperly adjusted.
9 Loose, worn or damaged steering components.
10 Damaged idler arm.
11 Worn balljoint.

78 Excessive pitching and/or rolling around corners or during braking

1 Defective shock absorbers. Replace as a set.
2 Broken or weak springs and/or suspension components.
3 Worn or damaged stabilizer bar or bushings.

79 Wandering or general instability

1 Improper tire pressures.
2 Worn or damaged upper and lower link or tension rod bushings.
3 Incorrect front end alignment.
4 Worn or damaged steering linkage or suspension components.
5 Improperly adjusted steering gear.
6 Out-of-balance wheels.
7 Loose wheel lug nuts.
8 Worn rear shock absorbers.
9 Fatigued or damaged rear leaf springs.

80 Excessively stiff steering

1 Lack of lubricant in power steering fluid reservoir (Chapter 1).
2 Incorrect tire pressures (Chapter 1).
3 Lack of lubrication at balljoints (Chapter 1).
4 Front end out of alignment.
5 Steering gear out of adjustment or lacking lubrication.
6 Improperly adjusted wheel bearings.
7 Worn or damaged steering gear.
8 Low tire pressures.
9 Worn or damaged balljoints.
10 See also Section 79.

81 Excessive play in steering

1 Loose wheel bearings (Chapter 1 or 8).
2 Excessive wear in suspension bushings (Chapter 1).
3 Steering gear worn.
4 Incorrect wheel alignment.
5 Steering gear mounting bolts loose.
6 Worn tie-rod ends.

82 Lack of power assistance

1 Steering pump drivebelt faulty or not adjusted properly (Chapter 1).
2 Fluid level low (Chapter 1).
3 Hoses or pipes restricting the flow. Inspect and replace parts as necessary.
4 Air in power steering system. Bleed system.
5 Defective power steering pump.

83 Steering wheel fails to return to straight-ahead position

1 Incorrect front end alignment.

2 Tire pressures low.
3 Steering gear worn or damaged.
4 Steering column out of alignment.
5 Worn or damaged balljoint.
6 Worn or damaged tie-rod end.
7 Lack of fluid in power steering pump.

84 Steering effort not the same in both directions (power system)

1 Leaks in steering gear.
2 Clogged fluid passage in steering gear.

85 Noisy power steering pump

1 Insufficient oil in pump.
2 Clogged hoses or oil filter in pump.
3 Loose pulley.
4 Improperly adjusted drivebelt (Chapter 1).
5 Defective pump.

86 Miscellaneous noises

1 Improper tire pressures.
2 Insufficiently lubricated balljoint or steering linkage.
3 Loose or worn steering gear, steering linkage or suspension components.
4 Defective shock absorber.
5 Defective wheel bearing.
6 Worn or damaged suspension bushings.
7 Damaged leaf spring.
8 Loose wheel lug nuts.
9 Worn or damaged rear axleshaft spline.
10 Worn or damaged rear shock absorber mounting bushing.
11 Excessive rear axle end play.
12 See also causes of noises at the rear axle and driveshaft.

87 Excessive tire wear (not specific to one area)

1 Incorrect tire pressures.
2 Tires out of balance. Have them balanced on the vehicle.
3 Wheels damaged. Inspect and replace as necessary.
4 Suspension or steering components worn (Chapter 1).

88 Excessive tire wear on outside edge

1 Incorrect tire pressure.
2 Excessive speed in turns.
3 Front end alignment incorrect (excessive toe-in).

89 Excessive tire wear on inside edge

1 Incorrect tire pressure.
2 Front end alignment incorrect (toe-out).
3 Loose or damaged steering components (Chapter 1).

90 Tire tread worn in one place

1 Tires out of balance. Have them balanced on the vehicle.
2 Damaged or buckled wheel. Inspect and replace if necessary.
3 Defective tire.

Section

Reference to other Chapters

CHECK ENGINE light on - See Chapter 6

1

TUNE-UP AND MAINTENANCE

Engine compartment component locations - 3.4L V6 engine

1	PCV. valve	5	Fuse box	9	Engine oil dipstick	12	Windshield washer fluid reservoir
2	Throttle body	6	Battery	10	Radiator cap	13	Air filter housing
3	Brake fluid reservoir	7	Coolant reservoir	11	Power steering fluid reservoir		
4	Clutch fluid reservoir	8	Engine oil filler cap				

Engine compartment component locations - 4.7L V8 engine

1	Brake fluid reservoir	7	Diagnostic connector	13	Windshield (and back window on Sequoia) washer fluid reservoir	16	Automatic transmission fluid dipstick (2004 and earlier models)
2	PCV valve	8	Battery			17	Ignition coils (spark plugs underneath) - right bank
3	Engine oil filler cap	9	Coolant reservoir	14	Air filter housing		
4	Main fuse	10	Radiator cap	15	Power steering fluid reservoir		
5	Fuse box	11	Upper radiator hose				
6	Engine oil dipstick	12	Drivebelt				

Typical engine compartment underside component locations

1	Lower radiator hose	6	Front differential check/fill plug
2	Radiator drain valve	7	Shock absorber/coil spring assembly
3	Drivebelt	8	Brake caliper
4	Engine oil drain plug	9	Driveaxle boot (inner CV joint)
5	Engine oil filter	10	Lower balljoint
		11	Tie-rod end
		12	Steering gear boot
		13	Automatic transmission fluid drain plug
		14	Front driveshaft

Typical rear underside component locations

1	Muffler	3	Driveshaft universal joint	5	Shock absorber	7	Differential drain plug
2	Fuel tank	4	Leaf spring	6	Brake drum	8	Parking brake cable

1 Toyota Tundra and Sequoia Maintenance schedule

EVERY 250 MILES OR WEEKLY, WHICHEVER COMES FIRST

Check the engine oil level (Section 4)
Check the engine coolant level (Section 4)
Check the windshield washer fluid level (Section 4)
Check the brake and clutch fluid level (Section 4)
Check the tires and tire pressures (Section 5)

EVERY 3000 MILES OR 3 MONTHS, WHICHEVER COMES FIRST

All items listed above, plus . . .
Check the automatic transmission fluid level (Section 6)
Check the power steering fluid level (Section 7)
Change the engine oil and filter (Section 8)

EVERY 7500 MILES OR 6 MONTHS, WHICHEVER COMES FIRST

Check and service the battery (Section 9)
Check the cooling system (Section 10)
Inspect and replace, if necessary, all underhood hoses (Section 11)
Inspect and replace, if necessary, the windshield wiper blades (Section 12)
Rotate the tires (Section 13)
Inspect the suspension and steering components (Section 14)
Lubricate the driveshaft and body components (Section 15)*
Inspect the exhaust system (Section 16)
Check the manual transmission lubricant (Section 17)
Check the transfer case lubricant level (Section 18)
Check the differential lubricant level (Section 19)
Check the seat belts (Section 20)

EVERY 15,000 MILES OR 12 MONTHS, WHICHEVER COMES FIRST

All items listed above, plus . . .
Check the driveaxle boots (Section 21)
Check the Positive Crankcase Ventilation (PCV) system (Section 22)
Replace the air filter (Section 23)*
Check the engine drivebelt(s) (Section 24)
Inspect the fuel system (Section 25)

Check the brakes (Section 26)*
Check the brake and clutch pedals for proper height and freeplay (Section 27)

EVERY 30,000 MILES OR 24 MONTHS, WHICHEVER COMES FIRST

All items listed above, plus . . .
Replace the spark plugs (non-platinum or iridium type) (Section 28)
Inspect the spark plug wires (V6 engine) (Section 29)
Service the cooling system (drain, flush and refill) (Section 30)
Change the automatic transmission fluid and filter (Section 31)**
Change the manual transmission lubricant (Section 32)
Change the transfer case lubricant (Section 33)
Change the differential lubricant (Section 34)*

EVERY 60,000 MILES OR 48 MONTHS, WHICHEVER COMES FIRST

Replace the fuel filter (Section 35)
Replace the spark plugs (platinum or iridium type) (Section 28)
Check and adjust if necessary, the valve clearance (Section 36)
Inspect the evaporative emissions control system (Section 37)
Replace the timing belt (Chapter 2A or 2B)

EVERY 80,000 MILES

Replace the oxygen sensors (see Chapter 6)
This item is affected by "severe" operating conditions, as described below. If the vehicle is operated under severe conditions, perform all maintenance indicated with an asterisk (*) at 5000 mile/four-month intervals. Severe conditions exist if you mainly operate the vehicle . . .

in dusty areas
towing a trailer
idling for extended periods and/or driving at low speeds when outside temperatures remain below freezing and most trips are less than four miles long

**If operated under one or more of the following conditions, change the automatic transmission fluid every 15,000 miles:*

in heavy city traffic where the outside temperature regularly reaches 90-degrees F or higher
in hilly or mountainous terrain
frequent trailer pulling

2 Introduction

This Chapter is designed to help the home mechanic maintain the Toyota Tundra and Sequoia for peak performance, economy, safety and long life.

Included is a master maintenance schedule, followed by Sections dealing specifically with each item on the schedule. Visual checks, adjustments, component replacement and other helpful items are included. Refer to the accompanying illustrations of the engine compartment and the underside of the vehicle for the location of various components.

Servicing your vehicle in accordance with the mileage/time maintenance schedule and the following Sections will provide it with a planned maintenance program that should result in a long and reliable service life. This is a comprehensive plan, so maintaining some items but not others at the specified service intervals will not produce the same results.

As you service your vehicle, you will discover that many of the procedures can, and should, be grouped together because of the nature of the particular procedure you're performing or because of the close proximity of two otherwise unrelated components to one another.

For example, if the vehicle is raised for any reason, you should inspect the exhaust, suspension, steering and fuel systems while you're under the vehicle. When you're rotating the tires, it makes good sense to check the brakes and wheel bearings since the wheels are already removed.

Finally, let's suppose you have to borrow or rent a torque wrench. Even if you only need to tighten the spark plugs, you might as well check the torque of as many critical fasteners as time allows.

The first step of this maintenance program is to prepare yourself before the actual work begins. Read through all Sections pertinent to the procedures you're planning to do, then make a list of and gather together all the parts and tools you will need to do the job. If it looks as if you might run into problems during a particular segment of some procedure, seek advice from your local auto parts store or dealer service department.

OWNER'S MANUAL AND VECI LABEL INFORMATION

Your vehicle Owner's Manual was written for your year and model and contains very specific information on component locations, specifications, fuse ratings, part numbers, etc. The Owner's Manual is an important resource for the do-it-yourselfer to have; if one was not supplied with your vehicle, it can generally be ordered from a dealer parts department.

Among other important information, the Vehicle Emissions Control Information (VECI) label contains specifications and procedures for tune-up adjustments (if applicable) and, in some instances, spark plugs (see Chapter 6 for more information on the VECI label). The information on this label is the exact maintenance data recommended by the manufacturer. This data often varies by intended operating altitude, local emissions regulations, month of manufacture, etc.

This Chapter contains procedural details, safety information and more ambitious maintenance intervals than you might find in the manufacturer's literature. However, you may also find procedures or specifications in your Owner's Manual or VECI label that differ with what's printed here. In these cases, the Owner's Manual or VECI label can be considered correct, since it is specific to your particular vehicle.

3 Tune-up general information

The term tune-up is used in this manual to represent a combination of individual operations rather than one specific procedure.

If, from the time the vehicle is new, the routine maintenance schedule is followed closely and frequent checks are made of fluid levels and high wear items, as suggested throughout this manual, the engine will be kept in relatively good running condition and the need for additional work will be minimized.

More likely than not, however, there will be times when the engine is running poorly due to lack of regular maintenance. This is even more likely if a used vehicle, which has not received regular and frequent maintenance checks, is purchased. In such cases, an engine tune-up will be needed outside of the regular routine maintenance intervals.

The first step in any tune-up or diagnostic procedure to help correct a poor running engine is a cylinder compression check. A compression check (see Chapter 2, Part C) will help determine the condition of internal engine components and should be used as a guide for tune-up and repair procedures. If, for instance, the compression check indicates serious internal engine wear, a conventional tune-up won't improve the performance of the engine and would be a waste of time and money. Because of its importance, the compression check should be done by someone with the right equipment and the knowledge to use it properly.

The following procedures are those most often needed to bring a generally poor running engine back into a proper state of tune.

MINOR TUNE-UP

Check all engine related fluids (Section 4)
Clean, inspect and test the battery (Section 9)
Check the cooling system (Section 10)
Check all underhood hoses (Section 11)
Check the PCV valve (Section 22)
Check the air filter (Section 23)
Check the drivebelt(s) (Section 24)
Replace the spark plugs (Section 28)
Inspect the spark plug wires (V6 engine) (Section 29)

MAJOR TUNE-UP

All items listed under Minor tune-up, plus . . .
Replace the PCV valve (Section 22)
Replace the air filter (Section 23)
Check the fuel system (Section 25)
Replace the spark plug wires (V6 engine) (Section 29)
Replace the fuel filter (Section 35)
Check and adjust the valve clearances (Section 36)
Check the ignition system (Chapter 5)
Check the charging system (Chapter 5)

4 Fluid level checks (every 250 miles or weekly)

→Note: The following are fluid level checks to be done on a 250 mile or weekly basis. Additional fluid level checks can be found in specific maintenance procedures which follow. Regardless of intervals, be alert to fluid leaks under the vehicle which would indicate a fault to be corrected immediately.

1 Fluids are an essential part of the lubrication, cooling, brake and windshield washer systems. Because the fluids gradually become depleted and/or contaminated during normal operation of the vehicle, they must be periodically replenished. See *Recommended lubricants and fluids* at the end of this Chapter before adding fluid to any of the following components.

→Note: The vehicle must be on level ground when fluid levels are checked.

ENGINE OIL

▶ Refer to illustrations 4.2, 4.4 and 4.6

2 The engine oil level is checked with a dipstick that extends through a tube and into the oil pan at the bottom of the engine (see illustration).

3 The oil level should be checked before the vehicle has been driven, or about 5 minutes after the engine has been shut off. If the oil is checked immediately after driving the vehicle, some of the oil will remain in the upper engine components, resulting in an inaccurate reading on the dipstick.

4 Pull the dipstick out of the tube and wipe all the oil from the end with a clean rag or paper towel. Insert the clean dipstick all the way back into the tube, then pull it out again. Note the oil at the end of the dipstick. Add oil as necessary to keep the level between the upper and lower marks on the dipstick (see illustration).

5 Do not overfill the engine by adding too much oil since this may result in oil-fouled spark plugs, oil leaks or oil seal failures.

6 Oil is added to the engine after removing the threaded cap from the valve cover (see illustration). A funnel may help to reduce spills.

7 Checking the oil level is an important preventive maintenance step. A consistently low oil level indicates oil leakage through damaged seals, defective gaskets or past worn rings or valve guides. If the oil looks milky or has water droplets in it, the cylinder head gasket(s) may be blown or the head(s) or block may be cracked. The engine should be checked immediately. The condition of the oil should also be checked. Whenever

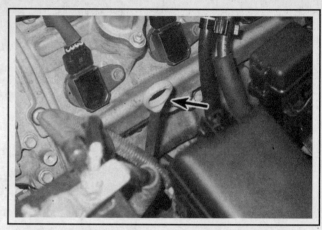

4.2 On the V8 engine, the oil dipstick is located on the left side of the engine (on the V6 it's at the front of the engine)

you check the oil level, slide your thumb and index finger up the dipstick before wiping off the oil. If you see small dirt or metal particles clinging to the dipstick, the oil should be changed (see Section 8).

ENGINE COOLANT

▶ Refer to illustration 4.8

❋❋ WARNING:

Do not allow antifreeze to come in contact with your skin or painted surfaces of the vehicle. Flush contaminated areas immediately with plenty of water. Don't store new coolant or leave old coolant lying around where it's accessible to children or pets – they're attracted by its sweet smell. Ingestion of even a small amount of coolant can be fatal! Wipe up garage floor and drip pan spills immediately. Keep antifreeze containers covered and repair cooling system leaks as soon as they're noticed.

8 All vehicles covered by this manual are equipped with a pressurized coolant recovery system. A coolant reservoir which is located in the

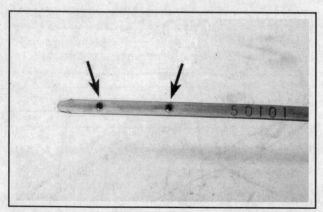

4.4 The oil level must be maintained between the marks at all times - it takes one quart of oil to raise the level from the lower mark to the upper mark

4.6 Oil is added to the engine after removing the twist-off cap located on the valve cover (V8 model shown, V6 models similar)

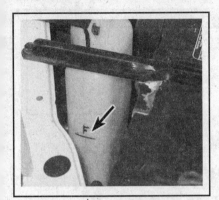

4.8 Check the coolant level in the reservoir with the engine hot - it should be visible through the translucent reservoir

4.14 The windshield washer fluid reservoir is located in the right front corner of the engine compartment

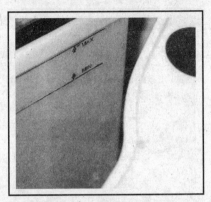

4.15 On some models the battery electrolyte level can be visually checked through the translucent battery case

left front corner of the engine compartment (next to the radiator) is connected by a hose to the base of the coolant filler cap (see illustration). If the coolant gets too hot during engine operation, coolant can escape through a pressurized filler cap, then through a connecting hose into the reservoir. As the engine cools, the coolant is automatically drawn back into the cooling system to maintain the correct level.

9 The coolant level should be checked regularly. It must be between the FULL and LOW lines on the tank. The level will vary with the temperature of the engine. When the engine is cold, the coolant level should be at or slightly above the LOW mark on the tank. Once the engine has warmed up, the level should be at or near the FULL mark. If it isn't, allow the fluid in the tank to cool, then remove the cap from the reservoir and add coolant to bring the level up to the FULL line. Use only the type of coolant and water in the mixture ratio recommended at the end of this Chapter. Do not use supplemental inhibitor additives. If only a small amount of coolant is required to bring the system up to the proper level, water can be used. However, repeated additions of water will dilute the recommended antifreeze and water solution. In order to maintain the proper ratio of antifreeze and water, it is advisable to top up the coolant level with the correct mixture.

10 If the coolant level drops within a short time after replenishment, there may be a leak in the system. Inspect the radiator, hoses, engine coolant filler cap, drain plugs and water pump. If no leak is evident, have the radiator cap pressure tested by your dealer.

❈❈ WARNING:

Never remove the radiator cap or the coolant recovery reservoir cap when the engine is running or has just been shut down, because the cooling system is hot. Escaping steam and scalding liquid could cause serious injury.

11 If it is necessary to open the radiator cap, wait until the system has cooled completely, then wrap a thick cloth around the cap and turn it to the first stop. If any steam escapes, wait until the system has cooled further, then remove the cap.

12 When checking the coolant level, always note its condition. It should be relatively clear. If it is brown or rust colored, the system should be drained, flushed and refilled. Even if the coolant appears to be normal, the corrosion inhibitors wear out with use, so it must be replaced at the specified intervals.

13 Do not allow antifreeze to come in contact with your skin or painted surfaces of the vehicle. Flush contacted areas immediately with plenty of water.

WINDSHIELD WASHER FLUID

▶ **Refer to illustration 4.14**

14 Fluid for the windshield washer system is located on the right (passenger) side of the engine compartment (see illustration). Sequoia models are equipped with a rear window washer system built into the front washer system. In milder climates, plain water can be used to top up the reservoir, but the reservoir should be kept no more than two-thirds full to allow for expansion should the water freeze. In colder climates, the use of a specially designed windshield washer fluid, available at your dealer and any auto parts store, will help lower the freezing point of the fluid. Mix the solution with water in accordance with the manufacturer's directions on the container. Do not use regular antifreeze. It will damage the vehicle's paint.

BATTERY ELECTROLYTE

▶ **Refer to illustration 4.15**

15 On models not equipped with a sealed battery, unscrew the filler/vent cap and check the electrolyte level. It must be between the upper and lower levels. On some batteries you can view the level through the translucent case (see illustration). If the level is low, remove the caps and add distilled water. Install and securely retighten the caps.

❈❈ CAUTION:

Overfilling the cells may cause electrolyte to spill over during periods of heavy charging, causing corrosion or damage.

BRAKE AND CLUTCH FLUID

▶ **Refer to illustrations 4.17a and 4.17b**

16 The brake master cylinder is mounted on the front of the power booster unit in the engine compartment. The hydraulic clutch master cylinder used on manual transmission vehicles is located next to the brake master cylinder.

17 To check the fluid level of the brake and clutch master cylinders, simply look at the MAX and MIN marks on the reservoir (see illustra-

4.17a The fluid level inside the brake fluid reservoir can be checked by observing the level from the outside

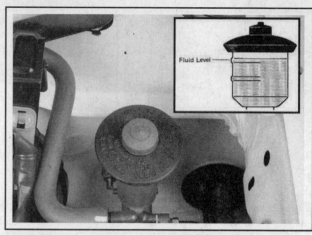

4.17b The fluid level inside the clutch reservoir can easily be checked by observing the level from the outside - fluid can be added to the reservoir after the cover is removed by pulling it straight up

tions). The level should be 1/4 inch below the maximum fill line.

18 If the level is low, wipe the top of the reservoir cover with a clean rag to prevent contamination of the brake system before lifting the cover off.

19 Add only the specified brake fluid to the brake and clutch reservoirs (refer to *Recommended lubricants and fluids* at the end of this Chapter or to your owner's manual). Mixing different types of brake fluid can damage the system.

✳✳ WARNING:

Use caution when filling either reservoir - brake fluid can harm your eyes and damage painted surfaces. Do not use brake fluid that has been opened for more than one year or has been left open. Brake fluid absorbs moisture from the air. Excess moisture can cause a dangerous loss of braking.

20 While the reservoir cover is removed, inspect the reservoir for contamination. If deposits, dirt particles or water droplets are present, the system should be drained and refilled.

21 After filling the reservoir to the proper level, make sure the cover is properly seated to prevent fluid leakage.

22 The fluid in the brake master cylinder will drop slightly as the brake pads at each wheel wear down during normal operation. If either master cylinder requires repeated replenishing to keep it at the proper level, this is an indication of leakage in the brake or clutch system, which should be corrected immediately. If the brake system shows an indication of leakage check all brake lines and connections, along with the calipers, wheel cylinders and booster (see Section 26 for more information). If the hydraulic clutch system shows an indication of leakage check all clutch lines and connections, along with the clutch release cylinder (see Chapter 8 for more information).

23 If, upon checking the brake or clutch master cylinder fluid level, you discover one or both reservoirs empty or nearly empty, the systems should be bled (see Chapter 9).

5 Tire and tire pressure checks (every 250 miles or weekly)

▶ Refer to illustrations 5.2, 5.3, 5.4a, 5.4b and 5.8

1 Periodic inspection of the tires may spare you the inconvenience of being stranded with a flat tire. It can also provide you with vital information regarding possible problems in the steering and suspension systems before major damage occurs.

2 The original tires on this vehicle are equipped with 1/2-inch wide wear bands that will appear when tread depth reaches 1/16-inch, at which point the tires can be considered worn out. Tread wear can be monitored with a simple, inexpensive device known as a tread depth gauge (see illustration).

3 Note any abnormal tread wear (see illustration). Tread pattern irregularities such as cupping, flat spots and more wear on one side than the other are indications of front end alignment and/or balance problems. If any of these conditions are noted, take the vehicle to a tire shop or service station to correct the problem.

4 Look closely for cuts, punctures and embedded nails or tacks. Sometimes a tire will hold air pressure for a short time or leak down very slowly after a nail has embedded itself in the tread. If a slow leak

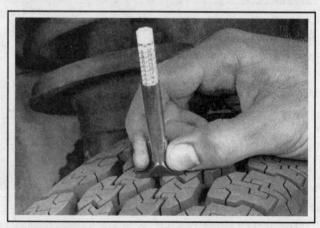

5.2 Use a tire tread depth gauge to monitor tire wear - they are available at auto parts stores and service stations and cost very little

UNDERINFLATION

CUPPING

Cupping may be caused by:
- Underinflation and/or mechanical irregularities such as out-of-balance condition of wheel and/or tire, and bent or damaged wheel.
- Loose or worn steering tie-rod or steering idler arm.
- Loose, damaged or worn front suspension parts.

OVERINFLATION

INCORRECT TOE-IN OR EXTREME CAMBER

FEATHERING DUE TO MISALIGNMENT

5.3 This chart will help you determine the condition of the tires, the probable cause(s) of abnormal wear and the corrective action necessary

persists, check the valve stem core to make sure it's tight (see illustration). Examine the tread for an object that may have embedded itself in the tire or for a "plug" that may have begun to leak (radial tire punctures are repaired with a rubber plug that's installed in the hole). If a puncture is suspected, it can be easily verified by spraying a solution of soapy water onto the puncture area (see illustration). The soapy solution will bubble if there's a leak. Unless the puncture is unusually large, a tire shop or service station can usually repair the tire.

5   Carefully inspect the inner sidewall of each tire for evidence of brake fluid leakage. If you see any, inspect the brakes immediately.

6   Correct air pressure adds miles to the lifespan of the tires, improves mileage and enhances overall ride quality. Tire pressure cannot be accurately estimated by looking at a tire, especially if it's a radial. A tire pressure gauge is essential. Keep an accurate gauge in the vehicle. The pressure gauges attached to the nozzles of air hoses at gas stations are often inaccurate.

7   Always check tire pressure when the tires are cold. Cold, in this case, means the vehicle has not been driven over a mile in the three hours preceding a tire pressure check. A pressure rise of four to eight pounds is not uncommon once the tires are warm.

8   Unscrew the valve cap protruding from the wheel or hubcap and push the gauge firmly onto the valve stem (see illustration). Note the reading on the gauge and compare the figure to the recommended tire pressure shown on the placard on the driver's side door jamb. Be sure to reinstall the valve cap to keep dirt and moisture out of the valve stem mechanism. Check all four tires and, if necessary, add enough air to bring them up to the recommended pressure.

9   Don't forget to keep the spare tire inflated to the specified pressure (consult your owner's manual).

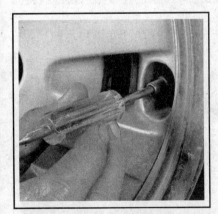
5.4a If a tire loses air on a steady basis, check the valve core first to make sure it's snug (special inexpensive wrenches are commonly available at auto parts stores)

5.4b If the valve core is tight, raise the corner of the vehicle with the low tire and spray a soapy water solution onto the tread as the tire is turned slowly - leaks will cause small bubbles to appear

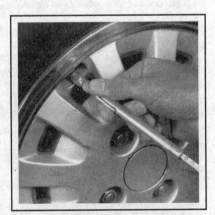
5.8 To extend the life of the tires, check the air pressure at least once a week with an accurate gauge (don't forget the spare)

6 Automatic transmission fluid level check (every 3000 miles or 3 months)

▶ **Refer to illustrations 6.4 and 6.6**

1   The level of the automatic transmission fluid should be carefully maintained. Low fluid level can lead to slipping or loss of drive, while overfilling can cause foaming, loss of fluid and transmission damage.

2   The transmission fluid level should only be checked when the transmission is hot (at its normal operating temperature). If the vehicle has just been driven over 10 miles (15 miles in a frigid climate), and the fluid temperature is 160 to 175-degrees F, the transmission is hot.

✳✳ CAUTION:

If the vehicle has just been driven for a long time at high speed or in city traffic in hot weather, or if it has been pulling a trailer, an accurate fluid level reading cannot be obtained. Allow the fluid to cool down for about 30 minutes.

3   If the vehicle has not been driven, park the vehicle on level ground, set the parking brake, then start the engine and bring it to operating temperature. While the engine is idling, depress the brake pedal and move the selector lever through all the gear ranges, beginning and ending in Park.

2004 AND EARLIER MODELS

4   With the engine still idling, remove the dipstick from its tube (see illustration). Check the level of the fluid on the dipstick and note its condition. Refer to the underhood photographs at the beginning of this chapter for the exact location of the automatic transmission dipstick.

5   Wipe the fluid from the dipstick with a clean rag and reinsert it back into the filler tube until the cap seats.

6   Pull the dipstick out again and note the fluid level (see illustration). If the transmission is cold, the level should be in the COOL range on the dipstick. If it is hot, the fluid level should be in the HOT range. If the level is at the low side of either range, add the specified automatic transmission fluid through the dipstick tube with a funnel.

7   Add just enough of the recommended fluid to fill the transmission to the proper level. It takes about one pint to raise the level from the low mark to the high mark when the fluid is hot, so add the fluid a little at a time and keep checking the level until it is correct. Proceed to Step 12.

2005 AND LATER MODELS

➡**Note 1: These models are not equipped with an automatic transmission fluid dipstick.**

➡**Note 2: The vehicle must be level for this check. If there is not enough room to crawl under the vehicle, raise both ends and support securely on jackstands.**

8   Turn the engine off. Remove the "overflow" plug from the bottom of the transmission fluid pan.

➡**Note: The overflow plug is located along the right side of the fluid pan; the drain plug is located at the back of the pan (don't remove the drain plug).**

9   If the fluid runs out of the hole, allow it to drip until it stops. If no fluid comes out of the hole, remove the refill plug, located on the side of the transmission housing, near the rear.

10 Add the proper type of transmission fluid (see this Chapter's Specifications) until fluid flows from the overflow hole in the bottom of the pan. When the flow of fluid slows to a trickle, install the overflow plug and tighten it to the torque listed in this Chapter's Specifications.

11 Tighten the refill plug to the torque listed in this Chapter's Specifications.

ALL MODELS

12 The condition of the fluid should also be checked along with the level. If the fluid at the end of the dipstick is black or a dark reddish brown color, or if it emits a burned smell, the fluid should be changed (see Section 31). If you are in doubt about the condition of the fluid, purchase some new fluid and compare the two for color and smell.

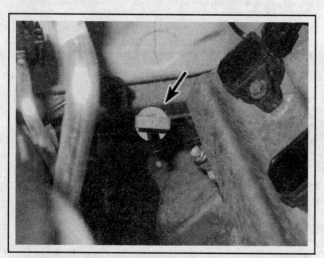

6.4 The automatic transmission fluid dipstick is located at the rear of the engine compartment (2004 and earlier models)

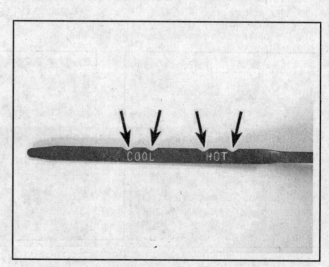

6.6 The automatic transmission fluid level must be maintained between the notches at the indicated operating temperature (2004 and earlier models)

7 Power steering fluid level check (every 3000 miles or 3 months)

♦ **Refer to illustrations 7.4 and 7.5**

1   On V6 models the fluid reservoir for the power steering pump is located at the right front corner of the engine, on the power steering pump. On V8 models the reservoir for the power steering pump is remotely mounted on the right side of the engine compartment.

2   The power steering fluid level can be checked with the engine either hot or cold.

V6 MODELS

3   With the engine off, use a rag to clean the reservoir cap and the area around the cap. This will help prevent foreign material from falling into the reservoir when the cap is removed.

4   Turn and pull out the reservoir cap, which has a dipstick attached to it. Wipe the fluid at the bottom of the dipstick with a clean rag. Reinstall the cap to get a fluid level reading. Remove the cap again and note the fluid level. It should be at the appropriate mark on the dipstick in relation to the engine temperature (see illustration).

V8 MODELS

5   On these models the power steering fluid reservoir is translucent and the level can be checked without removing the cap (see illustration).

ALL MODELS

6   If additional fluid is required, pour the specified type fluid (see *Recommended lubricants and fluids* at the end of this Chapter or your owner's manual) directly into the reservoir using a funnel to prevent spills.

7   If the reservoir requires frequent topping up, all power steering hoses, hose connections, the power steering pump and the steering gear should be carefully examined for leaks.

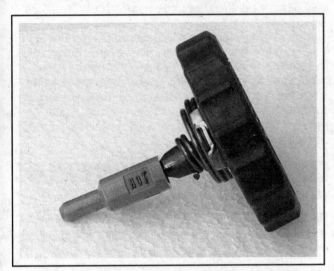

7.4 The marks on the dipstick indicate the safe fluid range (V6 models)

7.5 On V8 models the power steering fluid level can be viewed through the translucent reservoir

8 Engine oil and filter change (every 3000 miles or 3 months)

♦ **Refer to illustrations 8.2, 8.6, 8.7, 8.13a, 8.13b and 8.15**

1   Frequent oil changes are the best preventive maintenance the home mechanic can give the engine, because aging oil becomes diluted and contaminated, which leads to premature engine wear.

2   Make sure that you have all the necessary tools before you begin this procedure (see illustration). You should also have plenty of rags or newspapers handy for mopping up any spills.

3   Access to the underside of the vehicle is greatly improved if the vehicle can be lifted on a hoist, driven onto ramps or supported by jackstands.

4   If this is your first oil change, get under the vehicle and familiarize yourself with the location of the oil drain plug. The engine and exhaust components will be warm during the actual work, so try to anticipate any potential problems before the engine and accessories are hot.

5   Park the vehicle on a level spot. Start the engine and allow it to reach its normal operating temperature (the needle on the temperature gauge should be at least above the bottom mark). Warm oil and contaminates will flow out more easily. Turn off the engine when it's warmed up. Remove the filler cap on the valve cover.

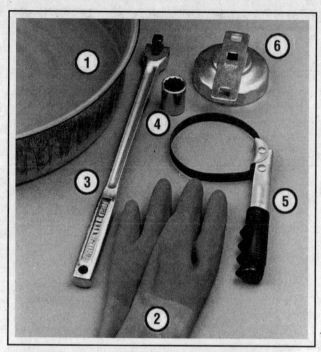

8.2 These tools are required when changing the engine oil and filter

1 **Drain pan** - It should be fairly shallow in depth, but wide to prevent spills

2 **Rubber gloves** - When removing the drain plug and filter, you will get oil on your hands (the gloves will prevent burns)

3 **Breaker bar** - Sometimes the oil drain plug is tight, and a long breaker bar is needed to loosen it

4 **Socket** - To be used with the breaker bar or a ratchet (must be the correct size to fit the drain plug)

5 **Filter wrench** - This is a metal band-type wrench, which requires clearance around the filter to be effective

6 **Filter wrench** - This type fits on the bottom of the filter and can be turned with a ratchet or breaker bar (different size wrenches are available for different types of filters)

6 Raise the vehicle and support it on jackstands.

✳✳ WARNING:

To avoid personal injury, never get beneath the vehicle when it is supported by only by a jack. The jack provided with your vehicle is designed solely for raising the vehicle to remove and replace the wheels. Always use jackstands to support the vehicle when it becomes necessary to place your body underneath the vehicle.

Remove the under-vehicle splash shields (see illustration).

7 Being careful not to touch the hot exhaust components, place the drain pan under the drain plug in the bottom of the pan and remove the plug (see illustration). You may want to wear gloves while unscrewing the plug the final few turns if the engine is really hot.

8 Allow the old oil to drain into the pan. It may be necessary to move the pan farther under the engine as the oil flow slows to a trickle. Inspect the old oil for the presence of metal shavings and chips.

9 After all the oil has drained, wipe off the drain plug with a clean rag. Even minute metal particles clinging to the plug would immediately contaminate the new oil.

10 Clean the area around the drain plug opening, reinstall the plug and tighten it securely, but do not strip the threads.

11 Move the drain pan into position under the oil filter.

12 Remove all tools, rags, etc. from under the vehicle, being careful not to spill the oil in the drain pan, then lower the vehicle.

13 Loosen the oil filter by turning it counterclockwise with the filter wrench (see illustration). Any standard filter wrench should work. Once the filter is loose, use your hands to unscrew it from the block. Just as the filter is detached from the block, immediately tilt the open end up to prevent the oil inside the filter from spilling out.

✳✳ WARNING:

The engine exhaust manifold may still be hot, so be careful.

14 With a clean rag, wipe off the mounting surface on the block. If a residue of old oil is allowed to remain, it will smoke when the block is

8.6 To gain access to various components under the front of the vehicle, the splash shields must be removed; they can be removed as a single assembly by unscrewing these bolts (2000 Tundra shown, other models similar)

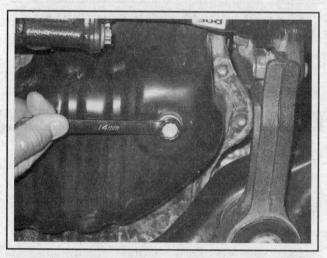

8.7 The engine oil drain plug is located on the bottom of the oil pan - it is usually very tight, so use the proper size box-end wrench or socket to avoid rounding it off

heated up. It will also prevent the new filter from seating properly. Also make sure that the none of the old gasket remains stuck to the mounting surface. It can be removed with a scraper if necessary.

15 Compare the old filter with the new one to make sure they are the same type. Smear some engine oil on the rubber gasket of the new filter (see illustration). Attach the filter to the engine, following the tightening directions printed on the filter canister or packing box. Most filter manufacturers recommend against using a filter wrench due to the possibility of overtightening and damage to the seal.

16 Add new oil to the engine through the oil filler cap in the valve cover. Use a spout or funnel to prevent oil from spilling onto the top of the engine. Pour three quarts of fresh oil into the engine. Wait a few minutes to allow the oil to drain into the pan, then check the level on the oil dipstick (see Section 4 if necessary). If the oil level is at or near the H mark, install the filler cap hand tight, start the engine and allow the new oil to circulate.

17 Allow the engine to run for about a minute. While the engine is running, look under the vehicle and check for leaks at the oil pan drain plug and around the oil filter. If either is leaking, stop the engine and tighten the plug or filter slightly.

18 Wait a few minutes to allow the oil to trickle down into the pan, then recheck the level on the dipstick and, if necessary, add enough oil to bring the level to the upper mark.

19 During the first few trips after an oil change, make it a point to check frequently for leaks and proper oil level.

20 The old oil drained from the engine cannot be reused in its present state and should be disposed of. Check with your local auto parts store, disposal facility or environmental agency to see if they will accept the oil for recycling. After the oil has cooled it can be drained into a container (capped plastic jugs, topped bottles, milk cartons, etc.) for transport to one of these disposal sites. Don't dispose of the oil by pouring it on the ground or down a drain!

8.13a The oil filter is usually on very tight and will require a special wrench for removal - DO NOT use the wrench to tighten the new filter!

8.13b On 2005 V6 models, the oil filter is located on the timing chain cover, so it's easier to access than on other models. But when you unscrew it, some oil will run out before you can tilt the open end up. The shield catches any spilled oil, and by removing a rubber plug from the underside of the shield, you can drain spilled oil through a small drain hole onto a shop rag

8.15 Lubricate the oil filter gasket with clean engine oil before installing the filter on the engine

9 Battery check, maintenance and charging (every 7500 miles or 6 months)

▶ Refer to illustrations 9.1, 9.6a, 9.6b, 9.7a and 9.7b

✳✳ WARNING:

Certain precautions must be followed when checking and servicing the battery. Hydrogen gas, which is highly flammable, is always present in the battery cells, so keep lighted tobacco and all other open flames and sparks away from the battery. The electrolyte inside the battery is actually dilute sulfuric acid, which will cause injury if splashed on your skin or in your eyes. It will also ruin clothes and painted surfaces. When removing the battery cables, always detach the negative cable first and hook it up last!

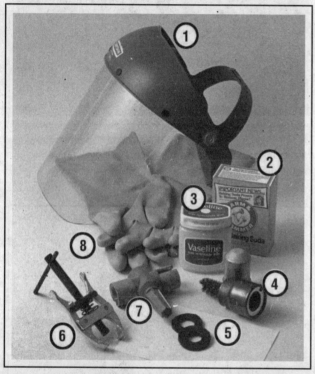

9.1 Tools and materials required for battery maintenance

1 *Face shield/safety goggles* - *When removing corrosion with a brush, the acidic particles can easily fly up into your eyes*
2 *Baking soda* - *A solution of baking soda and water can be used to neutralize corrosion*
3 *Petroleum jelly* - *A layer of this on the battery posts will help prevent corrosion*
4 *Battery post/cable cleaner* - *This wire brush cleaning tool will remove all traces of corrosion from the battery posts and cable clamps*
5 *Treated felt washers* - *Placing one of these on each post, directly under the cable clamps, will help prevent corrosion*
6 *Puller* - *Sometimes the cable clamps are very difficult to pull off the posts, even after the nut/bolt has been completely loosened. This tool pulls the clamp straight up and off the post without damage*
7 *Battery post/cable cleaner* - *Here is another cleaning tool which is a slightly different version of Number 4 above, but it does the same thing*
8 *Rubber gloves* - *Another safety item to consider when servicing the battery; remember that's acid inside the battery!*

MAINTENANCE

1 A routine preventive maintenance program for the battery in your vehicle is the only way to ensure quick and reliable starts. But before performing any battery maintenance, make sure that you have the proper equipment necessary to work safely around the battery (see illustration).

2 There are also several precautions that should be taken whenever battery maintenance is performed. Before servicing the battery, always turn the engine and all accessories off and disconnect the cable from the negative terminal of the battery.

3 The battery produces hydrogen gas, which is both flammable and explosive. Never create a spark, smoke or light a match around the battery. Always charge the battery in a ventilated area.

4 Electrolyte contains poisonous and corrosive sulfuric acid. Do not allow it to get in your eyes, on your skin on your clothes. Never ingest it. Wear protective safety glasses when working near the battery. Keep children away from the battery.

5 Note the external condition of the battery. If the positive terminal and cable clamp on your vehicle's battery is equipped with a rubber protector, make sure that it's not torn or damaged. It should completely cover the terminal. Look for any corroded or loose connections, cracks in the case or cover or loose hold-down clamps. Also check the entire length of each cable for cracks and frayed conductors.

6 If corrosion, which looks like white, fluffy deposits (see illustration) is evident, particularly around the terminals, the battery should be removed for cleaning. Loosen the cable clamp bolts with a wrench, being careful to remove the ground cable first, and slide them off the terminals (see illustration). Then disconnect the hold-down clamp bolt and nut, remove the clamp and lift the battery from the engine compartment.

7 Clean the cable clamps thoroughly with a battery brush or a terminal cleaner and a solution of warm water and baking soda (see illustration). Wash the terminals and the top of the battery case with the same solution but make sure that the solution doesn't get into the battery. When cleaning the cables, terminals and battery top, wear safety goggles and rubber gloves to prevent any solution from coming in contact with your eyes or hands. Wear old clothes too - even diluted, sulfuric acid splashed onto clothes will burn holes in them. If the terminals

9.6a Battery terminal corrosion usually appears as light, fluffy powder

have been extensively corroded, clean them up with a terminal cleaner (see illustration). Thoroughly wash all cleaned areas with plain water.

8 Make sure that the battery tray is in good condition and the hold-down clamp bolts are tight. If the battery is removed from the tray, make sure no parts remain in the bottom of the tray when the battery is reinstalled. When reinstalling the hold-down clamp bolts, do not over-tighten them.

9 Any metal parts of the vehicle damaged by corrosion should be covered with a zinc-based primer, then painted.

10 Information on removing and installing the battery can be found in Chapter 5. Information on jump starting can be found at the front of this manual.

CHARGING

✳✳ WARNING:

When batteries are being charged, hydrogen gas, which is very explosive and flammable, is produced. Do not smoke or allow open flames near a battery. Wear eye protection when near the battery during charging. Also, make sure the charger is unplugged before connecting or disconnecting the battery from the charger.

➡**Note: The manufacturer recommends the battery be removed from the vehicle for charging because the gas that escapes during this procedure can damage the paint. Fast charging with the battery cables connected can result in damage to the electrical system.**

11 Slow-rate charging is the best way to restore a battery that's discharged to the point where it will not start the engine. It's also a good way to maintain the battery charge in a vehicle that's only driven a few miles between starts. Maintaining the battery charge is particularly important in the winter when the battery must work harder to start the engine and electrical accessorics that drain the battery are in greater use.

12 It's best to use a one or two-amp battery charger (sometimes called a "trickle" charger). They are the safest and put the least strain on the battery. They are also the least expensive. For a faster charge, you can use a higher amperage charger, but don't use one rated more than 1/10th the amp/hour rating of the battery. Rapid boost charges that claim to restore the power of the battery in one to two hours are hardest on the battery and can damage batteries not in good condition. This type of charging should only be used in emergency situations.

13 The average time necessary to charge a battery should be listed in the instructions that come with the charger. As a general rule, a trickle charger will charge a battery in 12 to 16 hours.

14 Remove all the cell caps (if equipped) and cover the holes with a clean cloth to prevent spattering electrolyte. Disconnect the negative battery cable and hook the battery charger cable clamps up to the battery posts (positive to positive, negative to negative), then plug in the charger. Make sure it is set at 12-volts if it has a selector switch.

15 If you're using a charger with a rate higher than two amps, check the battery regularly during charging to make sure it doesn't overheat. If you're using a trickle charger, you can safely let the battery charge overnight after you've checked it regularly for the first couple of hours.

16 If the battery has removable cell caps, measure the specific gravity with a hydrometer every hour during the last few hours of the charging cycle. Hydrometers are available inexpensively from auto parts stores - follow the instructions that come with the hydrometer. Consider the battery charged when there's no change in the specific gravity reading for two hours and the electrolyte in the cells is gassing (bubbling) freely. The specific gravity reading from each cell should be very close to the others. If not, the battery probably has a bad cell(s).

17 Some batteries with sealed tops have built-in hydrometers on the top that indicate the state of charge by the color displayed in the hydrometer window. Normally, a bright-colored hydrometer indicates a full charge and a dark hydrometer indicates the battery still needs charging.

18 If the battery has a sealed top and no built-in hydrometer, you can hook up a voltmeter across the battery terminals to check the charge. A fully charged battery should read 12.6 volts or higher after the surface charge has been removed.

19 Further information on the battery and jump starting can be found in Chapter 5 and at the front of this manual.

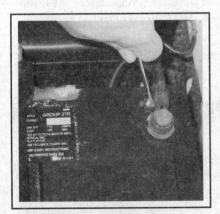

9.6b Removing the cable from a battery post with a wrench - sometimes special battery pliers are required for this procedure if corrosion has caused deterioration of the nut hex (always remove the ground cable first and hook it up last!)

9.7a When cleaning the cable clamps, all corrosion must be removed (the inside of the clamp is tapered to match the taper on the post, so don't remove too much material)

9.7b Regardless of the type of tool used on the battery posts, a clean, shiny surface should be the end result

10 Cooling system check (every 7500 miles or 6 months)

▶ **Refer to illustration 10.4**

1   Many major engine failures can be attributed to a faulty cooling system. If the vehicle is equipped with an automatic transmission, the cooling system also cools the transmission fluid and thus plays an important role in prolonging transmission life.

2   The cooling system should be checked with the engine cold. Do this before the vehicle is driven for the day or after it has been shut off for at least three hours.

3   Remove the radiator cap by turning it to the left until it reaches a stop. If you hear a hissing sound (indicating there is still pressure in the system), wait until this stops. Now press down on the cap with the palm of your hand and continue turning to the left until the cap can be removed. Thoroughly clean the cap, inside and out, with clean water. Also clean the filler neck on the radiator. All traces of corrosion should be removed. The coolant inside the radiator should be relatively transparent. If it is rust colored, the system should be drained and refilled (see Section 30). If the coolant level is not up to the top, add additional antifreeze/coolant mixture (see Section 4).

4   Carefully check the large upper and lower radiator hoses along with the smaller diameter heater hoses which run from the engine to the firewall. Inspect each hose along its entire length, replacing any hose which is cracked, swollen or shows signs of deterioration. Cracks may become more apparent if the hose is squeezed (see illustration). Regardless of condition, it's a good idea to replace hoses with new ones every two years.

5   Make sure all hose connections are tight. A leak in the cooling system will usually show up as white or rust colored deposits on the areas adjoining the leak. If wire-type clamps are used at the ends of the hoses, it may be a good idea to replace them with more secure screw-type clamps.

6   Use compressed air or a soft brush to remove bugs, leaves, etc. from the front of the radiator or air conditioning condenser. Be careful not to damage the delicate cooling fins or cut yourself on them.

7   Every other inspection, or at the first indication of cooling system problems, have the cap and system pressure tested. If you don't have a pressure tester, most gas stations and repair shops will do this for a minimal charge.

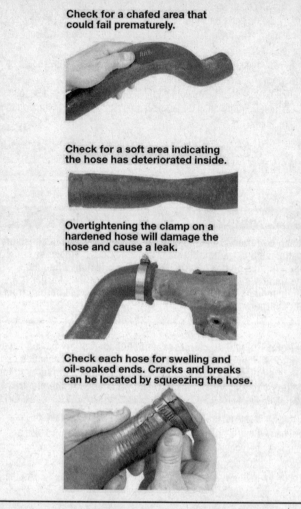

Check for a chafed area that could fail prematurely.

Check for a soft area indicating the hose has deteriorated inside.

Overtightening the clamp on a hardened hose will damage the hose and cause a leak.

Check each hose for swelling and oil-soaked ends. Cracks and breaks can be located by squeezing the hose.

10.4 Hoses, like drivebelts, have a habit of failing at the worst possible time - to prevent the inconvenience of a blown radiator or heater hose, inspect them carefully as shown here

11 Underhood hose check and replacement (every 7500 miles or 6 months)

GENERAL

❊❊ **WARNING:**

Replacement of air conditioning hoses must be left to a dealer service department or air conditioning shop that has the equipment to depressurize the system safely. Never remove air conditioning components or hoses until the system has been depressurized.

1   High temperatures in the engine compartment can cause the deterioration of the rubber and plastic hoses used for engine, accessory and emission systems operation. Periodic inspection should be made for cracks, loose clamps, material hardening and leaks. Information specific to the cooling system hoses can be found in Section 10.

2   Some, but not all, hoses are secured to the fittings with clamps. Where clamps are used, check to be sure they haven't lost their tension, allowing the hose to leak. If clamps aren't used, make sure the hose has not expanded and/or hardened where it slips over the fitting, allowing it to leak.

VACUUM HOSES

3   It's quite common for vacuum hoses, especially those in the emissions system, to be color coded or identified by colored stripes molded into them. Various systems require hoses with different wall thickness, collapse resistance and temperature resistance. When replacing hoses, be sure the new ones are made of the same material.

4   Often the only effective way to check a hose is to remove it completely from the vehicle. If more than one hose is removed, be sure to label the hoses and fittings to ensure correct installation.

5   When checking vacuum hoses, be sure to include any plastic T-fittings in the check. Inspect the fittings for cracks and the hose where it fits over the fitting for distortion, which could cause leakage.

6   A small piece of vacuum hose (1/4-inch inside diameter) can be used as a stethoscope to detect vacuum leaks. Hold one end of the hose to your ear and probe around vacuum hoses and fittings, listening for the "hissing" sound characteristic of a vacuum leak.

❄❄ WARNING:

When probing with the vacuum hose stethoscope, be very careful not to come into contact with moving engine components such as the drivebelt, cooling fan, etc.

FUEL HOSE

❄❄ WARNING:

Gasoline is extremely flammable, so take extra precautions when you work on any part of the fuel system. Don't smoke or allow open flames or bare light bulbs near the work area, and don't work in a garage where a gas-type appliance (such as a water heater or clothes dryer) is present. Since gasoline is carcinogenic, wear fuel-resistant gloves when there's a possibility of being exposed to fuel, and, if you spill any fuel on your skin, rinse it off immediately with soap and water. Mop up any spills immediately and do not store fuel-soaked rags where they could ignite. The fuel system is under constant pressure, so if any fuel lines are to be disconnected, the fuel pressure in the system must be relieved first (see Chapter 4). When you perform any kind of work on the fuel system, wear safety glasses and have a Class B Type fire extinguisher on hand.

7   Check all rubber fuel lines for deterioration and chafing. Check especially for cracks in areas where the hose bends and just before fittings, such as where a hose attaches to the fuel filter.

8   Only high quality fuel line, designed for high-pressure fuel injection systems, may be used for fuel line replacement. Never, under any circumstances, use standard fuel hose, unreinforced vacuum line, clear plastic tubing or water hose for fuel lines.

9   Spring-type clamps are commonly used on fuel lines. These clamps often lose their tension over a period of time, and can be "sprung" during removal. Replace all spring-type clamps with screw clamps whenever a hose is replaced.

METAL LINES

10 Sections of metal line are often used in the fuel system. Check carefully to be sure the line has not been bent or crimped and that cracks have not started in the line.

11 If a section of metal fuel line must be replaced, only seamless steel tubing should be used, since copper and aluminum tubing don't have the strength necessary to withstand normal engine vibration.

12 Check the metal brake lines where they enter the master cylinder and brake proportioning unit (if used) for cracks in the lines or loose fittings. Any sign of brake fluid leakage calls for an immediate thorough inspection of the brake system.

12 Wiper blade inspection and replacement (every 7500 miles or 6 months)

▶ **Refer to illustrations 12.3, 12.5 and 12.6**

1   The windshield wiper and blade assembly should be inspected periodically for damage, loose components and cracked or worn blade elements.

2   Road film can build up on the wiper blades and affect their efficiency, so they should be washed regularly with a mild detergent solution.

3   The action of the wiping mechanism can loosen bolts, nuts and fasteners, so they should be checked and tightened, as necessary (see illustration), at the same time the wiper blades are checked.

4   If the wiper blade elements are cracked, worn or warped, or no longer clean adequately, they should be replaced with new ones.

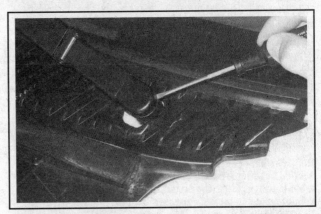

12.3 Pry open the trim cap and check the tightness of the wiper arm retaining nut

5   Lift the arm assembly away from the glass for clearance, press on the release lever, then slide the wiper blade assembly out of the hook in the end of the arm (see illustration).

6   Use needle-nose pliers to compress the blade element, then slide the element out of the frame and discard it (see illustration).

7   Installation is the reverse of removal.

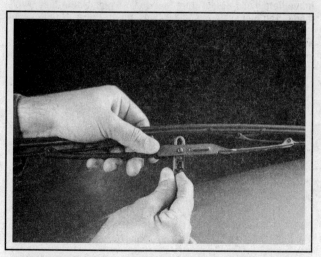

12.5 Press on the release tab, then push the blade assembly down and out of the hook in the arm

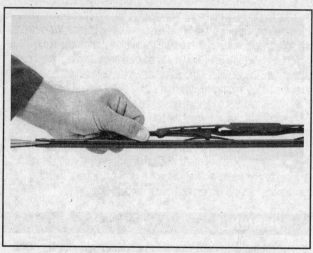

12.6 Use needle-nose pliers to compress the rubber element, then slide the element out - slide the new element in and lock the blade assembly fingers into the notches of the wiper element

13   Tire rotation (every 7500 miles or 6 months)

▶ **Refer to illustration 13.2**

1   The tires should be rotated at the specified intervals and whenever uneven wear is noticed. Since the vehicle will be raised and the tires removed anyway, check the brakes (see Section 26) at this time.

2   Radial tires must be rotated in a specific pattern (see illustration).

3   Refer to the information in *Jacking and towing* at the front of this manual for the proper procedures to follow when raising the vehicle and changing a tire. If the brakes are to be checked, do not apply the parking brake as stated. Make sure the tires are blocked to prevent the vehicle from rolling.

4   Preferably, the entire vehicle should be raised at the same time. This can be done on a hoist or by jacking up each corner and then lowering the vehicle onto jackstands placed under the frame rails. Always use four jackstands and make sure the vehicle is firmly supported.

5   After rotation, check and adjust the tire pressures as necessary and be sure to check the lug nut tightness.

6   For further information on the wheels and tires, refer to Chapter 10.

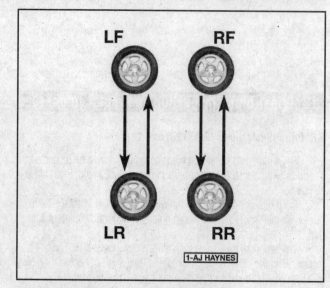

13.2 Tire rotation diagram for radial tires

14 Suspension and steering check (every 7500 miles or 6 months)

▶ Refer to illustrations 14.1, 14.6a, 14.6b and 14.7

➡Note: For detailed illustrations of the steering and suspension components, refer to Chapter 10.

WITH THE WHEELS ON THE GROUND

1   With the vehicle stopped and the front wheels pointed straight ahead, rock the steering wheel gently back and forth. If freeplay (see illustration) is excessive, a front wheel bearing, main shaft U-joint, intermediate shaft U-joint or tie rod end is worn or the steering gear is out of adjustment or broken. Refer to Chapter 10 for the appropriate repair procedure.

2   Other symptoms, such as excessive vehicle body movement over rough roads, swaying (leaning) around corners and binding as the steering wheel is turned, may indicate faulty steering and/or suspension components.

3   Check the shock absorbers by pushing down and releasing the vehicle several times at each corner. If the vehicle does not come back to a level position within one or two bounces, the shocks are worn and must be replaced. When bouncing the vehicle up and down, listen for squeaks and noises from the suspension components.

UNDER THE VEHICLE

4   Raise the vehicle with a floor jack and support it securely on jackstands. See *Jacking and towing* at the front of this book for proper jacking points.

5   Check the tires for irregular wear patterns and proper inflation. See Section 5 in this Chapter for information regarding tire wear.

6   Inspect the universal joint between the steering shaft and the steering rack. Check the steering rack and driveaxle boots for grease leakage. Check the steering linkage for looseness or damage. Check the tie-rod ends for excessive play. Look for loose bolts, broken or disconnected parts and deteriorated rubber bushings on all suspension and steering components (see illustrations). While an assistant turns the steering wheel from side to side, check the steering components

for free movement, chafing and binding. If the steering components do not seem to be reacting with the movement of the steering wheel, try to determine where the slack is located.

7   Check the wheel bearings. Do this by spinning the front wheels. Listen for any abnormal noises and watch to make sure the wheel spins true (doesn't wobble). Grab the top and bottom of the tire and pull in-and-out on it. Notice any movement which would indicate a loose wheel bearing assembly (see illustration). If the bearings are loose, they are in need of replacement. Refer to Chapter 10 for more information.

8   Check the steering knuckle, moving the knuckle up and down with a prybar to ensure that knuckle bearings have no play. If the bearings are suspect, they should be checked and repacked. Refer to Chapter 10 for more information.

9   Inspect the driveshafts for worn U-joints and for excessive play in the slip yoke and spline area (see Chapter 8).

10 Check the transfer case and differentials for evidence of fluid leakage.

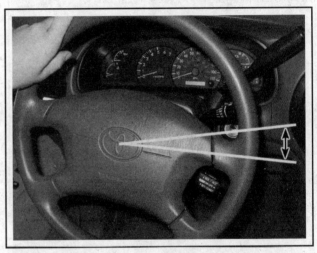

14.1 Steering wheel freeplay is the amount of travel between an initial steering input and the point at which the front wheels begin to turn (indicated by a slight resistance)

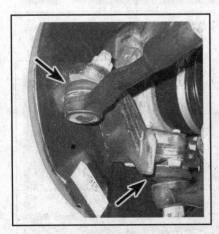

14.6a Inspect the suspension for deteriorated rubber bushings and torn grease seals

14.6b Check the stabilizer bar bushings for deterioration at the front and the rear of the vehicle

14.7 Grasp the tire as shown and check for endplay at the wheel bearings - if endplay is found the wheel bearings must be replaced

15 Driveshaft and body lubrication (every 7500 miles or 6 months)

♦ **Refer to illustrations 15.1, 15.5, 15.8 and 15.10**

1   Refer to *Recommended lubricants and fluids* at the end of this Chapter to obtain the necessary grease, etc. You will also need a grease gun (see illustration).

2   Look under the vehicle for grease fittings on the driveline components. They are normally found on the driveshaft universal joints and slip yokes.

3   For easier access under the vehicle, raise it with a jack and place jackstands under the frame. Make sure it is safely supported by the stands. If the wheels are to be removed at this interval for tire rotation or brake inspection, loosen the lug nuts slightly while the vehicle is still on the ground.

4   Before beginning, force a little grease out of the nozzle to remove any dirt from the end of the gun. Wipe the nozzle clean with a rag.

5   With the grease gun and plenty of clean rags, crawl under the vehicle and begin lubricating the driveshaft universal joints (see illustration).

6   Wipe the area around the grease fitting free of dirt, then squeeze the trigger on the grease gun to force grease into the component. Continue pumping grease into the fitting until it just oozes out of the bearing cup seals. If it escapes around the grease gun nozzle, the fitting is clogged or the nozzle is not completely seated on the fitting. Resecure the gun nozzle to the fitting and try again. If necessary, replace the fitting with a new one.

7   Wipe the excess grease from the components and the grease fitting. Repeat the procedure for the remaining fittings.

8   Lubricate the driveshaft slip yoke by pumping grease into the fitting until it can be seen coming out of the slip yoke seal (see illustration).

9   While you are under the vehicle, clean and lubricate the parking brake cable along with the cable guides and levers. This can be done by smearing some chassis grease onto the cable and its related parts with your fingers.

10 Lubricate the contact points on the steering knuckle stop and

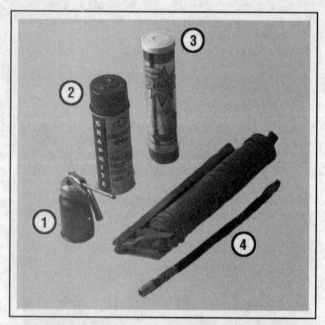

15.1 Materials required for chassis and body lubrication

1   *Engine oil - Light engine oil in a can like this can be used for door and hood hinges*
2   *Graphite spray - Used to lubricate lock cylinders*
3   *Grease - Grease, in a variety of types and weights, is available for use in a grease gun. Check the Specifications for your requirements*
4   *Grease gun - A common grease gun, shown here with a detachable hose and nozzle, is needed for chassis lubrication. After use, clean it thoroughly*

15.5 Pump grease into the universal joints until it can be seen coming out from the seals

15.8 The slip joint grease fitting is located on the yoke - pump grease into it until it comes out of the slip joint seal

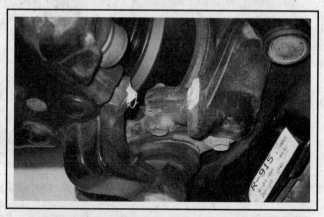

15.10 Use lithium base grease to lubricate the contact points on the steering knuckle stops

adjustment bolt if equipped (see illustration).

11 Open the hood and smear a little chassis grease on the hood latch mechanism. Have an assistant pull the hood release lever from inside the vehicle as you lubricate the cable at the latch.

12 Lubricate all the hinges (door, hood, etc.) with engine oil to keep them in proper working order.

13 The key lock cylinders can be lubricated with spray-on graphite or silicone lubricant, which is available at auto parts stores.

14 Lubricate the door weatherstripping with silicone spray. This will reduce chafing and retard wear.

16 Exhaust system check (every 7500 miles or 6 months)

▶ **Refer to illustrations 16.2a and 16.2b**

1 With the engine cold (at least three hours after the vehicle has been driven), check the complete exhaust system from the manifold to the end of the tailpipe. Be careful around the catalytic converter (if equipped), which may be hot even after three hours. The inspection should be done with the vehicle on a hoist to permit unrestricted access. If a hoist isn't available, raise the vehicle and support it securely on jackstands.

2 Check the exhaust pipes and connections for signs of leakage and/or corrosion indicating a potential failure. Make sure that all brack-ets and hangers are in good condition and tight (see illustrations).

3 Inspect the underside of the body for holes, corrosion, open seams, etc. which may allow exhaust gasses to enter the passenger compartment. Seal all body openings with silicone sealant or body putty.

4 Rattles and other noises can often be traced to the exhaust system, especially the hangers, mounts and heat shields. Try to move the pipes, mufflers and catalytic converter. If the components can come in contact with the body or suspension parts, secure the exhaust system with new brackets and hangers.

16.2a Check the exhaust pipes and connections for signs of leakage and corrosion

16.2b Check the exhaust system rubber hangers for cracks and damage

17 Manual transmission lubricant level check (every 7500 miles or 6 months)

▶ **Refer to illustration 17.2**

1 The manual transmission has a filler plug which must be removed to check the lubricant level. If the vehicle is raised to gain access to the plug, be sure to support it safely on jackstands - DO NOT crawl under a vehicle which is supported only by a jack! Be sure the vehicle is level or the check may be inaccurate.

2 Using a wrench, unscrew the fill plug from the transmission (see illustration) and use a finger to reach inside the housing to determine the lubricant level. The level should be at or near the bottom of the plug hole.

3 If it isn't, add the recommended lubricant through the plug hole with a pump or squeeze bottle.

4 Install and tighten the plug and check for leaks after the first few miles of driving.

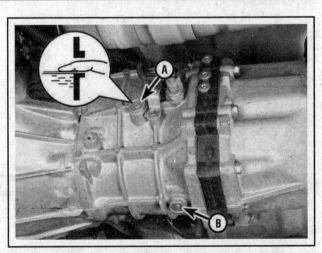

17.2 Locations of the manual transmission fill plug (A) and drain plug (B) - use your finger as a dipstick to check the lubricant level

18 Transfer case lubricant level check (4WD models only) (every 7500 miles or 6 months)

♦ **Refer to illustration 18.2**

1 The transfer case lubricant level is checked by removing the upper plug located in the back of the case.

2 Use a finger to reach inside the housing to determine the lubricant level. The lubricant level should be just at the bottom of the hole (see illustration). If not, add the appropriate lubricant through the opening.

3 Install and tighten the plug and check for leaks after the first few miles of driving.

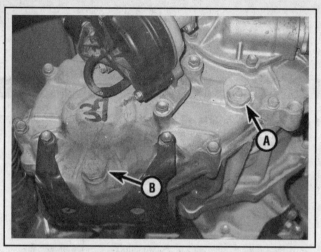

18.2 Typical locations of the transfer case fill plug (A) and drain plug (B) - use your finger as a dipstick to check the lubricant level

19 Differential lubricant level check (every 7500 miles or 6 months)

♦ **Refer to illustration 19.2**

➡ **Note: 4WD models covered by this manual have two differentials; be sure to check the lubricant level in both differentials.**

1 The differential has a check/fill plug which must be removed to check the lubricant level. If the vehicle must be raised to gain access to the plug, be sure to support it safely on jackstands - DO NOT crawl under the vehicle when it's supported only by the jack.

2 Remove the oil check/fill plug from the back of the rear differential or the front of the front differential (see illustration). On some models a tag is located in the area of the plug which gives information regarding lubricant type, particularly on models equipped with a limited-slip differential.

3 Use a finger to reach inside the housing to determine the lubricant level. The oil level should be at the bottom of the plug opening. If it isn't, use a hand pump (available at auto parts stores) to add the specified lubricant until it just starts to run out of the opening.

4 Install the plug and tighten it securely.

19.2 The differential fill plug (A) and drain plug (B) are located on the axle housing - use your finger as a dipstick to check the lubricant level

20 Seat belt check (every 7500 miles or 6 months)

1 Check the seat belts, buckles, latch plates and guide loops for any obvious damage or signs of wear.

2 Make sure the seat belt reminder light comes on when the key is turned on.

3 The seat belts are designed to lock up during a sudden stop or impact, yet allow free movement during normal driving. The retractors should hold the belt against your chest while driving and rewind the belt when the buckle is unlatched.

4 If any of the above checks reveal problems with the seat belt system, replace parts as necessary.

21 Driveaxle boot check (every 15,000 miles or 12 months)

♦ **Refer to illustration 21.2**

1 The driveaxle boots are important because they prevent dirt, water and foreign material from entering and damaging the constant velocity joints (CV). Oil and grease can cause the boot material to deteriorate prematurely, so it's a good idea to wash the boots with soap and water.

2 Inspect the boots for tears and cracks as well as loose clamps (see illustration). If there is any evidence of cracks or leaking grease, they must be replaced as described in Chapter 8.

21.2 Check the condition of the driveaxle boot for signs of cracks and grease leaks

22 Positive Crankcase Ventilation (PCV) valve check and replacement (every 15,000 miles or 12 months)

♦ **Refer to illustration 22.2**

1 The PCV valve is usually located in the valve cover. To find the PCV valve, refer to Section 18 in Chapter 6.

2 With the engine idling at normal operating temperature, pull the valve (with hose attached) from the valve cover (see illustration).

3 Place your thumb over the valve opening. If there's no vacuum, check for a plugged hose, manifold port, or the valve itself. Replace any plugged or deteriorated hoses.

4 Turn off the engine and shake the PCV valve, listening for a rattle. If the valve doesn't rattle, replace it with a new one.

5 To replace the valve, pull it from the end of the hose, noting its installed position.

6 When purchasing a replacement PCV valve, make sure it's for your particular vehicle and engine size. Compare the old valve with the new one to make sure they're the same.

7 Push the valve into the end of the hose until it's seated.

8 Inspect all rubber hoses and grommets for damage and hardening. Replace them, if necessary.

9 Press the PCV valve and hose securely into position.

22.2 The PCV valve is located in the valve cover - with the engine running, put your finger over the end of the PCV valve; you should feel vacuum

23 Air filter check and replacement (every 15,000 miles or 12 months)

1 At the specified intervals, the air filter should be replaced with a new one. A thorough preventive maintenance schedule would also require the filter to be inspected between filter changes.

ALL MODELS EXCEPT 2005 V6 MODELS

♦ **Refer to illustrations 23.3 and 23.5**

2 The air filter housing is mounted in the right front corner of the engine compartment.

3 Unlatch the cover retaining clips (see illustration).

4 Detach all hoses that would interfere with the removal of the air filter cover from the housing. While the top cover is off, be careful not to drop anything down into the housing.

5 Lift the air filter element out of the housing (see illustration) and wipe out the inside of the housing with a clean rag.

6 Inspect the outer surface of the filter element. If it is dirty, replace it. If it is only moderately dusty, it can be reused by blowing it clean

from the back to the front surface with compressed air. Because it is a pleated paper type filter, it cannot be washed or oiled. If it cannot be cleaned satisfactorily with compressed air, discard and replace it.

✳✳ CAUTION:

Never drive the vehicle with the air cleaner removed. Excessive engine wear could result and backfiring could even cause a fire under the hood.

7 Place the new filter in the air filter housing, making sure it seats properly.
8 Installation of the cover is the reverse of removal.

2005 AND LATER V6 MODELS

▶ **Refer to illustrations 23.10 and 23.11**

9 The air filter housing is located on the right side of the intake manifold plenum.
10 Open the two retaining clips (see illustration) and swing open the air filter housing.
11 Remove the air filter element from the housing (see illustration), then wipe out the inside of the housing with a clean rag.
12 Inspect the outer surface of the filter element. If it's dirty, replace it. If it's only moderately dusty, blow it out from the back to the front surface with low-pressure compressed air. Because the filter element is a pleated paper type filter, it cannot be oiled or washed. If you can't clean the filter element satisfactorily with compressed air, replace it.

✳✳ CAUTION:

Never operate the engine with the air filter removed.

13 Installation is the reverse of removal.

23.3 The first step in removing the air filter is detaching the clips on the housing

23.5 After raising the cover, the filter element can be lifted out of the housing

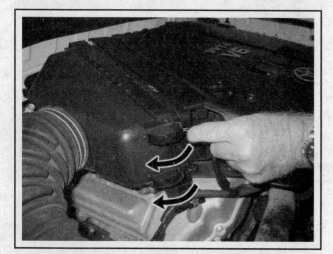

23.10 To remove the air filter housing, open these two clips and swing open the housing (the firewall end of the housing is hinged)

23.11 Note how the old air filter is installed, then pull it out of the air filter housing. The new filter must be installed in exactly the same way

24 Drivebelt check, adjustment and replacement (every 15,000 miles or 12 months)

�֎ WARNING:

Before checking, adjusting or replacing a drivebelt, make sure the ignition key is not in the ignition lock cylinder.

CHECK

▶ **Refer to illustrations 24.2, 24.3 and 24.4**

1   The drivebelts are located at the front of the engine and play an important role in the operation of the vehicle and its components. Due to their function and material makeup, belts are prone to failure after a period of time and should be inspected and adjusted periodically to prevent major damage. 2000 through 2004 V6 engines use multiple belts and require periodic adjustment. 2005 V6 engines and all V8 models use a single serpentine belt; no adjustment is necessary because an automatic tensioner is used.

2   With the engine turned off, open the hood and locate the drivebelt(s) at the front of the engine. Use a flashlight to carefully check for a severed core, separation of the adhesive rubber on both sides of the core and for core separation from the belt side. Inspect the ribs for separation from the adhesive rubber and for cracking or separation of the ribs, torn or worn ribs or cracks in the inner ridges of the ribs (see illustration). Also check for fraying and glazing, which gives the belt a shiny appearance. Inspect both sides of the belt by twisting the belt to check the underside. Use your fingers to feel the belt where you can't see it. If any of the above conditions are evident, replace the belt(s).

➡**Note: On some models drivebelt inspection can be made easier by removing the under-vehicle splash shield (see illustration 8.6).**

3   On V6 engines, the tension of each belt can be checked by pushing on it at a distance halfway between the pulleys. Apply about 10 pounds of force with your thumb and see how much the belt moves down (deflects). Measure the deflection with a ruler (see illustration). The belt should deflect about 1/4-inch if the distance between pulleys is between 7 and 11 inches and around 1/2-inch if the distance is between 12 and 16 inches.

4   On V8 models, the belt should be replaced when the pointer on the automatic tensioner falls outside of the acceptable range (see illustration).

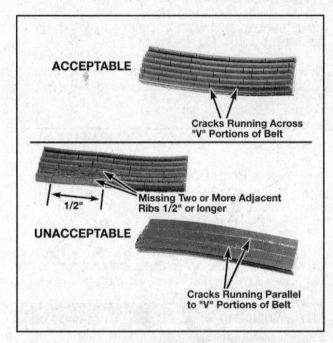

24.2 Small cracks in the underside of the V-ribbed belt are acceptable - lengthwise cracks or missing pieces are cause for replacement

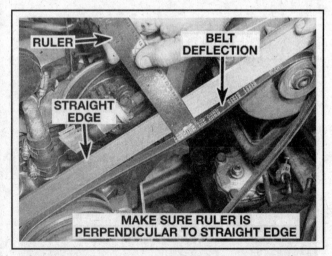

24.3 Measuring drivebelt deflection with a straightedge and ruler (typical)

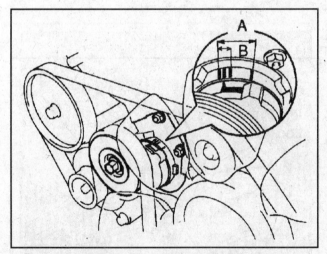

24.4 On V8 engines, drivebelt wear is checked on the automatic tensioner - if the pointer is outside of area A the belt must be replaced (when the belt is new, the pointer should be within area B)

ADJUSTMENT (2000 THROUGH 2004 V6 ENGINE)

♦ **Refer to illustrations 24.6a, 24.6b and 24.7**

5 If it is necessary to adjust the belt tension on these models, either to make the belt tighter or looser, it is done by moving an idler pulley or the component, depending on which belt is being adjusted.

6 The alternator and power steering pump drivebelts are adjusted by loosening a lock nut or bolt and a pivot bolt on the component, then turning an adjustment bolt to move the component in the required direction within its mounting bracket (see illustrations). One the desired tension has been obtained, tighten the lock bolt or nut securely, followed by the pivot bolt.

7 The air conditioning compressor drivebelt is adjusted with a idler pulley. Loosen the lock nut on the idler pulley, then turn the adjusting bolt until the desired tension is obtained (see illustration). Tighten the locknut securely.

REPLACEMENT

➡ **Note: Take the old belt(s) with you when purchasing new ones in order to make a direct comparison for length, width and design.**

2000 through 2004 V6 engine

♦ **Refer to illustration 24.9**

8 Follow the above procedures for drivebelt adjustment but slip the belt off the pulleys and remove it. Since belts tend to wear out more or less at the same time, it's a good idea to replace all of them at the same time.

9 When installing the belt(s), make sure the belt is centered on the pulleys (see illustration).

10 Adjust the belt(s) as described earlier in this Section.

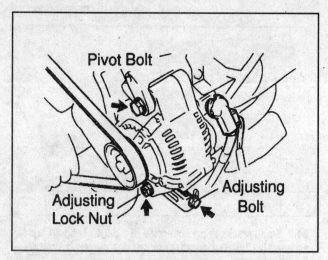

24.6a Alternator drivebelt adjustment details (2000 through 2004 V6 engine)

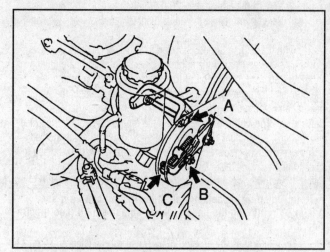

24.6b Power steering pump drivebelt adjustment details (2000 through 2004 V6 engine)

A	Pivot bolt	B	Lock nut	C	Adjusting bolt

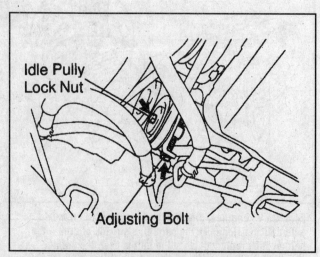

24.7 Air conditioning compressor drivebelt adjustment details (2000 through 2004 V6 engine)

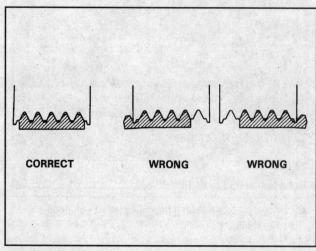

24.9 Make sure the belt is centered on all of the pulleys when installed

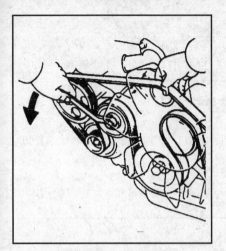

24.11 Rotate the belt tensioner counterclockwise to release tension on the belt (V8 engine)

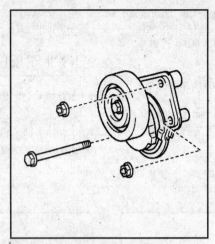

24.14 Drivebelt tensioner mounting details (V8 engine)

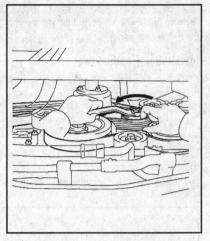

24.19 To release tension on the drivebelt, put a box wrench on the tensioner pulley bolt and rotate the tensioner counterclockwise (2005 and later V6 models)

V8 engine

Belt replacement

▶ Refer to illustration 24.11

11 The automatic tensioner must be released to allow drivebelt replacement. Check to make sure the key has been removed from the ignition lock cylinder, then place a wrench on the bolt in the center of the tensioner pulley and rotate it counterclockwise to release tension on the belt (see illustration). Remove the belt and slowly release the tensioner. Install the new belt and rotate the tensioner counterclockwise to allow the belt to slip over it, then release the tensioner slowly until it contacts the drivebelt. Make sure the drivebelt is centered on all of the pulleys (see illustration 24.9).

Tensioner replacement

▶ Refer to illustration 24.14

12 Be sure the key is not in the ignition lock cylinder, then remove the drivebelt as described in the previous Step.

13 Remove the under-vehicle splash shield (see illustration 8.6).

14 Working from above, remove the two tensioner mounting nuts (see illustration).

15 Working from below, remove the tensioner mounting bolt, then remove the tensioner and guide it out towards the bottom of the vehicle.

16 Installation is the reverse of the removal procedure. Tighten the fasteners to the torque listed in this Chapter's Specifications.

2005 and later V6 engine

Removal

▶ Refer to illustration 24.19

17 Disconnect the cable from the negative battery terminal.

18 Remove the four cooling fan mounting nuts and remove the cooling fan (see Chapter 3).

19 Put a box wrench on the belt tensioner pulley bolt, rotate the tensioner counterclockwise (see illustration) to release tension on the drivebelt and remove the belt from the tensioner.

Installation (method one)

▶ Refer to illustrations 24.20 and 24.21

20 Before installing the drivebelt, lock the tensioner as follows: turn the tensioner clockwise, align the two holes on the tensioner assembly and insert a 0.24-inch dowel pin or drill bit through the two holes (see illustration).

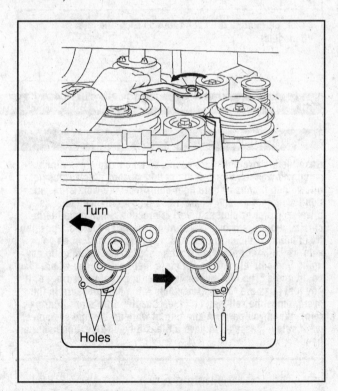

24.20 To lock the belt tensioner while installing the drivebelt, turn the tensioner counterclockwise until the two holes are aligned, then insert a 0.024-inch dowel pin or drill bit through the holes (2005 and later V6 models)

21 Install the drivebelt. Be sure to route it correctly (see illustration). Remove the dowel pin or drill bit and tension the belt. Make sure that the belt is centered on the all of the pulleys (see illustration 24.9).

Installation (method two)

22 If you have difficulty installing the drivebelt using method one, try this method instead.

23 Release tension on the belt tensioner pulley, then route the belt exactly as shown in illustration 24.21, except for the power steering pump pulley.

24 Release belt tension by rotating the belt tensioner counterclockwise, put the belt on the power steering pump pulley, then tension the belt. Make sure that the belt is centered on the all of the pulleys (see illustration 24.9).

25 Installation is otherwise the reverse of removal.

TENSIONER REPLACEMENT

◆ **Refer to illustration 24.27**

26 Remove the drivebelt (see Steps 17 through 19).
27 Remove the five tensioner mounting bolts (see illustration) and remove the tensioner.
28 Installation is the reverse of removal.

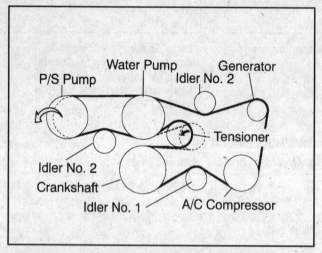

24.21 Serpentine drivebelt routing (2005 and later V6 models)

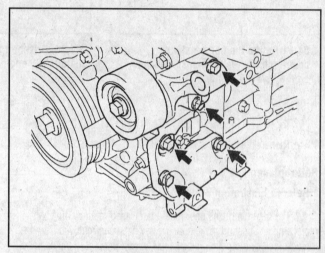

24.27 To detach the drivebelt tensioner, remove these five bolts (2005 and later V6 models)

25 Fuel system check (every 15,000 miles or 12 months)

✳✳ WARNING:

Gasoline is extremely flammable, so take extra precautions when you work on any part of the fuel system. Don't smoke or allow open flames or bare light bulbs near the work area, and don't work in a garage where a gas-type appliance (such as a water heater or clothes dryer) is present. Since gasoline is carcinogenic, wear fuel-resistant gloves when there's a possibility of being exposed to fuel, and, if you spill any fuel on your skin, rinse it off immediately with soap and water. Mop up any spills immediately and do not store fuel-soaked rags where they could ignite. The fuel system is under constant pressure, so, if any fuel lines are to be disconnected, the fuel pressure in the system must be relieved first (see Chapter 4 for more information). When you perform any kind of work on the fuel system, wear safety glasses and have a Class B type fire extinguisher on hand.

1 The fuel system is most easily checked with the vehicle raised on a hoist so the components underneath the vehicle are readily visible and accessible.

2 If the smell of gasoline is noticed while driving or after the vehicle has been in the sun, the system should be thoroughly inspected immediately.

3 Remove the fuel tank cap and check for damage, corrosion and an unbroken sealing imprint on the gasket. Replace the cap with a new one if necessary.

4 With the vehicle raised and safely supported, inspect the gas tank and filler neck for punctures, cracks and other damage. The connection between the filler neck and the tank is particularly critical. Sometimes a rubber filler neck will leak because of loose clamps or deteriorated rubber. These are problems a home mechanic can usually rectify.

✳✳ WARNING:

Do not, under any circumstances, try to repair a fuel tank (except rubber components). A welding torch or any open flame can easily cause fuel vapors inside the tank to explode.

5 Carefully check all rubber hoses and metal lines leading away from the fuel tank. Check for loose connections, deteriorated hoses, crimped lines and other damage. Follow the lines to the front of the vehicle, carefully inspecting them all the way to the carburetor or fuel injection system. Repair or replace damaged sections as necessary.

6 If a fuel odor is still evident after the inspection, refer to Chapter 6 and check the EVAP system.

26 Brake check (every 15,000 miles or 12 months)

✳✳ WARNING:

The dust created by the brake system is harmful to your health. Never blow it out with compressed air and don't inhale any of it. An approved filtering mask should be worn when working on the brakes. Do not, under any circumstances, use petroleum-based solvents to clean brake parts. Use brake system cleaner only! Try to use non-asbestos replacement parts whenever possible.

➡**Note: For detailed photographs of the brake system, refer to Chapter 9.**

1   In addition to the specified intervals, the brakes should be inspected every time the wheels are removed or whenever a defect is suspected. Any of the following symptoms could indicate a potential brake system defect: The vehicle pulls to one side when the brake pedal is depressed; the brakes make squealing or dragging noises when applied; brake pedal travel is excessive; the pedal pulsates; brake fluid leaks, usually onto the inside of the tire or wheel.

2   Loosen the wheel lug nuts.

3   Raise the vehicle and place it securely on jackstands.

4   Remove the wheels.

DISC BRAKES

▶ **Refer to illustrations 26.5 and 26.10**

5   There are two pads (an outer and an inner) in each caliper. The pads are visible after the wheels are removed (see illustration).

6   Measure the pad thickness. If the lining material is less than the minimum thickness listed in this Chapter's Specifications, replace the pads.

➡**Note: Keep in mind that the lining material is riveted or bonded to a metal backing plate and the metal portion is not included in this measurement.**

7   If it is difficult to determine the exact thickness of the remaining pad material by the above method, or if you are at all concerned about the condition of the pads, remove the pads for further inspection (see Chapter 9).

8   Once the pads are removed, clean them with brake cleaner and re-measure them.

9   Measure the disc thickness with a micrometer to make sure that it still has service life remaining. If any disc is thinner than the specified minimum thickness, replace it (refer to Chapter 9). Even if the disc has service life remaining, check its condition. Look for scoring, gouging and burned spots. If these conditions exist, remove the disc and have it resurfaced (see Chapter 9).

10   Before installing the wheels, check all brake lines and hoses for damage, wear, deformation, cracks, corrosion, leakage, bends and twists, particularly in the vicinity of the rubber hoses at the calipers (see illustration). Check the clamps for tightness and the connections for leakage. Make sure that all hoses and lines are clear of sharp edges, moving parts and the exhaust system. If any of the above conditions are noted, repair, reroute or replace the lines and/or fittings as necessary (see Chapter 9). Be sure to tighten the lug nuts to the torque listed in this Chapter's Specifications after the vehicle has been lowered.

DRUM BRAKES

▶ **Refer to illustrations 26.12 and 26.14**

11   Refer to Chapter 9 and remove the brake drums.

12   Note the thickness of the lining material on the brake shoes (see illustration) and look for signs of contamination by brake fluid and grease. If the lining material is within 1/16-inch of the recessed rivets or metal shoes, replace the brake shoes with new ones. The shoes should also be replaced if they are cracked, glazed (shiny lining surfaces) or contaminated with brake fluid or grease. See Chapter 9 for the replacement procedure.

13   Check the shoe return and hold-down springs and the adjusting mechanism to make sure they're installed correctly and in good condition (see Chapter 9). Deteriorated or distorted springs, if not replaced,

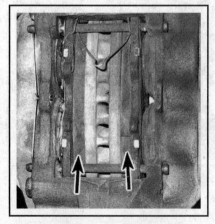

26.5 With the wheels removed, the brake pad lining can be inspected

26.10 Check for any sign of brake fluid leakage at the brake line fittings and the brake hoses

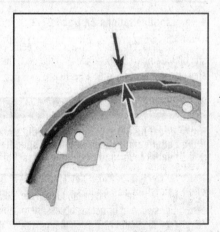

26.12 If the lining is bonded to the brake shoe, measure the lining thickness from the outer surface to the metal shoe, as shown here; if the lining is riveted to the shoe, measure from the lining outer surface to the rivet head

could allow the linings to drag and wear prematurely.

14 Check the wheel cylinders for leakage by carefully peeling back the rubber boots (see illustration). If brake fluid is noted behind the boots, the wheel cylinders must be replaced (see Chapter 9).

15 Check the drums for cracks, score marks, deep scratches and hard spots, which will appear as small discolored areas. If imperfections cannot be removed with emery cloth, the drums must be resurfaced by an automotive machine shop (see Chapter 9 for more detailed information).

16 Refer to Chapter 9 and install the brake drums.

17 Install the wheels and snug the wheel lug nuts finger tight.

18 Remove the jackstands and lower the vehicle.

19 Tighten the wheel lug nuts to the torque listed in this Chapter's Specifications.

26.14 Check for fluid leakage at both ends of the wheel cylinder dust covers

BRAKE BOOSTER CHECK

20 Sit in the driver's seat and perform the following sequence of tests.

21 With the brake fully depressed, start the engine - the pedal should move down a little when the engine starts.

22 With the engine running, depress the brake pedal several times- the travel distance should not change.

23 Depress the brake, stop the engine and hold the pedal in for about 30 seconds - the pedal should neither sink nor rise.

24 Restart the engine, run it for about a minute and turn it off. Then firmly depress the brake several times - the pedal travel should decrease with each application.

25 If your brakes do not operate as described above when the preceding tests are performed, the brake booster has failed. Refer to Chapter 9 for the replacement procedure.

PARKING BRAKE

26 Slowly pull up on the parking brake handle or depress the parking brake pedal, depending on design, and count the number of clicks you hear until maximum travel has been reached. The adjustment should be within the specified number of clicks listed in this Chapter's Specifications. If you hear more or fewer clicks, it's time to adjust the parking brake (see Chapter 9).

27 An alternative method of checking the parking brake is to park the vehicle on a steep hill with the parking brake set and the transmission in Neutral (be sure to stay in the vehicle during this check!). If the parking brake cannot prevent the vehicle from rolling, it is in need of adjustment (see Chapter 9).

27 Clutch/brake pedal height and freeplay check and adjustment (every 15,000 miles or 12 months)

PEDAL HEIGHT

▶ **Refer to illustrations 27.1 and 27.2**

1 The height of the clutch and brake pedal is the distance the pedal sits off the floor (see illustration). If the pedal height is not within the specified range, it must be adjusted.

❊❊ CAUTION:

Brake pedal height on Sequoia models should not be adjusted. If the pedal height is not within the specified range, check the pedal, pedal lever and brake pedal bracket for damage.

2 To adjust the clutch pedal, loosen the locknut and turn the stopper bolt or switch in or out, as necessary to achieve the proper pedal height (see illustration). Tighten the locknut securely.

3 Before measuring the brake pedal height make sure the pedal is in the fully returned position, then start the engine and depress the accelerator several times to activate the brake power booster. Measure the pedal height and adjust if necessary. To adjust the pedal height loosen the locknut and back the brake light switch out for clearance, then loosen the locknut on the brake pushrod. Turn the pushrod to

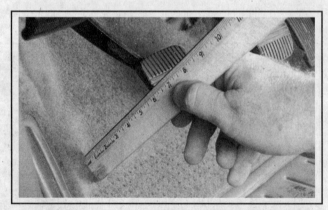

27.1 Pedal height is the distance between the pedal pad and the floor

adjust the pedal height in the middle of the specified range, then retighten the locknut.

4 Adjust the brake light switch or clutch pedal stopper bolt by turning it clockwise until the switch body or stopper bolt just contacts the pedal arm. If adjusting the brake light switch, turn the switch out until there is a clearance of 0.020 to 0.094-inch between the switch body and the pedal arm.

PEDAL FREEPLAY

▶ **Refer to illustration 27.5**

5   The freeplay is the pedal slack, or the distance the pedal can be depressed before it begins to have any effect on the clutch or brake system (see illustration). If the pedal freeplay is not within the specified range, it must be adjusted.

6   To adjust the clutch pedal freeplay, loosen the locknut on the clutch pushrod, then turn the pushrod to adjust the pedal freeplay to the specified range. Securely retighten the locknut.

7   Before measuring the brake pedal freeplay, turn off the engine and depress the brake pedal a half dozen times. Measure the pedal freeplay; if it isn't correct, adjust the brake light switch as described in Step 4. If the freeplay is still incorrect, troubleshoot the brake system.

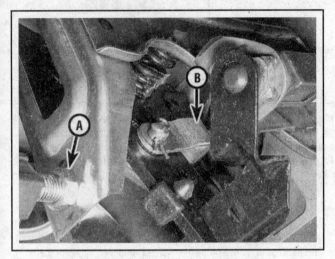

27.2 Pedal adjustment details

A   Stopper bolt or switch
B   Pushrod (locknut behind clevis)

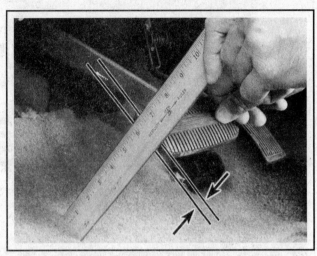

27.5 Pedal freeplay is the distance from the natural resting point of the pedal to the point at which resistance is felt

28 Spark plug replacement (see Maintenance schedule for service interval)

▶ **Refer to illustrations 28.1, 28.4a, 28.4b and 28.4c**

1   Spark plug replacement requires a spark plug socket, extension and ratchet. This socket is lined with a rubber grommet to protect the porcelain insulator of the spark plug and to hold the plug while you insert it into the spark plug hole. You will also need a wire-type feeler gauge to check and adjust the spark plug gap and a torque wrench to tighten the new plugs to the specified torque (see illustration).

2   If you are replacing the plugs, purchase the new plugs, adjust them to the proper gap and replace each plug one at a time.

➡ **Note: When buying new spark plugs, it's essential that you obtain the correct plugs for your specific vehicle. This information can be found in the Specifications Section at the end of this Chapter, on the Vehicle Emissions Control Information (VECI) label located on the underside of the hood or in the owner's manual. If these sources specify different plugs, purchase the spark plug type specified on the VECI label because that information is provided specifically for your engine.**

3   Inspect each of the new plugs for defects. If there are any signs of cracks in the porcelain insulator of a plug, don't use it.

4   Check the electrode gaps of the new plugs. Check the gap by inserting the wire gauge of the proper thickness between the electrodes at the tip of the plug (see illustration). The gap between the electrodes should be identical to that listed in this Chapter's Specifications or on

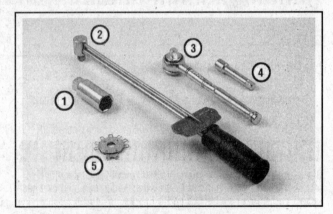

28.1 Tools required for changing spark plugs

*1   **Spark plug socket** - This will have special padding inside to protect the spark plug's porcelain insulator*
*2   **Torque wrench** - Although not mandatory, using this tool is the best way to ensure the plugs are tightened properly*
*3   **Ratchet** - Standard hand tool to fit the spark plug socket*
*4   **Extension** - Depending on model and accessories, you may need special extensions and universal joints to reach one or more of the plugs*
*5   **Spark plug gap gauge** - This gauge for checking the gap comes in a variety of styles. Make sure the gap for your engine is included*

28.4a Spark plug manufacturers recommend using a wire type gauge when checking the gap - if the wire does not slide between the electrodes with a slight drag, adjustment is required

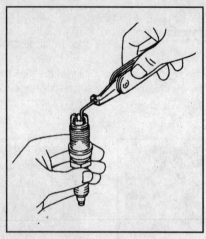

28.4b Some spark plugs have dual ground electrodes; on these plugs check the gap between each ground electrode and the center electrode

28.4c To change the gap, bend the side electrode only, as indicated by the arrows, and be very careful not to crack or chip the porcelain insulator surrounding the center electrode

the VECI label. If the gap is incorrect, use the notched adjuster on the feeler gauge body to bend the curved side electrode slightly (see illustration).

✳✳ CAUTION:

Platinum and iridium spark plugs generally come pre-gapped. If you check the gap, treat them very gently and do not scratch the coating on the electrodes. Also, don't attempt to adjust the gap on used platinum or iridium spark plugs.

5   If the side electrode is not exactly over the center electrode, use the notched adjuster to align them.

REMOVAL

▶ **Refer to illustrations 28.7a, 28.7b, 28.7c, 28.8 and 28.9**

6   If compressed air is available, blow any dirt or foreign material away from the coil/spark plug area before proceeding.

7   2000 through 2004 V6 engines have ignition coils mounted directly over the spark plugs on the right cylinder bank, and spark plug wires leading from these coils over to the three cylinders on the left cylinder bank. To avoid confusion, work on one spark plug at a time. 2005 and later V6 and all V8 engines have a separate coil mounted over each spark plug. Remove the ignition coil or spark plug wire, as applicable (see illustrations).

✳✳ CAUTION:

When removing spark plug wires, pull only on the boot, using a twisting motion. Also, don't bend the wires sharply - the conductor inside the wire could be damaged.

8   Remove the spark plug (see illustration).
9   Whether you are replacing the plugs at this time or intend to

reuse the old plugs, compare each old spark plug with the chart shown (see illustration) to determine the overall running condition of the engine.

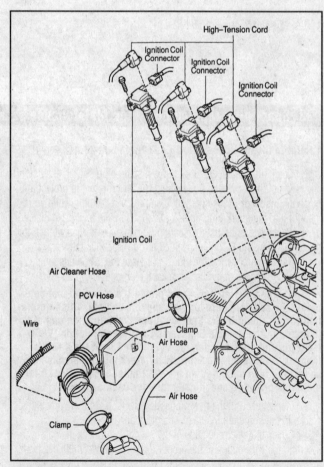

28.7a Ignition coil installation details - (2000 through 2004 V6 engine)

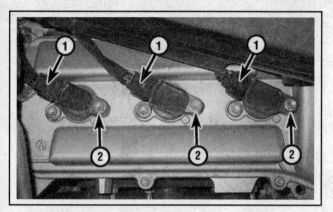

28.7b Ignition coil installation details (2005 and later V6 engines)

1   *Depress the release tab and pull off the electrical connector*
2   *Remove the coil mounting bolt and pull the coil off the spark plug*

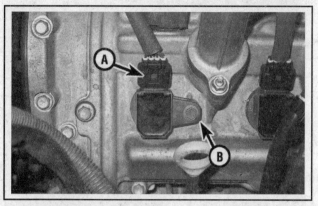

28.7c To remove an ignition coil from a V8, disconnect the electrical connector (A), then unscrew the mounting bolt (B) and twist the coil back-and-forth while pulling it up and out of the valve cover

INSTALLATION

▶ **Refer to illustrations 28.10a and 28.10b**

10 Prior to installation, apply a coat of anti-seize compound to the plug threads. It's often difficult to insert spark plugs into their holes without cross-threading them. To avoid this possibility, fit a short piece of rubber hose over the end of the spark plug (see illustrations). The flexible hose acts as a universal joint to help align the plug with the plug hole. Should the plug begin to cross-thread, the hose will slip on the spark plug, preventing thread damage. Tighten the plug to the torque listed in this Chapter's Specifications.

11 Attach the coil or spark plug wire to the new spark plug, again using a twisting motion until it is firmly seated on the end of the spark plug. When installing coils, tighten the mounting bolts securely.

12 Follow the above procedure for the remaining spark plugs, replacing them one at a time to prevent mixing up the spark plug wires.

28.8 Use a socket with a long extension to unscrew the spark plugs (V8 shown, V6 similar)

A **normally worn** spark plug should have light tan or gray deposits on the firing tip.

A **carbon fouled** plug, identified by soft, sooty, black deposits, may indicate an improperly tuned vehicle. Check the air cleaner, ignition components and engine control system.

An **oil fouled** spark plug indicates an engine with worn piston rings and/or bad valve seals allowing excessive oil to enter the chamber.

This spark plug has been **left in the engine too long,** as evidenced by the extreme gap. Plugs with such an extreme gap can cause misfiring and stumbling accompanied by a noticeable lack of power.

A **physically damaged** spark plug may be evidence of severe detonation in that cylinder. Watch that cylinder carefully between services, as a continued detonation will not only damage the plug, but could also damage the engine.

A **bridged or almost bridged** spark plug, identified by a build-up between the electrodes caused by excessive carbon or oil build-up on the plug.

28.9 Inspect the spark plug to determine engine running conditions

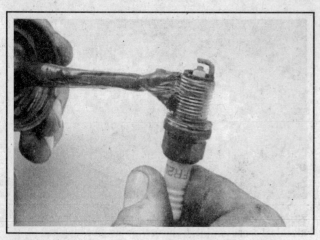

28.10a Apply a thin coat of anti-seize compound to the spark plug threads

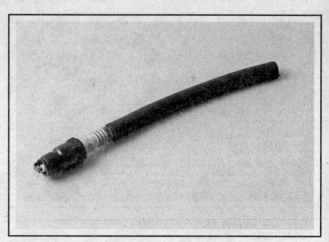

28.10b A length of rubber hose will save time and prevent damaged threads when installing the spark plugs

29 Spark plug wire check and replacement (V6 engine) (every 30,000 miles or 24 months)

1   The spark plug wires should be checked whenever new spark plugs are installed.

2   Begin this procedure by making a visual check of the spark plug wires while the engine is running. In a darkened garage (make sure there is adequate ventilation) start the engine and observe each plug wire. Be careful not to come into contact with any moving engine parts. If there is a break in the wire, you will see arcing or a small spark at the damaged area. If arcing is noticed, make a note to obtain new wires, then allow the engine to cool and check the distributor cap and rotor.

3   The spark plug wires should be inspected one at a time to prevent mixing up the order, which is essential for proper engine operation. Each original plug wire should be numbered to help identify its location. If the number is illegible, a piece of tape can be marked with the correct number and wrapped around the plug wire.

4   Disconnect the plug wire from the spark plug. Grasp the rubber boot, twist the boot and pull it free. Do not pull on the wire itself.

5   Check inside the boot for corrosion, which will look like a white crusty powder.

6   Push the wire and boot back onto the end of the spark plug. It should fit tightly onto the end of the plug. If it doesn't, remove the wire and use pliers to carefully crimp the metal connector inside the wire boot until the fit is snug.

7   Using a clean rag, wipe the entire length of the wire to remove built-up dirt and grease. Once the wire is clean, check for burns, cracks and other damage. Do not bend the wire sharply, because the conductor might break.

8   Disconnect the other end of the wire from the coil. To do this, lift up on the locking claw and pull the end of the wire off the coil. Check for corrosion and a tight fit. Reattach the wire to the coil, making sure the locking claw clips into place.

9   Inspect the remaining spark plug wires, making sure that each one is securely fastened at the coil and spark plug when the check is complete.

10 If new spark plug wires are required, purchase a set for your specific engine model. Remove and replace the wires one at a time to avoid mix-ups in the firing order.

30 Cooling system servicing (draining, flushing and refilling) (every 30,000 miles or 24 months)

❊❊ WARNING:

Do not allow antifreeze to come in contact with your skin or painted surfaces of the vehicle. Rinse off spills immediately with plenty of water. Antifreeze is highly toxic if ingested. Never leave antifreeze lying around in an open container or in puddles on the floor; children and pets are attracted by it's sweet smell and may drink it. Check with local authorities about disposing of used antifreeze. Many communities have collection centers which will see that antifreeze is disposed of safely.

1   Periodically, the cooling system should be drained, flushed and refilled to replenish the antifreeze mixture and prevent formation of rust and corrosion, which can impair the performance of the cooling system and cause engine damage. When the cooling system is serviced, all hoses and the radiator cap should be checked and replaced if necessary.

DRAINING

▶ **Refer to illustrations 30.3a, 30.3b and 30.3c**

2   Apply the parking brake and block the wheels.

❊❊ WARNING:

If the vehicle has just been driven, wait several hours to allow the engine to cool down before beginning this procedure.

Remove the under-vehicle splash shield (see illustration 8.6).

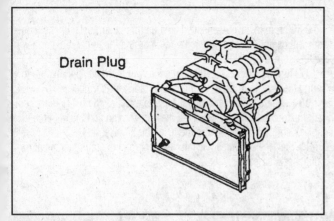

30.3a Coolant drain plug locations on the 2000 through 2004 V6 engine

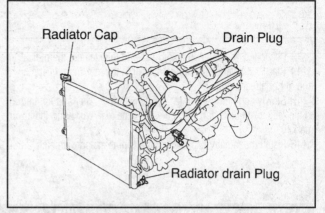

30.3b Coolant drain plug locations - 2005 V6 engine

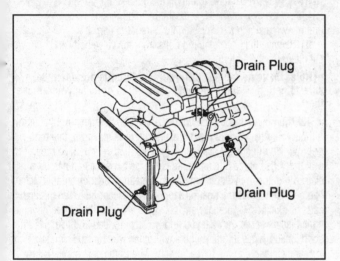

30.3c Coolant drain plug locations on the V8 engine

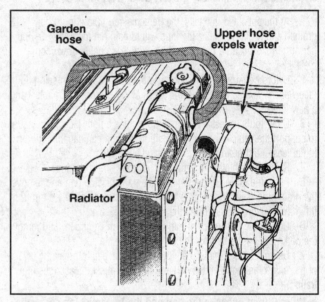

30.9 With the thermostat removed, disconnect the upper radiator hose and flush the radiator and engine block with a garden hose

3   Move a large container under the radiator drain to catch the coolant. The radiator drain plug is located on the left (driver's) side lower corner of the radiator (see illustrations). Unscrew the drain plug until coolant starts flowing from the drain hole (a pair of pliers may be required to turn it).

4   Remove the radiator cap and allow the radiator to drain, then, move the container under the engine. Loosen the engine block drain plug(s) and allow the coolant in the block to drain (see illustrations 30.3a and 30.3b).

➡**Note: Frequently, the coolant will not drain from the block after the plug is removed. This is due to a rust layer that has built up behind the plug. Insert a thin screwdriver or punch into the hole to break the rust barrier.**

5   While the coolant is draining, check the condition of the radiator hoses, heater hoses and clamps (refer to Section 10 if necessary).

6   Replace any damaged clamps or hoses. Close the drain plugs.

FLUSHING

▶ **Refer to illustration 30.9**

7   Once the system is completely drained, remove the thermostat from the engine (see Chapter 3), then reinstall the thermostat housing

without the thermostat. This will allow the system to be thoroughly flushed.

8   Turn the heating system controls to Hot, so that the heater core will be flushed at the same time as the rest of the cooling system.

9   Disconnect the upper radiator hose from the radiator, then place a garden hose in the upper radiator inlet and flush the system until the water runs clear at the upper radiator hose (see illustration).

10 In severe cases of contamination or clogging of the radiator, remove the radiator (see Chapter 3) and have a radiator repair facility clean and repair it if necessary.

11 Many deposits can be removed by the chemical action of a cleaner available at auto parts stores. Follow the procedure outlined in the manufacturer's instructions.

➡**Note: When the coolant is regularly drained and the system refilled with the correct antifreeze/water mixture, there should be no need to use chemical cleaners or descalers.**

12 Remove the overflow hose from the coolant recovery reservoir. Drain the reservoir and flush it with clean water, then reconnect the hose.

REFILLING

13 Reconnect the upper radiator hose and reinstall the thermostat.

14 Place the heater temperature control in the maximum heat position, if not already done.

15 Slowly add new coolant (a 50/50 mixture of water and antifreeze) to the radiator until it's full. Add coolant to the reservoir up to the lower mark.

16 Install the radiator cap and run the engine at approximately 2,000 to 2,500 rpm in a well-ventilated area until the thermostat opens (coolant will begin flowing through the radiator and the upper radiator hose will become hot).

17 Turn the engine off and let it cool. Add more coolant mixture to bring the level back up to the lip on the radiator filler neck.

18 Squeeze the upper radiator hose to expel air, then add more coolant mixture if necessary. Reinstall the radiator cap and the under-vehicle splash shield. Add coolant to the reservoir, if necessary.

19 Start the engine, allow it to reach normal operating temperature and check for leaks.

31 Automatic transmission fluid and filter change (every 30,000 miles or 24 months)

▶ **Refer to illustrations 31.5, 31.7, 31.9, 31.11 and 31.12**

1 At the specified intervals, the transmission fluid should be drained and replaced. Since the fluid will remain hot long after driving, perform this procedure only after the engine has cooled down completely.

2 Before beginning work, purchase the specified transmission fluid (see *Recommended lubricants and fluids* at the end of this Chapter) and a new filter.

3 Other tools necessary for this job include a floor jack, jackstands to support the vehicle in a raised position, a drain pan capable of holding at least eight quarts, newspapers and clean rags.

4 Raise the vehicle and support it securely on jackstands.

5 Place the drain pan underneath the transmission pan and remove the drain plug (see illustration). Allow the fluid to completely drain from the transmission, then reinstall the drain plug.

6 Detach the transmission pan rock shield (if equipped) and remove the pan mounting bolts from the outer edges of the pan.

7 Using a rubber mallet, carefully tap on the pan to break the layer of gasket sealant between the pan and the transmission case (see illustration).

➡ **Note: Prying between the pan and the transmission case with a screwdriver or similar tool may result in damage to the sealing surface on the transmission case.**

8 Lower the pan and separate the lower part of the dipstick tube from the upper part. If the union between the two tubes is stuck and won't separate, you'll have to unbolt the bracket at the upper end of the tube from the cylinder head. Once the pan and dipstick tube are free, drain any remaining transmission fluid from the pan.

9 Remove the filter retaining bolts from the valve body and remove the filter (see illustration).

➡ **Note: On some models different length bolts are used; be sure to note the length and locations of the bolts as you remove them.**

10 Thoroughly inspect the bottom of the pan, the filter and the fluid. Although normally bright red, transmission fluid may turn dark red or brown during normal use. If you find the fluid very dark colored, or if it smells burned, it usually indicates the transmission has been overheated. If you find small pieces of metal or clutch material in the pan or filter, it indicates wear or damage have occurred to the internal parts or clutches. If you have any concerns about the condition of your transmission based on what you find in the fluid, pan and filter, it's a good idea to take your vehicle to your dealer or a transmission shop for further evaluation.

11 Clean the pan with solvent and dry it. Use a gasket scraper to remove any traces of old gasket material remaining on the transmission case or valve body.

➡ **Note: Be very careful not to gouge the delicate aluminum gasket surfaces. Install new gaskets on the filter, then install the filter, tightening the bolts to the torque listed in this Chapter's Specifications (see illustration).**

31.5 The transmission drain plug is located on the bottom of the transmission pan (typical)

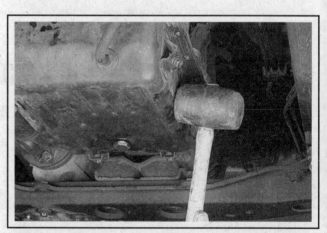

31.7 Using a rubber mallet, carefully tap on the pan to break the gasket seal between the pan and the transmission case

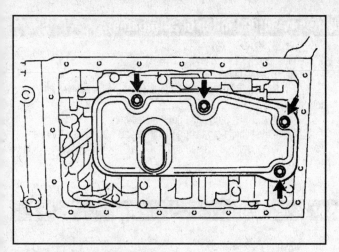

31.9 Remove the filter bolts

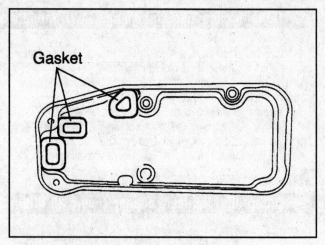

31.11 Be sure to install new gaskets on the filter

Gasket

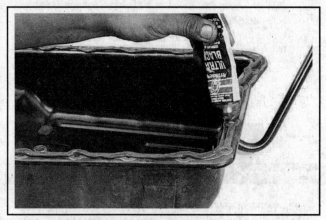

31.12 Apply a bead of RTV sealant all the way around the pan mating surface, inboard of the bolt holes

12 Be sure the gasket surface on the transmission pan is clean, then install a bead of RTV sealant to the pan (see illustration). Reinsert the filler tube onto the dipstick tube and place the pan against the transmis-

sion case. Working around the pan, tighten each bolt a little at a time to the torque listed in this Chapter's Specifications.

2004 AND EARLIER MODELS

13 Lower the vehicle and add approximately 1-1/2 quarts of the specified type of automatic transmission fluid through the filler tube (see Section 6).

14 With the transmission in Park and the parking brake set, start the engine.

15 Move the gear selector through each range and back to Park. Check the fluid level and add fluid, a little at a time, until the level is within the correct range on the dipstick. Between each application of fluid, move the selector lever through each range and back to Park.

2005 AND LATER MODELS

16 Refer to Section 6 for the refilling procedure.

17 Check under the vehicle for leaks during the first few trips. Check the fluid level again when the transmission is hot (see Section 6).

32 Manual transmission lubricant change (every 30,000 miles or 24 months)

1 Raise the vehicle and support it securely on jackstands.

2 Move a drain pan, rags, newspapers and wrenches under the transmission.

3 Remove the transmission drain plug at the bottom of the case and allow the lubricant to drain into the pan (see illustration 17.2).

4 After the lubricant has drained completely, reinstall the plug and tighten it securely.

5 Remove the fill plug from the side of the transmission case. Using a hand pump, syringe or funnel, fill the transmission with the specified lubricant until it is level with the lower edge of the filler hole. Reinstall the fill plug and tighten it securely.

6 Lower the vehicle.

7 Drive the vehicle for a short distance, then check the drain and fill plugs for leakage.

33 Transfer case lubricant change (4WD models only)(every 30,000 miles or 24 months)

1   Drive the vehicle for at least 15 minutes to warm the lubricant in the case. Perform this warm-up procedure with 4WD engaged, if possible. Use all gears, including Reverse, to ensure the lubricant is sufficiently warm to drain completely.

2   Raise the vehicle and support it securely on jackstands.

3   Remove the drain plug from the lower part of the case and allow the old lubricant to drain completely (see illustration 18.2).

4   After the lubricant has drained completely, reinstall the plug and

tighten it securely.

5   Remove the filler plug from the case.

6   Fill the case with the specified lubricant until it is level with the lower edge of the filler hole.

7   Install the filler plug and tighten it securely.

8   Drive the vehicle for a short distance and recheck the lubricant level. In some instances a small amount of additional lubricant will have to be added.

34 Differential lubricant change (every 30,000 miles or 24 months)

➡Note: The following procedure applies to the front and rear differentials.

1   Drive the vehicle for several miles to warm up the differential oil, then raise the vehicle and support it securely on jackstands.

2   Move a drain pan, rags, newspapers and the proper tools under the vehicle.

3   With the drain pan under the differential, use a socket and ratchet to loosen the drain plug. It's the lower of the two plugs (see illustration 19.2).

4   Once loosened, carefully unscrew it with your fingers until you can remove it from the case.

5   Allow all of the oil to drain into the pan, then replace the drain

plug and tighten it securely.

6   Feel with your hands along the bottom of the drain pan for any metal bits that may have come out with the oil. If there are any, it's a sign of excessive wear, indicating that the internal components should be carefully inspected in the near future.

7   Remove the differential check/fill plug (see Section 19). Using a hand pump, syringe or funnel, fill the differential with the correct amount and grade of oil (see the Specifications) until the level is just at the bottom of the plug hole.

8   Reinstall the plug and tighten it securely.

9   Lower the vehicle. Check for leaks at the drain plug after the first few miles of driving.

35 Fuel filter replacement (every 60,000 miles or 48 months)

▸ Refer to illustrations 35.2a and 35.2b

❋❋ WARNING:

Gasoline is extremely flammable, so take extra precautions when you work on any part of the fuel system. Don't smoke or allow open flames or bare light bulbs near the work area, and don't work in a garage where a gas-type appliance (such as a water heater or clothes dryer) is present. Since gasoline is carcinogenic, wear fuel-resistant gloves when there's a possibility of being exposed to fuel, and, if you spill any fuel on your skin, rinse it off immediately with soap and water. Mop up any spills immediately and do not store fuel-soaked rags where they could ignite. The fuel system is under constant pressure, so if any fuel lines are to be disconnected, the fuel pressure in the system

must be relieved first (see Chapter 4). When you perform any kind of work on the fuel system, wear safety glasses and have a Class B Type fire extinguisher on hand.

1   The fuel filter is located in a bracket mounted to the left side frame rail. Place a metal container, rags or newspapers under the filter to catch spilled fuel. Relieve the fuel system pressure (see Chapter 4).

2   On standard threaded fittings, use a flare-nut wrench to loosen the fittings on both ends of the fuel filter (see illustration). Hold the stationary fittings on the fuel filter with another wrench to prevent bending the filter bracket or twisting the lines. Some models may have quick-connect fittings; to disconnect this type of fitting, turn the end of the fitting to unlock it, then pull the flexible line off the filter (see illustration).

35.2a Unscrew the fittings from the fuel filter (A), then remove the bracket bolt (B) and separate the filter from the bracket

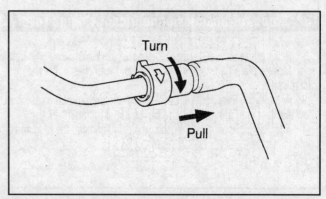

35.2b To disconnect a quick-connect fitting, turn the end of the fitting and pull

3   Remove the filter clamp bolt and separate the old filter from the bracket. Note how the filter is installed; most filters have an arrow on them indicating the direction of fuel flow.

4   Install the new filter in the bracket and connect the fuel line fittings, starting the threads by hand.

5   Tighten the bracket bolt securely, then tighten the fuel line fittings. If equipped with a quick-connect fitting, simply push the fitting onto the filter until it clicks into place. Pull on the hose to confirm proper engagement.

36 Valve clearance check and adjustment (every 60,000 miles or 48 months)

1   Refer to Chapter 2A or 2B and position the number 1 piston at TDC on the compression stroke.

2   Disconnect the cable from the negative terminal of the battery.

3   Disconnect the coil (and, on V6 engines, the spark plug wires) and remove any other components that will interfere with valve cover removal.

4   Blow out the recessed area around the spark plug openings with compressed air, if available, to remove any debris that might fall into the cylinders, then remove the spark plugs (see Section 28).

5   Remove the valve cover (see Chapter 2A or 2B).

CHECK

V6 engine

▶ **Refer to illustrations 36.6a, 36.6b, 36.7a and 36.7b**

6   Measure the clearances of the indicated valves with feeler gauges (see illustrations). Record the measurements which are out of specification. They will be used later to determine the required replacement shims.

7   Turn the crankshaft 2/3 revolution (240 degrees) and check the indicated valves (see illustration). Turn the crankshaft 2/3 revolution again and check the next group of valves (see illustration). Make notes of the cylinder number and the valve type (intake or exhaust) that need to be adjusted.

36.6a Check the clearance of each valve with a feeler gauge of the specified thickness - if the clearance is correct, you should feel a slight drag on the gauge as you pull it out

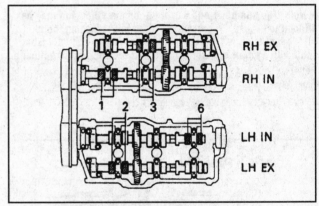

36.6b With the No. 1 piston at TDC on the compression stroke, check the indicated valves (V6 engine)

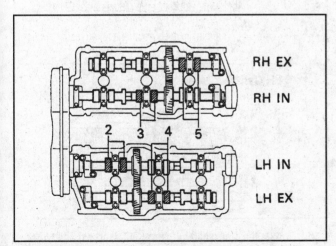

36.7a Rotate the crankshaft 2/3 turn (240 degrees) and check the indicated valves (V6 engine)

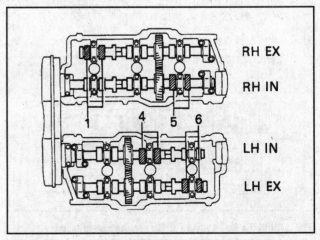

36.7b Rotate the crankshaft 2/3 turn (240 degrees) and check the remaining valves indicated (V6 engine)

V8 engine

▶ **Refer to illustrations 36.8 and 36.9**

8   Measure the clearances of the indicated valves with feeler gauges (see illustration). Record the measurements which are out of specification. They will be used later to determine the required replacement shims.

9   Turn the crankshaft one revolution (360 degrees) and check the remaining valves (see illustration).

ADJUSTMENT

2000 through 2004 V6 engines

▶ **Refer to illustrations 36.11a, 36.11b, 36.11c, 36.12 and 36.13**

10   After all the valves have been measured, turn the crankshaft pulley until the camshaft lobe above the first valve which you intend to adjust is pointing upward, away from the shim.

11   Position the notch in the valve lifter toward the spark plug. Then depress the valve lifter with the special valve lifter tools (see illustrations). Place the special valve lifter tool in position as shown, with the longer jaw of the tool gripping the lower edge of the cast lifter boss and the upper, shorter jaw gripping the upper edge of the lifter itself. Depress the valve lifter by squeezing the handles of the valve lifter tool together, then hold the lifter down with the smaller tool and remove the larger one.

➡ **Note: The number 1 and 6 cylinder on the V6 engine may be difficult to access. To prevent the shim from pinching the tool, tilt the tool slightly from the intake side. Remove the adjusting shim with a small screwdriver or a pair of tweezers (see illustrations).**

Note that the wire hook on the end of some valve lifter tool handles can be used to clamp both handles together to keep the lifter depressed while the shim is removed.

12   Measure the thickness of the shim with a micrometer (see illustration). To calculate the correct thickness of a replacement shim that will place the valve clearance within the specified value, use the following formula:

$N = T + (A - V)$

T = thickness of the old shim
A = valve clearance measured
N = thickness of the new shim
V = desired valve clearance (see this Chapter's Specifications)

13   Select a shim with a thickness as close as possible to the valve clearance calculated. Shims, which are available in 17 sizes in increments of 0.0020-inch (0.050 mm), range in size from 0.0984-inch (2.500 mm) to 0.1299-inch (3.300 mm) (see illustration).

➡ **Note: Through careful analysis of the shim sizes needed to bring the out-of-specification valve clearance within specification, it is often possible to simply move a shim that has to come out anyway to another valve lifter requiring a shim of that particular size, thereby reducing the number of new shims that must be purchased.**

14   Place the special valve lifter tool in position as shown in illustration 36.11a, with the longer jaw of the tool gripping the lower edge of the cast lifter boss and the upper, shorter jaw gripping the upper edge of the lifter itself, press down the valve lifter by squeezing the handles of the valve lifter tool together and install the new adjusting shim (note that the wire hook on the end of one valve lifter tool handle can be used to clamp the handles together to keep the lifter depressed while the shim is inserted. Measure the clearance with a feeler gauge to make sure that your calculations are correct.

15   Repeat this procedure until all the valves which are out of clearance have been corrected.

16   Reinstall the valve cover, spark plugs and coils/plug wires (V6 models) and any other components which were removed.

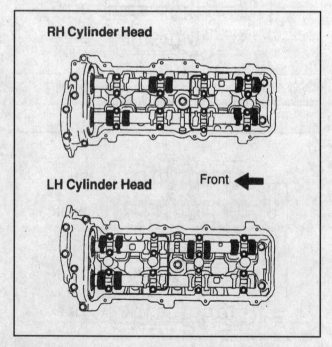

36.8 With the No. 1 piston at TDC on the compression stroke, check the valves indicated by the blackened cam lobes (V8 engine)

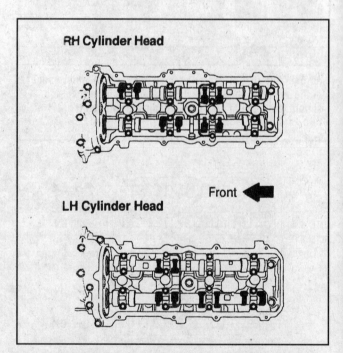

36.9 Rotate the crankshaft one turn (360 degrees) and check the remaining valves indicated by the blackened cam lobes (V8 engine)

36.11a Install the valve lifter tool as shown and squeeze the handles together to depress the valve lifter, then hold the lifter down with the smaller tool so the shim can be removed

36.11b Keep pressure on the lifter with the smaller tool and remove the shim with a small screwdriver . . .

36.11c . . . a pair of tweezers or a magnet as shown here (2000 through 2004 engines)

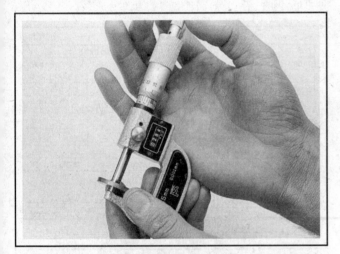

36.12 Measure the shim thickness with a micrometer (2000 through 2004 engines)

New shim thickness		mm (in.)	
Shim No.	Thickness	Shim No.	Thickness
1	2.500 (0.0984)	10	2.950 (0.1161)
2	2.550 (0.1004)	11	3.000 (0.1181)
3	2.600 (0.1024)	12	3.050 (0.1201)
4	2.650 (0.1043)	13	3.100 (0.1220)
5	2.700 (0.1063)	14	3.150 (0.1240)
6	2.750 (0.1083)	15	3.200 (0.1260)
7	2.800 (0.1102)	16	3.250 (0.1280)
8	2.850 (0.1122)	17	3.300 (0.1299)
9	2.900 (0.1142)		

36.13 Valve adjusting shim thickness chart (2000 through 2004 V6 engines)

2005 and later V6 engines

▶ **Refer to illustrations 36.18 and 36.20**

17 Remove the camshaft(s) for the valve(s) that you intend to adjust (see Chapter 2A).

18 Remove and measure each lifter (whose clearance is not correct) with a micrometer (see illustration). Put each lifter back into its bore in the cylinder head before moving on to the next lifter. Record the measurement for each lifter.

19 To calculate the correct thickness of a replacement lifter that will put the valve clearance within the specified range, use the following formula:

$N = T + (A - V)$, where:

N = thickness of the new lifter
T = thickness of the old lifter
A = measured valve clearance
V = specified valve clearance (see this Chapter's Specifications)

20 Select a lifter with a thickness as close as possible to the calculated valve clearance. Lifters for this V6 engine are available in 35 sizes in increments of 0.008-inch (0.020 mm), and range in size from

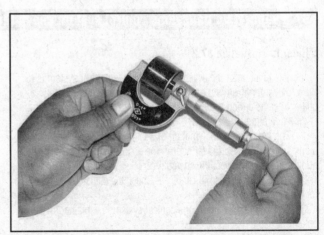

36.18 On 2005 and later V6 engines, measure the thickness of each lifter head with a micrometer

0.1992-inch (5.060 mm) to 0.2260-inch (5.740 mm) (see illustration).

21 Install the camshaft(s) (see Chapter 2A).

22 After the camshafts are reinstalled, check the valve clearances again to verify that they're now within the range of clearance listed in this Chapter's Specifications.

23 Installation of the valve cover(s), spark plugs, ignition coils, air filter housing and/or intake manifold is the reverse of removal.

V8 engines

▸ **Refer to illustration 36.28**

24 Remove the camshafts (see Chapter 2B).

➡**Note: It's only necessary to remove the camshaft(s) over any valve(s) whose clearance is incorrect.**

25 Remove the valve lifter(s) from the valves whose clearances were out of specification. Be sure to keep the lifters in order; they must be returned to the same bore they were removed from.

26 Remove the shim from the underside of the lifter. Clean the shim then measure its thickness with a micrometer (see illustration 36.12).

27 Calculate the required thickness of the new shim by using the formula given in Step 12.

28 Select a shim with a thickness as close as possible to the valve clearance calculated. Shims, which are available in 41 sizes in increments of 0.0008-inch (0.020 mm), range in size from 0.0787-inch (2.000 mm) to 0.1102-inch (2.800 mm) (see illustration).

➡**Note: Through careful analysis of the shim sizes needed to bring the out-of-specification valve clearance within specification, it is often possible to simply move a shim that has to come out anyway to another valve lifter requiring a shim of that particular size, thereby reducing the number of new shims that must be purchased.**

29 Once the proper shims have been selected, apply a thin coat of engine assembly lube to the shim and stick it in place on the underside of the lifter.

30 Repeat this procedure until all the valves which are out of clearance have been corrected.

31 Reinstall the lifters and camshafts following the procedures outlined in Chapter 2B.

New lifter thickness   mm (in.)

Lifter No.	Thickness	Lifter No.	Thickness	Lifter No.	Thickness
06	5.060 (0.1992)	30	5.300 (0.2087)	54	5.540 (0.2181)
08	5.080 (0.2000)	32	5.320 (0.2094)	56	5.560 (0.2189)
10	5.100 (0.2008)	34	5.340 (0.2102)	58	5.580 (0.2197)
12	5.120 (0.2016)	36	5.360 (0.2110)	60	5.600 (0.2205)
14	5.140 (0.2024)	38	5.380 (0.2118)	62	5.620 (0.2213)
16	5.160 (0.2031)	40	5.400 (0.2126)	64	5.640 (0.2220)
18	5.180 (0.2039)	42	5.420 (0.2134)	66	5.660 (0.2228)
20	5.200 (0.2047)	44	5.440 (0.2142)	68	5.680 (0.2236)
22	5.220 (0.2055)	46	5.460 (0.2150)	70	5.700 (0.2244)
24	5.240 (0.2063)	48	5.480 (0.2157)	72	5.720 (0.2252)
26	5.260 (0.2071)	50	5.500 (0.2165)	74	5.740 (0.2260)
28	5.280 (0.2079)	52	5.520 (0.2173)		

36.20 Valve lifter thickness chart (2005 V6 models)

New shim thickness   mm (in.)

Shim No.	Thickness	Shim No.	Thickness	Shim No.	Thickness
00	2.000 (0.0787)	28	2.280 (0.0898)	56	2.560 (0.1008)
02	2.020 (0.0795)	30	2.300 (0.0906)	58	2.580 (0.1016)
04	2.040 (0.0803)	32	2.320 (0.0913)	60	2.600 (0.1024)
06	2.060 (0.0811)	34	2.340 (0.0921)	62	2.620 (0.1031)
08	2.080 (0.0819)	36	2.360 (0.0929)	64	2.640 (0.1039)
10	2.100 (0.0827)	38	2.380 (0.0937)	66	2.660 (0.1047)
12	2.120 (0.0835)	40	2.400 (0.0945)	68	2.680 (0.1055)
14	2.140 (0.0843)	42	2.420 (0.0953)	70	2.700 (0.1063)
16	2.160 (0.0850)	44	2.440 (0.0961)	72	2.720 (0.1071)
18	2.180 (0.0858)	46	2.460 (0.0969)	74	2.740 (0.1079)
20	2.200 (0.0866)	48	2.480 (0.0976)	76	2.760 (0.1087)
22	2.220 (0.0874)	50	2.500 (0.0984)	78	2.780 (0.1094)
24	2.240 (0.0882)	52	2.520 (0.0992)	80	2.800 (0.1102)
26	2.260 (0.0890)	54	2.540 (0.1000)		

36.28 Valve adjusting shim thickness chart - V8 engine

37 Evaporative emissions control system check (every 60,000 miles or 48 months)

▸ **Refer to illustration 37.2**

1 The function of the evaporative emissions control system is to draw fuel vapors from the gas tank and fuel system, store them in a charcoal canister and route them to the intake manifold during normal engine operation.

2 The most common symptom of a fault in the evaporative emissions system is a strong fuel odor in the engine compartment. If a fuel odor is detected, inspect the charcoal canister, located in the front of the engine compartment (see illustration). Check the canister and all hoses for damage and deterioration.

3 The evaporative emissions control system is explained in more detail in Chapter 6.

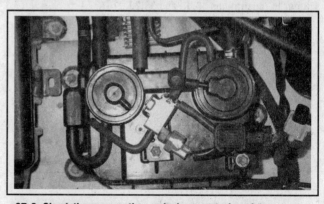

37.2 Check the evaporative emissions control canister and the hose connections for cracks and damage. On 2000 through 2002 models, the EVAP canister is mounted on the left (driver's) side of the engine compartment. On later models, it's mounted underneath the vehicle, near the fuel tank

Specifications

Recommended lubricants and fluids

Engine oil	
Type	API "certified for gasoline engines"
Viscosity	See accompanying chart
Coolant	
Type	Mixture of ethylene glycol based antifreeze and demineralized water
Concentration	
Down to -31 degrees F	50-percent antifreeze
Down to -58 degrees F	60-percent antifreeze

➡ **Note: Never exceed a concentration of 70-percent antifreeze to water.**

Brake fluid	DOT 3 brake fluid
Clutch fluid	DOT 3 brake fluid
Power steering fluid	DEXRON II or III automatic transmission fluid
Automatic transmission fluid	
2000 through 2002	DEXRON II or III ATF
2003 and 2004	TOYOTA Genuine ATF Type T-IV
2005 and later	TOYOTA Genuine ATF WS
Manual transmission lubricant	API GL-4 or GL-5 SAE 75W-90 gear oil
Transfer case lubricant	API GL-4 or GL-5 SAE 75W-90 gear oil
Differential lubricant	
Front	API GL-5 SAE 75W-90 hypoid gear oil
Rear	
Above 0 degrees F	API GL-5 SAE 90 hypoid gear oil
Below 0 degrees F	API GL-5 SAE 80W or 80W-90 hypoid gear oil

➡ **Note: On models with a limited slip rear differential, be sure to use hypoid gear oil for limited slip differentials.**

Chassis grease	NLGI No. 2 lithium base chassis grease

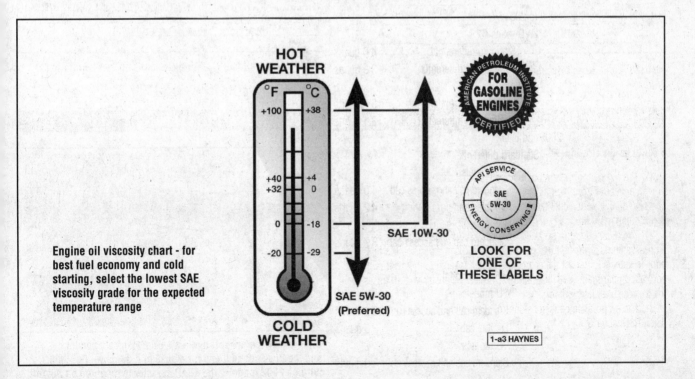

HOT WEATHER

FOR GASOLINE ENGINES CERTIFIED — AMERICAN PETROLEUM INSTITUTE

SAE 10W-30

SAE 5W-30 (Preferred)

API SERVICE — SAE 5W-30 — ENERGY CONSERVING II

LOOK FOR ONE OF THESE LABELS

Engine oil viscosity chart - for best fuel economy and cold starting, select the lowest SAE viscosity grade for the expected temperature range

COLD WEATHER

1-a3 HAYNES

Specifications (continued)

Capacities*

Engine oil (with filter change)

V6 engine	5.5 quarts
V8 engine	6.5 quarts

Cooling system

Tundra models

With manual transmission	10.6 quarts

With automatic transmission

V6 engine	10.5 quarts
V8 engine	12.3 quarts

Sequoia models

With rear heater	14.1
Without rear heater	12.4

Automatic transmission (drain and refill) (see Section 31 for refilling procedure)

2000 through 2004	2.1 quarts
2005 and later	3.2 quarts

Manual transmission

2000 through 2004

2WD	2.7 quarts
4WD	2.2 quarts
2005 and later	1.9 quarts

Transfer case

Tundra models	1.1 quarts
Sequoia models	1.3 quarts

Differential

Front	1.2 quarts

Rear

2000 to 2004

2WD models

Standard differential	4.0 quarts
Limited slip differential	3.3 quarts

4WD models

Standard differential	3.7 quarts
Limited slip differential	3.1 quarts

2005 and later

Standard differential

2WD models

Standard cab and access cab	4.0 quarts
Double cab	4.2 quarts

4WD models

Standard cab and access cab	3.7 quarts
Double cab	4.2 quarts

Limited slip differential

2WD models

Standard cab and access cab	3.3 quarts
Double cab	3.5 quarts

4WD models

Standard cab and access cab	3.0 quarts
Double cab	3.5 quarts

*All capacities approximate. Add as necessary to bring to appropriate level.

Spark plug type and gap

V6 engine
    2000 through 2004
        Type   Denso K16TR11 or NGK BKR5EKB-11
        Gap   0.043 inch
    2005 and later
        Type   Denso K20HR-U11 or NGK LFR6C-11
        Gap   0.039 to 0.043 inch
V8 engine
    2000 through 2004
        Type   Denso K20R-U or NGK BKR63YA
        Gap   0.031 inch
    2005 and later
        Type   Denso SK20R11 or NGK IFR6A11
        Gap   0.039 to 0.043 inch

Firing order

V6 engine   1-2-3-4-5-6
V8 engine   1-8-4-3-6-5-7-2

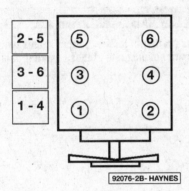

Cylinder and coil
terminal location
diagram - V6 engine
(cylinders 2, 4 and 6
are connected to the
coils on cylinders 1, 3
and 5 by spark
plug wires)

92076-2B- HAYNES

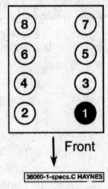

Cylinder location
diagram - V8 engine

Front

36060-1-specs.C HAYNES

Valve clearances (engine cold)

V6 engine
    Intake valve   0.006 to 0.009 inch
    Exhaust valve   0.011 to 0.014 inch
V8 engine
    Intake valve   0.006 to 0.010 inch
    Exhaust valve   0.010 to 0.014 inch

➡Note: Use the information printed on the Vehicle Emissions Control Information label, if different than the Specifications listed here.

Clutch pedal

Height   6.1 to 6.5 inches
Freeplay   0.197 to 0.591 inch

Brakes

Disc brake pad lining thickness (minimum)   1/16 inch
Drum brake shoe lining thickness (minimum)   1/16 inch
Brake pedal
    Height
        Tundra models
            2000 through 2003   6.46 to 6.85 inches

Brake pedal (continued)
    Height
        Tundra models
            2004 on

Standard and access cab (with or without VSC)	
2004	6.46 to 6.85 inches
2005 and later	6.44 to 6.83 inches
Double cab (with or without VSC)	5.95 to 6.50 inches
Sequoia models	5.949 to 6.500 inches
Freeplay	0.04 to 0.24 inch
Parking brake pedal adjustment	
All 2000 through 2004 models	6 to 9 clicks
2005 and later	
Sequoia models	6 to 9 clicks
Tundra models	8 to 10 clicks

Suspension and steering

Steering wheel freeplay limit	1.2 inches
Balljoint allowable movement	0.020 inch

Torque specifications   Ft-lbs (unless otherwise indicated)

➡**Note: One foot-pound (ft-lb) of torque is equivalent to 12 inch-pounds (in-lbs) of torque. Torque values below approximately 15 ft-lbs are expressed in inch-pounds, since most foot-pound torque wrenches are not accurate at these smaller values.**

Automatic transmission filter bolts	84 in-lbs
Automatic transmission fluid pan bolts	
2000 through 2004	65 in-lbs
2005 and later	39 in-lbs
Automatic transmission drain plug	15
Automatic transmission overflow	
plug (2005 and later models)	15
Automatic transmission refill	
plug (2005 and later models)	29
Drivebelt tensioner mounting bolt and nuts (V8 engine)	
2000 through 2004	144 in-lbs
2005 and later	19
Transfer case filler plug and drain plug	27
Front differential	
Drain plug	48
Fill plug	29
Rear differential fill/drain plug	
Tundra models	39
Sequoia models	36
Spark plugs	
V6 engine	
2000 through 2004	168 in-lbs
2005 and later	156 in-lbs
V8 engine	156 in-lbs
Engine oil drain plug	
V6 engine	
2000 through 2004	28
2005 and later	30
V8 engine	29
Wheel lug nuts	83

2A
V6 ENGINES

Section

Reference to other Chapters

1 General information

This Part of Chapter 2 is devoted to in-vehicle repair procedures for 3.4L (5VZ-FE) and 4.0L (1GR-FE) engines.

All information concerning engine removal and installation and engine block and cylinder head overhaul can be found in Part C of this Chapter.

The following repair procedures are based on the assumption that the engine is installed in the vehicle. If the engine has been removed from the vehicle and mounted on a stand, many of the steps outlined in this Part of Chapter 2 will not apply.

The Specifications included in this Part of Chapter 2 apply only to the procedures contained in this Part. Additional specifications can be found in Chapter 2 Part C

2 Repair operations possible with the engine in the vehicle

1   Many major repair operations can be accomplished without removing the engine from the vehicle. Clean the engine compartment and the exterior of the engine with some type of degreaser before any work is done. It will make the job easier and help keep dirt out of the internal areas of the engine.

2   Depending on the components involved, it may be helpful to remove the hood to improve access to the engine as repairs are performed (refer to Chapter 11 if necessary). Cover the fenders to prevent damage to the paint. Special pads are available, but an old bedspread or blanket will also work.

3   If vacuum, exhaust, oil or coolant leaks develop, indicating a need for gasket or seal replacement, the repairs can generally be made with the engine in the vehicle. The intake and exhaust manifold gaskets, oil pan gasket, crankshaft oil seals and cylinder head gaskets are all accessible with the engine in place.

4   Exterior engine components, such as the intake and exhaust manifolds, the oil pan, the oil pump, the water pump, the starter motor, the alternator and the fuel system components can be removed for repair with the engine in place.

➡Note: On 4WD models, it will be necessary to either remove the front axle/differential assembly (see Chapter 8) or remove the engine from the vehicle (see Chapter 2C) in order to access the oil pump and oil pan gasket.

5   Since the cylinder heads can be removed without pulling the engine, valve component servicing can also be accomplished with the engine in the vehicle. Replacement of the camshafts, timing belt and pulleys is also possible with the engine in the vehicle.

6   In extreme cases caused by a lack of necessary equipment, repair or replacement of piston rings, pistons, connecting rods and rod bearings is possible with the engine in the vehicle. However, this practice is not recommended because of the cleaning and preparation work that must be done to the components involved.

3 Top Dead Center (TDC) for number one piston - locating

♦ Refer to illustration 3.7

➡Note: The following procedure is based on the assumption that the distributor is correctly installed (3.0L engine only). If you are trying to locate TDC to install the distributor correctly, piston position must be determined by feeling for compression at the number one spark plug hole, then aligning the ignition timing marks as described in Step 8.

1   Top Dead Center (TDC) is the highest point in the cylinder that each piston reaches as it travels up the cylinder bore. Each piston reaches TDC on the compression stroke and again on the exhaust stroke, but TDC generally refers to piston position on the compression stroke.

2   Positioning the piston(s) at TDC is an essential part of many procedures such as valve timing, camshaft and timing belt/pulley removal and distributor removal.

3   Before beginning this procedure, remove the Number One spark plug (see Chapter 1, if necessary). Also, be sure to place the transmission in Neutral and apply the parking brake or block the rear wheels. Disable the ignition system by disconnecting the primary (small-diameter wires) electrical connector from each ignition coil pack. Also, disable the fuel pump (see Chapter 4, Section 2).

4   In order to bring any piston to TDC, the crankshaft must be turned using one of the methods outlined below. When looking at the front of the engine, normal crankshaft rotation is clockwise.

a) The preferred method is to turn the crankshaft with a socket and ratchet attached to the bolt threaded into the front of the crankshaft. Turn the crankshaft in a clockwise direction only.

b) A remote starter switch, which may save some time, can also be used. Follow the instructions included with the switch. Once the piston is close to TDC, use a socket and ratchet as described in the previous paragraph.

c) If an assistant is available to turn the ignition switch to the Start position in short bursts, you can get the piston close to TDC without a remote starter switch. Make sure your assistant is out of the vehicle, away from the ignition switch, then use a socket and ratchet as described in Paragraph (a) to complete the procedure.

5   This engine does not have a distributor, but rather a separate coil for each pair of companion cylinders. If not already done, disconnect the primary wires from the coil packs and remove the Number One spark plug (see Chapter 1).

6   Install a compression pressure gauge in the number one spark plug hole (refer to Chapter 2C). It should be a gauge with a screw-in fitting and a hose at least six inches long.

7   Rotate the crankshaft using one of the methods described above while observing the compression gauge. When TDC for the compression stroke of number one cylinder is reached, compression pressure will show on the gauge as the marks on the crankshaft pulley are beginning to line up (see illustration). If you go past the marks, release the gauge pressure and rotate the crankshaft around two more revolutions.

➡**Note: The most positive method for finding TDC is to examine the crankshaft timing mark and camshaft sprocket timing marks (see Section 7).**

8   After the number one piston has been positioned at TDC on the compression stroke, TDC for the next cylinder can be located by turning the crankshaft another 120-degrees (refer to the firing order in this Chapter's Specifications).

3.7 Turn the crankshaft until the notch in the pulley aligns with the zero on the timing plate (typical)

4   Valve covers - removal and installation

2000 THROUGH 2004 MODELS

▶ **Refer to illustrations 4.5 and 4.7**

1   Disconnect the cable from the negative battery terminal.

2   Remove the air intake plenum from the intake air connector (see Section 5). This will require a large amount of disassembly so read the procedure carefully and mark any harness connectors or vacuum lines with tape to insure proper installation.

3   Remove the ignition coils or spark plug wires from the spark plugs, depending on which side is being disassembled. Be sure to mark each spark plug wire using tape or other marking device to insure proper reassembly. See Chapter 5 for additional details.

4   Disconnect any vacuum lines or wires that may interfere with the removal process. Mark them carefully to insure proper reassembly.

5   Remove the retaining bolts (see illustration), then detach the cover(s). If the cover is stuck to the head, bump the end with a wood block and a hammer to jar it loose. If that doesn't work, try to slip a flexible putty knife between the head and cover to break the seal.

✳✳ CAUTION:

Don't pry at the cover-to-head joint or damage to the sealing surfaces may occur, leading to oil leaks after the cover is reinstalled.

6   The mating surfaces of the cylinder head and cover must be clean when the cover is installed. Use a gasket scraper to remove all traces of sealant and old gasket material, then clean the mating surfaces with lacquer thinner or acetone. If there's residue or oil on the mating surfaces when the cover is installed, oil leaks may develop.

7   Position the semi-circular seals in the cylinder head cutouts, sealing them with RTV sealant, then apply a thin, uniform layer of RTV sealant to the gasket/seal joints (see illustration).

8   Position a new gasket on the valve cover, then install the cover, sealing washers and nuts.

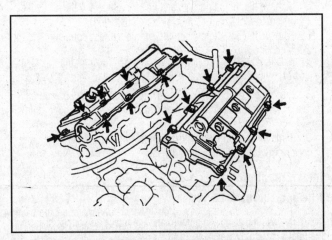

4.5 Valve cover bolt locations (2000 through 2004 models)

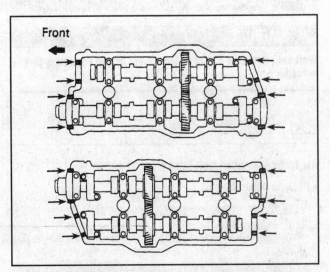

4.7 Be sure to use RTV sealant at the shaded areas (2000 through 2004 models)

➡Note: Be sure to install new seals on the spark plug tubes.

9   Tighten the nuts to the torque listed in this Chapter's Specifications in three or four equal steps.

10 Reinstall the remaining parts, run the engine and check for oil leaks.

2005 AND LATER MODELS

▸ **Refer to illustrations 4.12 and 4.16**

11 Disconnect the cable from the negative battery terminal.

12 Remove the engine cover (see illustration).

13 Remove the air intake duct and the air filter housing (see *Air filter housing - removal and installation* in Chapter 4).

14 If you're going to remove the left valve cover, remove the upper intake manifold (see Section 5).

15 Disconnect the PCV ventilation hose from the valve cover.

16 Remove the ignition coils (see illustration).

17 Remove the valve cover retaining bolts and nuts and remove the valve cover.

18 Remove and discard the valve cover gasket.

19 Installation is the reverse of removal. Be sure to use a new valve cover gasket and tighten the valve cover nuts and bolts to the torque listed in this Chapter's Specifications.

20 When you're done, start the engine and check for oil leaks around the edges of the valve cover.

4.12 To detach the engine cover from a 2005 or later V6 engine, remove these two nuts

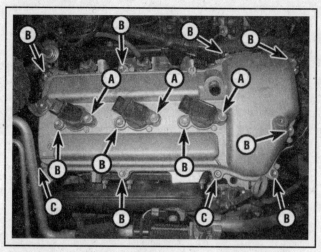

4.16 Valve cover details (2005 and later models)

A Ignition coil bolts	C Valve cover nuts
B Valve cover bolts	

5   Intake manifold - removal and installation

❋❋ WARNING:

Wait until the engine is completely cool before beginning this procedure.

2000 THROUGH 2004 MODELS

Air intake plenum and intake air connector

▸ **Refer to illustration 5.5**

1   Disconnect the negative cable from the battery.

2   Drain the coolant into a clean container (see Chapter 1).

3   Remove the air filter housing, throttle body and fuel injectors (see Chapter 4).

4   Clearly label, then detach all remaining wires, hoses and brackets still attached to the air intake plenum.

5   Remove the mounting nuts and bolts, then detach the air intake plenum from the intake air connector (see illustration).

6   Remove the bolts that retain the engine harness connectors to the intake air connector.

7   Detach the two fuel return hoses from the pipes on the intake air connector.

8   Disconnect the brake booster vacuum hose, the fuel pressure regulator vacuum hose and the ground strap from the intake air connector.

9   Remove the mounting bolts and nuts and separate the air intake connector from the intake manifold.

Intake manifold

▸ **Refer to illustration 5.12**

10 Remove the air intake plenum and the intake air connector from the intake manifold (see Steps 1 through 9).

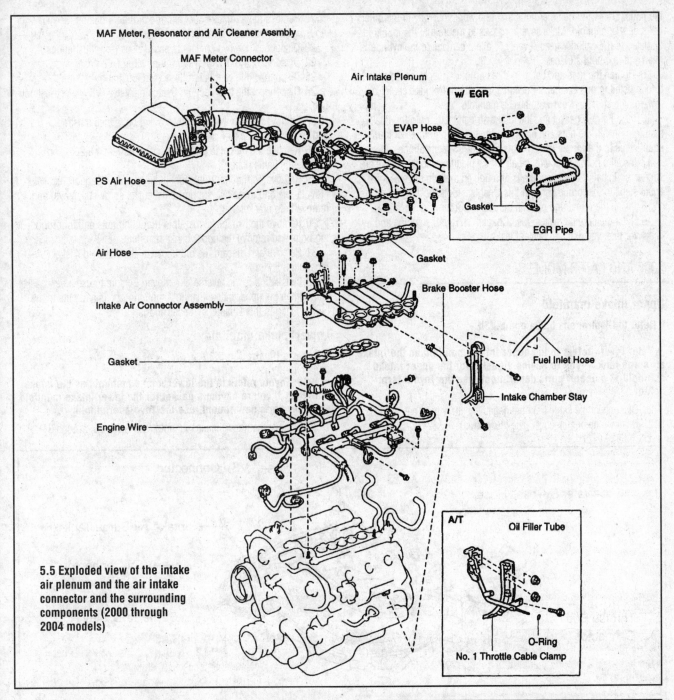

MAF Meter, Resonator and Air Cleaner Assmbly

MAF Meter Connector

Air Intake Plenum

EVAP Hose

w/ EGR

Gasket

EGR Pipe

PS Air Hose

Air Hose

Gasket

Intake Air Connector Assembly

Brake Booster Hose

Fuel Inlet Hose

Gasket

Intake Chamber Stay

Engine Wire

A/T

Oil Filler Tube

5.5 Exploded view of the intake air plenum and the air intake connector and the surrounding components (2000 through 2004 models)

O-Ring

No. 1 Throttle Cable Clamp

11 Clearly label, then detach all remaining wires, hoses and brackets still attached to the intake manifold and coolant outlets.

12 Remove the mounting nuts and bolts, then detach the intake manifold from the engine (see illustration). Start with the outer bolts first and work your way to the inner bolts on the manifold. Make several passes to insure that the manifold is separated from the cylinder heads evenly to avoid damage to the cylinder head and manifold surfaces. If it's stuck, don't pry between the gasket mating surfaces or damage may result.

13 Use a scraper to remove all traces of old gasket material and sealant from the manifold and cylinder heads, then clean the mating surfaces with lacquer thinner or acetone.

14 Install new gaskets, then position the intake manifold on the engine. Make sure the gaskets haven't shifted and install the nuts and

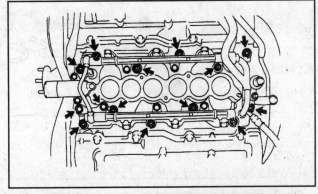

5.12 Intake manifold bolt locations (2000 through 2004 models)

the bolts. Start with the inner bolts first and work your way to the outer bolts on the manifold. Make several passes to insure that the manifold is mated to the cylinder heads evenly to avoid damage to the cylinder head and manifold surfaces.

15 Tighten the nuts and bolts, in three or four equal steps, to the torque listed in this Chapter's Specifications. Work from the center out towards the ends to avoid warping the manifold.

16 Install a new gasket and the intake air connector onto the intake manifold. Install the nuts and bolts and tighten them in three of four equal steps to the torque listed in this Chapter's Specifications.

17 Install a new gasket and the air intake plenum onto the intake air connector. Install the nuts and bolts and tighten them in three of four equal steps to the torque listed in this Chapter's Specifications.

18 Install the remaining parts in the reverse order of removal.

19 Refill the cooling system (see Chapter 1). Run the engine and check for fuel, vacuum and coolant leaks.

2005 AND LATER MODELS

Upper intake manifold

♦ Refer to illustrations 5.23a and 5.23b

➡Note: Toyota refers to the upper intake manifold as the intake air surge tank. If you're buying a gasket for the upper intake manifold at a dealer parts department, use the Toyota terminology.

20 Disconnect the cable from the negative battery terminal.

21 Remove the engine cover (see illustration).

22 Remove the air intake duct and the air filter housing (see *Air filter housing - removal and installation* in Chapter 4).

23 Disconnect the two coolant bypass hoses (see illustrations).

24 Disconnect the (EVAP system) fuel vapor feed hose.

25 Disconnect the ventilation hose (see illustration 5.23a).

26 Disconnect the two Vacuum Switching Valve (VSV) electrical connectors.

27 Remove the two throttle body bracket bolts (see illustration 5.23a) and remove the bracket.

28 Remove the oil baffle plate bolt (see illustration 5.23a) and remove the baffle plate.

29 Remove both bolts from each intake manifold support bracket (Toyota refers to these two support brackets as surge tank stays) and remove both brackets.

30 Remove the two intake manifold mounting nuts and four mounting bolts and remove the upper intake manifold.

31 Remove and discard the old upper intake manifold gasket (see illustration 5.23b).

32 Installation is the reverse of removal. Be sure to use a new gasket and tighten the upper intake manifold mounting bolts and nuts to the torque listed in this Chapter's Specifications.

Lower intake manifold

♦ Refer to illustration 5.36

➡Note: Toyota refers to the lower intake manifold as the intake manifold. If you're buying a gasket for the lower intake manifold at dealer parts department, use the Toyota terminology.

33 Remove the upper intake manifold (see Steps 20 through 31).

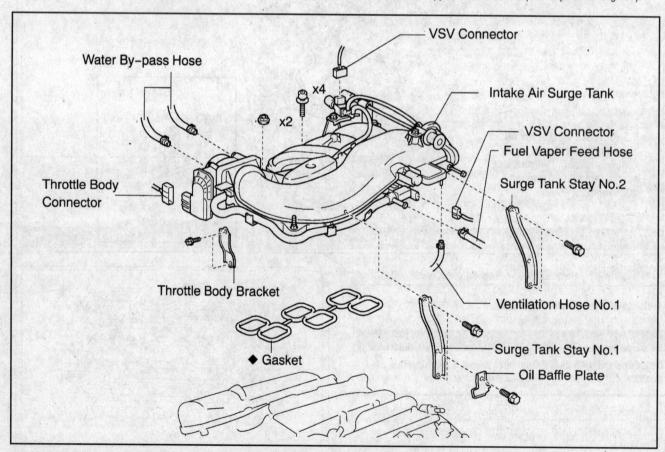

5.23a An exploded view of the upper intake manifold assembly (left side)

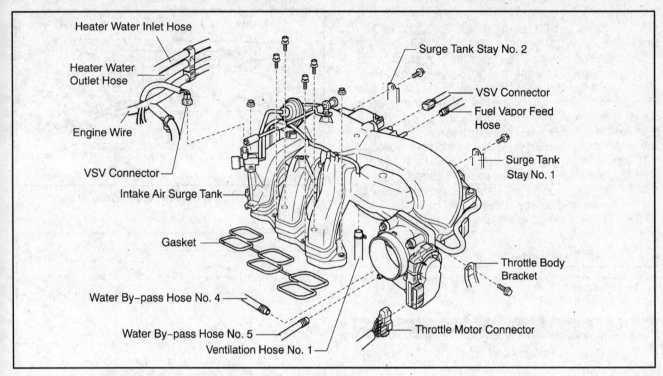

5.23b An exploded view of the upper intake manifold assembly (right side)

34 If you're removing the lower intake manifold to replace it, remove the fuel rail now (see Chapter 4).

35 If you're just removing the lower intake manifold to replace the gaskets, it's not necessary to remove the fuel rail, but you'll have to disconnect the fuel supply and return line connections (see Chapter 4).

36 Remove the lower intake manifold bolts and remove the lower intake manifold (see illustration).

➡**Note: If you're simply removing the lower intake manifold to replace the gaskets, you can remove it without detaching fuel rail assembly. Be sure to remove and discard the old gaskets.**

37 Remove and discard the old lower intake manifold gaskets.

38 Installation is the reverse of removal. Be sure to use new gaskets and tighten the lower intake manifold bolts to the torque listed in this Chapter's Specifications.

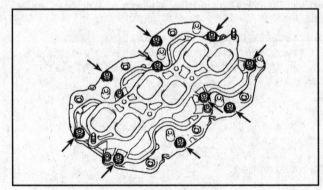

5.36 To detach the lower intake manifold, remove these 10 mounting bolts (fuel rail removed for clarity)

6   Exhaust manifolds - removal and installation

2000 THROUGH 2004 MODELS

◆ **Refer to illustrations 6.3 and 6.5**

❋ **WARNING:**

The engine must be completely cool before beginning this procedure.

1   Disconnect the cable from the negative terminal of the battery.

2   Spray penetrating oil on the exhaust manifold fasteners and allow it to soak in.

3   Remove the exhaust crossover pipe from the back of the cylinder heads (see illustration).

6.3 Exhaust crossover pipe installation details (2000 through 2004 models)

4 Remove the EGR pipe from the exhaust manifold (see Chapter 6).

5 Unbolt the exhaust manifolds from the cylinder heads (see illustration).

6 Carefully inspect the manifolds and fasteners for cracks and damage.

7 Use a scraper to remove all traces of old gasket material and carbon deposits from the manifolds and cylinder head mating surfaces. If the gasket was leaking, have the manifolds checked for warpage at an automotive machine shop and resurfaced if necessary.

8 Position new gaskets over the cylinder head studs.

9 Install the manifolds and thread the mounting nuts into place.

10 Working from the center out, tighten the nuts to the torque listed in this Chapter's Specifications in three or four equal steps.

11 Reinstall the remaining parts in the reverse order of removal. Use new gaskets when connecting the exhaust pipes.

12 Run the engine and check for exhaust leaks.

2005 AND LATER MODELS

♦ Refer to illustration 6.20

❊❊ WARNING:

The engine must be completely cool before beginning this procedure.

➠Note: The following procedure applies to either exhaust manifold.

13 Disconnect the cable from the negative battery terminal.

14 Raise the front of the vehicle and place it securely on jackstands.

15 Unbolt and disconnect the front exhaust pipe from the exhaust manifold.

16 Disconnect the electrical connector from the upstream oxygen sensors.

17 Remove the three bolts that secure the exhaust manifold stay (support bracket) and remove the stay.

18 Remove the six nuts that secure the exhaust manifold and remove the exhaust manifold.

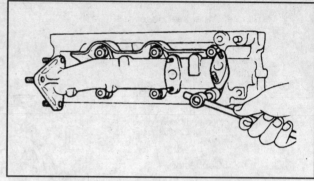

6.5 Exhaust manifold installation details (2000 through 2004 models)

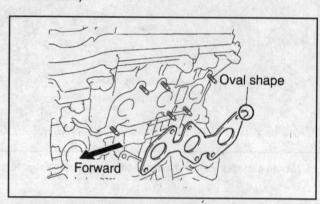

6.20 When installing the gasket for the left manifold (shown), make sure that the oval-shaped tip is facing to the rear; when installing the gasket for the right manifold, make sure that the oval-shaped tip is facing forward

19 Remove and discard the old exhaust manifold gasket.

20 Installation is the reverse of removal. Be sure to use a new gasket and be sure to install it with the oval-shaped tip facing in the correct direction (see illustration). Tighten the exhaust manifold nuts to the torque listed in this Chapter's Specifications.

7   Timing belt and sprockets - removal, inspection and installation (2000 through 2004 models)

❊❊ WARNING:

The engine must be completely cool before starting this procedure.

REMOVAL

❊❊ CAUTION ❊❊

The timing system is complex. Severe engine damage will occur if you make any mistakes. Do not attempt this procedure unless you are highly experienced with this type of repair. If you are at all unsure of your abilities, consult an expert. Double-check all your work and be sure everything is correct before you attempt to start the engine.

♦ Refer to illustrations 7.11, 7.14, 7.15, 7.22 and 7.27

1 Disconnect the negative cable from the battery.

2 Remove the under-vehicle splash shield (see illustration 8.6 in Chapter 1).

3 Remove the fan shroud and the fan assembly (see Chapter 3).

4 Drain the coolant and remove the radiator from the engine compartment (see Chapter 3).

5 Remove the drivebelts from the alternator and power steering pump (see Chapter 1).

6 Remove the spark plug wires (see Chapter 1).

7 Raise the front of the vehicle and support it securely on jackstands. Apply the parking brake and block the rear wheels.

8 Remove the power steering pump without disconnecting the hoses and secure it aside (see Chapter 10).

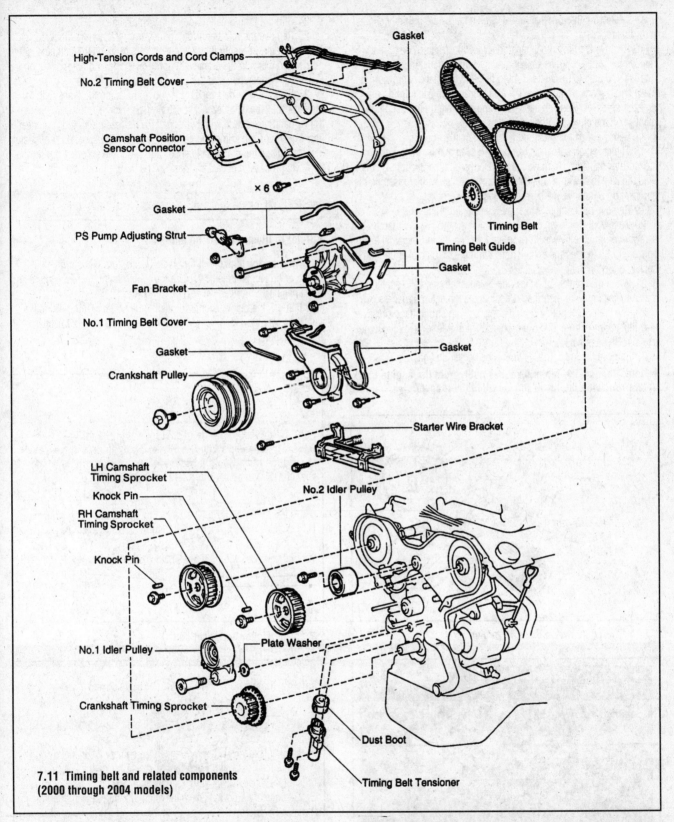

7.11 Timing belt and related components (2000 through 2004 models)

Labels (top to bottom, left to right):
- Gasket
- High-Tension Cords and Cord Clamps
- No.2 Timing Belt Cover
- Camshaft Position Sensor Connector
- × 6
- Gasket
- PS Pump Adjusting Strut
- Fan Bracket
- Timing Belt
- Timing Belt Guide
- Gasket
- No.1 Timing Belt Cover
- Gasket
- Crankshaft Pulley
- Gasket
- Starter Wire Bracket
- LH Camshaft Timing Sprocket
- No.2 Idler Pulley
- Knock Pin
- RH Camshaft Timing Sprocket
- Knock Pin
- No.1 Idler Pulley
- Plate Washer
- Crankshaft Timing Sprocket
- Dust Boot
- Timing Belt Tensioner

9   Remove the air conditioning compressor and the bracket from the engine (see Chapter 3).

10   Remove the spark plugs (see Chapter 1).

11   Remove the upper (no. 2) timing belt cover and gasket (see illustration). Also unbolt and remove the fan bracket.

➡**Note: Be sure to disconnect the camshaft position sensor harness connector and position it off to the side (see Chapter 6).**

12   Position the number one piston at TDC (see Section 3).

13   Check to see if there are installation marks on the timing belt - If you intend to re-use the belt and the marks have been obscured, make new ones.

14 Make sure the camshaft pulley timing marks are properly aligned (see illustration).

15 Remove the timing belt tensioner (see illustration). Be sure to remove the rubber boot as well; it may stick in the tensioner recess.

16 If you plan to re-use the timing belt and the marks were worn off, place a new mark on the belt at the exact location of each timing mark on each sprocket. This will allow easy and accurate timing belt alignment when the old timing belt is reused.

17 Remove the timing belt from the camshaft sprockets.

18 The camshaft sprockets can be re-moved at this point, if they are worn or damaged. Remove the valve cover(s) (see Section 4) and hold the camshaft with a wrench on the cast-in hex while loosening the sprocket bolt. Remove the bolt and detach the sprocket.

19 Remove the number 2 idler pulley (see illustration 7.11).

20 Remove the crankshaft (drivebelt) pulley bolt. Wedge a large screwdriver into the flywheel/driveplate ring gear teeth or against a converter bolt to keep the engine from turning. Use a breaker bar and socket to loosen the pulley bolt.

21 When the crankshaft pulley bolt is loosened, the TDC position of the crankshaft may be disturbed. Check and align again, if necessary.

22 The crankshaft pulley should come off with strong hand pressure (see illustration); if not, use two prybars behind it to lever it off. Do not use a jaw-type puller.

23 Remove the lower (no. 1) timing belt cover and gasket.

➡Note: Remove the fan bracket assembly from the engine block before removing the lower (no. 1) timing belt cover.

24 Slip the timing belt guide off the crankshaft.

25 Remove the number 1 idler pulley from the engine block.

26 If you're re-using the belt, check for a mark on the belt adjacent to the drilled mark on the crankshaft sprocket. If the original mark is gone, make a new one, then slip the belt off the sprocket.

27 If it's worn or damaged, or if you're replacing the crankshaft front oil seal, the crankshaft sprocket can now be removed. If it won't come off by hand, lever it off with two screwdrivers (see illustration). A steering wheel type puller may be needed to remove the sprocket. Be careful not to damage the crankshaft sensor portion of the sprocket during the removal process. If necessary, remove the lower sprocket retainer.

INSPECTION

⬧ **Refer to illustrations 7.30 and 7.31**

28 Inspect the timing belt for cracks, tears, torn belt strands, cut edges or broken belt teeth. Replace the timing belt if there are any signs of damage or prolonged wear (high mileage).

29 Check the belt tensioner for visible oil leakage. If there's only a faint trace of oil on the pushrod side, the tensioner seal is in satisfactory condition.

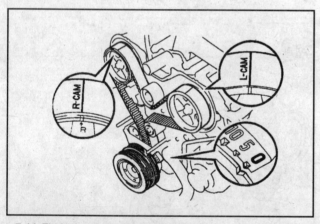

7.14 **Timing marks on the 3.4L engine**

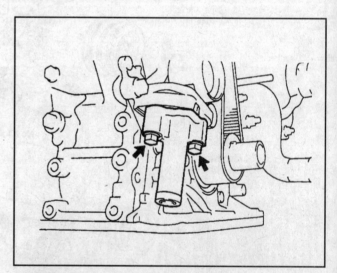

7.15 **Tensioner bolt locations**

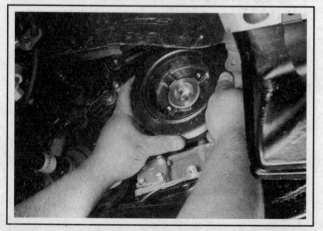

7.22 **A puller should not be necessary to remove the crankshaft pulley - if it is stuck, use two prybars behind it**

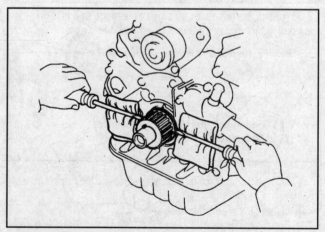

7.27 **Pad the front of the engine when prying off the crankshaft sprocket**

30 Hold the tensioner in both hands and push it forcefully against an immovable object (see illustration). If the pushrod moves, replace the tensioner.

31 Measure the protrusion of the pushrod from the housing end (see illustration). Compare your measurement to this Chapter's Specifications. If the protrusion is not as specified, replace the tensioner.

32 Check that the idler pulleys turn smoothly.

INSTALLATION

> ✳✳ **CAUTION** ✳✳
>
> **Before starting the engine, carefully rotate the crankshaft by hand through at least two full revolutions (use a socket and breaker bar on the crankshaft pulley centerbolt). If you feel any resistance, STOP! There is something wrong - most likely, valves are contacting the pistons. You must find the problem before proceeding. Check your work and see if any updated repair information is available.**

▶ **Refer to illustrations 7.34, 7.43 and 7.49**

33 Remove all dirt, oil and grease from the timing belt area at the front of the engine.

34 Align the crankshaft timing sprocket keyway with the crankshaft key and install the sprocket with the flange side up against the engine. Be careful not to damage the crankshaft sensor portion of the crankshaft sprocket. Check the alignment of the TDC marks on the sprocket and the oil pump housing, then reinstall the retainer and bolt (see illustration).

35 Apply thread locking compound to the first two or three threads on the number 1 idler pulley bolt, then position the idler pulley and washer and install the bolt. Tighten the bolt to the torque listed in this Chapter's Specifications.

36 Install the timing belt, starting at the crankshaft sprocket. If you're re-using the original belt, align the marks on the belt with the marks on the sprockets and covers. Install the belt over the lower number 1 idler and water pump pulleys.

37 Slip the belt guide over the crankshaft with the cupped side facing out.

38 Install the lower (no. 1) timing belt cover and gasket (see illustration 7.11).

39 Slip the crankshaft (drivebelt) sprocket onto the crankshaft, aligning the pulley keyway with the crankshaft key. Install the bolt and tighten it to the torque listed in this Chapter's Specifications. Use the method described in Step 20 to keep the crankshaft from turning.

40 Install the upper (no. 2) idler pulley. Tighten the bolt to the torque listed in this Chapter's Specifications. Make sure the pulley turns smoothly.

41 Install the front (left-hand) camshaft sprocket (if it was removed) on the camshaft with the flange side facing OUT. Align the pin hole in the sprocket with the pin in the end of the camshaft.

42 Install the retaining bolt and tighten it to the torque listed in this Chapter's Specifications. Use the method described in Step 18 to keep the camshaft from turning.

43 Recheck the timing marks to be sure the crankshaft hasn't turned (see illustration 7.14). If you're re-using the original belt, the installation mark should line up as it did in Step 14. If not, change the position of the timing belt on the crankshaft sprocket. The mark on the front (left-hand) camshaft sprocket should be at the top (12 o'clock position), aligned with the mark on the rear (no. 3) timing cover (see illustration).

7.30 Check the tensioner for signs of leakage and test for leakdown by forcing it against an immovable object

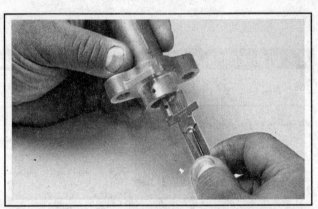

7.31 Measure the tensioner pushrod protrusion and compare it to the Specifications

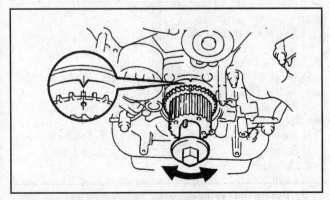

7.34 Timing gear marks on the 3.4L engine

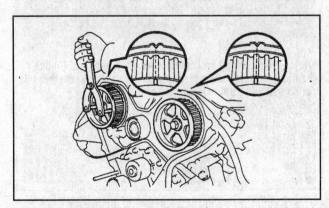

7.43 Alignment marks on the 3.4L engine

44 The rear (right-hand) camshaft knock pin hole should be at the top (12 o'clock position). If necessary, remove the valve cover and turn the camshaft slightly with a wrench to align the sprocket with the mark on the rear (no. 3) cover.

45 Turn the front (left-hand) camshaft sprocket clockwise slightly (about one tooth) with a pin spanner. If the special tool isn't available, grip the hex on the camshaft with a wrench and turn it. If you're re-using the original belt, align the installation mark with the camshaft timing mark. Slip the belt onto the sprocket, then turn the camshaft counterclockwise, back to its original position. There should now be slight tension on the belt.

46 Install the rear (right-hand) camshaft sprocket (if it was removed) on the camshaft with the flange side facing IN. Align the pin hole in the sprocket with the knock pin in the end of the camshaft.

47 Install the retaining bolt and tighten it to the torque listed in this Chapter's Specifications. Use the method described in Step 18 to keep the camshaft from turning. Be sure the timing mark is still aligned with the rear (no. 3) cover.

48 Slip the belt onto the sprocket. If you're re-using the original belt, align the installation marks.

49 Using a press or vise, slowly compress the timing belt tensioner pushrod (see illustration). Insert a metal pin, drill bit or Allen wrench through the holes in the pushrod and housing. Release the pressure from the press or vise.

50 Install the timing belt tensioner and tighten the bolts to the torque listed in this Chapter's Specifications. Remove the retaining pin.

51 Using a socket and breaker bar on the crankshaft pulley bolt, turn the crankshaft slowly through two complete revolutions (720-degrees). Recheck the timing marks.

✳✳ CAUTION:

If the timing marks are not aligned exactly as shown, repeat the timing belt installation procedure. DO NOT start the engine until you're absolutely certain that the timing belt is installed correctly. Serious and costly engine damage could occur if the belt is installed wrong.

52 Install the upper (no. 2) timing belt cover and gasket.

53 Reinstall the remaining parts in the reverse order of removal.

54 Refill the cooling system (See Chapter 1). Start the engine and check for proper operation and leaks.

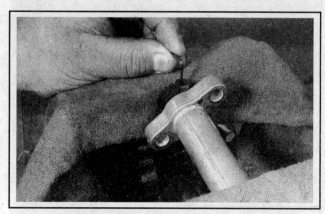

7.49 Restrain the tensioner pushrod by compressing the unit in a vise and inserting a pin approximately 0.050-inch (1.27 mm) in diameter - make sure the rubber boot is in place

8 Timing chain and sprockets - removal, inspection and installation (2005 and later models)

REMAL ▸ REMOVAL

▸ **Refer to illustrations 8.16, 8.23a, 8.23b, 8.24a, 8.24b, 8.25a, 8.25b and 8.31**

✳✳ CAUTION ✳✳

The timing system is complex. Severe engine damage will occur if you make any mistakes. Do not attempt this procedure unless you are highly experienced with this type of repair. If you are at all unsure of your abilities, consult an expert. Double-check all your work and be sure everything is correct before you attempt to start the engine.

➡ **Note: The following procedure is not for beginners. Please read the entire procedure carefully before deciding whether this is a job that you want to tackle at home.**

1 Disconnect the cable from the negative battery terminal.

2 Drain the engine oil and coolant (see Chapter 1).

3 Remove the battery (see Chapter 5).

4 Remove the engine cover (see illustration 4.12).

5 Remove the air filter housing (see Chapter 1).

6 Remove the radiator (see Chapter 3). Remove the cooling fan mounting nuts and remove the cooling fan (see illustrations 5.7a and 5.7b in Chapter 3).

7 Remove the serpentine drivebelt (see Chapter 1).

8 Remove the upper intake manifold (see Section 5).

9 Remove the ignition coils (see Chapter 5) and the valve covers (see Section 4).

10 Remove the Variable Valve Timing (VVT) sensor (see *Variable Valve Timing (VVT) system - description and component replacement* in Chapter 6).

11 Remove the dipstick, then remove the dipstick tube retaining bolt and the dipstick tube. Remove and discard the old dipstick tube O-ring.

12 Disconnect the electrical connector from the power steering pressure switch, then detach the power steering pump (see Chapter 10) and set it aside. Do NOT disconnect the power steering fluid hoses from the pump!

13 Remove the alternator (see Chapter 5).

14 Detach the air conditioning compressor (see Chapter 3). Do NOT disconnect the air conditioning refrigerant hoses!

15 Unbolt the serpentine drivebelt tensioner (see *Drivebelt check, adjustment and replacement* in Chapter 1).

16 Remove the idler pulley bolts and remove the two idler pulleys (see illustration).

17 Remove the crankshaft pulley (see Chapter 2B, Section 7).

18 Remove the two oil pans (see Section 12) and the oil pump pickup tube (see Section 13).

19 Remove the two oil cooler hoses (see illustration 8.16).

20 Disconnect the two radiator hoses from the water inlet (see illustration 8.16).

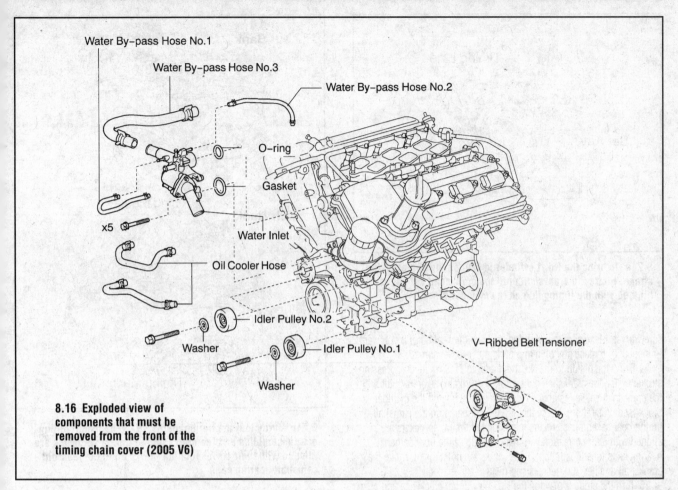

8.16 Exploded view of components that must be removed from the front of the timing chain cover (2005 V6)

Water By-pass Hose No.1
Water By-pass Hose No.3
Water By-pass Hose No.2
O-ring
Gasket
Water Inlet
x5
Oil Cooler Hose
Idler Pulley No.2
Washer
Idler Pulley No.1
Washer
V-Ribbed Belt Tensioner

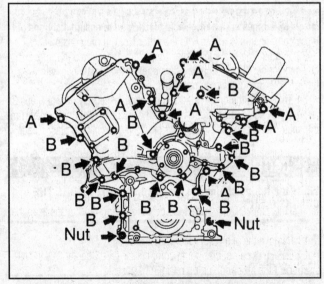
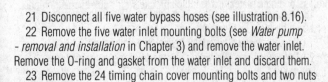

8.23a To detach the timing chain cover from a 2005 V6, remove these 24 bolts and two nuts. When installing the cover be sure to note the A (shorter) and B (longer) bolt holes

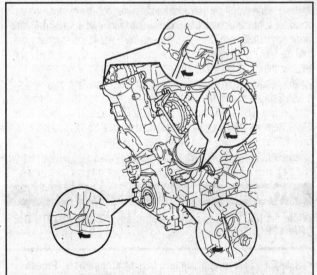

8.23b To remove the timing chain cover from a 2005 V6, pry between the cover and the engine only at these four points

21 Disconnect all five water bypass hoses (see illustration 8.16).
22 Remove the five water inlet mounting bolts (see *Water pump - removal and installation* in Chapter 3) and remove the water inlet. Remove the O-ring and gasket from the water inlet and discard them.
23 Remove the 24 timing chain cover mounting bolts and two nuts

(see illustration) and remove the timing chain cover. There are only four spots where you can pry between the timing chain cover and the engine (see illustration). Do NOT pry the timing chain cover loose at any other spot or you will damage the sealing surface of the cover. After removing the timing chain cover, carefully pry out the old crankshaft seal with a

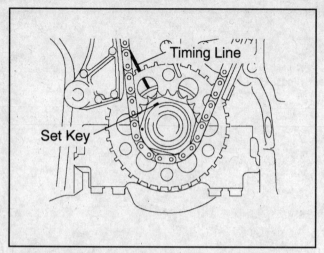

8.24a To bring the No. 1 cylinder to TDC on the compression stroke, rotate the crankshaft until the crankshaft set key is aligned with the timing line on the cylinder block

screwdriver. Make sure that you don't scratch the seal bore. If you want to inspect or replace any oil pump parts, refer to Section 13.

24 Bring the piston in the No. 1 cylinder to TDC on its compression stroke: Install the crankshaft pulley bolt, then rotate the crankshaft until the crankshaft set key is aligned with the timing line on the cylinder block (see illustration). Verify that the timing marks on the camshaft timing gear assemblies and the marks on the timing sprockets are aligned with their corresponding marks on top of the front camshaft bearing caps (see illustration). If the marks are not aligned, rotate the crankshaft another 360 degrees and recheck the marks.

25 Turn the stopper plate on the No. 1 tensioner clockwise and push in the tensioner plunger (see illustrations). To lock the plunger in this position, turn the stopper plate counterclockwise and insert a drill bit or punch (0.138-inch diameter) through the holes in the stopper plate and the tensioner body. Remove the two tensioner mounting bolts and remove the No. 1 tensioner.

26 Remove the chain tensioner slipper.

27 Using a 10 mm hex bit, unscrew idle gear shaft No. 2, then remove idle gear No. 1 and idle gear shaft No. 1 (see illustration 8.25a).

28 Remove the two chain vibration dampers No. 2 (see illustration 8.25a).

29 Remove the No. 1 timing chain (the long chain).

✳✳ **CAUTION:**

While the No. 1 timing chain is removed, DO NOT ROTATE THE CRANKSHAFT!

30 Remove the crankshaft timing chain sprocket.

31 Raise chain tensioner No. 2 and insert a drill bit or punch (0.039-inch diameter) into the hole (see illustration).

32 Immobilize the hex on the exhaust camshaft with an adjustable wrench and loosen and remove the two bolts that secure the camshaft timing gear and camshaft timing sprocket to the camshafts. Remove the timing gear, the timing sprocket and timing chain No. 2 as a single assembly. Keep the timing gear assembly, timing sprocket and timing

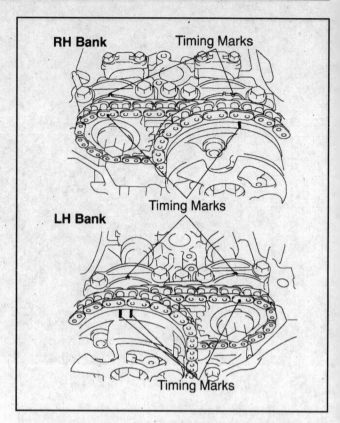

8.24b Verify that the timing marks on the camshaft timing gear assemblies and the marks on the timing sprockets are aligned with their corresponding marks on top of the front camshaft bearing caps

chain No. 2 together in a resealable plastic bag to ensure that none of these components is mixed with the other No. 2 timing chain set.

33 Remove the chain tensioner No. 2 mounting bolt and remove chain tensioner No. 2. Store the tensioner in the plastic bag with the other No. 2 timing chain components.

34 To remove the other No. 2 timing chain assembly, repeat Steps 31 through 33. Again, store the timing gear assembly, timing sprocket, timing chain No. 2 and chain tensioner No. 2 in a resealable plastic bag.

✳✳ **CAUTION:**

While the timing chains are removed, DO NOT ROTATE THE CRANKSHAFT!

INSPECTION

▶ **Refer to illustrations 8.36, 8.37 and 8.39**

35 Inspect all parts of the timing chain assembly for wear and damage. Inspect the three timing chains for loose pins, cracks, and worn rollers and side plates. Inspect the sprockets for hook-shaped, chipped and/or broken teeth.

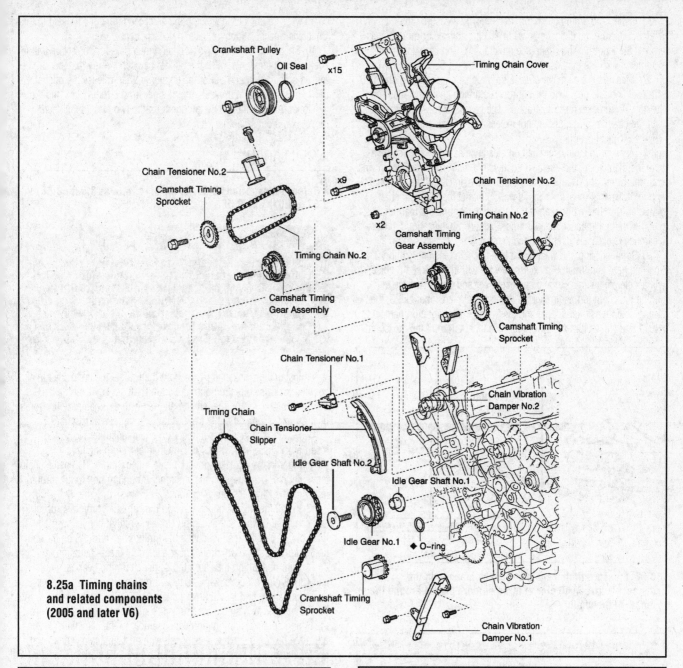

8.25a Timing chains and related components (2005 and later V6)

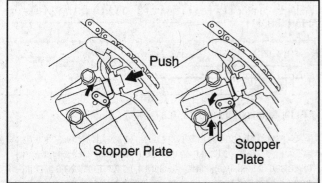

8.25b To lock the tensioner in the retracted position, rotate the stopper plate clockwise and push the plunger in, then rotate the stopper plate counterclockwise and insert a pin through the hole in the stopper plate and the tensioner body

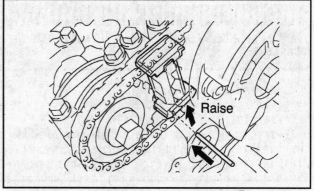

8.31 To lock tensioner no. 2 in the retracted position, push the plunger up and insert a pin through the hole in the tensioner

36 Inspect the timing chain for stretching. To measure timing chain stretch, measure the distance between 15 pins at three or more places around the length of the chain (see illustration). Compare your measurements with the distance listed in this Chapter's Specifications.

37 Measure the diameter of each timing chain gear, timing chain sprocket and idle gear with the appropriate timing chain installed on the gear or sprocket (see illustration). The gear or sprocket diameter, with the chain in place, should not exceed the dimensions listed in this Chapter's Specifications.

38 Measure the idle gear shaft oil clearance as follows. First, measure the diameter of the idle gear shaft with a micrometer and jot down your measurement. Then measure the inside diameter of the idle gear and jot down that measurement as well. Subtract the idle gear shaft diameter from the inside diameter of the idle gear and compare the result with the clearance listed in this Chapter's Specifications. If the clearance is excessive, replace the idle gear.

39 Some scoring and wear of the timing chain tensioners and vibration dampers is normal. But excessive wear will increase chain noise, will accelerate chain, gear and sprocket wear and could damage the engine if a chain jumps timing. So inspect chain tensioner No. 2, the timing chain tensioner slipper and the timing chain vibration dampers (No. 1 and No. 2) for excessive wear (see illustration). If the measured chain wear for any of these components exceeds the depth listed in this Chapter's Specifications, replace the component.

40 Check the No. 1 and No. 2 chain tensioners for correct operation. On the No. 1 tensioner, raise the ratchet pawl and verify that the plunger moves smoothly in and out of the tensioner, then release the ratchet pawl and verify that it prevents the plunger from sliding back into the tensioner. Also verify that the plungers on the two No. 2 tensioners move in and out smoothly.

INSTALLATION

▶ **Refer to illustrations 8.41, 8.43, 8.44, 8.52a, 8.52b, 8.56, 8.57a, 8.57b and 8.58**

> ※ **CAUTION** ※
> Before starting the engine, carefully rotate the crankshaft by hand through at least two full revolutions (use a socket and breaker bar on the crankshaft pulley centerbolt). If you feel any resistance, STOP! There is something wrong - most likely, valves are contacting the pistons. You must find the problem before proceeding. Check your work and see if any updated repair information is available.

41 Install the crankshaft pulley bolt, then rotate the crankshaft in a counterclockwise direction until the crankshaft set key is at the 270-degree position, i.e. on the left and aligned with an imaginary horizontal line, as you're looking at the front of the engine (see illustration).

42 Push in the tensioner plunger on chain tensioner No. 2 and insert a drill bit or punch (0.039 inch diameter) into the hole to lock the plunger in the retracted position (see illustration 8.31). Install chain tensioner No. 2 and tighten the tensioner mounting bolt to the torque listed in this Chapter's Specifications.

43 Install the No. 2 timing chain on the camshaft timing sprocket and camshaft timing gear assembly. Make sure that the yellow mark links on the chain are aligned with the timing marks (dots) on the cam timing sprocket and cam timing gear assembly (see illustration).

44 Align the timing marks on the camshaft timing gear assembly and cam timing sprocket with the timing marks on the bearing caps (see illustration) and install timing chain No. 2, the timing sprocket and the timing gear as a single assembly. Install the two bolts that secure the timing sprocket and timing gear assembly to the camshafts. Immobilize the hex on the exhaust camshaft with an adjustable wrench and tighten these two bolts to the torque listed in this Chapter's Specifications. Remove the drill bit or punch that you used to lock the tensioner

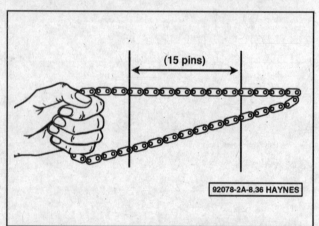

8.36 Measure timing chain stretch by measuring the distance between 15 pins at three or more places around the length of the chain

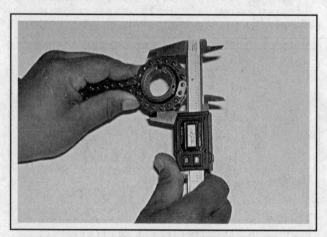

8.37 Wrap the chain around each of the timing sprockets and measure the diameter of the sprockets across the chain rollers. If the measurement is less than the minimum sprocket diameter, replace the chain and the timing sprockets

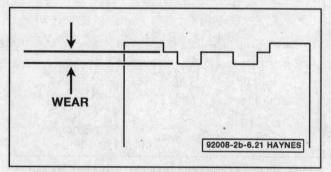

8.39 When inspecting chain tensioner No. 2, the chain tensioner slipper and the No. 1 and No. 2 chain vibration dampers, measure timing chain wear from the top of the chain contact surface to the bottom of the wear grooves

plunger in its retracted position and verify that the plunger tensions the chain.

45 Install the other No. 2 timing chain (repeat Steps 42 through 44).

46 Install chain vibration damper No. 1 and tighten the bolts to the torque listed in this Chapter's Specifications.

47 Install the crankshaft timing sprocket on the crankshaft. Be sure to align the timing sprocket keyway with the key on the crankshaft, and make sure that the sprocket end of the crank timing sprocket faces in, toward the engine.

48 Install the chain tensioner slipper.

49 Turn the stopper plate on the No. 1 tensioner clockwise and push in the tensioner plunger (see illustration 8.25b). To lock the plunger in this position, turn the stopper plate counterclockwise and insert a drill bit or punch (0.138-inch diameter) through the holes in the stopper plate and the tensioner. Install the tensioner and tighten the two tensioner mounting bolts to the torque listed in this Chapter's Specifications.

50 Verify that the timing marks on the camshaft timing gear assemblies and the marks on the camshaft timing sprockets are aligned with their corresponding marks on top of the front camshaft bearing caps (see illustration 8.24b).

51 Using the crankshaft pulley bolt, rotate the crankshaft clockwise until the crankshaft set key is aligned with the timing line on the cylinder block (see illustration 8.24a).

52 Install the long timing chain (No. 1) on the camshaft timing gear

assemblies and on the crankshaft timing sprocket. Make sure that the yellow mark link is aligned with the timing mark (dot) on the crankshaft timing sprocket (see illustration) and that the orange mark links are aligned with the timing marks on the camshaft timing gear assembly and the camshaft timing sprocket (see illustration).

53 Apply a light coat of engine oil to the bearing surface of idle gear

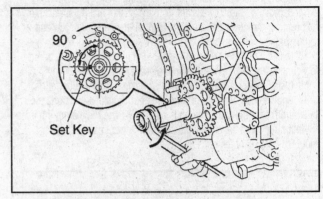

8.41 Install the crankshaft pulley bolt, then rotate the crankshaft in a counterclockwise direction until the crankshaft set key is at the 270-degree position, i.e. on the left and aligned with an imaginary horizontal line, as you're looking at the front of the engine

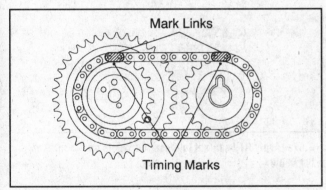

8.43 When installing the No. 2 timing chain on the camshaft timing sprocket and camshaft timing gear assembly, make sure that the yellow mark links on the chain are aligned with the timing marks (dots) on the cam timing sprocket and cam timing gear assembly

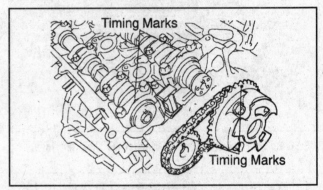

8.44 When installing the No. 2 timing chain, camshaft timing gear assembly and camshaft timing sprocket on the cylinder head, be sure align the timing marks on the cam timing gear and cam timing sprocket with the timing marks on the bearing caps

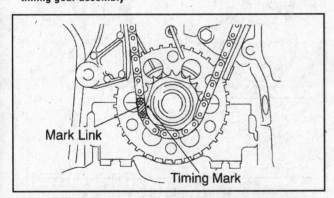

8.52a When installing timing chain No. 1 (the long timing chain) on the camshaft timing gear assemblies and the crankshaft timing sprocket, make sure that the yellow mark link is aligned with the timing mark (dot) on the crankshaft timing sprocket . . .

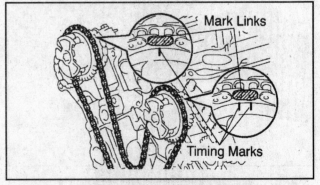

8.52b . . . and the orange mark links are aligned with the timing marks on the camshaft timing gear assembly and the camshaft timing sprocket

shaft No. 1. Install idle gear shaft No. 1 and idle gear No. 1. Make sure that the sprocket teeth on the idle gear are facing forward, then secure the idle gear No. 1 and idle gear shaft No. 1 with idle gear shaft No. 2 (the large bolt with the wide flat head with a 10 mm hex recess). Tighten idle gear shaft No. 2 to the torque listed in this Chapter's Specifications. Remove the drill bit or punch that you inserted into chain tensioner No. 1 (in step 49) and verify that it tensions timing chain No. 1.

54 Remove all old RTV sealant from the gasket mating surfaces of the timing chain cover and from the front of the cylinder heads and engine block.

55 Install a new crankshaft oil seal in the timing cover. Use a large socket and carefully tap the seal into place.

56 Install a new O-ring on the left cylinder head (see illustration).

57 Apply beads of RTV sealant to the four indicated locations (see illustration) and apply a continuous bead of RTV to the mating surfaces of the timing chain cover (see illustration). These beads should be about 0.12 to 0.16 inch-wide.

✳✳ CAUTION:

Once you have installed the RTV sealant on the engine and timing chain cover you have three minutes to install the cover. If you take longer than that, the sealant might not set up properly, so you'll have to remove the sealant and re-apply it.

58 Rotate the keyway on the oil drive rotor about 15-degrees to the right of vertical to align it with the square part of the crankshaft timing gear (see illustration) and slide the timing chain cover into place.

59 Install all 24 timing cover mounting bolts, install both nuts (see illustration 8.23a) and tighten all fasteners gradually and evenly, in a criss-cross fashion until they're all snug. Note that there are nine shorter bolts and 15 longer bolts. Do not put long bolts in short holes or vice versa. When all of the fasteners are snug, tighten them to the torque listed in this Chapter's Specifications.

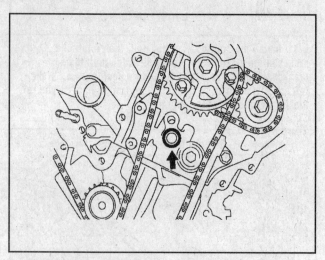

8.56 Be sure to install a new O-ring for the chain tensioner

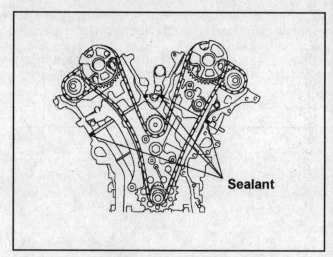

8.57a Apply RTV sealant to these four spots on the front of the engine . . .

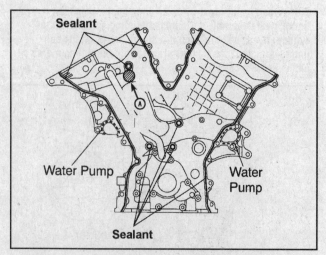

8.57b . . . and apply a continuous bead of sealant to the mating surfaces of the timing chain cover and to the indicated spots as well, including the mating surfaces around the two water pump passages. But make sure that you DON'T apply RTV to area A

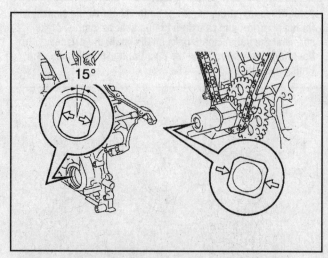

8.58 To align the keyway on the oil driver rotor with the square part of the crankshaft timing gear, rotate the keyway about 15 degrees to the right of vertical

60 Install the water inlet O-rings and the water inlet housing (see *Water pump - removal and installation* in Chapter 3). Install the five water bypass hoses and the two radiator hoses (see illustration 8.16).

61 Install the two oil cooler hoses (see illustration 8.16).

62 Install the oil pans (see Section 12).

63 Install the crankshaft pulley and tighten the pulley retaining bolt to the torque listed in this Chapter's Specifications.

64 Install the No. 1 and No.2 idler pulleys (see illustration 8.16). Tighten the idler pulley mounting bolts to the torque listed in this Chapter's Specifications.

65 Install the drivebelt belt tensioner (see *Drivebelt check, adjustment and replacement* in Chapter 1).

66 Install the air conditioning compressor (see Chapter 3).

67 Install the alternator (see Chapter 5).

68 Install the power steering pump (see Chapter 10) and reconnect the power steering pressure switch.

69 Install the dipstick tube. Be sure to use a new O-ring. Coat the O-ring with a light coat of engine oil, then push the dipstick tube into its guide hole in the oil pan. Install the dipstick tube mounting bracket bolt and tighten it securely.

70 Install the valve covers (see Section 4).

71 Install the spark plugs (see Chapter 1) and the ignition coils (see Chapter 5).

72 Install the upper intake manifold (see Section 5).

73 Install the serpentine drivebelt (see Chapter 1).

74 Install the cooling fan (see illustrations 5.7a and 5.7b in Chapter 3) and the radiator (see Chapter 3).

75 Install the air filter housing (see Chapter 1).

76 Install the engine cover (see illustration 4.12).

77 Install the battery (see Chapter 5).

78 Refill the engine with oil and coolant (see Chapter 1).

79 Reconnect the cable to the negative battery terminal.

80 Start the engine check for leaks.

9 Crankshaft front oil seal and camshaft oil seals replacement

➡**Note: This section applies to the crankshaft and camshaft seals used on 2000 through 2004 engines, which use a timing belt. 2005 and later V6 engines, which use a timing chain, do not have camshaft seals, and the crankshaft seal is in the timing chain cover. If you need to replace the crankshaft front oil seal, refer to Section 8.**

CRANKSHAFT FRONT OIL SEAL

▶ **Refer to illustrations 9.3 and 9.5**

1 Remove the timing belt and crankshaft pulley and the timing belt sprocket (see Section 7).

2 Note how far the seal is recessed in the bore, then cut away the seal lip with a razor knife.

3 Carefully pry the seal out of the engine with a screwdriver or seal removal tool (see illustration). If you use a screwdriver, wrap tape around the tip - don't scratch the housing bore or damage the crankshaft (if the crankshaft is damaged, the new seal will end up leaking).

4 Clean the bore in the engine and coat the outer edge of the new seal with engine oil or multi-purpose grease. Apply the same grease to the seal lip.

5 Using a socket with an outside diameter slightly smaller than the outside diameter of the seal, carefully drive the new seal into place with a hammer (see illustration). Make sure it's installed squarely and driven in to the same depth as the original. If a socket isn't available, a short section of large diameter pipe will also work. Check the seal after installation to make sure the spring didn't pop out of place.

6 Reinstall the crankshaft timing sprocket and timing belt (see Section 7). Be careful not to scratch the crankshaft sensor portion of the sprocket.

7 Run the engine and check for oil leaks at the front seal.

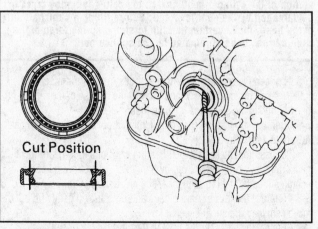

9.3 Cut away the crankshaft seal lip, wrap a screwdriver tip with tape and pry out the seal

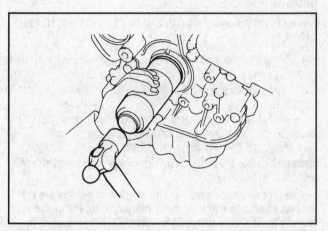

9.5 Lubricate the seal lip and drive the new crankshaft seal into place with a large socket or piece of pipe and a hammer

CAMSHAFT OIL SEALS

▶ **Refer to illustration 9.12**

8 Remove the timing belt and camshaft sprocket(s) (see Section 7).

9 Remove the bolts and detach the rear (no. 3) timing belt cover.

10 Note how far the seal is seated in the bore, then carefully pry it out with a screwdriver. Wrap the screwdriver tip with tape - don't scratch the bore or damage the camshaft (if the camshaft is damaged, the new seal will end up leaking).

11 Clean the bore and coat the outer edge of the new seal with engine oil or multi-purpose grease. Apply multi-purpose grease to the seal lip.

12 Using a socket with an outside diameter slightly smaller than the outside diameter of the seal, carefully drive the new seal into place with a hammer. Make sure it's installed squarely and driven in to the same depth as the original. If a socket isn't available, a short section of pipe will also work (see illustration).

13 Reinstall the rear timing belt cover and tighten the bolts.

14 Reinstall the camshaft sprocket(s) and timing belt (see Section 7).

15 Run the engine and check for oil leaks at the camshaft seal.

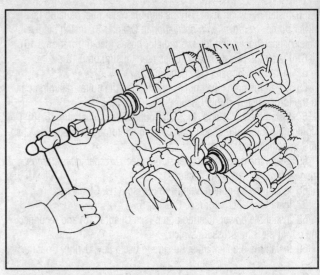

9.12 Lubricate the seal lip and tap the new camshaft seal into place with a large socket or piece of pipe and a hammer

---

10 Camshafts and lifters - removal, inspection and installation

➡**Note: Before beginning this procedure, obtain two 6 x 1.0 mm bolts 16 to 20 mm long. They will be referred to as service bolts in the text.**

2000 THROUGH 2004 MODELS

Removal

▶ **Refer to illustrations 10.1, 10.3, 10.4, 10.10, 10.12, 10.13 and 10.14**

1 Remove the valve covers (see Section 4) and the timing belt (see Section 7) (see illustration).

2 Make sure the engine is positioned on TDC for number 1 piston (see Section 3).

3 The following steps apply to the removal of each of the four camshafts. On each head, the exhaust camshaft subgear is secured first, the exhaust cam removed, then the intake camshaft. Align the cam timing marks on the drive and driven gears (see illustration). Turn the camshaft with a wrench if necessary.

➡**Note: For reference purposes, the outside camshafts are the exhaust camshafts while the inner camshafts are the intake camshafts. The right side cylinder head (passenger's side) is called the right bank while the left side cylinder head (driver's side) is called the left bank.**

4 Secure the exhaust camshaft sub-gear to the driven gear with a service bolt installed in the threaded hole (see illustration).

✳✳ **CAUTION:**

Since the camshaft thrust clearance is minimal, the camshafts must be held level as they are being removed. If they aren't, the portion of the cylinder head next to the cam gears may crack or be damaged by the gear leverage. Before lifting a camshaft out of the head, make certain that the torsional spring force of the sub-gear has been eliminated by the service bolt.

5 Loosen the camshaft bearing cap bolts in 1/4-turn increments until they can be removed by hand. Follow the reverse of the recommended tightening sequence (see illustration 10.34).

6 Remove the bearing caps and gently lift out the exhaust camshaft. Be sure to keep it level.

7 Loosen the intake camshaft bearing cap bolts in 1/4-turn increments until they can be removed by hand. Follow the reverse of the recommended tightening sequence (see illustration 10.29).

8 Remove the intake bearing caps and oil seal and gently lift out the intake camshaft. Be sure to keep it level.

9 Repeat the steps for the left-hand (front) cylinder head.

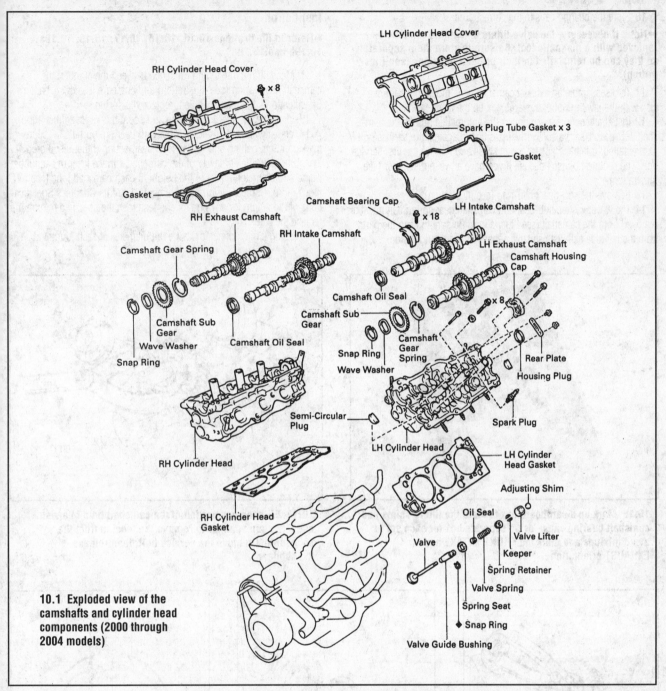

10.1 Exploded view of the camshafts and cylinder head components (2000 through 2004 models)

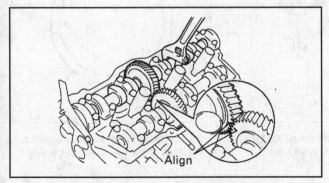

10.3 Align the timing marks (circled) on the camshaft gears

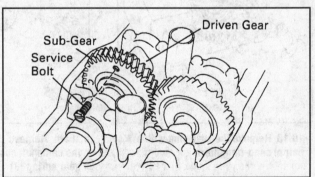

10.4 Install a service bolt through the sub-gear into the driven gear

10 Store the bearing caps in the correct order.

➡**Note: If necessary, the valve lifters and shims can now be removed with a magnetic tool. Be sure to store them separately so they can be reinstalled in their original locations (see illustration).**

11 To disassemble an exhaust camshaft gear, mount the cam in a vise with the jaws gripping the large hex on the shaft.

12 Install a second service bolt in the unthreaded hole in the camshaft sub-gear. Using a screwdriver positioned against the service bolt just installed, rotate the sub-gear clockwise and remove the first service bolt. The second bolt isn't needed if you have a two-pin spanner (see illustration).

13 Remove the sub-gear snap-ring (see illustration).

14 The wave washer, sub-gear and camshaft gear spring can now be removed from the camshaft (see illustration). Be sure to keep the parts from the left side exhaust camshaft separate from the right side.

Inspection

▸ **Refer to illustrations 10.15, 10.16, 10.17, 10.18, 10.19a, 10.19b and 10.20**

15 Measure the free length (distance between the ends) of the camshaft gear spring (see illustration) and compare it to this Chapter's Specifications. If not as specified, replace the spring.

16 Inspect each lifter for scuffing and score marks (see illustration).

17 Visually examine the cam lobes and bearing journals for score marks, pitting, galling and evidence of overheating (blue, discolored areas). Look for flaking away of the hardened surface layer of each lobe. Using a micrometer, measure the height of each camshaft lobe (see illustration). Compare your measurements with this Chapter's Specifications. If the height for any one lobe is less than the specified minimum, replace the camshaft.

18 Using a micrometer, measure the diameter of each journal at

10.10 Mark up a cardboard box to store the lifters/shims and camshaft bearing caps - use a separate box for each set to avoid mix-ups and mark the FRONT, INTAKE and EXHAUST orientation

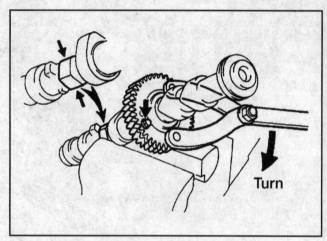

10.12 With the hex portion of the camshaft held in a vise, use a two-pin spanner to remove the tension from the subgear and remove the service bolt, then release the subgear

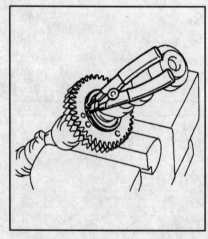

10.13 Remove the snap-ring with a pair of snap-ring pliers

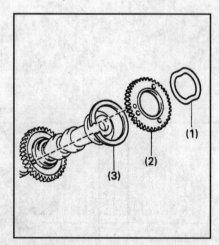

10.14 Remove the wave washer (1), the camshaft subgear (2) and the gear spring (3)

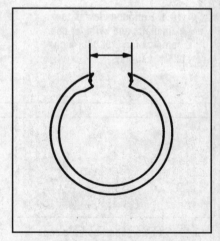

10.15 Measure the distance between the ends of the camshaft gear spring

several points (see illustration). Compare your measurements with this Chapter's Specifications. If the diameter of any one journal is less than specified, replace the camshaft.

19 Check the oil clearance for each camshaft journal as follows:

a) Clean the bearing caps and the camshaft journals with lacquer thinner or acetone.

b) Carefully lay the camshaft(s) in place in the cylinder head. Don't install the lifters or intake camshaft sub-gear and don't use any lubrication.

c) Lay a strip of Plastigage on each journal.

d) Install the bearing caps with the arrows pointing toward the front (timing chain end) of the engine (see illustration).

e) Tighten the bolts to the torque listed in this Chapter's Specifications in 1/4-turn increments.

➡ **Note: Don't turn the cam-shaft while the Plastigage is in place.**

f) Remove the bolts and detach the caps.

g) Compare the width of the crushed Plastigage (at its widest point) to the scale on the Plastigage envelope (see illustration).

h) If the clearance is greater than specified, replace the camshaft and/or cylinder head.

i) Scrape off the Plastigage with your fingernail or the edge of a credit card - don't scratch or nick the journals or bearing caps.

20 With the caps reinstalled temporarily, use a dial indicator to measure the backlash between the two camshaft gears. Hold one camshaft from turning (using a wrench on the hex portion) while measuring the movement in the other camshaft gear, and compare the results to Specifications (see illustration). If the backlash is beyond Specifications, replace both camshafts.

Installation

▶ **Refer to illustrations 10.26, 10.27, 10.28, 10.29 and 10.34**

21 Reassemble the exhaust camshaft gear(s) by installing the camshaft gear spring, sub-gear, wave washer and snap-ring.

10.16 Inspect each lifter for wear and scuffing

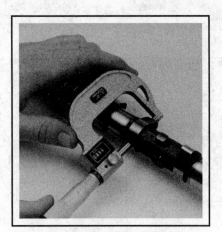

10.17 Measure the lobe heights on each camshaft - if any lobe height is less than the specified allowable minimum, replace that camshaft

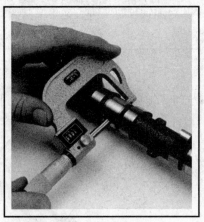

10.18 Measure each journal diameter with a micrometer (if any journal measures less than the specified limit, replace the camshaft)

10.19a The camshaft bearing caps are numbered with an arrow facing the front of the engine

10.19b Compare the width of the crushed Plastigage to the scale on the envelope to determine the oil clearance

10.20 Position a dial indicator as shown here to measure gear backlash - hold one camshaft steady with a wrench while moving the other camshaft with another wrench

22 Mount the camshaft in a padded vise.

23 Insert a service bolt into the unthreaded hole in the camshaft sub-gear. Using a screwdriver, align the holes of the camshaft driven gear and sub-gear by turning the camshaft sub-gear clockwise. Install a second service bolt in the threaded hole, tightening it to clamp the gears together. Remove the service bolt from the unthreaded hole. Repeat the procedure for the other camshaft.

24 Apply moly-base grease or engine assembly lube to the lifters, then install them in their original locations in the cylinder heads. Make sure the valve adjustment shims are in place in the lifters, and that all lifters are installed in their original bores.

Intake camshaft

→Note: For reference purposes, the outside camshafts are the exhaust camshafts while the inner camshafts are the intake camshafts. The right side cylinder head (passenger) is called the right bank while the left side cylinder head (driver) is called the left bank.

25 Apply moly-base grease or engine assembly lube to the camshaft lobes, bearing journals and gear thrust faces.

26 Set the intake camshaft in place in the right cylinder head with the timing marks (two dots) facing the exhaust camshaft side of the head (see illustration).

27 Apply a thin coat of RTV sealant to the outer edges of the front bearing cap cylinder head mating surfaces (see illustration).

28 Install the bearing caps in numerical order with the arrows pointing toward the front (timing belt end) of the engine (see illustration).

29 Tighten the bearing cap bolts in 1/4-turn increments to the torque listed in this Chapter's Specifications. Follow the recommended sequence (see illustration).

30 Refer to Section 9 and install a new camshaft oil seal.

Exhaust camshaft

31 Apply moly-base grease or engine assembly lube to the camshaft lobes, bearing journals and gear thrust faces.

32 Set the exhaust camshaft in place in the right cylinder head with the timing marks (two dots) aligned with the intake camshaft timing marks (see illustration 10.3).

33 Install the bearing caps in numerical order with the arrows pointing toward the front (timing-belt end) of the engine.

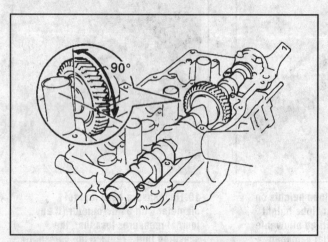

10.26 Set the right cylinder head intake camshaft into place with the two dots at the 3 o'clock position (facing the exhaust camshaft)

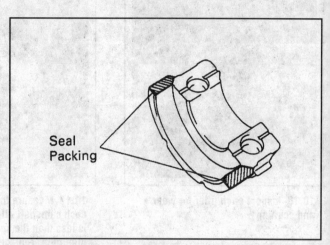

10.27 Apply RTV sealant to the shaded areas on the bearing cap

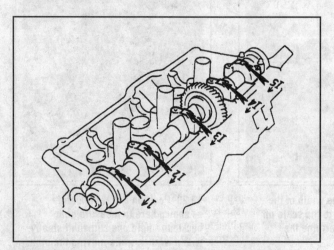

10.28 Install the intake camshaft bearing caps as shown with the arrows pointing toward the timing belt end of the engine

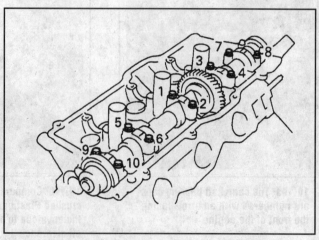

10.29 Tightening sequence for the intake camshaft bearing caps

34 Tighten the bearing cap bolts in 1/4-turn increments to the torque listed in this Chapter's Specifications. Follow the recommended sequence (see illustration).

35 Remove the service bolt.

36 Repeat Steps 21 through 35 for the camshafts on the left-bank cylinder head.

37 Reinstall the timing belt (see Section 7).

38 Reinstall the remaining components in the reverse order of removal.

39 Before reinstalling the valve covers, use RTV sealant in the areas indicated (see illustration 4.7). Clean the rubber half-circle plugs for the back of the heads and reinstall them with RTV sealant.

40 The remainder of installation is the reverse of removal.

41 Run the engine, then check for leaks and proper operation.

2005 AND LATER MODELS

Removal

▶ **Refer to illustration 10.45**

➡**Note 1: The following procedure is not for beginners. Please read the entire procedure carefully before deciding whether this is a job that you want to tackle at home.**

➡**Note 2: The following procedure applies to the camshafts for both cylinder heads.**

42 Disconnect the cable from the negative battery terminal.

43 Drain the engine oil and coolant (see Chapter 1).

44 Remove the timing chains (see Section 8).

45 To remove the camshafts from the right cylinder head, rotate the camshafts counterclockwise so that the nose of the intake cam lobe for the No. 1 cylinder is facing toward 7 o'clock, and the nose of the exhaust cam lobe is facing 12 o'clock, or straight up (see illustration). Use an open-end wrench on the integral hex cast into each camshaft to rotate the cams. (This step is NOT necessary for the camshafts on the

left cylinder head.)

46 Gradually and evenly loosen all 16 camshaft bearing cap bolts in the opposite order of the tightening sequence (see illustrations 10.53b and 10.54c).

47 Remove all eight bearing caps and remove the intake and exhaust camshafts.

48 Store the bearing caps in the correct order. One way to do this is to put them in a box and label the cap numbers with a utility marker pen (see illustration 10.10).

49 Remove the lifters with a magnet and put them in the same box with the cam bearing caps. Again, be sure to label each lifter with a marker. You can also write the lifter number right on top of the lifter.

50 If you're removing the camshafts from the other cylinder head, repeat Steps 46 through 49. Don't forget that Step 45 applies only to camshaft removal for the right cylinder.

✳✳ CAUTION:

While the timing chains and camshafts are removed, DO NOT ROTATE THE CRANKSHAFT!

Inspection

51 Refer to the inspection procedures in Steps 16 through 20. (Step 15 does not apply to the camshafts used on this engine.) Compare your measurements to the camshaft specifications for the 2005 V6 listed in this Chapter's Specifications.

Installation

▶ **Refer to illustrations 10.53a, 10.53b, 10.53c, 10.54a, 10.54b and 10.54c**

52 Lightly lubricate the lifter bores and the lifters with clean engine, then install the lifters in the same bores from which they were removed. Verify that each lifter rotates smoothly in its bore.

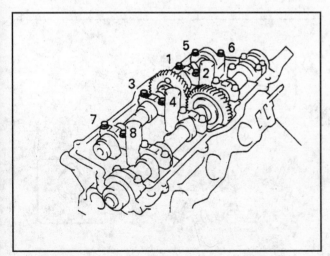

10.34 Tightening sequence for the exhaust camshaft bearing caps

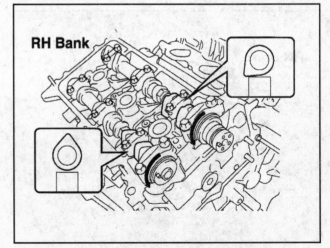

10.45 To remove the camshafts from the right cylinder head, rotate the camshafts counterclockwise so that the nose of the intake cam lobe for the No. 1 cylinder is facing toward 7 o'clock, and the nose of the exhaust cam lobe No. 1 is facing 12 o'clock, or straight up

53 Lightly lubricate the camshaft journals with clean engine oil. Install the camshafts on the right cylinder head so that the nose of the intake cam lobe for the No. 1 cylinder is facing toward 7 o'clock, and the nose of the exhaust cam lobe for No. 1 is facing 12 o'clock, or straight up (see illustration 10.45). Apply a light coat of engine oil to the upper bearings, then install the eight bearing caps in their correct locations (see illustration). Apply a light coat of engine oil to the threads of the bearing cap bolts and install the bolts. Then gradually and evenly tighten the bearing cap bolts, in the indicated sequence (see illustration) to the torque listed in this Chapter's Specifications. Rotate the camshafts 90 degrees clockwise until the knock pins on the front end of the camshafts are at a 90-degree position in relation to the cylinder head mating surface (see illustration).

54 Lightly lubricate the camshaft journals with clean engine oil.

Install the camshaft on the left cylinder head so that the nose of the intake cam lobe for the No. 2 cylinder is facing toward 7 o'clock, and the nose of the exhaust cam lobe for No. 2 is facing 12 o'clock, or straight up (see illustration). Apply a light coat of engine oil to the upper bearings, then install the eight bearing caps in their correct locations (see illustration). Apply a light coat of engine oil to the threads of the bearing cap bolts and install the bolts. Then gradually and evenly tighten the bearing cap bolts, in the indicated sequence (see illustration) to the torque listed in this Chapter's Specifications.

55 The remainder of installation is the reverse of removal. Install the timing chains, the timing chain cover and all of the components attached to the timing cover (see Section 8). Then refill the engine with oil and coolant (see Chapter 1), reconnect the cable to the negative battery terminal, start the engine and check for leaks.

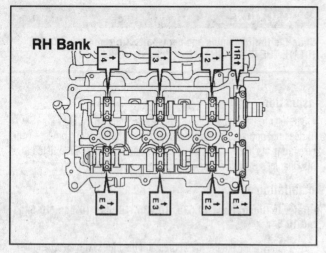

10.53a When installing the bearing caps on the right cylinder head, make sure that they're in the correct position and oriented so that the arrow is facing toward the timing chain end of the engine

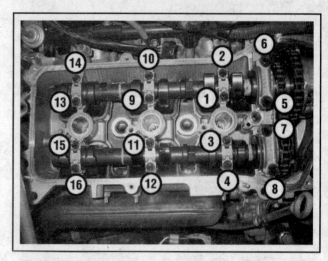

10.53b Camshaft bearing cap bolt tightening sequence (right cylinder head)

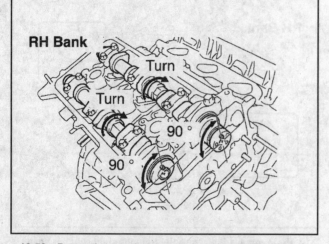

10.53c Rotate the camshafts on the right cylinder head 90 degrees clockwise until the knock pins on the front ends of the cams are at a 90-degree position in relation to cylinder head mating surface

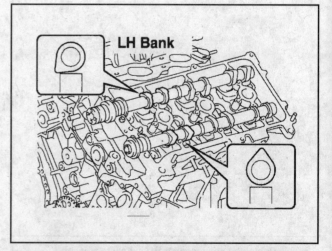

10.54a When installing the camshafts on the left cylinder head, position the cams so that the nose of the intake lobe for cylinder No. 2 is at 7 o'clock, and the nose of the exhaust cam for No. 2 is at 12 o'clock, or perpendicular, in relation to the lifter

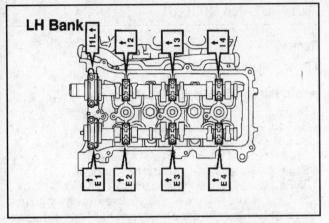

10.54b When installing the bearing caps on the left cylinder head, make sure that they're in the correct position and oriented so that the arrow is facing toward the timing chain end of the engine

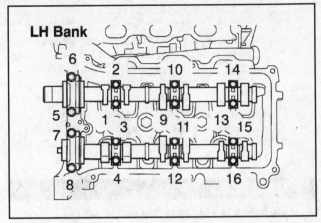

10.54c Camshaft bearing cap bolt tightening sequence (left cylinder head)

11 Cylinder heads - removal and installation

※※ WARNING:

The engine must be completely cool before starting this procedure.

2000 THROUGH 2004 MODELS

Removal

▶ **Refer to illustration 11.10**

1 Disconnect the negative cable from the battery.

2 Drain the cooling system, including the block (see Chapter 1).

3 Remove the air intake plenum (see Section 5), the fuel delivery pipes and injectors (see Chapter 4).

4 Remove the exhaust manifold(s) (see Section 6).

5 Remove the alternator (see Chapter 5).

6 Remove the intake manifold (see Section 5).

7 Remove the timing belt, camshaft sprockets and upper idler pulley (see Section 7).

8 Remove the upper timing belt cover (no. 3).

9 Remove the camshafts from the head you intend to remove (see Section 10).

10 Using an 8 mm hex bit or Allen wrench, remove the recessed head bolts (one in each head) (see illustration).

11 Using a 12-point socket, loosen the cylinder head bolts in 1/4-turn increments until they can be removed by hand. Follow the reverse order of the factory recommended tightening sequence (see illustration 11.22).

12 Lift the cylinder head off the engine block. If the head is stuck, place a wood block against it and strike the wood with a hammer.

※※ CAUTION:

Don't pry between the head and block. The gasket surfaces may be damaged and leaks could result.

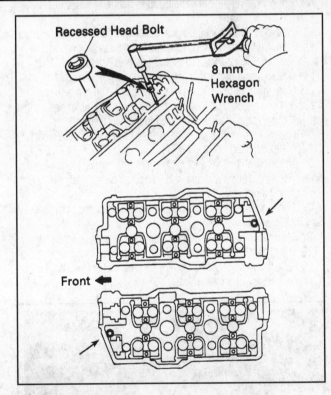

11.10 Remove the recessed bolts with an 8 mm hex bit socket

13 Repeat the procedure for the other head.

Installation

▶ **Refer to illustrations 11.19 and 11.22**

14 The mating surfaces of the cylinder heads and block must be perfectly clean when the heads are installed.

15 Use a gasket scraper to remove all traces of carbon and old gas-

ket material, then clean the mating surfaces with lacquer thinner or acetone. If there's oil on the mating surfaces when the head is installed, the gasket may not seal correctly and leaks could develop. When working on the block, stuff the cylinders with clean shop rags to keep out debris. Use a vacuum cleaner to remove material that falls into the cylinders.

16 Check the block and head mating surfaces for nicks, deep scratches and other damage. If damage is slight, it can be removed with a file; if it's excessive, machining may be the only alternative.

17 Use a tap of the correct size to chase the threads in the cylinder head bolt holes, then clean the holes with compressed air - make sure that nothing remains in the holes.

❋❋ WARNING:

Wear eye protection when using compressed air!

18 Mount each bolt in a vise and run a die down the threads to remove corrosion and restore the threads. Dirt, corrosion, sealant and damaged threads will affect torque readings.

19 Position the new gaskets over the dowel pins in the block (see illustration).

20 Carefully set the head on the block without disturbing the gasket.

21 Before installing the head bolts, apply a small amount of clean engine oil to the threads.

22 Install the bolts in their original locations and tighten them finger tight. Following the recommended sequence, tighten the bolts to the torque listed in this Chapter's Specifications (see illustration). Don't tighten the recessed bolt at this time.

23 Mark the front of each bolt head with paint. You can also mark the socket you are using. Place the socket over the 12-point bolt so that you can observe the mark, and proceed to the next step.

24 Following the same sequence, tighten each bolt an additional 1/4-turn (90-degrees).

25 Tighten each bolt yet another 1/4-turn (90-degrees) following the same sequence. The paint marks should now all be 180-degrees from the starting point.

26 Tighten the recessed bolt to the torque listed in this Chapter's Specifications.

27 Repeat the entire procedure to install the other cylinder head.

28 The remaining installation steps are the reverse of removal.

29 Refill the cooling system, change the oil and filter (see Chapter 1), run the engine and check for leaks.

11.19 Be sure the new head gaskets are positioned right side up (check all holes and coolant passages for correct alignment) and over the block dowels

2005 AND LATER MODELS

Removal

♦ **Refer to illustration 11.42**

➡**Note: This procedure applies to either cylinder head.**

30 Disconnect the cable from the negative battery terminal.

31 Drain the engine oil and coolant (see Chapter 1).

32 Remove the engine cover (see Section 4).

33 Remove the air filter housing (see Chapter 4).

34 Remove the upper intake manifold (see Section 5).

35 Remove the valve cover(s) (see Section 4).

36 Disconnect the fuel supply and return lines from the fuel rail (see *Fuel rail and injectors - removal and installation* in Chapter 4). Remove the lower intake manifold and the fuel rail as a single assembly (see Section 5).

37 Locate the coolant bypass housing on the back of the engine. The bypass housing is the casting that spans and is bolted to the upper rear part of the block and is attached to both cylinder heads by a pair of nuts at each end. Disconnect the Engine Coolant Temperature (ECT) sensor from the coolant bypass housing (see *Engine Coolant Temperature sensor - replacement* in Chapter 6). Disconnect the heater hose from the coolant bypass housing. Remove the two bolts and four nuts and remove the coolant bypass housing. Remove and discard the old coolant bypass housing gaskets. Remove the old O-ring from the coolant outlet pipe and discard it.

38 Disconnect the front exhaust pipe from the exhaust manifold and remove the exhaust manifold (see Section 6).

39 Remove the timing chain assembly (see Section 8).

40 Remove the camshaft timing oil control valve, the oil control

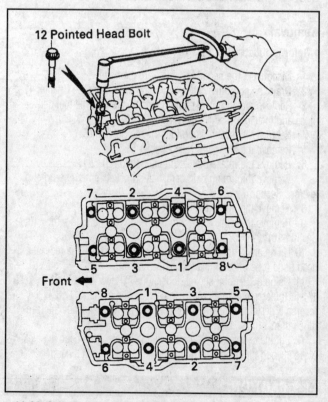

11.22 Cylinder head bolt TIGHTENING sequence

valve filter and the VVT-i sensor (see *Variable Valve Timing-intelligent (VVT-i) - description and component replacement* in Chapter 6).

41 Remove the intake and exhaust camshafts (see Section 10).

42 If you're removing the left cylinder head, remove the two front bolts in the indicated sequence (see illustration).

43 Remove the cylinder head bolts gradually and evenly, in the order opposite that of the tightening sequence (see illustrations 11.49a and 11.49b), then remove the cylinder head from the engine block. Discard the old cylinder head bolts and buy new ones. (The cylinder head bolts must be angle-torqued, which means that they're slightly stretched, so you will not be able to retighten them to the correct torque.)

44 Remove and discard the old cylinder head gasket.

45 If you're removing both cylinder heads, repeat this procedure for the other cylinder head.

INSTALLATION

▶ Refer to illustrations 11.47, 11.49a and 11.49b

➥Note: This procedure applies to either cylinder head.

46 Refer to Steps 14 through 18.

47 Apply two small beads of RTV sealant to the two indicated areas of the new cylinder head gasket (see illustration). Apply the sealant to the top and the bottom of these two spots.

✳✳ CAUTION:

Once you have applied the RTV sealant to the cylinder head gasket you must install the cylinder head within three minutes, and you must tighten and torque the cylinder head bolts within 15 minutes. If you don't, you must remove the RTV sealant and reapply it.

48 Position the cylinder head gasket on the engine block so that the lot number stamp is on the center upper edge of the gasket (next to the intake manifold), facing up. Carefully place the cylinder head on the head gasket.

49 Apply a light coat of oil to the new cylinder head bolts, then install all eight bolts - don't forget the washers! - and tighten them, gradually and evenly, in the proper sequence (see illustrations), to the initial torque listed in this Chapter's Specifications.

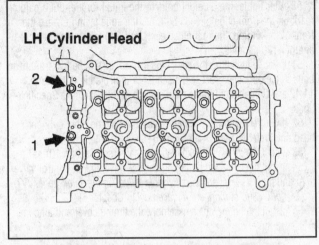

11.42 If you're removing the left cylinder head, remove these two bolts first, in this order. When installing the left head, install these two bolts, in reverse order, AFTER you have installed the rest of the cylinder head bolts

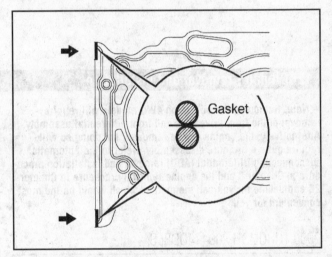

11.47 Apply small beads of sealant to the top and bottom of the head gasket at these two points

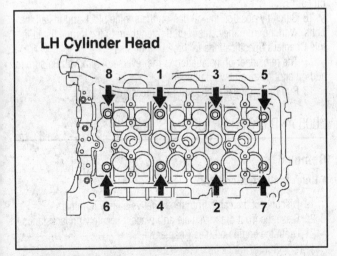

11.49a Cylinder head bolt tightening sequence - left cylinder head

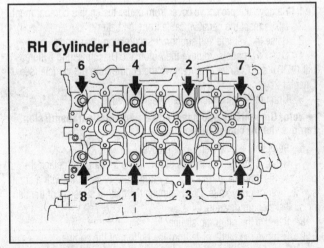

11.49b Cylinder head bolt tightening sequence - right cylinder head

50 After tightening all eight bolts to the initial torque, put a paint mark on the front edge of each bolt, i.e. the edge facing toward the front of the engine, then retighten each bolt, in the same sequence, another 180 degrees.

51 If you're installing the left cylinder head, install the two front head bolts in the opposite order to the sequence indicated in illustration 11.42 and tighten them to the torque listed in this Chapter's Specifications.

52 If you removed both cylinder heads, install the other cylinder head now (refer to Steps 47 through 51).

53 Install the intake and exhaust camshafts (see Section 10).

54 Install the camshaft timing oil control valve, the oil control valve filter and the VVT-i sensor (see *Variable Valve Timing-intelligent (VVT-i) - description and component replacement* in Chapter 6).

55 Install the timing chain assembly, the timing cover and all components attached to the cover (see Section 8).

56 Install the exhaust manifold and reconnect the front exhaust pipe from the exhaust manifold and (see Section 6).

57 Install the lower intake manifold and the fuel rail as a single assembly (see Section 5).

58 Reconnect the fuel supply and return lines to the fuel rail (see *Fuel rail and injectors - removal and installation* in Chapter 4).

59 Install the valve cover(s) (see Section 4).

60 Install the upper intake manifold (see Section 5).

61 Install the air filter housing (see Chapter 4).

62 Install the engine cover (see Section 4).

63 Refill the engine with oil and coolant (see Chapter 1).

64 Reconnect the cable to the negative battery terminal, start the engine and check for leaks.

12 Oil pan - removal and installation

→Note: **Removing the oil pan on 4WD models will require removal of the front driveaxles and front differential assembly. Alternatively, the engine can be removed, but doing so will require quite a bit more disassembly. Refer to the Automatic Disconnecting Differential (ADD) removal and installation procedure in Chapter 8 and the engine removal procedure in Chapter 2C and decide for yourself the method which would be the most convenient for you.**

2000 THROUGH 2004 MODELS

Removal

♦ **Refer to illustration 12.7**

1   Remove the protective cover from under the engine compartment.
2   Disconnect the negative cable from the battery.
3   Raise the vehicle and support it securely on jackstands.
4   On 4WD models, remove the Automatic Disconnecting Differential (ADD) (see Chapter 8) or remove the engine from the vehicle (see Chapter 2C).
5   Drain the engine oil and remove the oil filter.

→Note: **On vehicles equipped with an automatic transmission, remove the oil cooler tube.**

6   Remove the flywheel housing cover.
7   Remove the bolts and nuts securing the oil pan and detach the pan (see illustration). If it's stuck, pry it loose very carefully with a small screwdriver or putty knife. Don't damage the mating surfaces of the pan or oil leaks could develop.
8   Remove the oil pump strainer.
9   Remove the baffle plate from the bottom of the engine.

Installation

10   Use a scraper to remove all traces of old sealant from the block and oil pan. Clean the mating surfaces with lacquer thinner or acetone.
11   Make sure the threaded bolt holes in the block are clean.
12   Check the flange of the oil pan for distortion, particularly around the bolt holes. If necessary, place the pan on a wood block and use a hammer to flatten and restore the gasket surface.
13   If the baffle has been removed, reinstall it now.
14   Inspect the oil pump pick-up/strainer assembly for cracks and a blocked strainer. If the pick-up was removed, clean it with solvent or thinner and install it now, using a new gasket. Tighten the fasteners to the torque listed in this Chapter's Specifications.
15   Apply a 3 to 4 mm wide bead of RTV sealant to the flange of the oil pan.
16   Carefully position the oil pan on the engine block and install the bolts. Working from the center out, tighten them to the torque listed in this Chapter's Specifications in three or four steps.
17   The remainder of installation is the reverse of removal. Be sure to add oil and install a new oil filter.
18   Run the engine and check for oil pressure and leaks.

2005 AND LATER V6 MODELS

Removal

♦ **Refer to illustration 12.22**

19   Disconnect the cable from the negative battery terminal.
20   Raise the front of the vehicle and place it securely on jackstands.
21   Drain the engine oil (see Chapter 1).

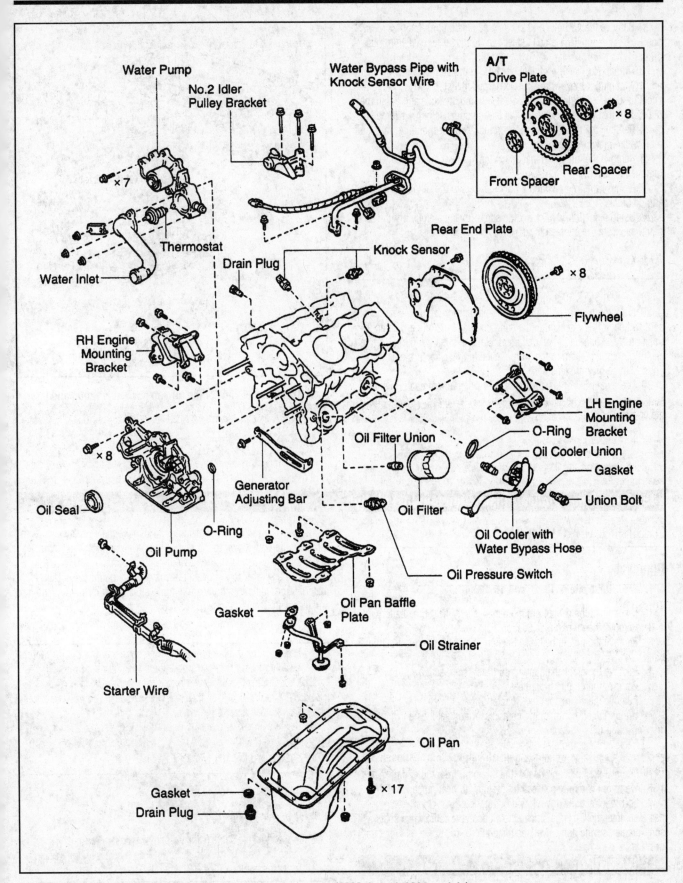

12.7 Exploded view of the oil pan and surrounding components (2000 through 2004 models)

22 Remove the 15 bolts and 2 nuts (see illustration) that secure oil pan No. 2 (the smaller stamped steel pan) to the larger cast aluminum oil pan.

23 Oil pan No. 2 will probably be stuck to the oil pan with RTV sealant. Try tapping it loose with a rubber-tipped mallet. If you're unable to knock it loose, carefully cut the sealant with a putty knife and a hammer. Make sure that you don't damage the mating surfaces of the two pans.

24 Remove the two oil pump pickup tube/strainer mounting nuts and remove the pickup/strainer assembly.

25 Remove the four bolts that attach the flywheel housing cover and remove the cover.

26 Remove the 17 bolts and 2 nuts that secure the oil pan to the engine block.

27 Using a small prybar or a large screwdriver, carefully pry the oil pan loose from the engine block.

INSTALLATION

28 Refer to Steps 10 through 16.

29 After you have installed the aluminum portion of the oil pan, apply a bead of RTV sealant to the flange of the No. 2 oil pan, carefully position No. 2 pan on the oil pan and install the bolts. Working from the center out, tighten them to the torque listed in this Chapter's specifications in three or four steps.

30 The remainder of installation is the reverse of removal. Be sure to add oil and install a new filter (see Chapter 1). When you're done, run the engine and check for leaks.

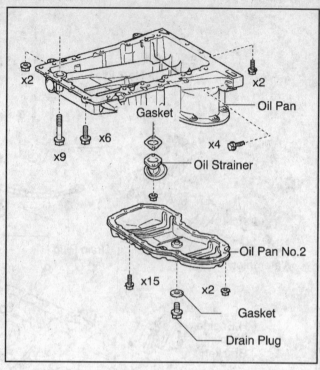

12.22 Exploded view of the oil pan assembly (2005 and later V6 models)

---

13 Oil pump - removal, inspection and installation

2000 THROUGH 2004 MODELS

Removal

▶ **Refer to illustrations 13.11 and 13.12**

1 Disconnect the cable from the negative terminal of the battery.

2 Remove the oil pan (see Section 12).

3 Remove the timing belt (see Section 7) and lower timing belt idler pulley.

4 Remove the crankshaft timing sprocket (see Section 7).

5 Remove the oil pick-up tube.

6 Remove the alternator (see Chapter 5).

7 Unbolt the air conditioning compressor and set it aside without disconnecting the refrigerant lines.

8 Remove the compressor bracket.

9 Remove the lower timing belt idler pulley.

10 Remove the bolts and detach the oil pump from the engine. You may have to pry carefully between the front main bearing cap and the pump body with a screwdriver.

11 Remove the O-ring. Remove the oil pressure relief valve snapring, retainer, spring and valve (see illustration).

✳✳ WARNING:

The spring is tightly compressed - be careful and wear eye protection.

12 Use a large Phillips screwdriver to remove the eight screws retaining the body cover to the rear of the oil pump (see illustration).

13 Lift the cover off and remove the pump rotors.

14 Use a scraper to remove all traces of sealant and old gasket material from the pump body and engine block, then clean the mating surfaces with lacquer thinner or acetone.

13.11 Remove the oil pressure relief plug, spring and valve

13.12 Use a large Phillips screwdriver or bit to remove the screws retaining the pump cover

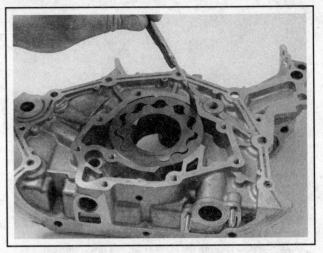

13.17a Measure the driven rotor-to-body clearance with a feeler gauge

Inspection

▶ **Refer to illustrations 13.17a, 13.17b and 13.17c**

15 Clean all components with solvent, then inspect them for wear and damage.

16 Check the oil pressure relief valve sliding surface and valve spring. If either the spring or the valve is damaged, they must be replaced as a set.

17 Check the clearance of the following components with a feeler gauge and compare the measurements to this Chapter's Specifications (see illustrations):

   a) Driven rotor-to-oil pump body
   b) Rotor side clearance
   c) Rotor tip clearance

Installation

▶ **Refer to illustration 13.26**

18 Pry the old crankshaft seal out with a screwdriver.

19 Apply multi-purpose grease or engine oil to the outer edge of the new seal and carefully drive it into place with a deep socket and a hammer. Also apply multi-purpose grease to the seal lip.

20 Place the drive and driven rotors into the pump body with the marks facing out (see illustration 13.17c).

21 Pack the pump cavities with petroleum jelly and install the cover. Tighten the screws securely following a criss-cross pattern.

22 Lubricate the oil pressure relief valve with engine oil and install the valve components in the pump body.

23 Use acetone or lacquer thinner and a clean rag to remove all traces of oil from the gasket surfaces.

24 Apply a 2 to 3 mm wide bead of RTV sealant to the oil pump. Avoid using an excessive amount of sealant, especially around oil passages and bolt holes. Assembly must be completed within five minutes of sealant application, otherwise the material must be removed and reapplied.

25 Position a new O-ring on the block.

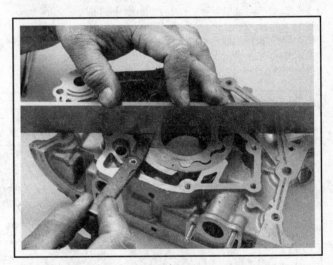

13.17b Measure the rotor side clearance with a precision straightedge and feeler gauge

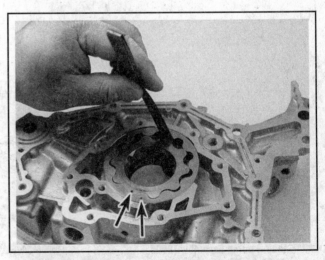

13.17c Measure the rotor tip clearance with a feeler gauge - note the rotor marks are facing out (when the pump body cover is installed, the marks will be against the cover)

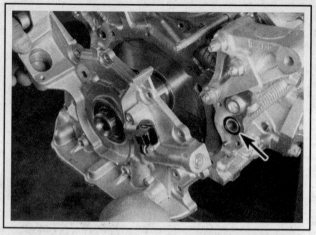

13.26 Be sure to align the drive rotor and the crankshaft as the oil pump is installed, and install a new O-ring (arrow) on the block

26 Engage the spline teeth on the oil pump drive rotor with the large teeth on the crankshaft and slide the pump into place (see illustration).

27 Install the oil pump mounting bolts in their original locations and tighten them to the torque listed in this Chapter's Specifications in a criss-cross pattern.

28 Using a new gasket, install the oil pick-up tube and tighten the fasteners to the torque listed in this Chapter's Specifications.

29 Reinstall the remaining parts in the reverse order of removal.

30 Add oil, start the engine and check for oil leaks.

31 Recheck the engine oil level.

2005 AND LATER MODELS

Removal

▶ **Refer to illustration 13.33**

32 Remove the timing chain cover (see Section 8).

33 Remove the oil pipe mounting bolts (see illustration) and remove the oil pipe. Remove and discard the two old oil pipe O-rings.

34 Remove the seven oil pump cover bolts and remove the oil pump cover. Remove the oil pump driver rotor and driven rotor.

35 Remove the oil pressure relief plug, valve spring and valve from the oil pump cover.

Inspection

36 Refer to steps 15 through 17 and to the illustrations accompanying those steps. The procedure for inspecting this oil pump is the same procedure used to inspect an oil pump on a 2000 through 2004 model.

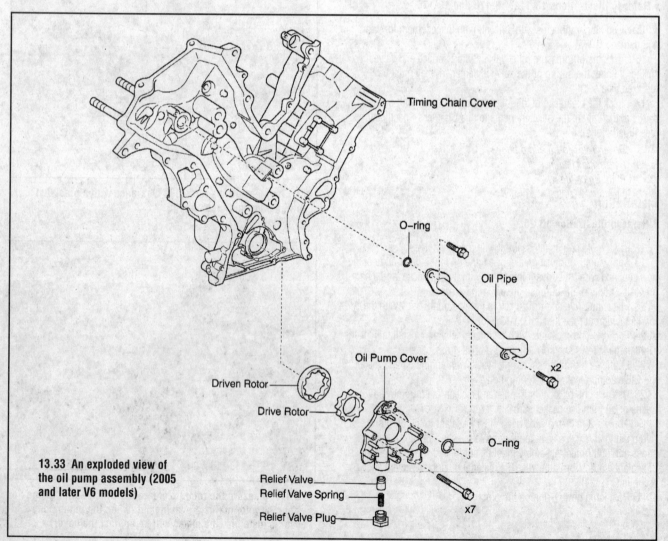

13.33 An exploded view of the oil pump assembly (2005 and later V6 models)

Installation

37 Pry the old crankshaft seal out of the timing chain cover with a screwdriver.

38 Apply multi-purpose grease or engine oil to the outer edge of the new crank seal and carefully drive it into place with a deep socket and a hammer. Then apply multi-purpose grease or engine oil to the seal lip.

39 Apply a coat of petroleum jelly to the pump drive and driven rotors, then place the two rotors into position in the timing chain cover. Make sure that the pump marks (dimples) are facing out, toward the pump cover (see illustration 13.17c).

40 Pack the pump cavity with petroleum jelly and install the cover. Tighten the cover bolts to the torque listed in this Chapter's Specifications.

41 Lubricate the oil pressure relief valve with clean engine oil and insert the valve, then the spring, into the pump cover. Screw in the plug and tighten it to the torque listed in this Chapter's Specifications.

42 Install the timing chain cover (see Section 8).

14 Flywheel/driveplate - removal and installation

REMOVAL

1 Disconnect the negative cable from the battery. Raise the vehicle and support it securely on jackstands, then refer to Chapter 7 and remove the transmission.

2 If you're working on a manual transmission-equipped, remove the pressure plate and clutch disc (see Chapter 8).

3 Make alignment marks on the flywheel/driveplate and crankshaft to ensure correct alignment during reinstallation.

4 Remove the bolts securing the flywheel/driveplate to the crankshaft. If the crankshaft turns, wedge a screwdriver in the ring gear teeth to hold the flywheel/driveplate.

5 Remove the flywheel/driveplate from the crankshaft. Since the flywheel is fairly heavy, be sure to support it while removing the last bolt. Automatic transmission equipped vehicles have spacers on both sides of the driveplate. Keep them with the driveplate.

✷✷ WARNING:

The ring-gear teeth may be sharp, wear gloves to protect your hands.

INSTALLATION

6 Clean the flywheel/driveplate to remove grease and oil. Inspect the surface for cracks, rivet grooves, burned areas and score marks. Light scoring can be removed with emery cloth. Check for cracked or broken ring gear teeth. Lay the flywheel/driveplate on a flat surface and use a straightedge to check for warpage.

7 Clean and inspect the mating surfaces of the flywheel/driveplate and the crankshaft. If the crankshaft rear seal is leaking, replace it before reinstalling the flywheel/driveplate (see Section 15).

8 Position the flywheel/driveplate against the crankshaft. Be sure to align the marks made during removal. Note that some engines have an alignment dowel or staggered bolt holes to ensure correct installation. Before installing the bolts, apply thread-locking compound to the threads.

9 Wedge a screwdriver in the ring gear teeth to keep the flywheel/driveplate from turning and tighten the bolts to the torque listed in this Chapter's Specifications. Follow a criss-cross pattern and work up to the final torque in three or four steps.

10 The remainder of installation is the reverse of the removal procedure.

15 Rear main oil seal - replacement

▶ Refer to illustrations 15.2, 15.5, 15.6 and 15.7

1 Remove the transmission (see Chapter 7). Remove the rear end plate.

2 The seal can be replaced without removing the oil pan or seal retainer. However, this method is not recommended because the lip of the seal is quite stiff and it's possible to cock the seal in the retainer bore or damage it during installation. If you want to take the chance, pry out the old seal with a screwdriver (see illustration). Apply multi-purpose grease to the crankshaft seal journal and the lip of the new seal and carefully press the new seal into place. The lip is stiff so carefully work it onto the seal journal of the crankshaft with a smooth object like the end of an extension as you tap the seal into place. Don't rush it or you may damage the seal.

3 The following method is recommended but requires removing the seal retainer and resealing the rear of the oil pan (see Section 11).

4 After removing the two rearmost oil pan-to-seal retainer bolts, break the seal between the rear of the oil pan and the bottom of the seal retainer with a putty knife. Remove the retainer-to-engine block bolts, detach the seal retainer and remove all the old gasket material. remove

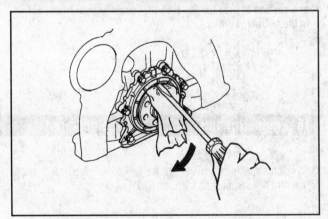

15.2 The rear main oil seal may be pried out with a screwdriver - when installing, lubricate the crankshaft journal and the lip of the new seal with multi-purpose grease and press the new seal into place - the seal lip is very stiff and can be easily damaged during installation if you're not careful

the sealant from the top of the oil pan flange.

➡Note: Cover the open area of the oil pan with clean rags to keep debris out while bracing the pan flange.

5   Position the seal and retainer assembly between two wood blocks on a workbench and drive the old seal out from the back side with a screwdriver (see illustration).

6   Drive the new seal into the retainer with a wood block (see illustration) or a section of pipe slightly smaller in diameter than the outside diameter of the seal.

7   Lubricate the crankshaft seal journal and the lip of the new seal with multi-purpose grease. Note that the engine doesn't have a gasket between the seal retainer and the engine block. Instead, apply a 2 to 3 mm wide bead of RTV sealant to the retainer flange (see illustration) before attaching the retainer to the block.

8   Slowly and carefully push the seal and retainer onto the crankshaft. The seal lip is stiff, so work it onto the crankshaft with a smooth object such as the end of an extension as you push the retainer against the block.

9   Install and tighten the retainer bolts to the torque listed in this Chapter's Specifications.

10   The remainder of installation is the reverse of removal.

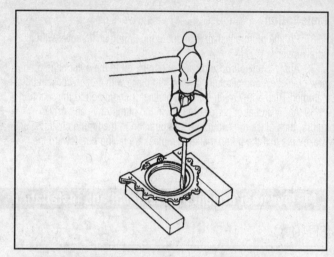

15.5 After removing the retainer assembly from the engine block, support it between two wooden blocks and drive out the old seal with a screwdriver and hammer

15.6 Drive the new seal into the retainer with a wood block or a section of pipe - make sure that you don't cock the seal in the retainer bore

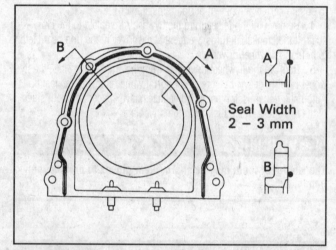

15.7 Apply RTV sealant to the oil seal retainer-to-block mating surface (2000 through 2004 model shown, later models similar)

16   Engine mounts - check and replacement

1   Engine mounts seldom require attention, but broken or deteriorated mounts should be replaced immediately or the added strain placed on the driveline components may cause damage or wear.

CHECK

▸ Refer to illustration 16.4

2   During the check, the engine must be raised slightly to remove the weight from the mounts.

3   Raise the vehicle and support it securely on jackstands, then position a jack under the engine oil pan. Place a large wood block between the jack head and the oil pan, then carefully raise the engine just enough to take the weight off the mounts. Do not position the wood block under the drain plug.

❋❋ WARNING:

DO NOT place any part of your body under the engine when it's supported only by a jack!

4   Check the mounts (see illustration) to see if the rubber is cracked, hardened or separated from the metal plates. Sometimes the rubber will split down the center.

5   Check for relative movement between the mount plates and the engine or frame (use a large screwdriver or pry bar to attempt to move the mounts). If movement is noted, lower the engine and tighten the mount fasteners.

6   Rubber preservative should be applied to the mounts to slow deterioration.

REPLACEMENT

♦ **Refer to illustration 16.8**

7   Disconnect the negative battery cable from the battery, then raise the vehicle and support it securely on jackstands (if not already done). Support the engine as described in Step 3.

8   To remove an engine mount, remove the fasteners (see illustration), raise the engine and detach the mount.

9   Installation is the reverse of removal. Use thread locking compound on the mount bolts/nuts and be sure to tighten them securely.

10 See Chapter 7 for transmission mount replacement.

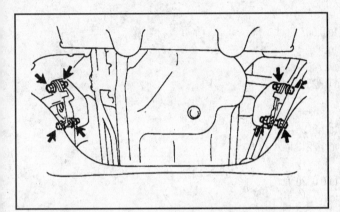

16.4 Engine mount bolt locations (2000 through 2004 models - typical)

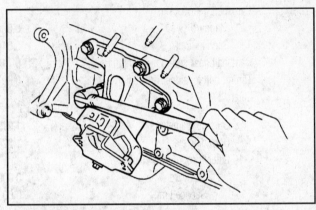

16.8 Removing a typical engine mount

Specifications

3.4L V6 engine (2000 through 2004)

General

Engine designation	5VZ-FE
Displacement	3.4 liters
Cylinder numbers (timing belt end-to-transmission end)	
Right (passenger) side	1-3-5
Left (driver) side	2-4-6
Firing order	1-2-3-4-5-6

Warpage limits

Cylinder head	0.0039 inch
Intake manifold	0.0039 inch
Exhaust manifolds	0.0394 inch

Camshaft and related components

Valve clearance (engine cold)	See Chapter 1
Bearing journal diameter	1.0610 to 1.0616 inches
Bearing oil clearance	
Standard	0.0014 to 0.0028 inch
Service limit	0.0039 inch

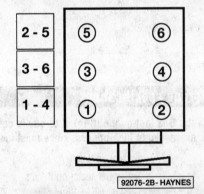

92076-2B- HAYNES

Cylinder numbering and coil terminal locations

Lobe height	
Intake	
Standard	1.6657 to 1.6697 inches
Service Limit	1.6598 inches
Exhaust	
Standard	1.6520 to 1.6559 inches
Service Limit	1.6461 inches
Thrust clearance (endplay)	
Standard	0.0013 to 0.0031 inch
Service limit	0.0047 inch
Runout limit (total indicator reading)	0.0024 inch
Camshaft gear backlash	
Standard	0.0008 to 0.0079 inch
Service limit	0.0188 inch
Camshaft gear spring free length	0.712 to 0.740 inch
Timing belt tensioner protrusion	0.394 to 0.425 inch
Lifters	
Outside diameter	1.2191 to 1.2195 inch
Bore diameter	1.2205 to 1.2212 inch
Lifter-to-bore (oil) clearance	
Standard	0.0009 to 0.0020 inch
Service limit	0.003 inch

Oil pump

Driven rotor-to-pump body clearance	
Standard	0.0039 to 0.0069 inch
Service limit	0.0118 inch
Rotor tip clearance	
Standard	0.0043 to 0.0094 inch
Service limit	0.0138 inch
Rotor side clearance	
Standard	0.0012 to 0.0035 inch
Service limit	0.0059 inch

Torque specifications   Ft-lbs (unless otherwise indicated)

➡ **Note: One foot-pound (ft-lb) of torque is equivalent to 12 inch-pounds (in-lbs) of torque. Torque values below approximately 15 ft-lbs are expressed in inch-pounds, since most foot-pound torque wrenches are not accurate at these smaller values.**

Intake manifold bolts/nuts	156 in-lbs
Intake air connector bolts/nuts	156 in-lbs
Air intake plenum bolts/nuts	156 in-lbs
Exhaust manifold nuts	30
Exhaust crossover pipe	33
Crankshaft pulley bolt	217
Timing belt cover bolts (no. 3)	80 in-lbs
Idler pulley bolts*	
No. 1	26
No. 2	30
Timing belt tensioner bolts	20
Valve cover nuts	53 in-lbs
Camshaft pulley bolts	81
Camshaft bearing cap bolts	144 in-lbs

Torque specifications (continued)   Ft-lbs (unless otherwise indicated)

➡Note: One foot-pound (ft-lb) of torque is equivalent to 12 inch-pounds (in-lbs) of torque. Torque values below approximately 15 ft-lbs are expressed in inch-pounds, since most foot-pound torque wrenches are not accurate at these smaller values.

Cylinder head bolts (in sequence - see illustration 11.22)	
12-point head bolts	
Step 1	25
Step 2	Turn an additional 90-degrees (1/4-turn)
Step 3	Turn an additional 90-degrees (1/4-turn)
Recessed head bolts**	156 in-lbs
Oil pan bolts	52 in-lbs
Oil pump body cover bolts	96 in-lbs
Oil pump mounting bolts	
14 mm bolt head	15
12 mm bolt head	31
Oil pick-up tube mounting bolts	66 in-lbs
Flywheel bolts (manual transmission) *	63
Driveplate bolts (automatic transmission) *	61
Rear crankshaft oil seal retainer mounting bolts	69 in-lbs

Apply thread locking compound to the threads prior to installation

* Recessed head bolts are torqued once and are NOT torqued again at a 90-degree turn like the 12-point head bolts*

4.0L V6 engine

General

Engine designation	1GR-FE
Displacement	4.0 liters
Cylinder numbers (timing chain end-to-transmission end)	
Right (passenger) side	1-3-5
Left (driver) side	2-4-6
Firing order	1-2-3-4-5-6

Warpage limits

Cylinder head	0.0039 inch
Intake manifold	
Intake plenum side	0.031 inch
Cylinder head side	0.008 inch
Exhaust manifolds	0.028 inch

Camshaft and related components

Valve clearance (engine cold)	See Chapter 1
Bearing journal diameter	
No. 1 journal	1.4162 to 1.4167 inch
Other journals	0.9030 to 0.9045 inch
Bearing oil clearance	
Standard	
Right side No. 1 (intake)	0.0003 to 0.0015 inch
Right side No. 1 (exhaust)	0.0016 to 0.0031 inch
Left side (intake)	0.0016 to 0.0031 inch
Others	0.0010 to 0.0024 inch

Camshaft and related components (continued)

Bearing oil clearance (continued)
  Service limit
    Right side No. 1 (intake) — 0.0028 inch
    Others — 0.0039 inch
Lobe height
  Intake
    Standard — 1.7389 to 1.7428 inch
    Service limit — 1.7330 inch
  Exhaust
    Standard — 1.7551 to 1.7591 inch
    Service limit — 1.7492 inch
Thrust clearance (endplay)
  Standard — 0.016 to 0.035 inch
  Service limit — 0.0043 inch
Runout limit (total indicator reading) — 0.0024 inch
Lifters
  Lifter outside diameter — 1.2191 to 1.2195 inch
  Lifter bore diameter — 1.2208 to 1.2215 inch
  Lifter-to-bore (oil) clearance
    Standard — 0.0013 to 0.0023 inch
    Service limit — 0.0031 inch

Timing chain

Timing chain stretch limit (15 pins)
  (No. 1 and both No. 2 chains) — 5.780 inches
Timing gear assembly wear limits
  Larger camshaft timing gear
    (with No. 1 chain installed) — 4.547 inches
  Smaller camshaft timing gear
    (with No. 2 chain installed) — 2.878 inches
Timing sprocket wear limits
  Camshaft timing sprocket
    (with No. 2 chain installed) — 2.878 inches
  Crankshaft timing sprocket
    (with No. 1 chain installed) — 2.402 inches
Idle gear wear limits
  With No. 1 chain installed) — 2.402 inches
  Idle gear shaft diameter — 0.9050 to 0.9055 inch
  Idle gear inside diameter — 0.9063 to 0.90967 inch
  Oil clearance
    Standard — 0.0008 to 0.0017 inch
    Maximum — 0.0037 inch
Chain tensioner No. 2 wear limit — 0.039 inch
Chain tensioner slipper wear limit — 0.039 inch
Vibration damper No. 1 and No. 2 wear limit — 0.039 inch

Oil pump

Driven rotor-to-pump body clearance
  Standard — 0.0098 to 0.00128 inch
  Service limit — 0.0128 inch

Rotor tip clearance
    Standard   0.0024 to 0.0063 inch
    Service limit   0.0063 inch
Rotor side clearance
    Standard   0.0012 to 0.0035 inch
    Service limit   0.0035 inch

Torque specifications   Ft-lbs (unless otherwise indicated)

➡ **Note: One foot-pound (ft-lb) of torque is equivalent to 12 inch-pounds (in-lbs) of torque. Torque values below approximately 15 ft-lbs are expressed in inch-pounds, since most foot-pound torque wrenches are not accurate at these smaller values.**

Intake manifold assembly	
Upper intake manifold bolts/nuts	21
Lower intake manifold bolts	19
Valve cover bolts/nuts	80 in-lbs
Exhaust manifold nuts	32
Crankshaft pulley bolt	184
Timing chain cover bolts/nuts	17
Idler pulley bolts	
Idler pulley No. 1 bolt	40
Idler pulley No. 2 bolt	29
Drivebelt tensioner mounting bolts	27
Camshaft timing gear bolt	
(applies to timing gear and timing sprocket)	74
Camshaft bearing cap bolts	
10 mm bolts	80 in-lbs
12 mm bolts	18
Cylinder head bolts (in sequence - see illustrations 11.49a and 11.49b)	
Initial torque	27
Final torque	An additional 180 degrees
Left cylinder head (two front bolts)	22
Oil filter assembly	
Oil filter bracket mounting bolts	168 in-lbs
Oil cooler-to-oil filter bracket bolt	50
Oil pan bolts/nuts	
Oil pan No. 1	
All bolts & nuts except 2 bolts at	
flywheel/driveplate end of pan)	180 in-lbs
2 bolts at flywheel/driveplate end of pan	84 in-lbs
Oil pan No. 2	
Bolts	80 in-lbs
Nuts	84 in-lbs
Oil pick-up tube nuts	80 in-lbs
Oil pump assembly	
Oil pump cover bolts	80 in-lbs
Oil pipe flange bolts	80 in-lbs
Oil pump relief valve plug	36
Flywheel bolts (manual transmission)	Not available
Driveplate bolts (automatic transmission)	36

Torque specifications (continued)   Ft-lbs (unless otherwise indicated)

➡Note: One foot-pound (ft-lb) of torque is equivalent to 12 inch-pounds (in-lbs) of torque. Torque values below approximately 15 ft-lbs are expressed in inch-pounds, since most foot-pound torque wrenches are not accurate at these smaller values.

Engine rear oil seal retainer	
Bolts	84 in-lbs
Nuts	80 in-lbs
Timing chain assembly	
Timing chain tensioner bolts	
Chain tensioner No. 1	84 in-lbs
Chain tensioner No. 2	168 in-lbs
Timing chain vibration damper	
No. 1 mounting bolts	168 in-lbs

2B
V8 ENGINE

Reference to other Chapters

CHECK ENGINE light - See Chapter 6
Cylinder compression check - See Chapter 2C
Drivebelt check, adjustment and replacement - See Chapter 1
Engine oil and filter change - See Chapter 1
Engine overhaul - general information - See Chapter 2C
Engine - removal and installation - See Chapter 2C
Spark plug replacement - See Chapter 1
Water pump - removal and installation - See Chapter 3

1 General information

This Part of Chapter 2 is devoted to in-vehicle repair procedures for the 4.7L V8 (2UZ-FE) engine. The 4.7L V8 engine is designed with a Double-OverHead-Cam (DOHC) arrangement with four valves per cylinder (32 in all).

All information concerning engine removal and installation and engine block and cylinder head overhaul can be found in Part C of this Chapter.

The following repair procedures are based on the assumption that the engine is installed in the vehicle. If the engine has been removed from the vehicle and mounted on a stand, many of the steps outlined in this Part of Chapter 2 will not apply.

The Specifications included in this Part of Chapter 2 apply only to the procedures contained in this Part. Additional specifications can be found in Chapter 2 Part C.

2 Repair operations possible with the engine in the vehicle

1   Many major repair operations can be accomplished without removing the engine from the vehicle. Clean the engine compartment and the exterior of the engine with some type of degreaser before any work is done. It will make the job easier and help keep dirt out of the internal areas of the engine.

2   Depending on the components involved, it may be helpful to remove the hood to improve access to the engine as repairs are performed (refer to Chapter 11 if necessary). Cover the fenders to prevent damage to the paint. Special pads are available, but an old bedspread or blanket will also work.

3   If vacuum, exhaust, oil or coolant leaks develop, indicating a need for gasket or seal replacement, the repairs can generally be made with the engine in the vehicle. The intake and exhaust manifold gaskets, crankshaft oil seals and cylinder head gaskets are all accessible with the engine in place.

➡Note: It will be necessary to remove the engine from the vehicle in order to access the oil pump and oil pan.

4   Exterior engine components, such as the intake and exhaust manifolds, the oil pan, the oil pump, the water pump, the starter motor, the alternator and the fuel system components can be removed for repair with the engine in place.

5   Since the cylinder heads can be removed without pulling the engine, valve component servicing can also be accomplished with the engine in the vehicle. Replacement of the camshafts, timing belt and pulleys is also possible with the engine in the vehicle.

6   In extreme cases caused by a lack of necessary equipment, repair or replacement of piston rings, pistons, connecting rods and rod bearings is possible with the engine in the vehicle. However, this practice is not recommended because of the cleaning and preparation work that must be done to the components involved.

3 Top Dead Center (TDC) for number one piston - locating

➡Note: The most positive method for finding TDC is to examine the crankshaft timing marks and camshaft sprocket timing marks (see Section 7).

Refer to Chapter 2, Part A for this procedure, but note that the 4.7L V8 engine is equipped with different timing belt covers and accessory component locations. Otherwise, the TDC procedure is the same.

4 Valve covers - removal and installation

REMOVAL

▶ **Refer to illustration 4.5**

1   Disconnect the negative cable from the battery.

2   Disconnect the wire clamps from the right valve cover and set the harness off to the side.

3   Remove the igniter/coil assemblies from the spark plugs. Be sure to mark each igniter/coil assembly using tape or other marking device to insure proper reassembly. See Chapter 5 for additional details.

4   Disconnect any vacuum lines or wires that may interfere with the removal process: Mark them carefully to insure proper reassembly.

5   Remove the retaining bolts (see illustration), then detach the cover(s). If the cover is stuck to the head, bump the end with a wood block and a hammer to jar it loose. If that doesn't work, try to slip a

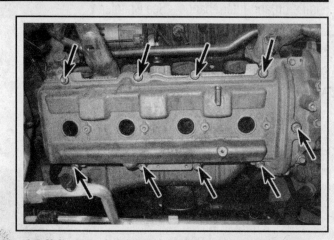

4.5 Valve cover bolt locations on the right bank valve cover

flexible putty knife between the head and cover to break the seal.

✳✳ CAUTION:

Don't pry at the cover-to-head joint or damage to the sealing surfaces may occur, leading to oil leaks after the cover is reinstalled.

INSTALLATION

▶ **Refer to illustration 4.7**

6   The mating surfaces of the cylinder head and cover must be clean when the cover is installed. Use a gasket scraper to remove all traces of sealant and old gasket material, then clean the mating surfaces with lacquer thinner or acetone. If there's residue or oil on the mating surfaces when the cover is installed, oil leaks may develop.

7   Position the semi-circular plugs in the cylinder head cutouts, sealing them with RTV sealant, then apply a thin, uniform layer of RTV sealant to the gasket/plug joints (see illustration).

8   Position a new gasket on the valve cover, then install the valve cover, sealing washers and nuts.

➡**Note: Be sure to install new spark plug tube seals into the valve cover.**

9   Tighten the nuts to the torque listed in this Chapter's Specifications in three or four equal steps.

10 Reinstall the remaining parts, run the engine and check for oil leaks.

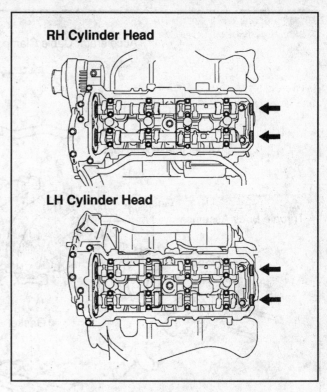

4.7 Apply RTV sealant to the area on the cylinder heads directly next to the semi-circular plugs

---

5	Intake manifold - removal and installation

✳✳ WARNING:

The engine must be completely cool before beginning this procedure.

REMOVAL

Upper intake manifold

▶ **Refer to illustration 5.7**

1   Disconnect the negative cable from the battery.

2   Drain the coolant into a clean container (see Chapter 1).

3   Disconnect the fuel inlet and return hoses at the intake manifold (see Chapter 4).

4   Remove the air filter housing, throttle body and fuel injectors (see Chapter 4).

5   Clearly label, then detach all remaining wires, hoses and brackets still attached to the upper intake manifold.

6   Remove the mounting bolt for the EVAP vacuum switching valve and separate the component from the upper intake manifold.

7   Remove the mounting nuts and bolts, then detach the upper intake manifold from the lower intake manifold (see illustration).

8   Make several passes to insure that the upper intake manifold is separated from the lower intake manifold evenly to avoid damage to the manifold surfaces. If it's stuck, don't pry between the gasket mating surfaces or damage may result.

Lower intake manifold

▶ **Refer to illustration 5.10**

9   Remove the upper intake manifold from the lower intake manifold (see Steps 1 through 8).

➡**Note: Some repair procedures do not require the upper intake manifold to be separated from the lower intake manifold. In these cases, it is possible to remove the upper and lower intake manifolds as a single assembly (but you'll still have to perform Steps 1, 2, 3, 5 and 6).**

10 Clearly label, then detach all remaining wires, hoses and brackets still attached to the lower intake manifold, coolant outlets and rear coolant bypass joint (see illustration).

11 Remove the mounting nuts and bolts, then detach the intake manifold from the cylinder heads. Start with the outer bolts first and work your way to the inner bolts on the manifold. Make several passes to insure that the manifold is separated from the cylinder heads evenly to avoid damage to the cylinder head and manifold surfaces. If it's stuck, don't pry between the gasket mating surfaces or damage may result.

INSTALLATION

▶ **Refer to illustration 5.13**

12 Use a scraper to remove all traces of old gasket material and sealant from the upper and lower intake manifolds and cylinder heads, then clean the mating surfaces with lacquer thinner or acetone.

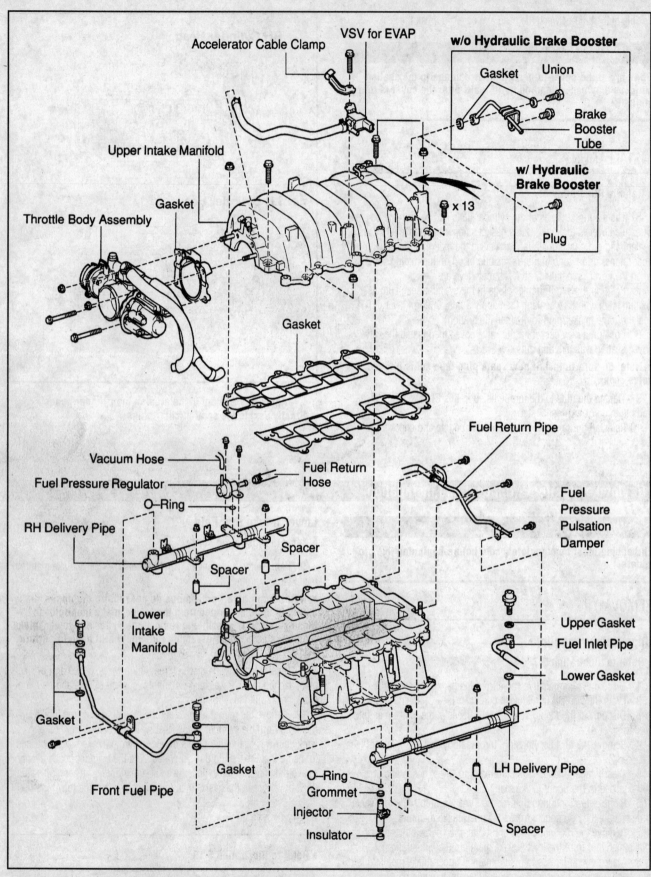

5.7 Exploded view of the upper and the lower intake manifolds with the surrounding components (2000 through 2004 model shown, later models similar)

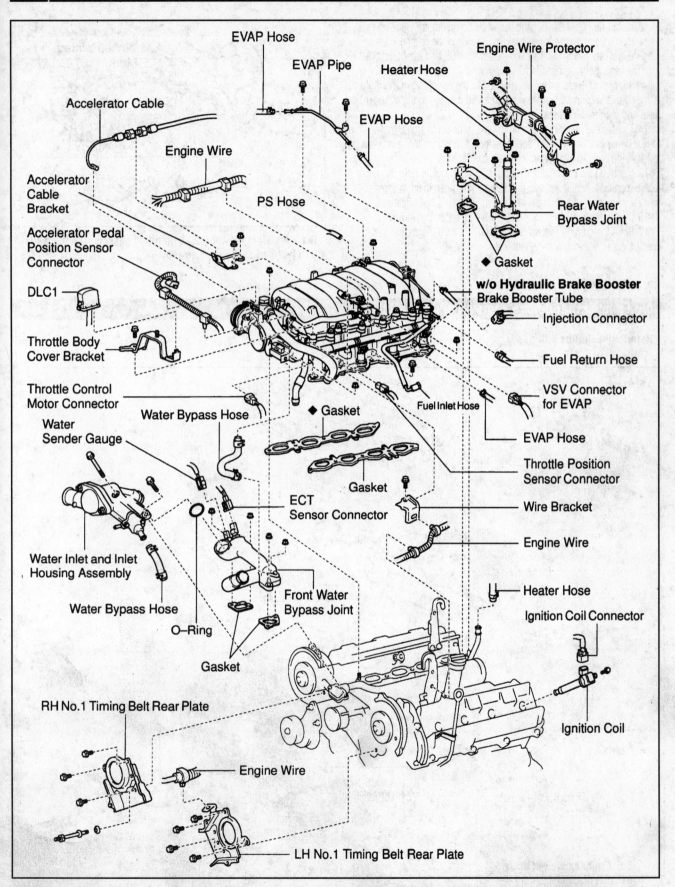

EVAP Hose

EVAP Pipe

Heater Hose

Engine Wire Protector

Accelerator Cable

EVAP Hose

Engine Wire

Accelerator Cable Bracket

PS Hose

Rear Water Bypass Joint

Accelerator Pedal Position Sensor Connector

◆ Gasket

DLC1

w/o Hydraulic Brake Booster
Brake Booster Tube

Injection Connector

Throttle Body Cover Bracket

Fuel Return Hose

Throttle Control Motor Connector

Water Bypass Hose

◆ Gasket

Fuel Inlet Hose

VSV Connector for EVAP

Water Sender Gauge

EVAP Hose

Gasket

Throttle Position Sensor Connector

ECT Sensor Connector

Wire Bracket

Engine Wire

Water Inlet and Inlet Housing Assembly

Water Bypass Hose

O–Ring

Front Water Bypass Joint

Heater Hose

Ignition Coil Connector

Gasket

RH No.1 Timing Belt Rear Plate

Ignition Coil

Engine Wire

LH No.1 Timing Belt Rear Plate

5.10 Exploded view of the upper and lower intake manifold bolted together as a single unit (2000 through 2004 model shown, later models similar)

13 Install new gaskets. Be sure the white painted marks face up (see illustration). Position the lower intake manifold on the engine. Make sure the gaskets haven't shifted and install the nuts and the bolts. Start with the inner bolts first and work your way to the outer bolts on the manifold. Make several passes to insure that the lower intake manifold is mated to the cylinder heads evenly to avoid damage to the cylinder head and manifold surfaces.

14 Tighten the nuts and bolts, in three or four equal steps, to the torque listed in this Chapter's Specifications. Work from the center out towards the ends to avoid warping the manifold.

15 Install a new gasket and the upper intake manifold onto the lower intake manifold. Install the nuts and bolts and tighten them in three of four equal steps to the torque listed in this Chapter's Specifications.

16 Install the remaining parts in the reverse order of removal.

17 Refill the cooling system (see Chapter 1). Run the engine and check for fuel, vacuum and coolant leaks.

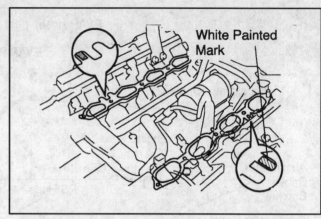

5.13 Install the lower intake manifold gaskets with the white painted marks facing UP as shown

---

6 Exhaust manifolds - removal and installation

▶ **Refer to illustration 6.5**

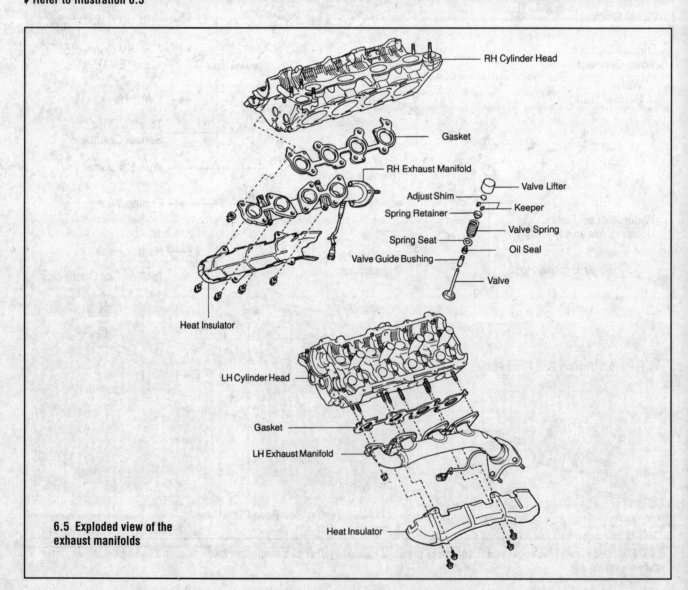

6.5 Exploded view of the exhaust manifolds

1 Disconnect the negative cable from the battery.
2 Spray penetrating oil on the exhaust manifold and exhaust flange fasteners and allow it to soak in.
3 Remove the exhaust flange nuts from the exhaust manifolds.
4 Remove the EGR pipe from the exhaust manifold (see Chapter 6).
5 Unbolt the exhaust manifolds from the cylinder heads (see illustration).

6 Carefully inspect the manifolds and fasteners for cracks and damage.
7 Use a scraper to remove all traces of old gasket material and carbon deposits from the manifolds and cylinder head mating surfaces. If the gasket was leaking, have the manifolds checked for warpage at an automotive machine shop and resurfaced if necessary.
8 Position new gaskets over the cylinder head studs.
9 Install the manifolds and thread the mounting nuts into place.
10 Working from the center out, tighten the nuts to the torque listed in this Chapter's Specifications in three or four equal steps.
11 Reinstall the remaining parts in the reverse order of removal. Use new gaskets when connecting the exhaust pipes.
12 Run the engine and check for exhaust leaks.

7   Timing belt and sprockets - removal, inspection and installation

REMOVAL

◆ **Refer to illustrations 7.7, 7.9a, 7.9b, 7.10, 7.11, 7.12, 7.13, 7.14a, 7.14b, 7.16, 7.17a, 7.17b, 7.17c, 7.19, 7.20a, 7.20b, 7.21, 7.22, 7.23a, 7.23b and 7.24**

1 Disconnect the negative cable from the battery.
2 Remove the under-vehicle splash shield (see illustration 8.6 in Chapter 1).
3 Remove the fan shroud and the fan assembly (see Chapter 3).
4 Remove the drivebelt (see Chapter 1).
5 Drain the coolant and remove the radiator from the engine compartment (see Chapter 3).
6 Remove the throttle body (see Chapter 5).
7 Remove the drivebelt idler pulley (see illustration).
8 Raise the front of the vehicle and support it securely on jackstands. Apply the parking brake and block the rear wheels.
9 Remove the left side number 3 timing belt cover (see illustrations).

7.7 Remove the drivebelt idler pulley mounting bolt

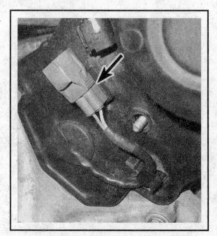

7.9a Disconnect the camshaft sensor from the left side number 3 timing belt cover

7.9b Remove the bolts from the left side number 3 timing belt cover

7.10 Remove the bolts from the right side number 3 timing belt cover (one bolt hidden from view)

7.11 Remove the number 2 timing belt cover mounting bolts

7.12 Remove the fan bracketmounting bolts

10 Remove the right side number 3 timing belt cover (see illustration).

11 Remove the number 2 timing belt cover (see illustration).

➡**Note: On models with air conditioning, remove (and reposition, without disconnecting the hoses) the air conditioning compressor (see Chapter 3) to access the number 2 timing belt cover.**

12 Remove the fan bracket (see illustration).

13 Remove the drivebelt tensioner mounting bolts (see illustration) and tensioner from the engine block.

14 Position the number one piston at TDC (see Section 3). Check the installation marks on the timing belt on the camshaft sprockets (see illustration) and the crankshaft sprocket (see illustration).

15 Check to see if there are installation marks on the timing belt - if you intend to re-use the belt and the marks have been obscured, make new ones. Place a new mark on the belt at the exact location of each timing mark on each sprocket. This will allow easy and accurate timing belt alignment when the old timing belt is reused.

16 Remove the crankshaft pulley. If air tools are not available, lock the crankshaft pulley with a special tool to prevent the engine from rotating and remove the crankshaft pulley bolt with a breaker bar and

socket (see illustration).

17 Remove the crankshaft pulley (see illustration) and the crankshaft sensor timing sprocket (see illustrations).

18 Double-check to make sure the camshaft sprockets and the crankshaft timing sprocket timing marks are properly aligned after removal of the crankshaft pulley. If the engine was rotated slightly, re-align the timing marks (see illustrations 7.14a and 7.14b).

19 Remove the timing belt tensioner (see illustration). Be sure to remove the rubber boot as well; it may stick in the tensioner recess.

20 Remove the timing belt from the camshaft sprockets. Using a special camshaft sprocket tool or a pin spanner (see illustration), release the tension between the right side and left side camshaft sprockets by carefully rotating the right side camshaft sprocket clockwise. This will allow slack on the timing belt. The camshaft sprocket should now be aligned with the T mark on the rear timing cover (see illustration).

21 Rotate the left side camshaft sprocket to release tension between the left side camshaft and the crankshaft sprocket. The camshaft sprocket should now be aligned with the T mark on the rear timing belt cover (see illustration). Remove the timing belt from the sprocket.

➡**Note this position for timing belt installation.**

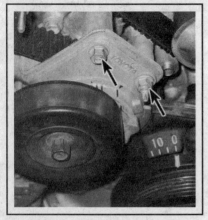

7.13 Remove the drivebelt tensioner mounting bolts (one bolt hidden from view)

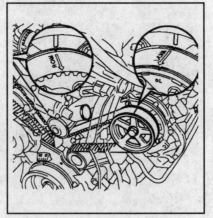

7.14a TDC number 1 locating marks at the 12 o'clock position on the camshaft sprockets

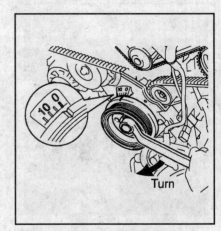

7.14b TDC number 1 locating mark on the crankshaft pulley

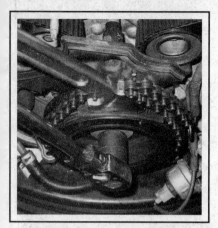

7.16 Use a chain wrench to lock the crankshaft pulley into place - note the piece of old drivebelt used to prevent damage to the pulley

7.17a Use a puller to remove the crankshaft pulley

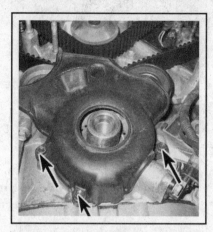

7.17b Remove the number 1 timing belt cover from the lower section of the engine

7.17c Remove the crankshaft sensor timing sprocket from the crankshaft

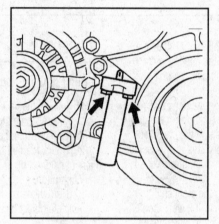

7.19 Remove the tensioner mounting bolts

7.20a Use a special tool to turn the right camshaft sprocket clockwise slightly

7.20b With the right side camshaft sprocket at TDC number 1 (1), rotate the sprocket clockwise until it is aligned with the T mark on the rear cover (2)

7.21 With the left side camshaft sprocket at TDC number 1 (1), rotate the sprocket clockwise until it is aligned with the T mark on the rear cover (2)

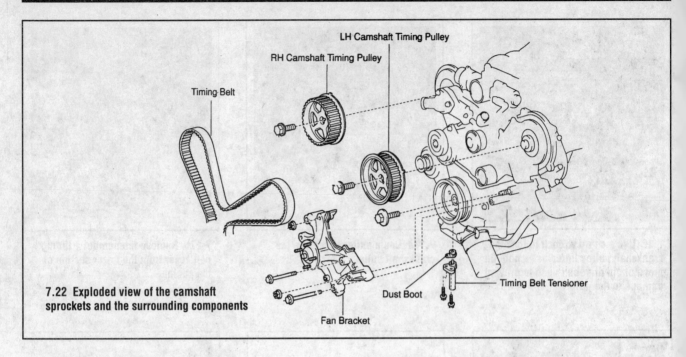

7.22 Exploded view of the camshaft sprockets and the surrounding components

22 The camshaft sprockets can be removed at this point if they are worn or damaged (see illustration). Use a tool like the one shown in illustration 7.20a to prevent the camshaft from turning as the bolt is removed and installed.

23 If you're re-using the belt, check for a mark on the belt adjacent to the drilled mark on the crankshaft sprocket (see illustration). If the original mark is gone, make a new one, then slip the belt off the sprocket. Remove the timing belt from the engine. If necessary, it is now possible to remove the number 1 and number 2 idler pulleys from the lower section of the engine (see illustration).

24 If it's worn or damaged, or if you're replacing the crankshaft front oil seal, the crankshaft sprocket can now be removed. If it won't come off by hand, lever it off with two screwdrivers. A steering wheel type puller may be needed to remove the sprocket (see illustration).

INSPECTION

▶ **Refer to illustrations 7.27 and 7.28**

25 Inspect the timing belt for cracks, tears, torn belt strands, cut edges or broken belt teeth. Replace the timing belt if there are any signs of damage or prolonged wear (high mileage).

26 Check the belt tensioner for visible oil leakage. If there's only a faint trace of oil on the pushrod side, the tensioner seal is in satisfactory condition.

27 Hold the tensioner in both hands and push it forcefully against an immovable object (see illustration). If the pushrod moves, replace the tensioner.

28 Measure the protrusion of the pushrod from the housing end (see illustration). Compare your measurement to this Chapter's Specifications. If the protrusion is not as specified, replace the tensioner.

29 Check that the idler pulleys turn smoothly.

7.23a Locate the drilled mark on the camshaft sprocket

7.23b Remove the number 1 and the number 2 idler pulleys from the lower section of the engine

7.24 Use a puller to remove the crankshaft sprocket from the crankshaft (also note the notch in the sprocket flange aligns with the small dot on the oil pump housing at TDC)

7.27 Check the tensioner for signs of leakage and test for leakdown by forcing it against an immovable object

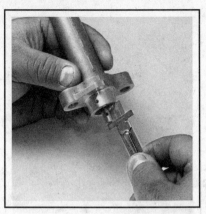

7.28 Measure the tensioner pushrod protrusion and compare it to the Specifications

INSTALLATION

> ✳✳ **CAUTION** ✳✳
>
> **Before starting the engine, carefully rotate the crankshaft by hand through at least two full revolutions (use a socket and breaker bar on the crankshaft pulley centerbolt). If you feel any resistance, STOP! There is something wrong - most likely, valves are contacting the pistons. You must find the problem before proceeding. Check your work and see if any updated repair information is available.**

▶ **Refer to illustrations 7.34, 7.41, 7.42a, 7.42b and 7.44**

30 Remove all dirt, oil and grease from the timing belt area at the front of the engine.

31 Install the number 1 and number 2 idler pulleys (see illustration 7.23b) onto the lower section of the engine block if they were removed.

32 Align the crankshaft timing sprocket keyway with the crankshaft key and install the sprocket with the flange side up against the engine.

33 Check the alignment of the TDC marks on the sprocket and the oil pump housing (see illustration 7.24).

34 Install the timing belt, starting at the crankshaft sprocket. If you're re-using the original belt, align the marks on the belt with the marks on the sprockets and covers (see illustration). Install the belt between the number 1 and number 2 idler pulleys. Tie the belt up with a length of string so it doesn't become disengaged from the camshaft sprocket.

35 Slip the crankshaft sensor timing sprocket over the crankshaft with the cupped side facing out.

36 Install the number 1 timing belt cover and gasket (see illustration 7.17b).

37 Slip the crankshaft (drivebelt) pulley onto the crankshaft, aligning the pulley keyway with the crankshaft key. Install the bolt and tighten it to the torque listed in this Chapter's Specifications. Use the method described in Step 16 to keep the crankshaft from turning.

38 Install the drivebelt tensioner (see illustration 7.13). Tighten the bolts to the torque listed in this Chapter's Specifications. Make sure the pulley turns smoothly.

39 Check the crankshaft timing marks for TDC number 1 (see Section 3).

40 Install the camshaft sprockets (if they were removed) onto the camshafts. Align the pin hole in the sprocket with the pin in the end of the camshaft.

41 Install the camshaft sprocket retaining bolts and tighten them to the torque listed in this Chapter's Specifications (see illustration).

7.34 Install the new timing belt with the crankshaft designation (CR) on the timing belt with the crankshaft sprocket alignment mark

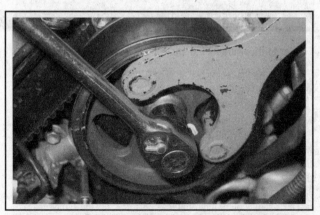

7.41 Use a special tool to retain the camshaft sprocket position when tightening the bolt

7.42a Align the timing belt designation L-CAM with TDC number 1 location mark for the left side camshaft sprocket

7.42b Align the timing belt designation R-CAM with TDC number 1 location mark for the right side camshaft sprocket

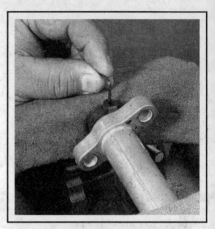

7.44 Restrain the tensioner pushrod by compressing the unit in a vise and inserting a pin approximately 0.050-inch (1.27 mm) in diameter - make sure the rubber boot is in place

42 Connect the timing belt to the left side camshaft sprocket. Be sure the alignment mark on the camshaft sprocket is aligned with the T mark on the rear timing belt cover. With the timing belt installed, turn the left side camshaft sprocket counterclockwise until there is tension on the timing belt between the left side camshaft sprocket and the crankshaft sprocket. The camshaft sprocket should now align with the TDC mark (see illustration). Connect the timing belt to the right side camshaft sprocket. Be sure the alignment marks on the right side camshaft sprocket is aligned with the T mark on the rear timing belt cover. Rotate the sprocket to apply tension to timing belt and check that the TDC marks are now aligned (see illustration).

43 Recheck the timing marks to be sure the crankshaft hasn't turned (see illustrations 7.14a and 7.14b). If you're re-using the original belt, the installation mark should line up as it did in Step 14. If not, change the position of the timing belt.

44 Using a press or vise, slowly compress the timing belt tensioner pushrod (see illustration). Insert a metal pin, drill bit or Allen wrench through the holes in the pushrod and housing. Release the pressure from the press or vise.

45 Install the timing belt tensioner and tighten the bolts to the torque listed in this Chapter's Specifications. Remove the retaining pin.

46 Using a socket and breaker bar on the crankshaft pulley bolt, turn the crankshaft slowly through two complete revolutions (720-degrees). Recheck the timing marks.

✳✳ CAUTION:

If the timing marks are not aligned exactly as shown, repeat the timing belt installation procedure. DO NOT start the engine until you're absolutely certain that the timing belt is installed correctly. Serious and costly engine damage will occur if the belt is installed wrong.

47 Install the fan bracket.

48 Install the air conditioning compressor, if equipped.

49 Install the upper (number 2) timing belt cover and gasket.

50 Install the right and left number 3 timing belt covers (see illustration 7.9a and 7.9b).

51 Reinstall the remaining parts in the reverse order of removal.

52 Refill the cooling system (see Chapter 1).

53 Start the engine and check for leaks and proper operation.

8   Crankshaft front oil seal - replacement

▶ **Refer to illustrations 8.3 and 8.5**

1 Remove the timing belt and crankshaft timing belt sprocket (see Section 7).

2 Note how far the seal is recessed in the bore, then cut away the seal lip with a razor knife.

3 Carefully pry the seal out of the engine with a screwdriver or seal removal tool (see illustration). If you use a screwdriver, wrap tape around the tip - don't scratch the housing bore or damage the crankshaft (if the crankshaft is damaged, the new seal will end up leaking).

4 Clean the bore in the engine and coat the outer edge of the new seal with engine oil or multi-purpose grease. Apply the same grease to the seal lip.

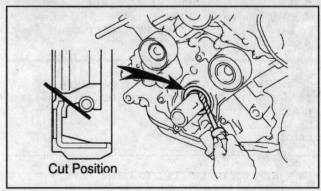

Cut Position

8.3 Cut away the crankshaft seal lip, wrap a screwdriver tip with tape and pry out the seal

5   Using a socket with an outside diameter slightly smaller than the outside diameter of the seal, carefully drive the new seal into place with a hammer (see illustration). Make sure it's installed squarely and driven in to the same depth as the original. If a socket isn't available, a short section of large diameter pipe will also work. Check the seal after installation to make sure the spring didn't pop out of place.

6   Reinstall the crankshaft timing sprocket and timing belt (see Section 7).

7   Run the engine and check for oil leaks at the front seal.

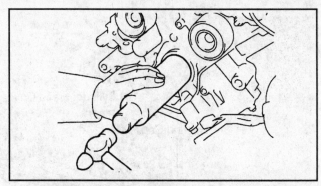

8.5 Lubricate the seal lip and drive the new crankshaft seal into place with a large socket or piece of pipe and a hammer

9   Camshaft oil seals - replacement

▶ **Refer to illustration 9.4**

1   Remove the timing belt and camshaft sprocket(s) (see Section 7).

2   Note how far the seal is seated in the bore, then carefully pry it out with a screwdriver. Wrap the screwdriver tip with tape - don't scratch the bore or damage the camshaft (if the camshaft is damaged, the new seal will end up leaking).

3   Clean the bore and coat the outer edge of the new seal with engine oil or multi-purpose grease.

4   Using a socket with an outside diameter slightly smaller than the outside diameter of the seal, carefully drive the new seal into place with a hammer. Make sure it's installed squarely and driven in to the same depth as the original. If a socket isn't available, a short section of pipe will also work (see illustration).

5   Reinstall the camshaft sprocket(s) and timing belt (see Section 7).

6   Run the engine and check for oil leaks at the camshaft seal.

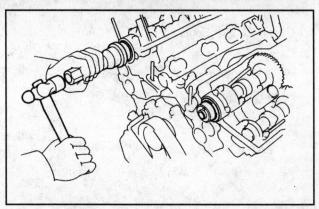

9.4 Lubricate the seal lip and tap the new camshaft seal into place with a large socket or piece of pipe and a hammer

10 Camshafts and lifters - removal, inspection and installation

※※ **WARNING:**

The engine must be completely cool before beginning this procedure.

※※ **CAUTION:**

Before removing the camshafts, be sure the TDC mark on the crankshaft pulley is aligned with the number 2 idler pulley bolt (see Step 2). If the procedure is not followed exactly, damage to the valves and lifters may occur.

➡**Note: Before beginning this procedure, obtain two 6 x 1.0 mm bolts 16 to 20 mm long. They will be referred to as service bolts in the text.**

REMOVAL

▶ **Refer to illustrations 10.1, 10.2, 10.3, 10.4, 10.7a, 10.7b, 10.10, 10.12 and 10.14**

1   Remove the valve covers (see Section 4) and the timing belt (see Section 7) (see illustration).

2   Make sure the TDC mark on the crank-shaft pulley is aligned with the number 2 idler pulley bolt (see illustration).

3   The following steps apply to the removal of the camshafts on each cylinder head. Start the removal process on the right bank cylinder head. Secure the exhaust camshaft sub-gear to the driven gear with a service bolt installed in the threaded hole (see illustration). Turn the camshaft with a wrench if necessary. Use the hexagonal portion of the exhaust camshaft.

➡**Note: For reference purposes, the outside camshafts are the exhaust camshafts while the inner camshafts are the intake camshafts. The right side cylinder head (passenger's side) is called the right bank while the left side cylinder head (driver's side) is called the left bank.**

4   Align the cam timing marks on the drive and driven gears at a 10-degree angle (see illustration).

5   Loosen the camshaft bearing cap bolts in 1/4-turn increments until they can be removed by hand. Follow the reverse of the recommended tightening sequence (see illustration 10.25a).

6   Remove the bearing caps and gently lift out the oil feed pipe and the camshafts. Be sure to keep it level.

➡**Note: The camshaft cap bolts vary in length. Be sure to mark each bolt carefully to avoid reassembly problems.**

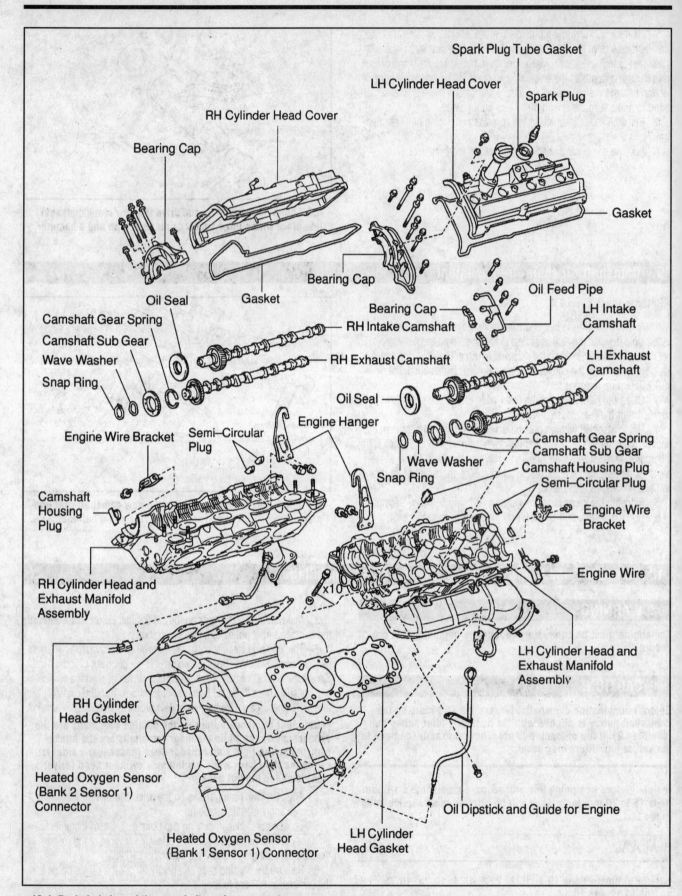

10.1 Exploded view of the camshafts and components

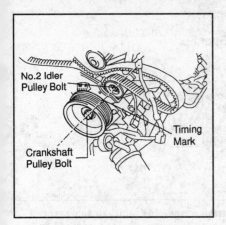

10.2 Make sure the TDC mark on the crankshaft pulley is aligned with the number 2 idler pulley bolt

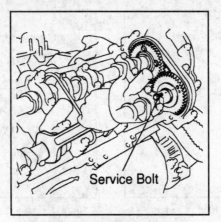

10.3 Install a service bolt through the sub-gear into the driven gear on the exhaust camshaft on the right bank cylinder head

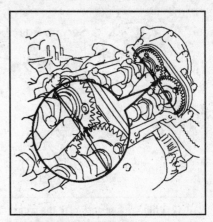

10.4 Align the timing marks (circled) on the camshaft gears at a 10-degree angle

❋❋ CAUTION:

Since the camshaft thrust clearance is minimal, the camshafts must be held level as they are being removed. If they aren't, the portion of the cylinder head next to the cam gears may crack or be damaged. Before lifting a camshaft out of the head, make certain that the torsional spring force of the sub-gear has been eliminated by the service bolt in the exhaust camshaft.

❋❋ CAUTION:

Since the camshaft thrust clearance is minimal, the camshafts must be held level as they are being removed. If they aren't, the portion of the cylinder head next to the cam gears may crack or be damaged. Before lifting a camshaft out of the head, make certain that the torsional spring force of the sub-gear has been eliminated by the service bolt in the exhaust camshaft.

7 Repeat the steps for the left-bank cylinder head (see illustration).

➥Note: Align the camshaft timing gears together on the left bank cylinder head camshafts (see illustration).

8 Loosen the camshaft bearing cap bolts in 1/4-turn increments until they can be removed by hand. Follow the reverse of the recommended tightening sequence (see illustration 10.33a).

9 Remove the bearing caps and gently lift out the oil feed pipe and the camshafts. Be sure to keep it level.

➥Note: The camshaft cap bolts vary in length. Be sure to mark each bolt carefully to avoid reassembly problems.

10 Store the bearing caps in the correct order.

➥Note: If necessary, the valve lifters and shims can now be removed with a magnetic tool. Be sure to store them separately so they can be reinstalled in their original locations (see illustration).

11 To disassemble an exhaust camshaft gear, mount the cam in a vise with the jaws gripping the large hex on the shaft.

12 Install a second service bolt in the unthreaded hole in the camshaft sub-gear. Using a screwdriver positioned against the service bolt just installed, rotate the sub-gear clockwise and remove the first service

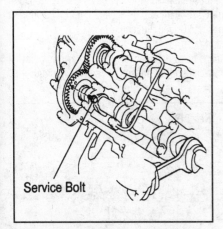

10.7a Install a service bolt through the sub-gear into the driven gear on the exhaust camshaft on the left bank cylinder head

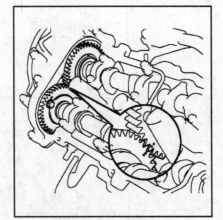

10.7b Align the timing marks (circled) on the camshaft gears directly even on the left bank acylinder head

10.10 Mark up a cardboard box to store the lifters/shims and camshaft bearing caps - use a separate box for each set to avoid mix-ups and mark the FRONT, INTAKE and EXHAUST orientation

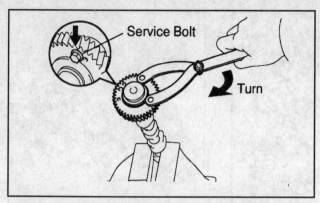

10.12 With the hex portion of the camshaft held in a vise, use a two-pin spanner to remove the tension from the sub-gear and remove the service bolt, then release the sub-gear

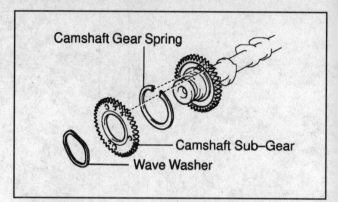

10.14 Remove the wave washer, the camshaft sub-gear and the gear spring

bolt. The second bolt isn't needed if you have a two-pin spanner (see illustration).

13 Remove the sub-gear snap-ring.

14 The wave washer, sub-gear and camshaft gear spring can now be removed from the camshaft (see illustration). Be sure to keep the parts from the left side camshaft separate from the right side.

INSPECTION

15 Refer to Chapter 2, Part A for camshaft, lifter and related component inspection procedures. Be sure to use the Specifications in this Part of Chapter 2.

INSTALLATION

▸ **Refer to illustrations 10.16, 10.22, 10.24, 10.25a, 10.25b, 10.30, 10.32, 10.33a and 10.33b**

16 Insert new camshaft plugs into the cylinder head (see illustration). Apply a small amount of RTV sealant to the grooves.

17 Reassemble the exhaust camshaft gear(s) by installing the cam-

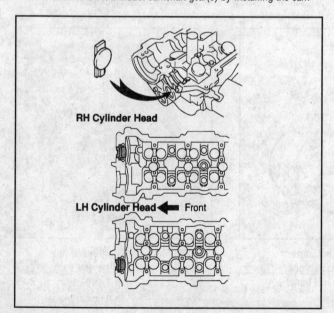

10.16 Location of the camshaft cylinder head plugs

shaft gear spring, sub-gear, wave washer and snap-ring.

18 Mount the camshaft in a padded vise. Insert a service bolt into the unthreaded hole in the camshaft subgear. Using a screwdriver, align the holes of the camshaft driven gear and sub-gear by turning the camshaft sub-gear clockwise. Install a second service bolt in the threaded hole, tightening it to clamp the gears together. Remove the service bolt from the unthreaded hole.

19 Apply moly-base grease or engine assembly lube to the lifters, then install them in their original locations in the cylinder heads. Make sure the valve adjustment shims are in place in the lifters, and that all lifters are installed in their original bores.

Right-bank cylinder head

➡ **Note: For reference purposes, the outside camshafts are the exhaust camshafts while the inner camshafts are the intake camshafts. The right side cylinder head (passenger) is called the right bank while the left side cylinder head (driver) is called the left bank.**

20 Apply moly-base grease or engine assembly lube to the camshaft lobes, bearing journals and gear thrust faces.

21 Set the intake camshaft and exhaust camshaft in place in the cylinder head with the timing marks (two dots) at an 10 degree angle facing each camshaft (see illustration 10.4).

22 Apply a thin coat of RTV sealant to the edges of the front bearing cap cylinder head mating surfaces (see illustration).

23 Install the front bearing cap.

24 Install the bearing caps in numerical order with the arrows pointing toward the front (timing-belt end) of the engine (see illustration).

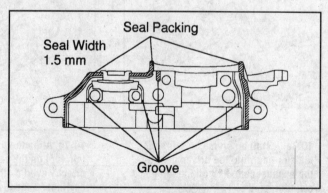

10.22 Apply RTV sealant to the shaded areas on the bearing cap (right bank)

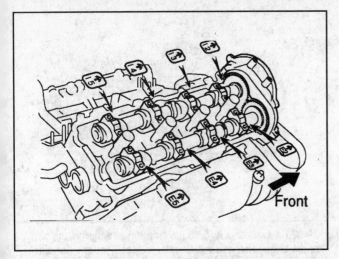

10.24 Install the camshaft bearing caps as shown with the arrows pointing toward the timing belt end of the engine (right bank)

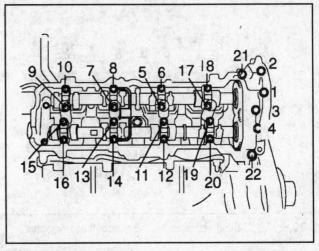

10.25a Tightening sequence for the camshaft bearing caps (right bank)

25 Tighten the bearing cap bolts in 1/4-turn increments to the torque listed in this Chapter's Specifications. Follow the recommended sequence (see illustrations).

➥**Note: Be sure to tighten Bolt A to the specified torque listed at the end of this Chapter.**

26 Remove the service bolt.

27 Refer to Section 9 and install a new camshaft oil seal.

Left-bank cylinder head

28 Apply moly-base grease or engine assembly lube to the camshaft lobes, bearing journals and gear thrust faces.

29 Set the intake camshaft and exhaust camshaft in place in the cylinder head with the timing marks aligned next to each other on each camshaft (see illustration 10.7b).

30 Apply a thin coat of RTV sealant to the edges of the front bearing cap cylinder head mating surfaces (see illustration).

31 Install the front bearing cap.

32 Install the bearing caps in numerical order with the arrows pointing toward the front (timing-belt end) of the engine (see illustration).

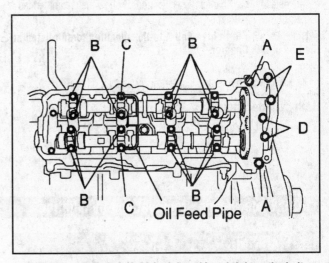

10.25b Locate Bolt A (0.98 inch length) and tighten the bolt to the torque listed in this Chapter's Specifications (right bank)

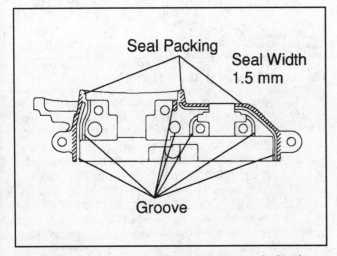

10.30 Apply RTV sealant to the shaded areas on the bearing cap (left bank)

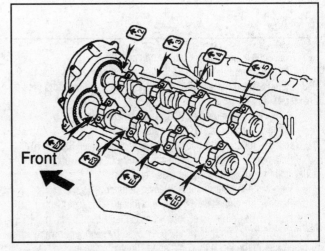

10.32 Install the camshaft bearing caps as shown with the arrows pointing toward the timing belt end of the engine (left bank)

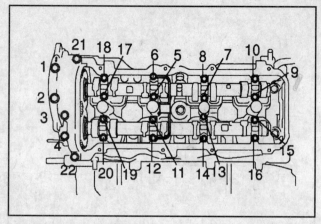

10.33a Tightening sequence for the camshaft bearing caps (left bank)

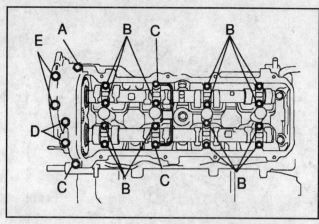

10.33b Locate Bolt A (0.98 inch length) and tighten the bolt to the torque listed in this Chapter's Specifications (left bank)

33 Tighten the bearing cap bolts in 1/4-turn increments to the torque listed in this Chapter's Specifications. Follow the recommended sequence (see illustration).

➡Note: Be sure to tighten Bolt A to the specified torque listed at the end of this Chapter.

34 Remove the service bolt.
35 Refer to Section 9 and install a new camshaft oil seal.

Both cylinder heads

36 Reinstall the timing belt (see Section 7).

37 Reinstall the remaining components in the reverse order of removal.
38 Adjust the valve clearances (see Chapter 1).
39 Before reinstalling the valve covers, use RTV sealant in the areas indicated (see illustration 4.7).
40 The remainder of installation is the reverse of removal. Refill the cooling system (see Chapter 1).
41 Run the engine, then check for leaks and proper operation.

11 Cylinder heads - removal and installation

✱✱ WARNING:

The engine must be completely cool before beginning this procedure.

✱✱ CAUTION:

Don't pry between the head and block. The gasket surfaces may be damaged and leaks could result.

12 Repeat the procedure for the other head.

INSTALLATION

REMOVAL

1 Disconnect the negative cable from the battery.
2 Drain the cooling system, including the block (see Chapter 1).
3 Remove the throttle body, fuel delivery pipes and injectors (see Chapter 4).
4 Remove the exhaust manifold(s) (see Section 6).
5 Remove the alternator (see Chapter 5).
6 Remove the upper and lower intake manifolds (see Section 5).
7 Remove the timing belt, camshaft sprockets, the drivebelt idler pulley and the drivebelt tensioner (see Section 7).
8 Remove the upper timing belt cover number 3.
9 Remove the camshaft(s) from the cylinder heads (see Section 10).
10 Using a 12-point socket, loosen the cylinder head bolts in 1/4-turn increments until they can be removed by hand. Follow the reverse order of the factory recommended tightening sequence (see illustration 11.21).
11 Lift the cylinder head off the engine block. If the head is stuck, place a wood block against it and strike the wood with a hammer.

▶ **Refer to illustrations 11.18 and 11.21**

13 The mating surfaces of the cylinder heads and block must be perfectly clean when the heads are installed.
14 Use a gasket scraper to remove all traces of carbon and old gasket material, then clean the mating surfaces with lacquer thinner or acetone. If there's oil on the mating surfaces when the head is installed, the gasket may not seal correctly and leaks could develop. When working on the block, stuff the cylinders with clean shop rags to keep out debris. Use a vacuum cleaner to remove material that falls into the cylinders.
15 Check the block and head mating surfaces for nicks, deep scratches and other damage. If damage is slight, it can be removed with a file; if it's excessive, machining may be the only alternative.
16 Use a tap of the correct size to chase the threads in the cylinder head bolt holes, then clean the holes with compressed air - make sure that nothing remains in the holes.

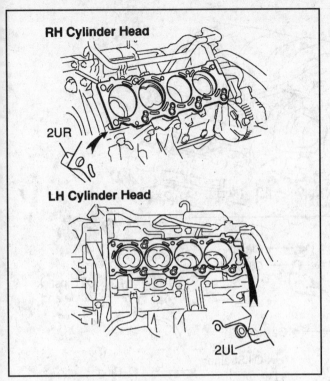

11.18 Be sure the new head gaskets are positioned right side up (check all holes and coolant passages for correct alignment) and over the block dowels

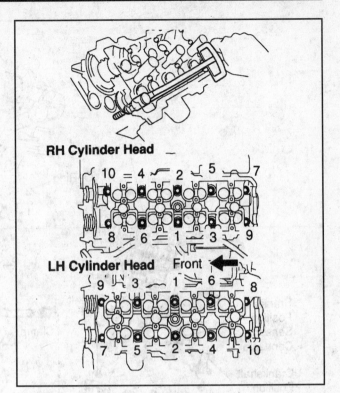

11.21 Cylinder head bolt TIGHTENING sequence for the 4.7L V8 engine

⁂ WARNING:

Wear eye protection when using compressed air!

17 Mount each bolt in a vise and run a die down the threads to remove corrosion and restore the threads. Dirt, corrosion, sealant and damaged threads will affect torque readings.

18 Position the new gaskets over the dowel pins in the block (see illustration).

19 Carefully set the head on the block without disturbing the gasket.

20 Before installing the head bolts, apply a small amount of clean engine oil to the threads.

21 Install the bolts in their original locations and tighten them finger tight. Following the recommended sequence, tighten the bolts to the torque listed in this Chapter's Specifications (see illustration).

22 Mark the front of each bolt head with paint. You can also mark the socket you are using. Place the socket over the 12-point bolt so that you can observe the mark.

23 Following the same sequence, tighten each bolt an additional 1/4-turn (90-degrees).

24 Tighten each bolt yet another 1/4-turn (90-degrees) following the same sequence. The paint marks should now all be 180-degrees from the starting point.

25 Repeat the entire procedure to install the other cylinder head.

26 The remaining installation steps are the reverse of removal.

27 Refill the cooling system, change the oil and filter (see Chapter 1), run the engine and check for leaks.

12 Oil pan - removal and installation

REMOVAL

⬩ **Refer to illustration 12.6**

1 Drain the engine oil and remove the oil filter.

2 Remove the engine from the vehicle (see Chapter 2C).

3 Remove the timing belt, the number 1 and number 2 idler pulleys, the crankshaft pulley (see Section 7) and crankshaft sensor (see Chapter 6).

➡**Note: This step won't be necessary if the oil pump isn't going to be removed.**

4 Remove the oil dipstick tube from the engine.

5 Remove the oil filter, the oil cooler and oil filter bracket assembly from the engine.

6 Remove the bolts and nuts securing the number 2 oil pan and detach the number 2 pan (see illustration). If it's stuck, pry it loose very carefully with a small screwdriver or putty knife. Don't damage the mating surfaces of the pan or oil leaks could develop.

7 Remove the baffle plate from the bottom of the engine.

8 Remove the bolts and nuts securing the number 1 oil pan and detach the number 1 pan (see illustration 12.6). If it's stuck, pry it loose very carefully with a small screwdriver or putty knife. Don't damage the mating surfaces of the pan or oil leaks could develop.

9 Remove the oil pump strainer.

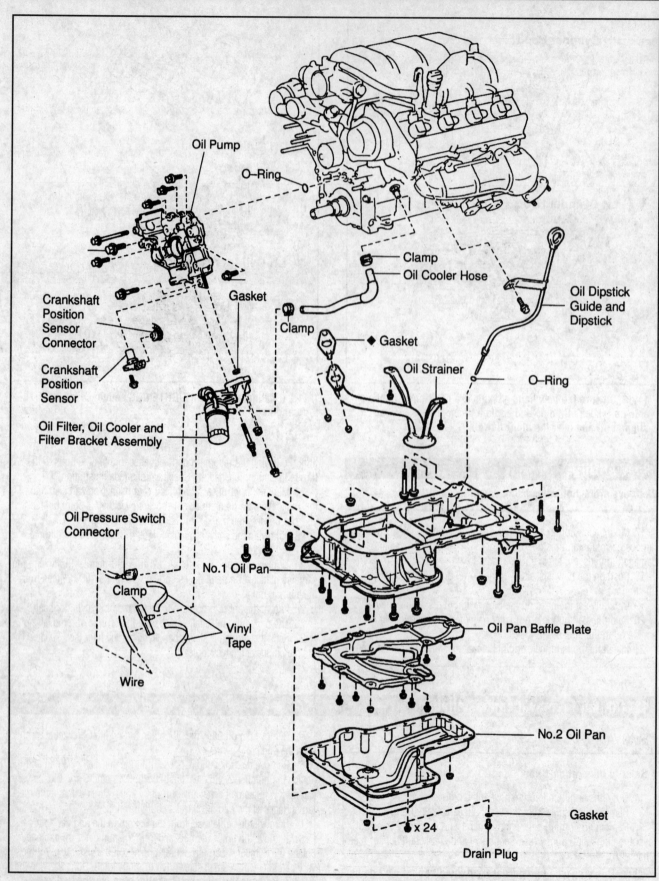

Oil Pump

O—Ring

Gasket

Crankshaft Position Sensor Connector

Crankshaft Position Sensor

Oil Filter, Oil Cooler and Filter Bracket Assembly

Clamp
Oil Cooler Hose

Clamp

◆ Gasket

Oil Strainer

Oil Dipstick Guide and Dipstick

O—Ring

Oil Pressure Switch Connector

Clamp

Vinyl Tape

Wire

No.1 Oil Pan

Oil Pan Baffle Plate

No.2 Oil Pan

Gasket

x 24

Drain Plug

12.6 Exploded view of the oil pan and surrounding components

INSTALLATION

10 Use a scraper to remove all traces of old sealant from the block, the number 1 oil pan and the number 2 oil pan. Clean the mating surfaces with lacquer thinner or acetone.

11 Make sure the threaded bolt holes in the engine block and the number 1 oil pan are clean.

12 Check the flange of the number 2 oil pan for distortion, particularly around the bolt holes. If necessary, place the pan on a wood block and use a hammer to flatten and restore the gasket surface.

13 Inspect the oil pump pick-up/strainer assembly for cracks and a blocked strainer. If the pick-up was removed, clean it with solvent or thinner and install it now, using a new gasket. Tighten the fasteners to the torque listed in this Chapter's Specifications.

14 Apply a 2 to 3 mm wide bead of RTV sealant to the flange of the number 1 oil pan.

15 Carefully position the number 1 oil pan on the engine block and install the bolts. Working from the center out, tighten them to the torque listed in this Chapter's Specifications in three or four steps.

16 Install the baffle plate.

17 Apply a 2 to 3 mm wide bead of RTV sealant to the flange of the number 2 oil pan.

18 Carefully position the number 2 oil pan on the engine block and install the bolts. Working from the center out, tighten them to the torque listed in this Chapter's Specifications in three or four steps.

19 The remainder of installation is the reverse of removal. Be sure to add oil and install a new oil filter, and refill the cooling system (see Chapter 1).

20 Run the engine and check for oil pressure and leaks.

13 Oil pump - removal, inspection and installation

REMOVAL

▶ **Refer to illustrations 13.6a, 13.6b and 13.7**

1 Remove the oil pan (see Section 12).

2 Remove the timing belt (see Section 7) and number 1 and number 2 idler pulleys.

3 Remove the crankshaft timing sprocket (see Section 7).

4 Remove the oil strainer.

5 Remove the air conditioning compressor bracket.

6 Remove the bolts and detach the oil pump from the engine (see illustration). Pry carefully only in the designated areas of the oil pump body with a screwdriver (see illustration).

7 Remove the O-ring from the engine block, Remove the oil pressure relief valve snap-ring, retainer, spring and valve (see illustration).

❋❋ WARNING:

The spring is tightly compressed - be careful and wear eye protection.

8 Use a large Phillips screwdriver to remove the screws retaining the body cover to the rear of the oil pump.

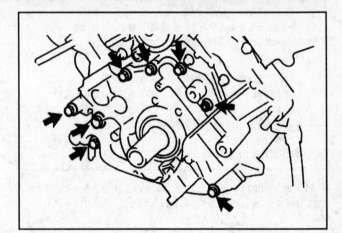

13.6a Oil pump bolt locations

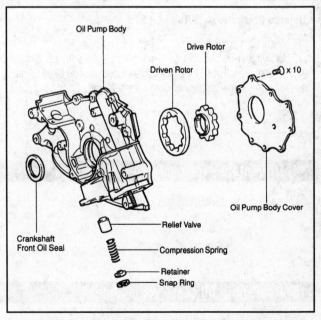

13.7 Remove the oil pressure relief plug, spring and valve

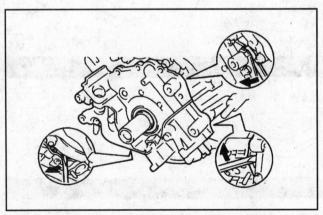

13.6b Pry the oil pump carefully at the designated areas to avoid damaging the oil pump surfaces

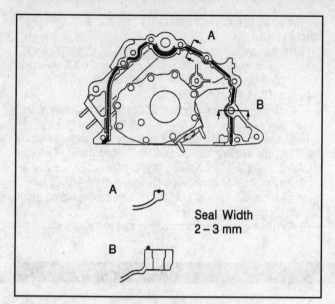

13.18 Apply a 2 to 3 mm bead of RTV sealant to the designated area on the oil pump body

9   Lift the cover off and remove the pump rotors.
10 Use a scraper to remove all traces of sealant and old gasket material from the pump body and engine block, then clean the mating surfaces with lacquer thinner or acetone.

INSPECTION

11 Refer to Chapter 2, Part A for this procedure, but be sure to use the clearance specifications in this Part of Chapter 2 for the 4.7L V8 engine.

INSTALLATION

▶ **Refer to illustrations 13.18 and 13.20**

12 Pry the old crankshaft seal out with a screwdriver.
13 Apply multi-purpose grease or engine oil to the outer edge of the new seal and carefully drive it into place with a deep socket and a hammer. Also apply multi-purpose grease to the seal lip.

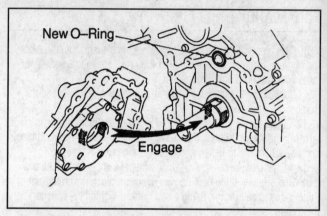

13.20 Be sure to align the drive rotor and the crankshaft as the oil pump is installed, and install a new O-ring on the block

14 Place the drive and driven rotors into the pump body with the marks facing out.
15 Pack the pump cavity with petroleum jelly and install the cover. Tighten the screws securely following a criss-cross pattern.
16 Lubricate the oil pressure relief valve with engine oil and install the valve components in the pump body.
17 Use acetone or lacquer thinner and a clean rag to remove all traces of oil from the gasket surfaces.
18 Apply a 2 to 3 mm wide bead of RTV sealant to the oil pump (see illustration). Avoid using an excessive amount of sealant, especially around oil passages and bolt holes. Assembly must be completed within five minutes of sealant application, otherwise the material must be removed and reapplied.
19 Position a new O-ring on the block.
20 Engage the spline teeth on the oil pump drive rotor with the large teeth on the crankshaft and slide the pump into place (see illustration).
21 Install the oil pump mounting bolts in their original locations and tighten them to the torque listed in this Chapter's Specifications in a criss-cross pattern.
22 Using a new gasket, install the oil pick-up tube and tighten the fasteners to the torque listed in this Chapter's Specifications.
23 Reinstall the remaining parts in the reverse order of removal. Install the engine
24 Refill the engine oil and coolant (see Chapter 1). Start the engine and check for oil leaks.

14 Driveplate - removal and installation

Refer to Chapter 2, Part A for this procedure, but be sure to use the torque specifications in this Part of Chapter 2 for the 4.7L V8 engine.

15 Rear main oil seal - replacement

▶ **Refer to illustrations 15.1a and 15.1b**

Refer to Chapter 2, Part A for this procedure. Apply a 2 to 3 mm

wide bead of RTV sealant to the retainer flange (see illustrations) before attaching the retainer to the block. Also, be sure to use the torque specifications in this Part of Chapter 2 for the 4.7L V8 engine.

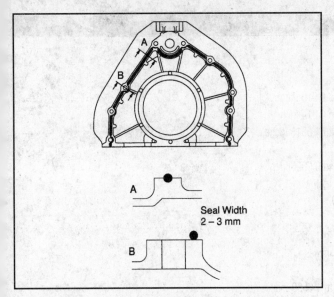

15.1a Apply RTV sealant to the designated area on the oil seal retainer-to-block mating surface

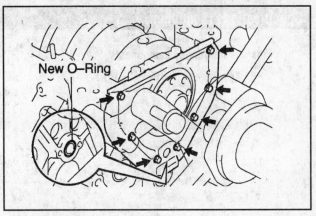

15.1b Be sure to install a new O-ring into the engine block

16 Engine mounts - check and replacement

⬧ **Refer to illustration 16.1**

Refer to Chapter 2, Part A, but note that the 4.7L V8 engine mounts are slightly different in ways that don't significantly affect the check and replacement procedures (see illustration).

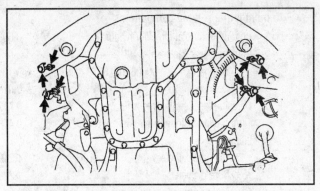

16.1 Location of the engine mount bolts

Specifications

General

Engine designation	2UZ-FE
Displacement	4.7 liters
Cylinder numbers (timing belt end-to-transmission end)	
Right (passenger) side	2-4-6-8
Left (driver) side	1-3-5-7
Firing order	1-8-4-3-6-5-7-2

Warpage limits

Cylinder head	0.0039 inch
Intake manifold	0.0059 inch
Exhaust manifolds	0.0197 inch

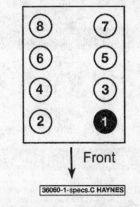

36060-1-specs.C HAYNES

Cylinder numbering diagram

Specifications (continued)

Camshaft and related components

Valve clearance (engine cold)

    Intake   0.006 to 0.010 inch

    Exhaust   0.010 to 0.014 inch

Bearing journal diameter   1.0612 to 1.0618 inches

Bearing oil clearance

    Standard   0.0012 to 0.0026 inch

    Service limit   0.0039 inch

Lobe height

    Intake

        Standard   1.6512 to 1.6551 inches

        Service Limit   1.6453 inches

    Exhaust

        Standard   1.6520 to 1.6559 inches

        Service Limit   1.6461 inches

Thrust clearance (endplay)

    Standard

        Intake   0.0016 to 0.0033 inch

        Exhaust   0.0011 to 0.0030 inch

    Service limit   0.0047 inch

Camshaft and related components

Runout limit (total indicator reading)   0.0031 inch

Camshaft gear backlash

    Standard   0.0008 to 0.0079 inch

    Service limit   0.0188 inch

Camshaft gear spring free length   0.712 to 0.740 inch

Timing belt tensioner protrusion   0.413 to 0.453 inch

Lifters

    Outside diameter   1.2192 to 1.2195 inch

    Bore diameter   1.2205 to 1.2211 inch

    Lifter-to-bore (oil) clearance

        Standard   0.0009 to 0.0020 inch

        Service limit   0.003 inch

Oil pump

Driven rotor-to-pump body clearance

    Standard   0.0039 to 0.0069 inch

    Service limit   0.0118 inch

Rotor tip clearance

    Standard   0.0043 to 0.0094 inch

    Service limit   0.0138 inch

Rotor side clearance

    Standard   0.0012 to 0.0035 inch

    Service limit   0.0059 inch

Torque specifications   Ft-lbs (unless otherwise indicated)

➡**Note: One foot-pound (ft-lb) of torque is equivalent to 12 inch-pounds (in-lbs) of torque. Torque values below approximately 15 ft-lbs are expressed in inch-pounds, since most foot-pound torque wrenches are not accurate at these smaller values.**

Intake manifold	
Upper intake manifold bolts/nuts	156 in-lbs
Lower intake manifold bolts/nuts	156 in-lbs
Exhaust manifold nuts	33
Crankshaft pulley bolt	181
Drivebelt idler pulley	29
Drivebelt tensioner bolts	144 in-lbs
Fan bracket bolts	
12 mm bolt head	144 in-lbs
14 mm bolt head	24
Driveplate bolts	
Step 1	36
Step 2	Turn an additional 90-degrees (1/4-turn)
Timing belt cover number 2 bolts	144 in-lbs
Timing belt cover (left and right) number 3 bolts	66 in-lbs
Idler pulley bolts*	
Number 1	25
Number 2	25
Timing belt tensioner bolts	19
Valve cover nuts	53 in-lbs
Camshaft timing pulley bolts	80
Camshaft bearing cap bolts	
Bolt A (see illustrations 10.25b and 10.33b)	69 in-lbs
All others	144 in-lbs
Cylinder head bolts (in sequence - see illustration 11.21)	
Step 1	24
Step 2	Turn an additional 90-degrees (1/4-turn)
Step 3	Turn an additional 90-degrees (1/4-turn)
Oil pump body cover bolts	84 in-lbs
Oil pump mounting bolts	
14 mm bolt head	22
All others	132 in-lbs
Oil pan number 1 bolts	
10 mm bolt head	66 in-lbs
12 mm bolt head	21
Oil pan number 2 bolts	66 in-lbs
Oil strainer mounting bolts	66 in-lbs
Rear crankshaft oil seal retainer mounting bolts	69 in-lbs

* Apply thread locking compound to the threads prior to installation

Notes

Section

Reference to other Chapters

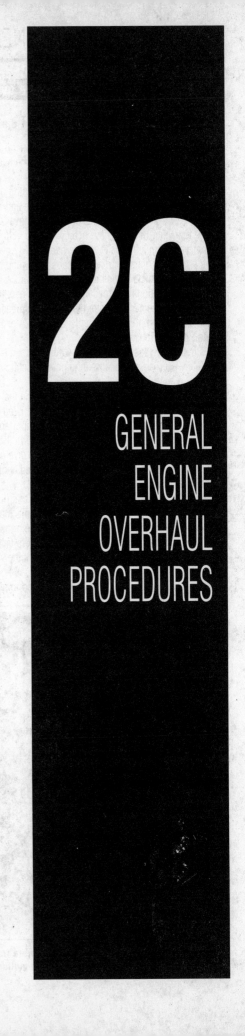

2C

GENERAL ENGINE OVERHAUL PROCEDURES

1   General information - engine overhaul

▶ **Refer to illustrations 1.2, 1.3, 1.4, 1.5, 1.6 and 1.7**

Included in this portion of Chapter 2 are general information and diagnostic testing procedures for determining the overall mechanical condition of your engine.

The information ranges from advice concerning preparation for an overhaul and the purchase of replacement parts and/or components to detailed, step-by-step procedures covering removal and installation.

The following Sections have been written to help you determine whether your engine needs to be overhauled and how to remove and install it once you've determined it needs to be rebuilt. For information concerning in-vehicle engine repair, see Chapter 2A or 2B.

The Specifications included in this Part are general in nature and include only those necessary for testing the oil pressure and checking the engine compression. Refer to Chapter 2A or 2B for additional engine Specifications.

It's not always easy to determine when, or if, an engine should be completely overhauled, because a number of factors must be considered.

High mileage is not necessarily an indication that an overhaul is needed, while low mileage doesn't preclude the need for an overhaul. Frequency of servicing is probably the most important consideration. An engine that's had regular and frequent oil and filter changes, as well as other required maintenance, will most likely give many thousands of miles of reliable service. Conversely, a neglected engine may require an overhaul very early in its service life.

Excessive oil consumption is an indication that piston rings, valve seals and/or valve guides are in need of attention. Make sure that oil leaks aren't responsible before deciding that the rings and/or guides are bad. Perform a cylinder compression check to determine the extent of the work required (see Section 3). Also check the vacuum readings under various conditions (see Section 4).

Check the oil pressure with a gauge installed in place of the oil pressure sending unit and compare it to this Chapter's Specifications (see Section 2). If it's extremely low, the bearings and/or oil pump are probably worn out.

Loss of power, rough running, knocking or metallic engine noises, excessive valve train noise and high fuel consumption rates may also point to the need for an overhaul, especially if they're all present at the same time. If a complete tune-up doesn't remedy the situation, major mechanical work is the only solution.

An engine overhaul involves restoring the internal parts to the specifications of a new engine. During an overhaul, the piston rings are replaced and the cylinder walls are reconditioned (rebored and/ or honed) (see illustrations 1.2 and 1.3). If a rebore is done by an automotive machine shop, new oversize pistons will also be installed. The main bearings, connecting rod bearings and camshaft bearings are generally replaced with new ones and, if necessary, the crankshaft may be reground to restore the journals (see illustration 1.4). Generally, the valves are serviced as well, since they're usually in less-than-perfect condition at this point. While the engine is being overhauled, other components, such as the distributor, starter and alternator, can be rebuilt as well. The end result should be a like-new engine that will give many trouble-free miles.

➡ **Note: Critical cooling system components such as the hoses, drivebelts, thermostat and water pump should be replaced with new parts when an engine is overhauled. The radiator should be checked carefully to ensure that it isn't clogged or leaking (see Chapter 3). If you purchase a rebuilt engine or short block, some rebuilders will not warranty their engines unless the radiator has been professionally flushed. Also, we don't recommend overhauling the oil pump - always install a new one when an engine is rebuilt.**

Overhauling the internal components on today's engines is a difficult and time-consuming task which requires a significant amount of specialty tools and is best left to a professional engine rebuilder

1.2 An engine block being bored. An engine rebuilder will use special machinery to recondition the cylinder bores

1.3 If the cylinders are bored, the machine shop will normally hone the engine on a machine like this

1.4 A crankshaft having a main bearing journal ground

1.5 A machinist checks for a bent connecting rod, using specialized equipment

1.6 A bore gauge being used to check the main bearing bore

(see illustrations 1.5, 1.6 and 1.7). A competent engine rebuilder will handle the inspection of your old parts and offer advice concerning the reconditioning or replacement of the original engine. Never purchase parts or have machine work done on other components until the block has been thoroughly inspected by a professional machine shop. As a general rule, time is the primary cost of an overhaul, especially since the vehicle may be tied up for a minimum of two weeks or more. Be aware that some engine builders only have the capability to rebuild the engine you bring them while other rebuilders have a large inventory of rebuilt exchange engines in stock. Also be aware that many machine shops could take as much as two weeks time to completely rebuild your engine depending on shop workload. Sometimes it makes more sense to simply exchange your engine for another engine that's already rebuilt to save time.

1.7 Uneven piston wear like this indicates a bent connecting rod

2   Oil pressure check

♦ **Refer to illustrations 2.2a, 2.2b and 2.2c**

1   Low engine oil pressure can be a sign of an engine in need of rebuilding. A "low oil pressure" indicator (often called an "idiot light") is not a test of the oiling system. Such indicators only come on when the oil pressure is dangerously low. Even a factory oil pressure gauge in the instrument panel is only a relative indication, although much better for driver information than a warning light. A better test is with a mechanical (not electrical) oil pressure gauge.

2   Locate the oil pressure indicator sending unit on the engine block.

a) *On 2000 through 2004 V6 models, the oil pressure sending unit (see illustration) is located on the engine block, next to the oil filter assembly.*

b) *On 2005 V6 models, the oil pressure sending unit (see illustration) is located on the timing chain cover, next to the oil filter.*

c) *On V8 models, the oil pressure sending unit (see illustration) is located on the side of the oil filter bracket next to the oil cooler.*

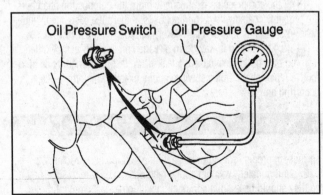

2.2a On 2000 through 2004 V6 models, the oil pressure sending unit is located on the left side of the engine block, next to the oil filter assembly

2.2b On 2005 and later V6 models, the oil pressure sending unit is located on the timing chain cover, next to the oil filter

3   Unscrew and remove the oil pressure sending unit and then screw in the hose for your oil pressure gauge. If necessary, install an adapter fitting. Use Teflon tape or thread sealant on the threads of the adapter and/or the fitting on the end of your gauge's hose.

4   Connect an accurate tachometer to the engine, according to the tachometer manufacturer's instructions.

5   Check the oil pressure with the engine running (normal operating temperature) at the specified engine speed, and compare it to this Chapter's Specifications. If it's extremely low, the bearings and/or oil pump are probably worn out.

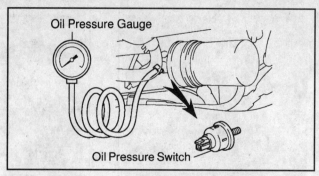

2.2c On V8 models, the oil pressure sending unit is located on the oil filter assembly next to the oil cooler

3   Cylinder compression check

▶ **Refer to illustration 3.6**

1   A compression check will tell you what mechanical condition the upper end of your engine (pistons, rings, valves, head gaskets) is in. Specifically, it can tell you if the compression is down due to leakage caused by worn piston rings, defective valves and seats or a blown head gasket.

➡**Note: The engine must be at normal operating temperature and the battery must be fully charged for this check.**

2   Begin by cleaning the area around the spark plugs before you remove them (compressed air should be used, if available). The idea is to prevent dirt from getting into the cylinders as the compression check is being done.

3   Remove all of the spark plugs from the engine (see Chapter 1).

4   Block the throttle wide open.

5   Disable the ignition system by disconnecting the electrical connector(s) from the igniter (V6 models) or the igniter/ignition coil assemblies (V8 models) (see Chapter 5). The fuel pump circuit should also be disabled by removing the circuit opening relay from the fuse/relay center in the engine compartment (see Chapter 4).

6   Install a compression gauge in the spark plug hole (see illustration).

7   Crank the engine over at least seven compression strokes and watch the gauge. The compression should build up quickly in a healthy engine. Low compression on the first stroke, followed by gradually increasing pressure on successive strokes, indicates worn piston rings. A low compression reading on the first stroke, which doesn't build up during successive strokes, indicates leaking valves or a blown head gasket (a cracked head could also be the cause). Deposits on the undersides of the valve heads can also cause low compression. Record the highest gauge reading obtained.

8   Repeat the procedure for the remaining cylinders and compare the results to this Chapter's Specifications.

9   Add some engine oil (about three squirts from a plunger-type oil

3.6 Use a compression gauge with a threaded fitting for the spark plug hole, not the type that requires hand pressure to maintain the seal

can) to each cylinder, through the spark plug hole, and repeat the test.

10 If the compression increases after the oil is added, the piston rings are definitely worn. If the compression doesn't increase significantly, the leakage is occurring at the valves or head gasket. Leakage past the valves may be caused by burned valve seats and/or faces or warped, cracked or bent valves.

11 If two adjacent cylinders have equally low compression, there's a strong possibility that the head gasket between them is blown. The appearance of coolant in the combustion chambers or the crankcase would verify this condition.

12 If one cylinder is slightly lower than the others, and the engine has a slightly rough idle, a worn lobe on the camshaft could be the cause.

13 If the compression is unusually high, the combustion chambers are probably coated with carbon deposits. If that's the case, the cylinder head(s) should be removed and decarbonized.

14 If compression is way down or varies greatly between cylinders, it would be a good idea to have a leak-down test performed by an automotive repair shop. This test will pinpoint exactly where the leakage is occurring and how severe it is.

4   Vacuum gauge diagnostic checks

▶ **Refer to illustrations 4.4 and 4.6**

1   A vacuum gauge provides inexpensive but valuable information about what is going on in the engine. You can check for worn rings or cylinder walls, leaking head or intake manifold gaskets, incorrect carburetor adjustments, restricted exhaust, stuck or burned valves, weak valve springs, improper ignition or valve timing and ignition problems.

2   Unfortunately, vacuum gauge readings are easy to misinterpret, so they should be used in conjunction with other tests to confirm the diagnosis.

3   Both the absolute readings and the rate of needle movement are important for accurate interpretation. Most gauges measure vacuum in

inches of mercury (in-Hg). The following references to vacuum assume the diagnosis is being performed at sea level. As elevation increases (or atmospheric pressure decreases), the reading will decrease. For every 1,000 foot increase in elevation above approximately 2000 feet, the gauge readings will decrease about one inch of mercury.

4   Connect the vacuum gauge directly to the intake manifold vacuum, not to ported (throttle body) vacuum (see illustration). Be sure no hoses are left disconnected during the test or false readings will result.

5   Before you begin the test, allow the engine to warm up completely. Block the wheels and set the parking brake. With the transmission in Park, start the engine and allow it to run at normal idle speed.

❊❊ WARNING:

Keep your hands and the vacuum gauge clear of the fans.

6   Read the vacuum gauge; an average, healthy engine should normally produce about 17 to 22 in-Hg with a fairly steady needle (see illustration). Refer to the following vacuum gauge readings and what they indicate about the engine's condition:

7   A low steady reading usually indicates a leaking gasket between the intake manifold and cylinder head(s) or throttle body, a leaky vacuum hose, late ignition timing or incorrect camshaft timing. Check ignition timing with a timing light and eliminate all other possible causes, utilizing the tests provided in this Chapter before you remove the timing chain cover to check the timing marks.

8   If the reading is three to eight inches below normal and it fluctuates at that low reading, suspect an intake manifold gasket leak at an intake port or a faulty fuel injector.

9   If the needle has regular drops of about two-to-four inches at a steady rate, the valves are probably leaking. Perform a compression check or leak-down test to confirm this.

10   An irregular drop or down-flick of the needle can be caused by a sticking valve or an ignition misfire. Perform a compression check or leak-down test and read the spark plugs.

11   A rapid vibration of about four in-Hg vibration at idle combined with exhaust smoke indicates worn valve guides. Perform a leak-down test to confirm this. If the rapid vibration occurs with an increase in engine speed, check for a leaking intake manifold gasket or head gasket, weak valve springs, burned valves or ignition misfire.

12   A slight fluctuation, say one inch up and down, may mean ignition problems. Check all the usual tune-up items and, if necessary, run the engine on an ignition analyzer.

13   If there is a large fluctuation, perform a compression or leak-down test to look for a weak or dead cylinder or a blown head gasket.

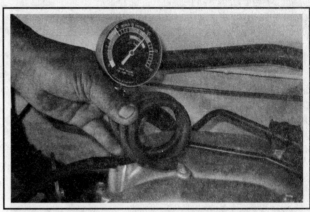

4.4 A simple vacuum gauge can be handy in diagnosing engine condition and performance

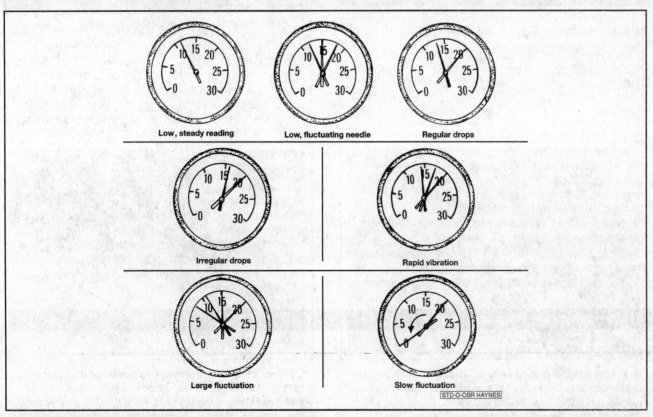

4.6 Typical vacuum gauge readings

14 If the needle moves slowly through a wide range, check for a clogged PCV system, incorrect idle fuel mixture, throttle body or intake manifold gasket leaks.

15 Check for a slow return after revving the engine by quickly snapping the throttle open until the engine reaches about 2,500 rpm and let it shut. Normally the reading should drop to near zero, rise above normal idle reading (about 5 in-Hg over) and then return to the previous idle reading. If the vacuum returns slowly and doesn't peak when the throttle is snapped shut, the rings may be worn. If there is a long delay, look for a restricted exhaust system (often the muffler or catalytic converter). An easy way to check this is to temporarily disconnect the exhaust ahead of the suspected part and redo the test.

5   Engine rebuilding alternatives

The do-it-yourselfer is faced with a number of options when purchasing a rebuilt engine. The major considerations are cost, warranty, parts availability and the time required for the rebuilder to complete the project. The decision to replace the engine block, piston/connecting rod assemblies and crankshaft depends on the final inspection results of your engine. Only then can you make a cost effective decision whether to have your engine overhauled or simply purchase an exchange engine for your vehicle.

Some of the rebuilding alternatives include:

Individual parts - If the inspection procedures reveal that the engine block and most engine components are in reusable condition, purchasing individual parts and having a rebuilder rebuild your engine may be the most economical alternative. The block, crankshaft and piston/connecting rod assemblies should all be inspected carefully by a machine shop first.

Short block - A short block consists of an engine block with a crankshaft and piston/connecting rod assemblies already installed. All new bearings are incorporated and all clearances will be correct. The existing camshafts, valve train components, cylinder head and external parts can be bolted to the short block with little or no machine shop work necessary.

Long block - A long block consists of a short block plus an oil pump, oil pan, cylinder head, valve cover, camshaft and valve train components, timing sprockets and chain or gears and timing cover. All components are installed with new bearings, seals and gaskets incorporated throughout. The installation of manifolds and external parts is all that's necessary.

Low mileage used engines - Some companies now offer low mileage used engines which is a very cost effective way to get your vehicle up and running again. These engines often come from vehicles which have been in totaled in accidents or come from other countries which have a higher vehicle turn over rate. A low mileage used engine also usually has a similar warranty like the newly remanufactured engines.

Give careful thought to which alternative is best for you and discuss the situation with local automotive machine shops, auto parts dealers and experienced rebuilders before ordering or purchasing replacement parts.

6   Engine removal - methods and precautions

▶ **Refer to illustrations 6.1, 6.2, 6.3, 6.4 and 6.5**

If you've decided that an engine must be removed for overhaul or major repair work, several preliminary steps should be taken. Read all removal and installation procedures carefully prior to committing this job. Some engines are removed by lowering to the floor and then raising the vehicle sufficiently to slide it out; this will require a vehicle hoist.

Locating a suitable place to work is extremely important. Adequate work space, along with storage space for the vehicle, will be needed. If a shop or garage isn't available, at the very least a flat, level, clean work surface made of concrete or asphalt is required. Cleaning the engine compartment and engine before beginning the removal procedure will help keep tools clean and organized (see illustrations 6.1 and 6.2).

An engine hoist or A-frame will also be necessary. Make sure the

6.1 **After tightly wrapping water-vulnerable components, use a spray cleaner on everything, with particular concentration on the greasiest areas, usually around the valve cover and lower edges of the block. If one section dries out, apply more cleaner**

6.2 **Depending on how dirty the engine is, let the cleaner soak in according to the directions and then hose off the grime and cleaner. Get the rinse water down into every area you can get at; then dry important components with a hair dryer or paper towels**

6.3 Get an engine hoist that's strong enough to easily lift your engine in and out of the engine compartment; an adapter, like the one shown here (arrow), can be used to change the angle of the engine as it's being removed or installed

6.4 Get an engine stand sturdy enough to firmly support the engine while you're working on it. Stay away from three-wheeled models: they have a tendency to tip over more easily, so get a four-wheeled unit

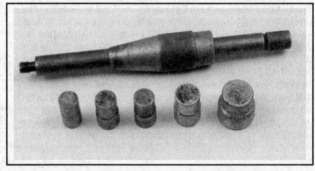

6.5 A clutch alignment tool is necessary if you plan to install a rebuilt engine mated to a manual transmission

equipment is rated in excess of the combined weight of the engine and transmission. Safety is of primary importance, considering the potential hazards involved in lifting the engine out of the vehicle.

If you're a novice at engine removal, get at least one helper. One person cannot easily do all the things you need to do to lift a big heavy engine out of the engine compartment. Also helpful is to seek advice and assistance from someone who's experienced in engine removal.

Plan the operation ahead of time. Arrange for or obtain all of the tools and equipment you'll need prior to beginning the job (see illustrations 6.3, 6.4 and 6.5). some of the equipment necessary to perform engine removal and installation safely and with relative ease are (in addition to an engine hoist) a heavy duty floor jack, complete sets of wrenches and sockets as described in the front of this manual, wooden

blocks, plenty of rags and cleaning solvent for mopping up spilled oil, coolant and gasoline. If the hoist must be rented, make sure that you arrange for it in advance and have everything disconnected and/or removed before bringing the hoist home. This will save you money and time.

Plan for the vehicle to be out of use for quite a while. A machine shop can do the work that is beyond the scope of the home mechanic. Machine shops often have a busy schedule, so before removing the engine, consult the shop for an estimate of how long it will take to rebuild or repair the components that may need work.

7 Engine - removal and installation

✳✳ WARNING 1:

DO NOT use a cheap engine hoist designed for lifting four-cylinder engines. Obtain a heavy-duty hoist designed for lifting heavy engines. And always be extremely careful when removing and installing the engine. Serious injury can result from careless actions.

✳✳ WARNING 2:

The models covered by this manual are equipped with Supplemental Restraint systems (SRS), more commonly known as airbags. Always disable the airbag system before working in the vicinity of any airbag system component to avoid the possibility of accidental deployment of the airbag, which could cause personal injury (see Chapter 12).

7.5 Place sturdy jackstands under the frame of the vehicle and set them both at uniform height

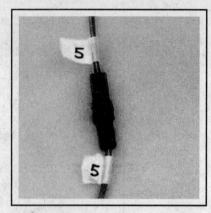

7.9 Label both ends of each wire and hose before disconnecting it - do the same for vacuum hoses

7.28a Engine lifting hanger on the right side of the engine block (V8 model shown, V6 models similar)

❋❋ WARNING 3:

Gasoline is extremely flammable, so take extra precautions when you work on any part of the fuel system. Don't smoke or allow open flames or bare light bulbs near the work area, and don't work in a garage where a gas-type appliance (such as a water heater or a clothes dryer) is present. Since gasoline is carcinogenic, wear latex gloves when there's a possibility of being exposed to fuel, and, if you spill any fuel on your skin, rinse it off immediately with soap and water. Mop up any spills immediately and do not store fuel-soaked rags where they could ignite. The fuel system is under constant pressure, so, if any fuel lines are to be disconnected, the fuel pressure in the system must be relieved first (see Chapter 4 for more information). When you perform any kind of work on the fuel system, wear safety glasses and have a Class B type fire extinguisher on hand.

❋❋ WARNING 4:

The air conditioning system is under high pressure, and refrigerant is expensive. Have a dealer service department or an automotive air conditioning shop discharge the system before beginning this procedure.

REMOVAL

▶ Refer to illustrations 7.5 and 7.9

1   On vehicles with air conditioning, have the air conditioning system discharged.

2   Relieve the fuel system pressure (see Chapter 4).

3   Disconnect the battery cables and. remove the battery (see Chapter 5).

4   Remove the hood (see Chapter 11) and cover the fenders and cowl. Special pads are available to protect the fenders, but an old bedspread or blanket will also work.

5   Raise the vehicle and place it securely on jackstands (see illustration).

➥Note: On 4WD models, and on models with large tires, this step may not be necessary, because some models already have sufficient ground clearance to allow disconnection of the exhaust system, the engine mounts, etc. from underneath the vehicle. Raising these vehicles any higher might even make

engine removal more difficult because it might position the vehicle too high to lift the engine out of the engine compartment with a hoist.

6   Drain the engine oil and remove the oil filter (see Chapter 1). Detach the engine oil dipstick tube bracket and remove the engine oil dipstick tube.

7   Drain the cooling system (see Chapter 1).

8   Remove the air cleaner assembly and the air intake duct (see Chapter 4).

9   Label all vacuum lines, emissions system hoses, wiring harness electrical connectors and ground straps to ensure correct reinstallation, then disconnect them. Pieces of masking tape with numbers or letters written on them work well (see illustration). So does colored electrical tape. If there's any possibility of confusion, make a sketch of the engine compartment and clearly label the lines, hoses and wires. You can also use an inexpensive disposable or digital camera to take photos of connectors, grounds, harness routing, etc.

10 Disconnect the fuel lines running from the engine to the chassis (see Chapter 4). Plug or cap all open fittings/lines.

❋❋ WARNING:

Gasoline is extremely flammable, so extra precautions must be taken when working on any part of the fuel system. DO NOT smoke or allow open flames or bare light bulbs near the vehicle. Also, don't work in a garage if a gas-type appliance is present.

4WD models

11 Remove the front exhaust pipes (see Chapter 4).

12 Remove the front and rear driveshafts (see Chapter 8).

13 Remove the front stabilizer bar (see Chapter 10).

14 Remove the transmission (see Chapter 7).

All models

▶ Refer to illustration, 7.28a, 7.28b, 7.28c and 7.31

15 Disconnect the MAF sensor electrical connector (see Chapter 6).

16 Disconnect the throttle cable from the throttle linkage on the throttle body, then remove the throttle body (see Chapter 4). Disconnect the fuel lines from the fuel rail, then remove the fuel rail and the injectors (see Chapter 4). Remove the intake manifold (see Chapter 2A or 2B). Also remove all fuel and/or emission control components that

7.28b Engine lifting hanger on the left side of the engine block (V8 model shown, V6 models similar)

7.28c Secure the engine hangers to the lifting device with heavy chain

might be damaged during engine removal (see Chapters 4 and 6).

17 Clearly label and disconnect all coolant and heater hoses. Remove the cooling fan, shroud and radiator (see Chapter 3).

18 Remove the accessory drivebelt(s) (see Chapter 1), then remove the alternator (see Chapter 5).

19 On vehicles with air conditioning, remove the compressor (see Chapter 3). Look carefully at the air conditioning system lines. If some section(s) of the air conditioning lines - particularly any section consisting of rigid metal lines - looks like it's going to impede engine removal and installation, detach it from the engine and/or vehicle and set it aside. Secure it with wire if necessary to make sure that it won't be damaged by the engine when the engine is lifted out of the engine compartment.

20 Disconnect the air ducts near the timing belt cover and remove them from the engine compartment (see Chapter 4).

21 Disconnect the vacuum lines from the intake manifold (see Chapters 2A or 2B).

22 Remove the transmission oil cooler pipes from the radiator and the transmission cooler (see Chapter 3).

23 Remove the radiator (see Chapter 3).

24 Remove the power steering pump from its mounting bracket (see Chapter 10) and secure it with wire so that it won't interfere with engine removal.

25 If you're going to be replacing the block, now is a good time to remove all large brackets such as the alternator, air conditioning compressor and power steering pump brackets (see Chapter 9).

26 Remove the part of the exhaust system that's routed underneath the engine, between the exhaust manifolds and the downstream catalytic converter (see Chapter 4). It's not absolutely necessary to remove the exhaust manifolds in order to remove the engine, but removing the manifolds will shave a little weight off the engine.

27 Remove the starter motor (see Chapter 5).

28 Locate the lifting brackets on the engine (see illustrations). Roll a heavy-duty hoist into position and attach it to the lifting brackets with a couple pieces of heavy-duty chain (see illustration). Take up the slack in the sling or chain, but don't lift the engine.

❋❋ WARNING:

DO NOT use a cheap hoist designed to lift four-cylinder engines. Obtain a heavy-duty hoist designed for lifting heavy engines. And DO NOT place any part of your body under the engine when it's supported only by a hoist or other lifting device.

7.31 Use long high-strength bolts (arrows) to hold the engine block on the engine stand - make sure they are tight before resting all the weight on the stand

29 Remove the engine mount fasteners (see the "Engine mounts - check and replacement" section in Chapter 2A).

30 Recheck to be sure nothing is still connecting the engine to the transmission or vehicle. Disconnect anything still remaining. Raise the engine slightly and inspect it thoroughly once more to make sure that nothing is still attached, then slowly lift the engine out of the engine compartment. Check carefully to make sure nothing is hanging up.

31 Remove the flywheel/driveplate (see Chapter 2A or 2B) and mount the engine on an engine stand (see illustration).

32 Inspect the engine and transmission mounts (see the "Engine mounts - check and replacement" Section in Chapter 2A). If they're worn or damaged, replace them.

INSTALLATION

33 Install the flywheel/driveplate (see Chapter 2A or 2B).

34 If you're working on a vehicle with a manual transmission, install the clutch and pressure plate (see Chapter 8). Now is a good time to install a new clutch.

35 Carefully lower the engine into the engine compartment, then reattach it to the engine mounts (see the "Engine mounts - check and replacement" Section in Chapter 2A).

4WD models

36 Install the transmission (see Chapter 7). If you're working on a manual transmission model, apply a dab of high-temperature grease to the input shaft and guide it into the crankshaft pilot bearing until the bellhousing is flush with the engine block. If you're working on a vehicle with an automatic transmission, guide the torque converter into the crankshaft following the procedure outlined in Chapter 7B. Install the transmission-to-engine bolts and tighten them securely.

❋❋ CAUTION:

DO NOT use the bolts to force the transmission and engine together!

37 Install the front exhaust pipes (see Chapter 4).
38 Install the front and rear driveshafts (see Chapter 8).
39 Install the front stabilizer bar (see Chapter 10).

All models

40 Reinstall the remaining components in the reverse order of removal.
41 Add coolant, oil and transmission fluid as needed.
42 Run the engine and check for leaks and proper operation of all accessories, then install the hood and test drive the vehicle.
43 Have the air conditioning system re-charged and leak tested, if it was discharged.

8 Engine overhaul - disassembly sequence

1 It's much easier to remove the external components if it's mounted on a portable engine stand. A stand can often be rented quite cheaply from an equipment rental yard. Before the engine is mounted on a stand, the flywheel/driveplate should be removed from the engine.

2 If a stand isn't available, it's possible to remove the external engine components with it blocked up on the floor. Be extra careful not to tip or drop the engine when working without a stand.

3 If you're going to obtain a rebuilt engine, all external components must come off first, to be transferred to the replacement engine. These components include:

Clutch and flywheel (models with manual transmission)
Driveplate (models with automatic transmission)
Ignition system components
Emissions-related components
Engine mounts and mount brackets
Engine rear cover (spacer plate between flywheel/driveplate and engine block)

Fuel injection components
Intake/exhaust manifolds
Oil filter
Spark plug wires and spark plugs
Thermostat and housing assembly
Water pump

➡**Note: When removing the external components from the engine, pay close attention to details that may be helpful or important during installation. Note the installed position of gaskets, seals, spacers, pins, brackets, washers, bolts and other small items.**

4 If you're going to obtain a short block (assembled engine block, crankshaft, pistons and connecting rods), then remove the timing belt, cylinder head, oil pan, oil pump pick-up tube, oil pump and water pump from your engine so that you can turn in your old short block to the rebuilder as a core. See *Engine rebuilding alternatives* for additional information regarding the different possibilities to be considered.

9 Pistons and connecting rods - removal and installation

REMOVAL

◆ **Refer to illustrations 9.1, 9.3, 9.4 and 9.6**

➡**Note: Prior to removing the piston/connecting rod assemblies, remove the cylinder head and oil pan (see Chapter 2A or 2B).**

1 Use your fingernail to feel if a ridge has formed at the upper limit of ring travel (about 1/4-inch down from the top of each cylinder). If carbon deposits or cylinder wear have produced ridges, they must be completely removed with a special tool (see illustration). Follow the manufacturer's instructions provided with the tool. Failure to remove the ridges before attempting to remove the piston/connecting rod assemblies may result in piston breakage.

2 After the cylinder ridges have been removed, turn the engine so the crankshaft is facing up.

3 Before the main bearing cap assembly and connecting rods are removed, check the connecting rod endplay with feeler gauges. Slide them between the first connecting rod and the crankshaft throw until the play is removed (see illustration). Repeat this procedure for each connecting rod. The endplay is equal to the thickness of the feeler gauge(s). Check with an automotive machine shop for the endplay service limit. If the play exceeds the service limit, new connecting rods will be required.

9.1 Before you try to remove the pistons, use a ridge reamer to remove the raised material (ridge) from the top of the cylinders

If new rods (or a new crankshaft) are installed, the endplay may fall under the minimum allowable clearance. If it does, the rods will have to be machined to restore it. If necessary, consult an automotive machine shop for advice.

4 Check the connecting rods and caps for identification marks (see illustration). If they aren't plainly marked, use a small center-

9.3 Checking the connecting rod endplay (side clearance)

9.4 If the connecting rods and caps are not marked, use a center punch or numbered impression stamps to mark the caps to the rods by cylinder number (for example, this would be the No. 4 connecting rod)

punch to make the appropriate number of indentations on each rod and cap (1, 2, 3, etc., depending on the cylinder they're associated with).

5　Loosen each of the connecting rod cap nuts or bolts 1/2-turn at a time until they can be removed by hand. Remove the number one connecting rod cap and bearing insert. Don't drop the bearing insert out of the cap.

6　Slip a short length of plastic or rubber hose over each connecting rod bolt to protect the crankshaft journal and cylinder wall as the rod is removed (see illustration).

7　Remove the bearing insert and push the connecting rod/piston assembly out through the top of the engine. Use a wooden or plastic hammer handle to push on the upper bearing surface in the connecting rod. If resistance is felt, double-check to make sure that all of the ridge was removed from the cylinder.

8　Repeat the procedure for the remaining cylinders.

➠**Note: If the connecting rod caps are secured by bolts (instead of nuts), discard the old rod cap bolts. Use new bolts when reassembling the engine.**

9　After removal, reassemble the connecting rod caps and bearing inserts in their respective connecting rods and install the cap bolts finger tight. Leaving the old bearing inserts in place until reassembly will help prevent the connecting rod bearing surfaces from being accidentally nicked or gouged.

10　The pistons and connecting rods are now ready for inspection and overhaul at an automotive machine shop.

PISTON RING INSTALLATION

▶ **Refer to illustrations 9.13, 9.14, 9.15, 9.19a, 9.19b, 9.21 and 9.22**

11　Before installing the new piston rings, the ring end gaps must be checked. It's assumed that the piston ring side clearance has been checked and verified correct.

12　Lay out the piston/connecting rod assemblies and the new ring sets so the ring sets will be matched with the same piston and cylinder during the end gap measurement and engine assembly.

13　Insert the top (number one) ring into the first cylinder and square it up with the cylinder walls by pushing it in with the top of the piston (see illustration). The ring should be near the bottom of the cylinder, at the lower limit of ring travel.

14　To measure the end gap, slip feeler gauges between the ends of the ring until a gauge equal to the gap width is found (see illustration). The feeler gauge should slide between the ring ends with a slight amount of

9.6 Push a short section of plastic or rubber hose over the connecting rod bolts to prevent damage to the crankshaft journals during piston/rod removal

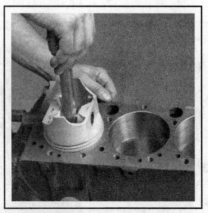

9.13 Install the piston ring into the cylinder then push it down into position using a piston so the ring will be square in the cylinder

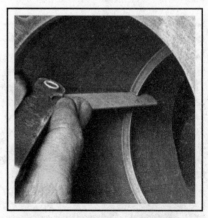

9.14 With the ring square in the cylinder, measure the ring end gap with a feeler gauge

9.15 If the ring end gap is too small, clamp a file in a vise as shown and file the piston ring ends - be sure to remove all raised material

9.19a Installing the spacer/expander in the oil ring groove

9.19b DO NOT use a piston ring installation tool when installing the oil control side rails

drag. Check with an automotive machine shop for the correct end gap for your engine. If the gap is larger or smaller than specified, double-check to make sure you have the correct rings before proceeding.

15 If the gap is too small, it must be enlarged or the ring ends may come in contact with each other during engine operation, which can cause serious damage to the engine. The end gap can be increased by filing the ring ends very carefully with a fine file. Mount the file in a vise equipped with soft jaws, slip the ring over the file with the ends contacting the file face and slowly move the ring to remove material from the ends. When performing this operation, file only by pushing the ring from the outside end of the file towards the vise (see illustration).

16 Excess end gap isn't critical unless it's greater than approximately 0.040-inch. Again, double-check to make sure you have the correct ring type and that you are referencing the correct section and category of specifications.

17 Repeat the procedure for each ring that will be installed in the first cylinder and for each ring in the remaining cylinders. Remember to keep rings, pistons and cylinders matched up.

18 Once the ring end gaps have been checked/corrected, the rings can be installed on the pistons.

19 The oil control ring (lowest one on the piston) is usually installed first. It's composed of three separate components. Slip the spacer/expander into the groove (see illustration). If an anti-rotation tang is used, make sure it's inserted into the drilled hole in the ring groove.

Next, install the upper side rail in the same manner (see illustration). Don't use a piston ring installation tool on the oil ring side rails, as they may be damaged. Instead, place one end of the side rail into the groove between the spacer/expander and the ring land, hold it firmly in place and slide a finger around the piston while pushing the rail into the groove. Finally, install the lower side rail.

20 After the three oil ring components have been installed, check to make sure that both the upper and lower side rails can be rotated smoothly inside the ring grooves.

21 The number two (middle) ring is installed next. It's usually stamped with a mark which must face up (see illustration), toward the top of the piston. Do not mix up the top and middle rings, as they have different cross-sections.

➡**Note: Always follow the instructions printed on the ring package or box - different manufacturers may require different approaches.**

22 Use a piston ring installation tool and make sure the identification mark is facing the top of the piston, then slip the ring into the middle groove on the piston (see illustration). Don't expand the ring any more than necessary to slide it over the piston.

23 Install the number one (top) ring in the same manner. Make sure the mark is facing up. Be careful not to confuse the number one and number two rings (see illustration 9.21).

24 Repeat the procedure for the remaining pistons and rings.

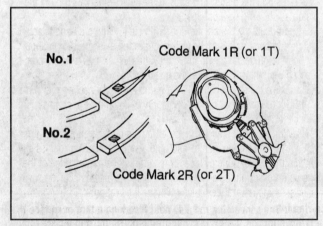

9.21 The number 1 piston ring (top) is marked 1R and the number 2 piston ring (second) is marked 2R. Install the piston ring with the mark UP

9.22 Use a piston ring installation tool to install the number 2 and the number 1 (top) rings - be sure the directional mark on the piston ring(s) is facing toward the top of the piston

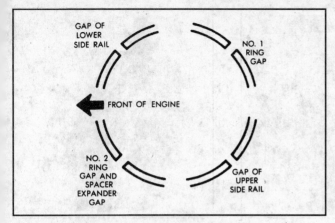

9.30 Position the piston ring end gaps as shown

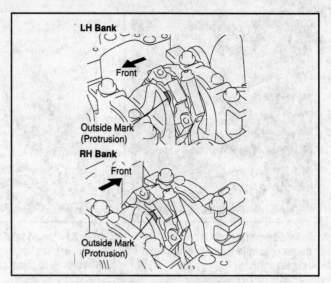

9.33b On V8 models, the left bank connecting rod marks face front and the right bank connecting rod marks face the rear of the engine

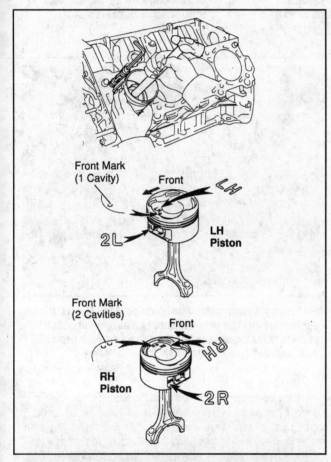

9.33a The pistons on the V8 models are designed specifically for the left bank and the right bank. The left bank pistons have a single cavity mark that faces forward while the right bank pistons have two cavity marks that face forward

INSTALLATION

25 Before installing the piston/connecting rod assemblies, the cylinder walls must be perfectly clean, the top edge of each cylinder bore must be chamfered, and the crankshaft must be in place.

26 Remove the cap from the end of the number one connecting rod (refer to the marks made during removal). Remove the original bearing inserts and wipe the bearing surfaces of the connecting rod and cap with a clean, lint-free cloth. They must be kept spotlessly clean.

Connecting rod bearing oil clearance check

▶ Refer to illustrations 9.30, 9.33a, 9.33b, 9.35, 9.37, 9.38 and 9.41

27 Clean the back side of the new upper bearing insert, then lay it in place in the connecting rod. Make sure the tab on the bearing fits into the recess in the rod. Don't hammer the bearing insert into place and be very careful not to nick or gouge the bearing face. Don't lubricate the bearing at this time.

28 Clean the back side of the other bearing insert and install it in the rod cap. Again, make sure the tab on the bearing fits into the recess in the cap, and don't apply any lubricant. It's critically important that the mating surfaces of the bearing and connecting rod are perfectly clean and oil free when they're assembled.

29 Install a short section of plastic or rubber hose over the connecting rod bolts to avoid damaging the cylinder wall or crankshaft journal (see illustration 9.6).

30 Position the piston ring gaps at 90-degree intervals around the piston as shown (see illustration).

31 Lubricate the piston and rings with clean engine oil and attach a piston ring compressor to the piston. Leave the skirt protruding about 1/4-inch to guide the piston into the cylinder. The rings must be compressed until they're flush with the piston.

32 Rotate the crankshaft until the number one connecting rod journal is at BDC (bottom dead center) and apply a liberal coat of engine oil to the cylinder walls.

33 With the mark (cavity) on top of the piston facing the front (timing belt end) of the engine, gently insert the piston/connecting rod assembly into the number one cylinder bore and rest the bottom edge of the ring compressor on the engine block. Install the pistons with the cavity mark(s) facing toward the timing belt (see illustration).

➡Note: The connecting rod also has a mark on it that must face the correct direction. On V6 models, the marks on the connecting rods face the front (timing belt) of the engine while on V8 models, the left bank connecting rod marks face front and the right bank connecting rod marks face the rear of the engine (see illustration).

9.35 Use a plastic or wooden hammer handle to push the piston into the cylinder

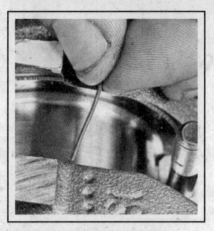

9.37 Place Plastigage on each connecting rod bearing journal parallel to the crankshaft centerline

9.38 Install the connecting rod cap making sure the cap and rod identification numbers match

34 Tap the top edge of the ring compressor to make sure it's contacting the block around its entire circumference.

35 Gently tap on the top of the piston with the end of a wooden or plastic hammer handle (see illustration) while guiding the end of the connecting rod into place on the crankshaft journal. The piston rings may try to pop out of the ring compressor just before entering the cylinder bore, so keep some downward pressure on the ring compressor. Work slowly, and if any resistance is felt as the piston enters the cylinder, stop immediately. Find out what's hanging up and fix it before proceeding. Do not, for any reason, force the piston into the cylinder - you might break a ring and/or the piston.

36 Once the piston/connecting rod assembly is installed, the connecting rod bearing oil clearance must be checked before the rod cap is permanently installed.

37 Cut a piece of the appropriate size Plastigage slightly shorter than the width of the connecting rod bearing and lay it in place on the number one connecting rod journal, parallel with the journal axis (see illustration).

38 Clean the connecting rod cap bearing face and install the rod cap. Make sure the mating mark on the cap is on the same side as the mark on the connecting rod (see illustration).

39 Install the old rod bolts or nuts, at this time, and tighten them to the torque listed in this Chapter's Specifications, working up to it in three steps.

➡Note: Use a thin-wall socket to avoid erroneous torque readings that can result if the socket is wedged between the rod cap and the bolt or nut. If the socket tends to wedge itself between the fastener and the cap, lift up on it slightly until it no longer contacts the cap. DO NOT rotate the crankshaft at any time during this operation.

40 Remove the fasteners and detach the rod cap, being very careful not to disturb the Plastigage. Discard the cap bolts at this time as they cannot be reused.

➡Note: You MUST use new connecting rod bolts.

41 Compare the width of the crushed Plastigage to the scale printed on the Plastigage envelope to obtain the oil clearance (see illustration). The connecting rod oil clearance is usually about 0.001 to 0.002 inch. Consult an automotive machine shop for the clearance specified for the rod bearings on your engine.

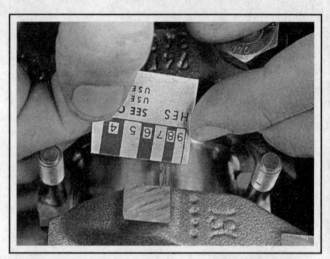

9.41 Use the scale on the Plastigage package to determine the bearing oil clearance - be sure to measure the widest part of the Plastigage and use the correct scale; it comes with both standard and metric scales

42 If the clearance is not as specified, the bearing inserts may be the wrong size (which means different ones will be required). Before deciding that different inserts are needed, make sure that no dirt or oil was between the bearing inserts and the connecting rod or cap when the clearance was measured. Also, recheck the journal diameter. If the Plastigage was wider at one end than the other, the journal may be tapered. If the clearance still exceeds the limit specified, the bearing will have to be replaced with an undersize bearing.

✳✳ CAUTION:

When installing a new crankshaft always use a standard size bearing.

Final installation

43 Carefully scrape all traces of the Plastigage material off the rod

journal and/or bearing face. Be very careful not to scratch the bearing - use your fingernail or the edge of a plastic card.

44 Make sure the bearing faces are perfectly clean, then apply a uniform layer of clean moly-base grease or engine assembly lube to both of them. You'll have to push the piston into the cylinder to expose the face of the bearing insert in the connecting rod.

✳✳ CAUTION:

If you're working on a V8 engine, carefully inspect the connecting rod bolts for distortion or signs of stretching. Replace the bolts if any undesirable conditions exist.

45 Slide the connecting rod back into place on the journal, install the rod cap, install the nuts or bolts and tighten them to the torque listed in this Chapter's Specifications. Again, work up to the torque in three steps.

46 Repeat the entire procedure for the remaining pistons/connecting rods.

47 The important points to remember are:
a) *Keep the back sides of the bearing inserts and the insides of the connecting rods and caps perfectly clean when assembling them.*
b) *Make sure you have the correct piston/rod assembly for each cylinder.*
c) *The mark on the piston must face the front (timing belt end) of the engine.*
d) *Lubricate the cylinder walls liberally with clean oil.*
e) *Lubricate the bearing faces when installing the rod caps after the oil clearance has been checked.*

48 After all the piston/connecting rod assemblies have been correctly installed, rotate the crankshaft a number of times by hand to check for any obvious binding.

49 As a final step, check the connecting rod endplay again. If it was correct before disassembly and the original crankshaft and rods were reinstalled, it should still be correct. If new rods or a new crankshaft were installed, the endplay may be inadequate. If so, the rods will have to be removed and taken to an automotive machine shop for resizing.

10 Crankshaft - removal and installation

REMOVAL

▸ **Refer to illustrations 10.1 and 10.3**

➡**Note: The crankshaft can be removed only after the engine has been removed from the vehicle. It's assumed that the flywheel or driveplate, crankshaft pulley, timing belt, oil pan, oil pump body, oil filter and piston/connecting rod assemblies have already been removed. The rear main oil seal retainer must be unbolted and separated from the block before proceeding with crankshaft removal.**

1 Before the crankshaft is removed, measure the endplay. Mount a dial indicator with the indicator in line with the crankshaft and touching the end of the crankshaft (see illustration).

2 Pry the crankshaft all the way to the rear and zero the dial indicator. Next, pry the crankshaft to the front as far as possible and check the reading on the dial indicator. The distance traveled is the endplay. A typical crankshaft endplay will fall between 0.003 to 0.010-inch. If it's greater than that, check the crankshaft thrust surfaces for wear after its removed. If no wear is evident, new main bearings should correct the endplay.

3 If a dial indicator isn't available, feeler gauges can be used. Gently pry the crankshaft all the way to the front of the engine. Slip feeler gauges between the crankshaft and the front face of the thrust bearing or washer to determine the clearance (see illustration).

4 Loosen the main bearing cap bolts 1/4-turn at a time each, until they can be removed by hand.

➡**Note: If you're working on a 2005 V6 engine, first remove the main bearing cap side bolts in the order opposite that of the tightening sequence (see illustration 10.19c)**

5 Gently tap the main bearing cap(s) (2005 V6 and all V8 models) or cap assembly (2004 and earlier V6 models) with a soft-face hammer. Pull the main bearing cap(s) straight up and off the cylinder block. Try not to drop the bearing inserts if they come out with the assembly.

6 Carefully lift the crankshaft out of the engine. It may be a good idea to have an assistant available, since the crankshaft is quite heavy and awkward to handle. With the bearing inserts in place inside the engine block and main bearing caps, reinstall the main bearing cap assembly onto the engine block and tighten the bolts finger tight. Make sure you install the main bearing cap(s) with the arrow facing the front end of the engine.

10.1 Checking crankshaft endplay with a dial indicator

10.3 Checking crankshaft endplay with feeler gauges at the thrust bearing journal

ENGINE BEARING ANALYSIS

Debris

Babbitt bearing embedded with debris from machinings

Microscopic detail of debris

Microscopic detail of gouges

Overplated copper alloy bearing gouged by cast iron debris

Aluminum bearing embedded with glass beads

Microscopic detail of glass beads

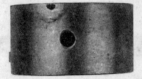

Damaged lining caused by dirt left on the bearing back

Misassembly

Result of a lower half assembled as an upper - blocking the oil flow

Excessive oil clearance is indicated by a short contact arc

Polished and oil-stained backs are a result of a poor fit in the housing bore

Result of a wrong, reversed, or shifted cap

Overloading

Damage from excessive idling which resulted in an oil film unable to support the load imposed

Damaged upper connecting rod bearings caused by engine lugging; the lower main bearings (not shown) were similarly affected

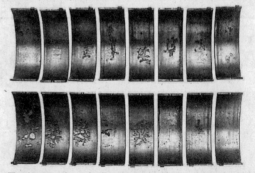

The damage shown in these upper and lower connecting rod bearings was caused by engine operation at a higher-than-rated speed under load

Misalignment

A warped crankshaft caused this pattern of severe wear in the center, diminishing toward the ends

A poorly finished crankshaft caused the equally spaced scoring shown

A tapered housing bore caused the damage along one edge of this pair

A bent connecting rod led to the damage in the "V" pattern

Lubrication

Result of dry start: The bearings on the left, farthest from the oil pump, show more damage

Result of a low oil supply or oil starvation

Severe wear as a result of inadequate oil clearance

Corrosion

Microscopic detail of corrosion

Corrosion is an acid attack on the bearing lining generally caused by inadequate maintenance, extremely hot or cold operation, or inferior oils or fuels

Microscopic detail of cavitation

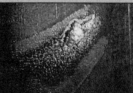

Example of cavitation - a surface erosion caused by pressure changes in the oil film

Damage from excessive thrust or insufficient axial clearance

Bearing affected by oil dilution caused by excessive blow-by or a rich mixture

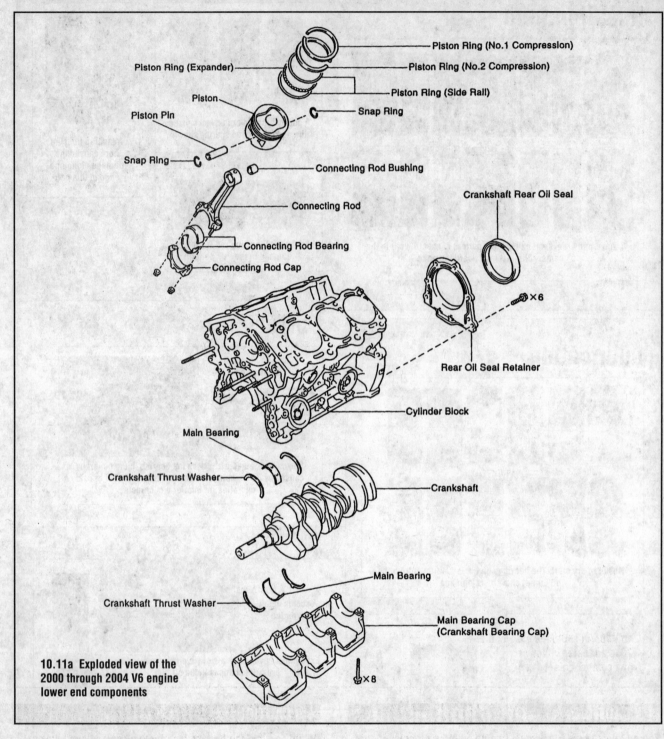

Piston Ring (Expander)

Piston Ring (No.1 Compression)

Piston Ring (No.2 Compression)

Piston Ring (Side Rail)

Piston

Piston Pin

Snap Ring

Snap Ring

Connecting Rod Bushing

Crankshaft Rear Oil Seal

Connecting Rod

Connecting Rod Bearing

Connecting Rod Cap

Rear Oil Seal Retainer

×6

Cylinder Block

Main Bearing

Crankshaft Thrust Washer

Crankshaft

Main Bearing

Crankshaft Thrust Washer

Main Bearing Cap
(Crankshaft Bearing Cap)

**10.11a Exploded view of the
2000 through 2004 V6 engine
lower end components**

×8

INSTALLATION

7   Crankshaft installation is the first step in engine reassembly. It's assumed at this point that the engine block and crankshaft have been cleaned, inspected and repaired or reconditioned.

8   Position the engine block with the bottom facing up.

9   Remove the mounting bolts and lift off the main bearing cap(s).

10   If they're still in place, remove the original bearing inserts from the block and from the main bearing cap(s). Wipe the bearing surfaces of the block and main bearing cap(s) with a clean, lint-free cloth. They must be kept spotlessly clean. This is critical for determining the correct bearing oil clearance.

MAIN BEARING OIL CLEARANCE CHECK

▸ **Refer to illustrations 10.11a, 10.11b, 10.11c, 10.17, 10.19a, 10.19b, 10.19c, 10.19d and 10.21**

11   Without mixing them up, clean the back sides of the new upper main bearing inserts (with grooves and oil holes) and lay one in each main bearing saddle in the block. Each upper bearing has an oil groove and oil hole in it.

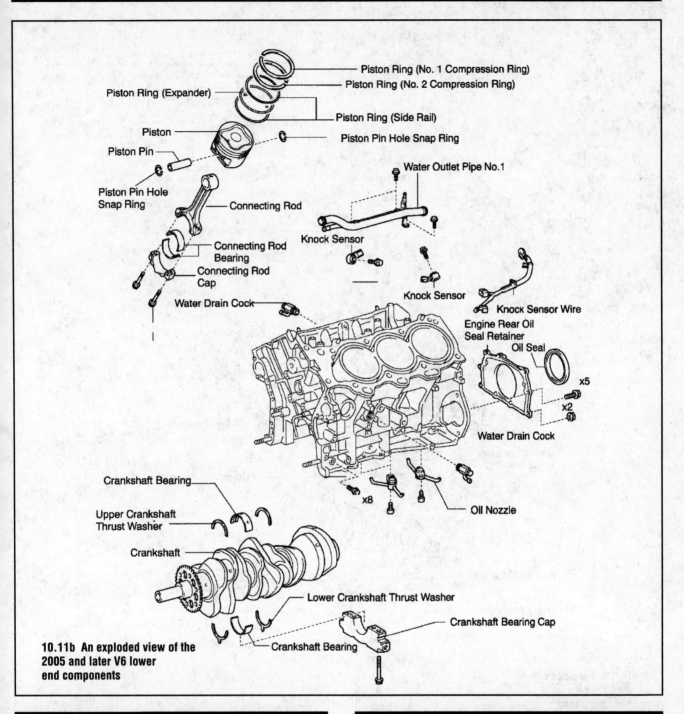

10.11b An exploded view of the 2005 and later V6 lower end components

---

✻✻ CAUTION:

The oil holes in the block must line up with the oil holes in the upper bearing inserts. The thrust washer on the V6 models is located on the number 2 crankshaft journal (see illustration).

✻✻ CAUTION:

Do not hammer the bearing insert into place and don't nick or gouge the bearing faces. DO NOT apply any lubrication at this time.

The thrust washer on the V8 models is located on the number 3 crankshaft journal (see illustration). Install the thrust washers with the grooved side facing out. Install the thrust washers so that one set is located in the block and the other set is with the main bearing cap assembly. Clean the back sides of the lower main bearing inserts (without grooves) and lay them in the corresponding location in the main bearing cap assembly. Make sure the tab on the bearing insert fits into the recess in the block or main bearing cap assembly.

12 Clean the faces of the bearing inserts in the block and the crankshaft main bearing journals with a clean, lint-free cloth.

13 Check or clean the oil holes in the crankshaft, as any dirt here can go only one way - straight through the new bearings.

14 Once you're certain the crankshaft is clean, carefully lay it in position in the cylinder block.

15 Before the crankshaft can be permanently installed, the main bearing oil clearance must be checked.

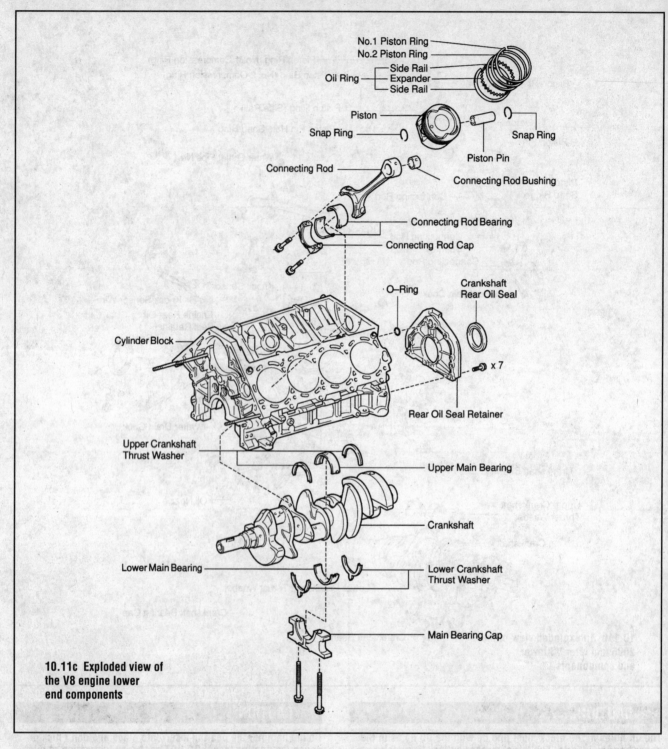

No.1 Piston Ring
No.2 Piston Ring
Side Rail
Oil Ring — Expander
Side Rail

Piston

Snap Ring

Snap Ring

Piston Pin

Connecting Rod

Connecting Rod Bushing

Connecting Rod Bearing

Connecting Rod Cap

O–Ring

Crankshaft Rear Oil Seal

Cylinder Block

x 7

Rear Oil Seal Retainer

Upper Crankshaft Thrust Washer

Upper Main Bearing

Crankshaft

Lower Main Bearing

Lower Crankshaft Thrust Washer

Main Bearing Cap

10.11c Exploded view of the V8 engine lower end components

16 Cut several strips of the appropriate size of Plastigage (they must be slightly shorter than the width of the main bearing journal).

17 Place one piece on each crankshaft main bearing journal, parallel with the journal axis (see illustration).

18 Clean the faces of the bearing inserts in the main bearing cap assembly. Hold the bearing inserts in place and install the assembly (or caps) onto the crankshaft and cylinder block. DO NOT disturb the Plastigage. Make sure you install the main bearing cap assembly with the arrow facing the front of the engine.

19 Apply clean engine oil to all bolt threads prior to installation, then install all bolts finger-tight. Tighten main bearing cap bolts in the sequence shown (see illustrations) progressing in two steps, to the torque listed in this Chapter's Specifications. DO NOT rotate the crankshaft at any time during this operation.

➡**Note: If you're working on a 2005 V6 engine, tighten the main bearing cap bolts first, then the main bearing cap side bolts.**

20 Remove the bolts in the reverse order of the tightening sequence and carefully lift the main bearing cap assembly straight up and off the block. Do not disturb the Plastigage or rotate the crankshaft. If the main bearing cap assembly is difficult to remove, tap it gently from side-to-side with a soft-face hammer to loosen it.

10.17 Place the Plastigage (arrow) onto the crankshaft bearing journal as shown

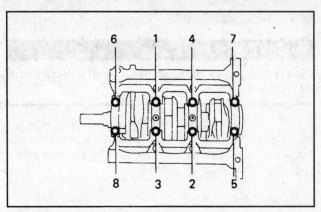

10.19a Main bearing cap bolt tightening sequence (2004 and earlier V6 models)

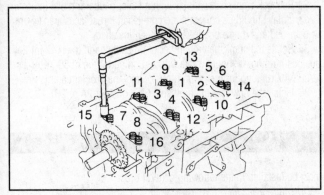

10.19b Main bearing cap bolt tightening sequence (2005 and later V6 engine)

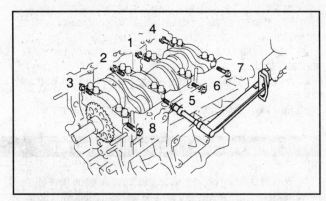

10.19c Main bearing cap side bolt tightening sequence (2005 and later V6 engine)

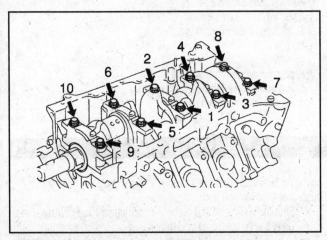

10.19d Main bearing cap bolt tightening sequence (V8 models)

10.21 Use the scale on the Plastigage package to determine the bearing oil clearance - be sure to measure the widest part of the Plastigage and use the correct scale; it comes with both standard and metric scales

21 Compare the width of the crushed Plastigage on each journal to the scale printed on the Plastigage envelope to determine the main bearing oil clearance (see illustration). A typical main bearing oil clearance should fall between 0.0015 to 0.0023-inch. Check with an automotive machine shop for the clearance specified for your engine.

22 If the clearance is not as specified, the bearing inserts may be the wrong size (which means different ones will be required). Before deciding if different inserts are needed, make sure that no dirt or oil was between the bearing inserts and the cap assembly or block when the clearance was measured. If the Plastigage was wider at one end than the other, the crankshaft journal may be tapered. If the clearance still exceeds the limit specified, the bearing insert(s) will have to be replaced

with an undersize bearing insert(s).

When installing a new crankshaft always install a standard bearing insert set.

23 Carefully scrape all traces of the Plastigage material off the main bearing journals and/or the bearing insert faces. Be sure to remove all residue from the oil holes. Use your fingernail or the edge of a plastic card - don't nick or scratch the bearing faces.

FINAL INSTALLATION

24 Carefully lift the crankshaft out of the cylinder block.

25 Clean the bearing insert faces in the cylinder block, then apply a thin, uniform layer of moly-base grease or engine assembly lube to each of the bearing surfaces. Be sure to coat the thrust faces as well as the journal face of the thrust bearing.

26 Make sure the crankshaft journals are clean, then lay the crankshaft back in place in the cylinder block.

27 Clean the bearing insert faces and then apply the same lubricant to them.

28 Hold the bearing inserts in place and install the main bearing caps on the crankshaft and cylinder block. Tap the bearing caps into place with a brass punch or a soft-face hammer.

29 Using NEW main bearing cap bolts, apply clean engine oil to the bolt threads, wipe off any excess oil and then install the bolts finger-tight.

30 Tighten the main bearing cap bolts in the indicated sequence (see illustration 10.19a, 10.19b, 10.19c or 10.19d), to 10 or 12 foot-pounds.

31 Push the crankshaft forward using a screwdriver or prybar to seat the thrust bearing. Once the crankshaft is pushed fully forward to seat the thrust bearing, leave the screwdriver in position so that pressure stays on the crankshaft until after all main bearing cap bolts have been tightened.

32 Tighten the main bearing cap bolts in two steps in the indicated sequence (see illustration 10.19a, 10.19b, 10.19c or 10.19d) and to the torque and angle listed in this Chapter's Specifications.

33 If you're working on a 2005 V6 engine, install and tighten the main bearing cap side bolts in the correct sequence (see illustration).

34 Recheck crankshaft endplay with a feeler gauge or a dial indicator. The endplay should be correct if the crankshaft thrust faces aren't worn or damaged and if new bearings have been installed.

35 Rotate the crankshaft a number of times by hand to check for any obvious binding. It should rotate with a running torque of 50 in-lbs or less. If the running torque is too high, correct the problem at this time.

36 Install the new rear main oil seal (see Chapter 2A or 2B).

11 Engine overhaul - reassembly sequence

1 Before beginning engine reassembly, make sure you have all the necessary new parts, gaskets and seals as well as the following items on hand:

Common hand tools
A 1/2-inch drive torque wrench
New engine oil
Gasket sealant
Thread locking compound

2 If you obtained a short block it will be necessary to install the cylinder head, the oil pump and pick-up tube, the oil pan, the water pump, the timing belt and timing cover, and the valve cover (see Chapter 2A or 2B). In order to save time and avoid problems, the external components must be installed in the following general order:

Thermostat and housing cover
Water pump
Intake and exhaust manifolds
Fuel injection components
Emission control components
Spark plug wires and spark plugs
Ignition coils
Oil filter
Engine mounts and mount brackets
Clutch and flywheel (manual transmission)
Driveplate (automatic transmission)

12 Initial start-up and break-in after overhaul

Have a fire extinguisher handy when starting the engine for the first time.

1 Once the engine has been installed in the vehicle, double-check the engine oil and coolant levels.

2 With the spark plugs out of the engine and the ignition system and fuel pump disabled, crank the engine until oil pressure registers on the gauge or the light goes out.

3 Install the spark plugs, hook up the plug wires and/or ignition coils and restore the ignition system and fuel pump functions.

4 Start the engine. It may take a few moments for the fuel system to build up pressure, but the engine should start without a great deal of effort.

5 After the engine starts, it should be allowed to warm up to normal operating temperature. While the engine is warming up, make a thorough check for fuel, oil and coolant leaks.

6 Shut the engine off and recheck the engine oil and coolant levels.

7 Drive the vehicle to an area with minimum traffic, accelerate from 30 to 50 mph, then allow the vehicle to slow to 30 mph with the throttle closed. Repeat the procedure 10 or 12 times. This will load the piston rings and cause them to seat properly against the cylinder walls. Check again for oil and coolant leaks.

8 Drive the vehicle gently for the first 500 miles (no sustained high speeds) and keep a constant check on the oil level. It is not unusual for an engine to use oil during the break-in period.

9 At approximately 500 to 600 miles, change the oil and filter.

10 For the next few hundred miles, drive the vehicle normally. Do not pamper it or abuse it.

11 After 2000 miles, change the oil and filter again and consider the engine broken in.

Specifications

General

Displacement	
V6 models	
2000 through 2004	207 cubic inches
2005 and later	241 cubic inches
V8 models	286 cubic inches
Bore and stroke	
V6 models	
2000 through 2004	3.68 x 3.23 inches
2005 and later	3.70 x 3.74 inches
V8 models	3.70 x 3.31 inches
Cylinder compression pressure	Lowest cylinder must be within 75 percent of highest cylinder
Oil pressure	
V6 models	
2000 through 2004	
At curb idle	4 to 6 psi
At 3000 rpm	36 to 75 psi
2005 and later	
At curb idle	4.3 psi minimum
At 3000 rpm	33 to 85 psi
V8 models	
At curb idle	4 to 5 psi
At 3000 rpm	43 to 85 psi

Torque specifications	Ft-lbs (unless otherwise indicated)
Driveplate-to-torque converter bolts	35
Connecting rod bearing cap bolts (all)	
Step 1	18
Step 2	90 degrees
Main bearing cap bolts (see illustration 10.19a for tightening sequence)	
V6 models	
Step 1	45
Step 2	90 degrees
V8 models (see illustration 10.19b for tightening sequence)	
Step 1	20
Step 2	90 degrees

GLOSSARY

B

Backlash - The amount of play between two parts. Usually refers to how much one gear can be moved back and forth without moving the gear with which it's meshed.

Bearing Caps - The caps held in place by nuts or bolts which, in turn, hold the bearing surface. This space is for lubricating oil to enter.

Bearing clearance - The amount of space left between shaft and bearing surface. This space is for lubricating oil to enter.

Bearing crush - The additional height which is purposely manufactured into each bearing half to ensure complete contact of the bearing back with the housing bore when the engine is assembled.

Bearing knock - The noise created by movement of a part in a loose or worn bearing.

Blueprinting - Dismantling an engine and reassembling it to EXACT specifications.

Bore - An engine cylinder, or any cylindrical hole; also used to describe the process of enlarging or accurately refinishing a hole with a cutting tool, as to bore an engine cylinder. The bore size is the diameter of the hole.

Boring - Renewing the cylinders by cutting them out to a specified size. A boring bar is used to make the cut.

Bottom end - A term which refers collectively to the engine block, crankshaft, main bearings and the big ends of the connecting rods.

Break-in - The period of operation between installation of new or rebuilt parts and time in which parts are worn to the correct fit. Driving at reduced and varying speed for a specified mileage to permit parts to wear to the correct fit.

Bushing - A one-piece sleeve placed in a bore to serve as a bearing surface for shaft, piston pin, etc. Usually replaceable.

C

Camshaft - The shaft in the engine, on which a series of lobes are located for operating the valve mechanisms. The camshaft is driven by gears or sprockets and a timing chain. Usually referred to simply as the cam.

Carbon - Hard, or soft, black deposits found in combustion chamber, on plugs, under rings, on and under valve heads.

Cast iron - An alloy of iron and more than two percent carbon, used for engine blocks and heads because it's relatively inexpensive and easy to mold into complex shapes.

Chamfer - To bevel across (or a bevel on) the sharp edge of an object.

Chase - To repair damaged threads with a tap or die.

Combustion chamber - The space between the piston and the cylinder head, with the piston at top dead center, in which air-fuel mixture is burned.

Compression ratio - The relationship between cylinder volume (clearance volume) when the piston is at top dead center and cylinder volume when the piston is at bottom dead center.

Connecting rod - The rod that connects the crank on the crankshaft with the piston. Sometimes called a con rod.

Connecting rod cap - The part of the connecting rod assembly that attaches the rod to the crankpin.

Core plug - Soft metal plug used to plug the casting holes for the coolant passages in the block.

Crankcase - The lower part of the engine in which the crankshaft rotates; includes the lower section of the cylinder block and the oil pan.

Crank kit - A reground or reconditioned crankshaft and new main and connecting rod bearings.

Crankpin - The part of a crankshaft to which a connecting rod is attached.

Crankshaft - The main rotating member, or shaft, running the length of the crankcase, with offset throws to which the connecting rods are attached; changes the reciprocating motion of the pistons into rotating motion.

Cylinder sleeve - A replaceable sleeve, or liner, pressed into the cylinder block to form the cylinder bore.

D

Deburring - Removing the burrs (rough edges or areas) from a bearing.

Deglazer - A tool, rotated by an electric motor, used to remove glaze from cylinder walls so a new set of rings will seat.

E

Endplay - The amount of lengthwise movement between two parts. As applied to a crankshaft, the distance that the crankshaft can move forward and back in the cylinder block.

F

Face - A machinist's term that refers to removing metal from the end of a shaft or the face of a larger part, such as a flywheel.

Fatigue - A breakdown of material through a large number of loading and unloading cycles. The first signs are cracks followed shortly by breaks.

Feeler gauge - A thin strip of hardened steel, ground to an exact thickness, used to check clearances between parts.

Free height - The unloaded length or height of a spring.

Freeplay - The looseness in a linkage, or an assembly of parts, between the initial application of force and actual movement. Usually perceived as slop or slight delay.

Freeze plug - See Core plug.

G

Gallery - A large passage in the block that forms a reservoir for engine oil pressure.

Glaze - The very smooth, glassy finish that develops on cylinder walls while an engine is in service.

H

Heli-Coil - A rethreading device used when threads are worn or damaged. The device is installed in a retapped hole to reduce the thread size to the original size.

I

Installed height - The spring's measured length or height, as installed on the cylinder head. Installed height is measured from the spring seat to the underside of the spring retainer.

J

Journal - The surface of a rotating shaft which turns in a bearing.

K

Keeper - The split lock that holds the valve spring retainer in position on the valve stem.

Key - A small piece of metal inserted into matching grooves machined into two parts fitted together - such as a gear pressed onto a shaft - which prevents slippage between the two parts.

Knock - The heavy metallic engine sound, produced in the combustion chamber as a result of abnormal combustion - usually detonation. Knock is usually caused by a loose or worn bearing. Also referred to as detonation, pinging and spark knock. Connecting rod or main bearing knocks are created by too much oil clearance or insufficient lubrication.

L

Lands - The portions of metal between the piston ring grooves.

Lapping the valves - Grinding a valve face and its seat together with lapping compound.

Lash - The amount of free motion in a gear train, between gears, or in a mechanical assembly, that occurs before movement can begin. Usually refers to the lash in a valve train.

Lifter - The part that rides against the cam to transfer motion to the rest of the valve train.

M

Machining - The process of using a machine to remove metal from a metal part.

Main bearings - The plain, or babbitt, bearings that support the crankshaft.

Main bearing caps - The cast iron caps, bolted to the bottom of the block, that support the main bearings.

O

O.D. - Outside diameter.

Oil gallery - A pipe or drilled passageway in the engine used to carry engine oil from one area to another.

Oil ring - The lower ring, or rings, of a piston; designed to prevent excessive amounts of oil from working up the cylinder walls and into the combustion chamber. Also called an oil-control ring.

Oil seal - A seal which keeps oil from leaking out of a compartment. Usually refers to a dynamic seal around a rotating shaft or other moving part.

O-ring - A type of sealing ring made of a special rubberlike material; in use, the O-ring is compressed into a groove to provide the sealing action.

Overhaul - To completely disassemble a unit, clean and inspect all parts, reassemble it with the original or new parts and make all adjustments necessary for proper operation.

P

Pilot bearing - A small bearing installed in the center of the flywheel (or the rear end of the crankshaft) to support the front end of the input shaft of the transmission.

Pip mark - A little dot or indentation which indicates the top side of a compression ring.

Piston - The cylindrical part, attached to the connecting rod, that moves up and down in the cylinder as the crankshaft rotates. When the fuel charge is fired, the piston transfers the force of the explosion to the connecting rod, then to the crankshaft.

Piston pin (or wrist pin) - The cylindrical and usually hollow steel pin that passes through the piston. The piston pin fastens the piston to the upper end of the connecting rod.

Piston ring - The split ring fitted to the groove in a piston. The ring contacts the sides of the ring groove and also rubs against the cylinder wall, thus sealing space between piston and wall. There are two types of rings: Compression rings seal the compression pressure in the combustion chamber; oil rings scrape excessive oil off the cylinder wall.

Piston ring groove - The slots or grooves cut in piston heads to hold piston rings in position.

Piston skirt - The portion of the piston below the rings and the piston pin hole.

Plastigage - A thin strip of plastic thread, available in different sizes, used for measuring clearances. For example, a strip of plastigage is laid across a bearing journal and mashed as parts are assembled. Then parts are disassembled and the width of the strip is measured to determine clearance between journal and bearing. Commonly used to measure crankshaft main-bearing and connecting rod bearing clearances.

Press-fit - A tight fit between two parts that requires pressure to force the parts together. Also referred to as drive, or force, fit.

Prussian blue - A blue pigment; in solution, useful in determining the area of contact between two surfaces. Prussian blue is commonly used to determine the width and location of the contact area between the valve face and the valve seat.

R

Race (bearing) - The inner or outer ring that provides a contact surface for balls or rollers in bearing.

Ream - To size, enlarge or smooth a hole by using a round cutting tool with fluted edges.

Ring job - The process of reconditioning the cylinders and installing new rings.

Runout - Wobble. The amount a shaft rotates out-of-true.

S

Saddle - The upper main bearing seat.

Scored - Scratched or grooved, as a cylinder wall may be scored by abrasive particles moved up and down by the piston rings.

Scuffing - A type of wear in which there's a transfer of material between parts moving against each other; shows up as pits or grooves in the mating surfaces.

Seat - The surface upon which another part rests or seats. For example, the valve seat is the matched surface upon which the valve face rests. Also used to refer to wearing into a good fit; for example, piston rings seat after a few miles of driving.

Short block - An engine block complete with crankshaft and piston and, usually, camshaft assemblies.

Static balance - The balance of an object while it's stationary.

Step - The wear on the lower portion of a ring land caused by excessive side and back-clearance. The height of the step indicates the ring's extra side clearance and the length of the step projecting from the back wall of the groove represents the ring's back clearance.

Stroke - The distance the piston moves when traveling from top dead center to bottom dead center, or from bottom dead center to top dead center.

Stud - A metal rod with threads on both ends.

T

Tang - A lip on the end of a plain bearing used to align the bearing during assembly.

Tap - To cut threads in a hole. Also refers to the fluted tool used to cut threads.

Taper - A gradual reduction in the width of a shaft or hole; in an engine cylinder, taper usually takes the form of uneven wear, more pronounced at the top than at the bottom.

Throws - The offset portions of the crankshaft to which the connecting rods are affixed.

Thrust bearing - The main bearing that has thrust faces to prevent excessive endplay, or forward and backward movement of the crankshaft.

Thrust washer - A bronze or hardened steel washer placed between two moving parts. The washer prevents longitudinal movement and provides a bearing surface for thrust surfaces of parts.

Tolerance - The amount of variation permitted from an exact size of measurement. Actual amount from smallest acceptable dimension to largest acceptable dimension.

U

Umbrella - An oil deflector placed near the valve tip to throw oil from the valve stem area.

Undercut - A machined groove below the normal surface.

Undersize bearings - Smaller diameter bearings used with re-ground crankshaft journals.

V

Valve grinding - Refacing a valve in a valve-refacing machine.

Valve train - The valve-operating mechanism of an engine; includes all components from the camshaft to the valve.

Vibration damper - A cylindrical weight attached to the front of the crankshaft to minimize torsional vibration (the twist-untwist actions of the crankshaft caused by the cylinder firing impulses). Also called a harmonic balancer.

W

Water jacket - The spaces around the cylinders, between the inner and outer shells of the cylinder block or head, through which coolant circulates.

Web - A supporting structure across a cavity.

Woodruff key - A key with a radiused backside (viewed from the side).

Notes

Section

Reference to other Chapters

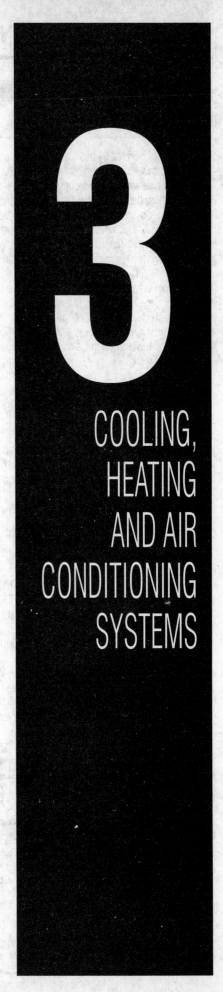

3

COOLING, HEATING AND AIR CONDITIONING SYSTEMS

1 General information

ENGINE COOLING SYSTEM

All vehicles covered by this manual employ a pressurized engine cooling system with thermostatically controlled coolant circulation. An impeller type water pump mounted on the front of the block pumps coolant through the engine. The coolant flows around each cylinder and toward the rear of the engine. Cast-in coolant passages direct coolant around the intake and exhaust ports, near the spark plug areas and in proximity to the exhaust valve guides. The water pump on both engines (V6 and V8 models) is mounted on the block and requires timing belt removal.

A wax-pellet type thermostat is located in the thermostat housing at the radiator end of the engine. During warm up, the closed thermostat prevents coolant from circulating through the radiator. When the engine reaches normal operating temperature, the thermostat opens and allows hot coolant to travel through the radiator, where it is cooled before returning to the engine.

The cooling system is sealed by a pressure-type radiator cap. This raises the boiling point of the coolant, and the higher boiling point of the coolant increases the cooling efficiency of the radiator. If the system pressure exceeds the cap pressure-relief value, the excess pressure in the system forces the spring-loaded valve inside the cap off its seat and allows the coolant to escape through the overflow tube into a coolant overflow reservoir. When the system cools, the excess coolant is automatically drawn from the reservoir back into the radiator.

The coolant reservoir serves as both the point at which fresh coolant is added to the cooling system to maintain the proper fluid level and as a holding tank for overheated coolant.

This type of cooling system is known as a closed design because coolant that escapes past the pressure cap is saved and reused.

TRANSMISSION COOLING SYSTEMS

V6 models with an automatic transmission are equipped with a transmission cooler, located inside the radiator, which cools the transmission fluid. The transmission is connected to the cooler by a pair of hoses: one delivers hot transmission fluid to the radiator and the other brings the cooled fluid back to the transmission. V8 models are also equipped with an auxiliary external cooler, located in front of the air conditioning condenser and the radiator.

For more information on transmission oil coolers, refer to Chapter 7B.

ENGINE OIL COOLING SYSTEM

Besides the engine and transmission cooling systems described above, engine heat is also dissipated through an external oil cooler that's integrated into the lubrication system. The oil cooler helps keep engine and oil temperatures within design limits under extreme load conditions.

The oil cooling system consists of a housing mounted between the oil filter and the filter adapter. Hoses connected to the oil cooler circulate coolant from the radiator through the oil cooler housing.

HEATING SYSTEM

The heating system consists of a blower fan and heater core located within the heater box under the dashboard, the inlet and outlet hoses connecting the heater core to the engine cooling system and the heater/air conditioning control head on the dashboard. Hot engine coolant is circulated through the heater core. When the heater mode is activated, a flap opens to expose the heater box to the passenger compartment. A fan switch on the control head activates the blower motor, which forces air through the core, heating the air.

Some Sequoia models are equipped with a rear heating and cooling system. This separate A/C and heating unit is mounted behind the right quarter trim panel.

AIR CONDITIONING SYSTEM

The air conditioning system consists of a condenser mounted in front of the radiator, an evaporator mounted adjacent to the heater core, a compressor mounted on the engine, a filter-drier which contains a high pressure relief valve and the plumbing connecting all of the above.

A blower fan forces the warmer air of the passenger compartment through the evaporator core (sort of a radiator-in-reverse), transferring the heat from the air to the refrigerant. The liquid refrigerant boils off into low pressure vapor, taking the heat with it when it leaves the evaporator. The compressor keeps refrigerant circulating through the system, pumping the warmed coolant through the condenser where it is cooled and then circulated back to the evaporator.

Some Sequoia models are equipped with a rear heating and cooling system. This separate A/C and heating unit is mounted behind the right quarter panel.

2 Antifreeze - general information

▶ Refer to illustration 2.5

➡Note: Non-toxic antifreeze is now manufactured and available at local auto parts stores, but even this type must be disposed of properly.

The cooling system should be filled with a water/ethylene glycol based antifreeze solution, which will prevent freezing down to at least -20-degrees F (even lower in cold climates). It also provides protection against corrosion and increases the coolant boiling point. The engines in these vehicles have aluminum cylinder heads. The manufacturer recommends that the correct type of coolant be used and strongly urges that coolant types not be mixed.

Drain, flush and refill the cooling system at least every other year (see Chapter 1). The use of antifreeze solutions for periods of longer than two years is likely to cause damage and encourage the formation of rust and scale in the system.

Before adding antifreeze to the system, inspect all hose connections. Antifreeze can leak through very minute openings.

The exact mixture of antifreeze to water, which you should use, depends on the relative weather conditions. The mixture should contain at least 50-percent antifreeze, but should never contain more than 70-percent anti-freeze. Consult the mixture ratio chart on the container before adding coolant.

Hydrometers are available at most auto parts stores to test the coolant (see illustration). Use antifreeze that meets Toyota specifications for engines with aluminum heads.

2.5 An inexpensive hydrometer can be used to test the condition of your coolant

3   Thermostat - check and replacement

✳✳ WARNING:

Do not remove the radiator cap, coolant or thermostat until the engine has cooled completely.

CHECK

1   Before proceeding, check the coolant level and the drivebelt tension (see Chapter 1) and then check the operation of the temperature gauge or temperature warning light circuit (see Chapter 12).

2   If the engine takes a long time to warm up, the thermostat is probably stuck open. Replace the thermostat.

3   If the engine runs hot, check the temperature of the upper radiator hose. If the hose isn't hot, the thermostat is probably stuck shut.

Replace the thermostat.

4   If the upper radiator hose is hot, then the coolant is circulating and the thermostat is open. Refer to the *Troubleshooting* Section at the front of this manual for the cause of overheating.

5   If the engine has overheated, it might have leaking cylinder head gaskets, scuffed pistons and/or warped or cracked cylinder heads.

REPLACEMENT

▶ **Refer to illustrations 3.7a, 3.7b, 3.7c, 3.8, 3.9 and 3.10**

6   Disconnect the negative cable from the battery. Drain the cooling system (see Chapter 1).

7   Detach the housing from the engine (see illustrations). Be prepared for some coolant to spill as the gasket seal is broken. The radiator

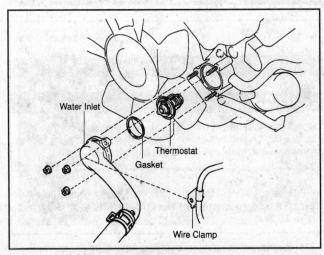

3.7a Thermostat housing details on V6 models

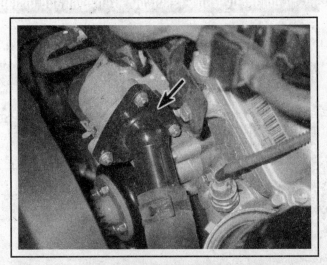

3.7b Thermostat location on 2005 and later V6 models

hose can be left attached to the housing, unless the housing itself is to be replaced.

8 Remove the thermostat, noting the direction in which it was installed in the housing, and thoroughly clean the sealing surfaces (see illustration).

9 Install a new gasket onto the thermostat (see illustration). Make sure it is evenly fitted all the way around.

10 Install the thermostat and housing, positioning the jiggle pin, if equipped, at the highest point (see illustration).

11 Tighten the housing fasteners to the torque listed in this Chapter's Specifications and reinstall the remaining components in the reverse order of removal.

12 Refill the cooling system (see Chapter 1), run the engine and check for leaks and proper operation.

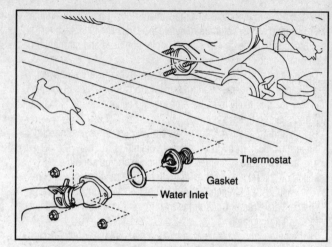

3.7c Thermostat housing details on V8 models

3.8 The thermostat is installed with the spring end towards the water inlet housing on V8 models

3.9 The thermostat gasket, which is actually a grooved sealing ring, fits around the edge of the thermostat

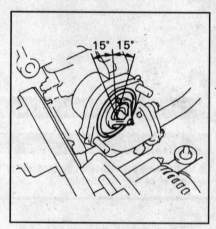

3.10 The jiggle valve must be within 15-degrees of the 12 o'clock position when installed

4 Coolant overflow reservoir - removal and installation

▶ Refer to illustration 4.1

1 Remove the hose clamp (see illustration) and then disconnect the overflow hose from the reservoir.

2 Remove the bolts from the coolant overflow reservoir and bracket.

3 Lift the coolant overflow reservoir from the engine compartment.

4 Pour the coolant into a container. Wash out and inspect the reservoir for cracks and chafing. Replace the reservoir if damaged.

5 Installation is the reverse of removal.

4.1 Disconnect the coolant overflow hose from the coolant overflow reservoir (typical)

5 Cooling fan and clutch - check, removal and installation

CHECK

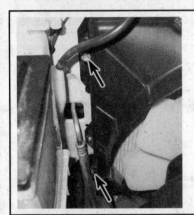

5.6a Location of the number 1 fan shroud bolts on V8 models - left side shown

5.6b Location of the number 1 fan shroud bolts on V8 models - right side shown

1   Symptoms of fan clutch failure are continuous noisy operation, looseness, vibration and/or silicone fluid leaking from the clutch.

Cold engine checks

2   Rock the fan back and forth by hand to check for excessive bearing play.

3   With the engine cold, turn the blades by hand. The fan should turn freely.

4   Visually inspect for substantial fluid leakage from the fan clutch assembly, a deformed bi-metal spring or grease leakage from the cooling fan bearing. If any of these conditions exist, replace the fan clutch.

Hot engine check

5   Start the engine and then allow it to warm up to its normal operating temperature. When the engine is fully warmed up, turn off the ignition switch. Turn the fan by hand. Some resistance should be felt. If the fan turns easily, replace the fan clutch.

REMOVAL

▶ **Refer to illustrations 5.6a, 5.6b, 5.7a, 5.7b and 5.8**

6   Remove the fan shroud mounting bolts (see illustrations).

➡**Note: It will be necessary to remove the lower fan shroud (number 2) first, before removing the main fan shroud (number 1) (see illustrations 6.10a and 6.10b).**

7   Remove the fan clutch-to-water pump mounting nuts (see illustration) and separate the fan from the water pump. Lift the fan and shroud from the engine compartment (see illustration).

8   To separate the fan blades from the viscous clutch hub, remove the mounting nuts (see illustration).

9   Installation is the reverse of removal.

5.7a Unscrew the four nuts that retain the fan to the water pump, detach the fan/clutch assembly . . .

5.7b . . . and remove the fan shroud and fan at the same time (V8 model shown, V6 models similar)

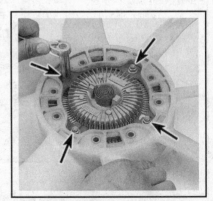

5.8 Remove these four nuts to separate the fan blades from the clutch (V8 model shown, V6 models similar)

6   Radiator - removal and installation

▶ Refer to illustrations 6.3, 6.9a and 6.9b

❋❋ WARNING:

Do not start this procedure until the engine is completely cool.

1   Disconnect the negative battery cable.

2   Drain the coolant into a container (see Chapter 1).

3   Detach the upper and lower radiator hoses from the radiator (see illustration). If the hoses stick, twist them with a pair of large pliers to break the bond, but be careful not to damage the fittings on the radiator.

4   Disconnect the reservoir hose from the radiator filler neck.

5   On V8 models, remove the intake air connector from the engine compartment (see Chapter 4).

6   Remove the cooling fan and fan shroud (see Section 4).

7   If equipped with an automatic transmission, disconnect the trans-

6.3   Use a pair of pliers to squeeze the hose clamp open, then slide the clamp back on the hose

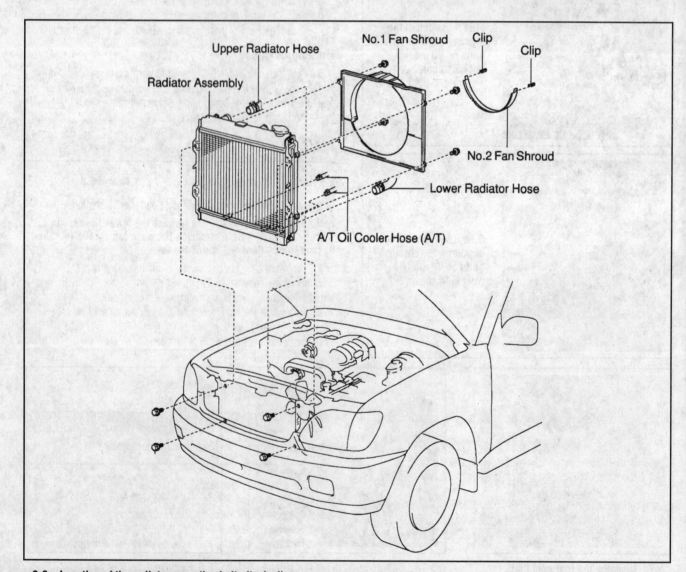

6.9a   Location of the radiator mounting bolts (typical)

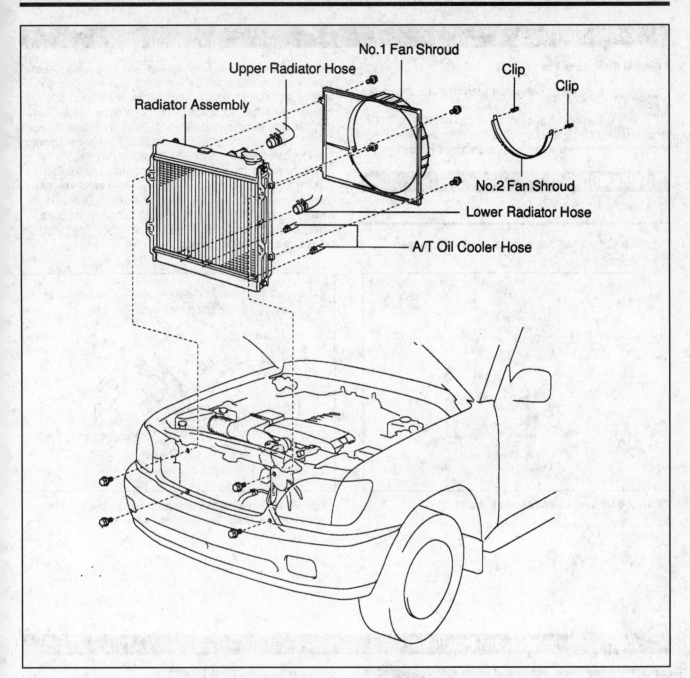

6.9b Location of the radiator mounting bolts - V8 models shown

mission cooler lines (see Chapter 7B). Place a drip pan to catch the fluid and cap the fittings.

8 Remove the radiator grille (see Chapter 11).

9 Remove the radiator mounting bolts (see illustrations).

10 Lift out the radiator. Be aware of dripping fluids and the sharp fins.

11 With the radiator removed, it can be inspected for leaks, damage and internal blockage. If in need of repairs, have a radiator shop or dealer service department perform the work as special techniques are required.

12 Bugs and dirt can be cleaned from the radiator with compressed air and a soft brush. Don't bend the cooling fins as this is done.

✳✳ WARNING:

Wear eye protection when using compressed air.

13 Installation is the reverse of the removal procedure.

14 After installation, fill the cooling system with the proper mixture of antifreeze and water. Refer to Chapter 1 if necessary.

15 Start the engine and check for leaks. Allow the engine to reach normal operating temperature, indicated by both radiator hoses becoming hot. Recheck the coolant level and add more if required.

16 On automatic transmission equipped models, check and add fluid as needed.

7 Water pump - check

▶ **Refer to illustrations 7.3a and 7.3b**

1   A failure in the water pump can cause serious engine damage due to overheating.

2   With the engine running and warmed to normal operating temperature, squeeze the upper radiator hose. If the water pump is working properly, a pressure surge should be felt as the hose is released.

✳✳ WARNING:

Keep hands away from fan blades!

3   Water pumps on all models, except 2005 V6 models, are equipped with weep or vent holes (see illustration). If a failure occurs in the pump seal, coolant will leak from this hole. In most cases it will be necessary to use a flashlight to find the hole on the water pump by looking through the space behind the pulley just below the water pump shaft. The water pump used on 2005 and later V6 engines has two weep holes; a "water hole" and an air hole (ventilation hole) (see illustration). If coolant leaks from either of these holes, replace the pump.

4   If the water pump shaft bearings fail there may be a howling sound at the front of the engine while it is running. Bearing wear can be felt if the water pump pulley is rocked up and down. Do not mistake drivebelt slippage, which causes a squealing sound, for water pump failure. Spray automotive drivebelt dressing on the belts to eliminate the belt as a possible cause of the noise.

7.3a Typical weep hole on a water pump

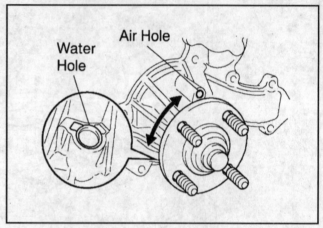

7.3b On 2005 and later V6 engines the water pump has two weep holes (a "water hole" and an "air hole")

8 Water pump - removal and installation

✳✳ WARNING 1:

Wait until the engine is completely cool before beginning this procedure.

✳✳ WARNING 2:

Do not allow antifreeze to come in contact with your skin or painted surfaces of the vehicle. Rinse off spills immediately with plenty of water. Antifreeze is highly toxic if ingested. Never leave antifreeze lying around in an open container or in puddles on the floor; children and pets are attracted by its sweet smell and may drink it. Check with local authorities on disposing of used antifreeze. Many communities have collection centers, which will see that antifreeze is disposed of safely. Never dump used antifreeze on the ground or into drains.

2004 AND EARLIER V6 MODELS

▶ **Refer to illustration 8.3**

1   Remove the oil dipstick and dipstick guide. Remove the timing belt (see Chapter 2A).

2   Remove the alternator adjuster bracket and bolt assembly and remove the thermostat housing and thermostat from the water pump (see Section 3).

3   Disconnect the oil cooler hose from the water pump, if equipped (see illustration).

4   Remove the water pump bolts and separate the pump from the engine. Disconnect the radiator inlet hose and bypass hose.

➡**Note: Be sure to mark the location of the bolts for proper reassembly.**

5   Thoroughly clean the gasket mating surface, removing all traces of oil with acetone or lacquer thinner and a clean rag.

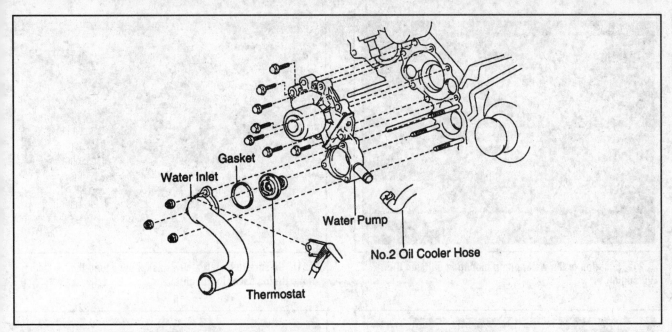

8.3 Exploded view of the water pump on 2004 and earlier V6 models

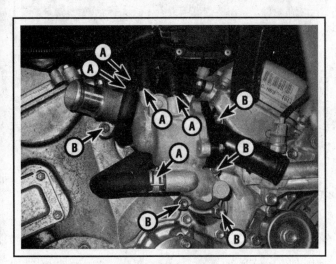

8.16 Water inlet assembly details on the 2005 and later V6 engine

A   Bypass hose clamp locations
B   Water inlet assembly mounting bolts

8.17 Remove the number 1 (A) and number 2 (B) idler pulleys

6   Apply a 3 mm bead of RTV sealant to the water pump. Be sure to install the water pump within five minutes of sealant application.

7   Installation is the reverse of removal. Tighten the water pump bolts to the torque listed in this Chapter's Specifications.

8   Refill the cooling system (see Chapter 1), run the engine and check for leaks.

2005 AND LATER V6 MODELS

▶ **Refer to illustrations 8.16, 8.17, 8.21a, 8.21b and 8.24**

9   Disconnect the cable from the negative terminal of the battery (see Chapter 5, Section 1).

10   Drain the cooling system (see Chapter 1).

11   Remove the engine cover.

12   Remove the engine splash shield (see illustration 8.6 in Chapter 1).

13   Refer to Chapter 1 and remove the drivebelt(s), then remove the engine cooling fan (see Section 5).

14   Remove the upper and lower radiator hoses (see Section 6).

15   Remove the air filter assembly (see Chapter 4).

16   Remove the water inlet assembly (see illustration).

17   Remove the number 1 and number 2 idler pulleys (see illustration).

18   Remove the alternator (see Chapter 5).

19   Remove the air conditioning compressor (see Section 15).

8.21a Location of the water pump mounting bolts on the V6 engine

8.21b Be sure to install a new gasket and clean the surface of the timing cover thoroughly

8.24 After installing the water pump, be sure to replace this O-ring with a new one

8.29 Remove the water inlet assembly mounting bolts

20 Remove the drivebelt tensioner (see Chapter 1).

21 Remove the water pump mounting bolts (see illustrations).

➡**Note: Be sure to mark the location of the stud bolts for proper reassembly.**

22 Thoroughly clean the gasket mating surfaces.

23 Install a new gasket and water pump. Tighten the water pump bolts to the torque listed in this Chapter's Specifications.

24 Install a new O-ring on the water pump (see illustration).

25 Install the water inlet assembly. Be sure to install new O-rings, then tighten the water inlet assembly mounting fasteners to the torque listed in this Chapter's Specifications.

26 The remainder of installation is the reverse of removal.

27 Refill the cooling system (see Chapter 1), run the engine and check for leaks.

V8 MODELS

▶ **Refer to illustrations 8.29, 8.30 and 8.34**

28 Remove the timing belt (see Chapter 2B). Remove the number 2

idler pulley (see Chapter 2B).

29 Disconnect the bypass hose from the water inlet assembly, remove the mounting bolts and separate the water inlet from the engine (see illustration).

30 Remove the water pump mounting bolts (see illustration) and separate the pump from the engine block.

31 Remove the O-ring from the coolant bypass pipe directly behind the water pump. Replace the O-ring with a new one

32 Install a new gasket and water pump. Tighten the water pump bolts to the torque listed in this Chapter's Specifications.

33 Thoroughly clean the gasket surface of the water inlet assembly.

34 Install a new O-ring, then apply a 3 mm bead of RTV sealant to the groove on the water inlet assembly (see illustration). Be sure to install the water pump within five minutes of sealant application. Tighten the water inlet assembly fasteners to the torque listed in this Chapter's Specifications.

35 The remainder of installation is the reverse of removal, with the following points:

36 Refill the cooling system (see Chapter 1), run the engine and check for leaks.

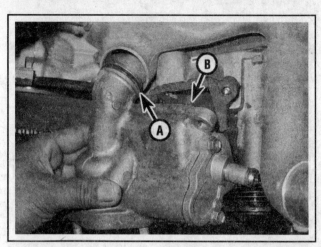

8.30 Location of the 5 bolts, the 2 stud bolts and the single nut on the V8 water pump

8.34 Install a new O-ring (A), then apply a 3 mm bead of RTV sealant to the groove on the backside of the water inlet assembly (B)

9   Coolant temperature sending unit - check and replacement

❊❊ WARNING:

Do not start this procedure until the engine is completely cool.

CHECK

◆ **Refer to illustration 9.3**

1   If the coolant temperature gauge is inoperative, check the fuses first (see Chapter 12).

2   If the temperature gauge indicates excessive temperature after running awhile, see the *Troubleshooting* section in the front of the manual.

3   If the temperature gauge indicates HOT as soon as the engine is started cold, disconnect the wire at the coolant temperature sender (see illustration). If the gauge reading drops, replace the sending unit. If the reading remains high, the wire to the gauge may be shorted to ground or the gauge is faulty.

➡**Note: The coolant temperature sending unit is located on the intake manifold (see Chapter 2A) near the left rear of the engine on 2000 through 2004 V6 engines or next to the engine coolant temperature (ECT) sensor (see Chapter 6) on V8 engines.**

On 2005 and later V6 engines, the Engine Coolant Temperature (ECT) sensor handles the coolant temperature sending unit's function. The ECT sensor is monitored by the Powertrain Control Module (PCM), which uses the ECT signal to calculate engine coolant temperature and outputs the appropriate commands to control engine temperature. The PCM displays the engine temperature to the driver via the coolant temperature gauge on the instrument cluster. When the PCM senses that the coolant temperature is excessive, it indicates an overheated engine

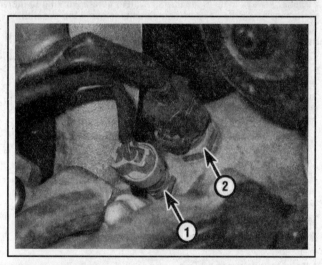

9.3 Location of the coolant temperature sending unit (1) and the ECT sensor (2) on V8 models

to the driver by moving the needle on the temperature gauge into the red zone. In the event that this happens to you, pull over immediately, turn off the engine and wait for it to cool down before checking the coolant level. The ECT sensor is located on the coolant bypass housing on the back of the engine, between the cylinder heads (see illustration 11.37 in Chapter 2A). To replace the ECT sensor, refer to Chapter 6.

4   If the coolant temperature gauge fails to show any indication after the engine has been warmed up, (approximately 10 minutes) and the fuses checked out OK, shut off the engine. Disconnect the wire at the sending unit and, using a jumper wire, connect the wire to a clean ground on the engine. Briefly turn on the ignition without starting the

engine. If the gauge now indicates Hot, replace the sending unit.

➡Note: The ECT sensor is mounted next to the coolant temperature sending unit. Do NOT try to test the ECT sensor as described in this step. Checking this sensor is beyond the scope of the home mechanic. Have the ECT sensor and circuit checked by a dealer service department. You can, however, replace the ECT sensor yourself (see Chapter 6).

5   If the gauge fails to respond, the circuit may be open or the gauge may be faulty - see Chapter 12 for additional information.

REPLACEMENT

6   Disconnect the electrical connector from the sending unit.
7   Using a deep socket or a wrench, remove the sending unit. Be prepared for coolant spillage.
8   Install the new unit and tighten it securely.
9   Reconnect the electrical connector, refill the cooling system (see Chapter 1) and check for coolant leakage and proper gauge function.

10 Air conditioning and heater blower motor and resistor - removal and installation

➡Note: Some Sequoia models are equipped with rear heating and air conditioning. The rear A/C unit is mounted behind the rear quarter trim panel. The rear A/C unit houses the heater core (heater radiator) and the evaporator (see Section 13).

BLOWER MOTOR

1   The blower unit is located in the cooling assembly under the dash, and behind the right, rear quarter trim panel on Sequoia models with rear heater units.

Front mounted blower motors

▶ Refer to illustration 10.2

2   Remove the blower motor connector, then remove the three mounting screws (see illustration).
3   Lower the blower assembly from the housing.

Rear mounted blower motors

▶ Refer to illustration 10.6

4   Remove the rear seat (see Chapter 11).
5   Remove the rear door scuff plate (see Chapter 11).
6   Remove the garnish from the right side window (see illustration).
7   Remove the right, rear quarter trim panel.
8   Remove the blower motor connector, then remove the three mounting screws.

9   Remove the blower assembly from the housing.

All models

▶ Refer to illustrations 10.10a and 10.10b

10 Remove the clamp from the fan (see illustrations) and separate the fan from the blower motor frame.
11 Installation is the reverse of removal.

RESISTOR

▶ Refer to illustration 10.12

12 Disconnect the resistor connector and remove the mounting screws (see illustration).
13 Remove the resistor from the vehicle.
14 Installation is the reverse of removal.

10.2 Blower fan mounting screws

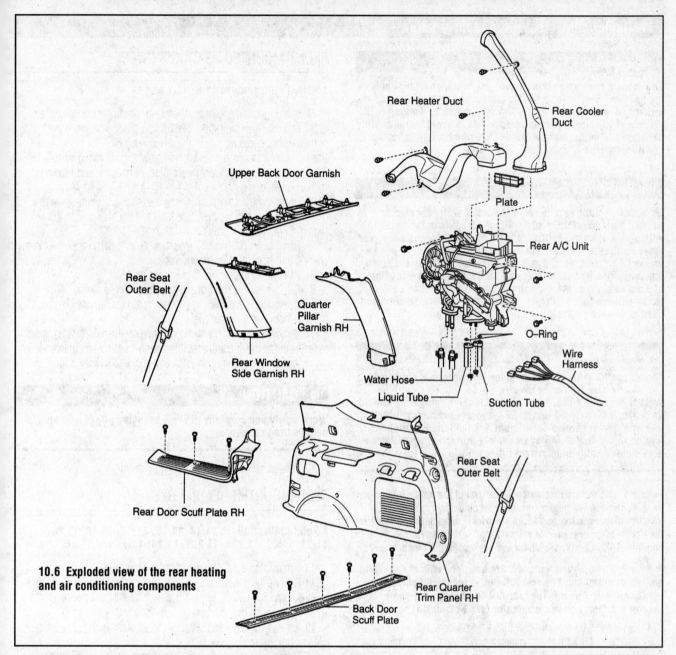

Rear Heater Duct

Rear Cooler Duct

Upper Back Door Garnish

Plate

Rear A/C Unit

Rear Seat Outer Belt

Quarter Pillar Garnish RH

O—Ring

Wire Harness

Rear Window Side Garnish RH

Water Hose

Liquid Tube

Suction Tube

Rear Seat Outer Belt

Rear Door Scuff Plate RH

10.6 Exploded view of the rear heating and air conditioning components

Rear Quarter Trim Panel RH

Back Door Scuff Plate

10.10a Remove the clamp from the shaft . . .

10.10b . . . and separate the blower fan from the blower motor

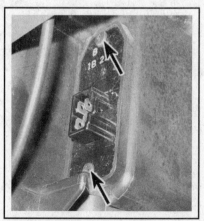

10.12 Location of the resistor mounting screws

11 Heater core - removal and installation

❋❋ WARNING 1:

The models covered by this manual are equipped with Supplemental Restraint systems (SRS), more commonly known as airbags. Always disable the airbag system before working in the vicinity of any airbag system components to avoid the possibility of accidental deployment of the airbag, which could cause personal injury (see Chapter 12).

❋❋ WARNING 2:

Do not allow antifreeze to come in contact with your skin or painted surfaces of the vehicle. Rinse off spills immediately with plenty of water. Antifreeze is highly toxic if ingested. Never leave antifreeze lying around in an open container or in puddles on the floor; children and pets are attracted by it's sweet smell and may drink it. Check with local authorities about disposing of used antifreeze. Many communities have collection centers which will see that antifreeze is disposed of safely. Never dump used antifreeze on the ground or pour it into drains.

❋❋ WARNING 3:

The air conditioning system is under high pressure. Do not loosen any hose fittings or remove any components until the system has been discharged. Air conditioning refrigerant should be properly discharged into an EPA-approved recovery/recycling unit by a dealer service department or an automotive air conditioning repair facility. Always wear eye protection when disconnecting air conditioning system fittings.

➡Note 1: The factory recommends removal of the entire instrument panel to remove the heater core on front mounted systems. This involves disconnecting numerous electrical connectors and there is the potential for breakage of delicate plastic tabs on various components. This is a difficult job for the average home mechanic.

➡Note 2: Some Sequoia models are equipped with rear heating and air conditioning. The rear A/C unit is mounted behind the rear quarter trim panel. The rear A/C unit houses the heater core (heater radiator) and the evaporator (see Section 13).

1 Disconnect the negative cable from the battery.

2 Wait until the engine is completely cool, then drain the cooling system (see Chapter 1).

3 Have the refrigerant discharged and recovered by an air conditioning technician.

FRONT-MOUNTED HEATING SYSTEM

◗ **Refer to illustrations 11.6, 11.7 and 11.10**

4 Working in the engine compartment, disconnect the heater hoses at the firewall. Push the rubber seal around the hoses toward the inside of the vehicle, releasing it from the sheetmetal. Plug the heater core (heater radiator) tubes to prevent leakage when it is removed.

5 Remove the instrument panel. Also, remove the instrument panel reinforcement brace.

6 Remove the cooling unit. Using a special air conditioning line disconnect tool, separate the liquid tube and the suction tube (see illustration) from the cooling unit.

7 Remove the cooling unit mounting bolts (see illustration). Lift the cooling unit from the passenger compartment.

8 Disconnect the control cables from the heater unit.

9 Remove the defroster nozzle and the heater to register duct.

10 Remove the heater unit mounting bolts (see illustration) and lift the assembly from the passenger compartment.

11 Remove the screws and clamps and pull the heater core (heater radiator) out of the housing (see illustration 11.10). Keep plenty of towels or rags on the carpeting to catch any coolant that may drip.

❋❋ CAUTION:

Work slowly and carefully to avoid breaking any plastic components during removal.

12 Installation is the reverse order of removal.

REAR-MOUNTED HEATING SYSTEM

◗ **Refer to illustrations 11.14, 11.16, 11.17, 11.19, 11.20, 11.21, 11.22, 11.23a, 11.23b, 11.24, 11.25 and 11.26**

13 Remove the right, rear quarter trim panel (see Section 10).

14 Remove the bolts from the air conditioning pressure lines and separate the liquid tube and the suction tube from the A/C unit (see illustration).

15 Remove the heater hoses from the coolant pipes.

16 Remove the clips and separate the air ducts from the vehicle (see illustration).

17 Remove the mounting bolts and then remove the A/C unit case from the rear of the vehicle (see illustration).

11.6 Disconnect the liquid tube and the suction tube from the cooling unit

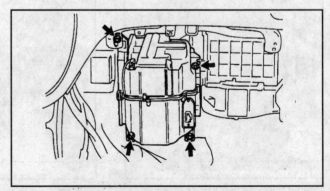

11.7 Location of the cooling unit mounting bolts

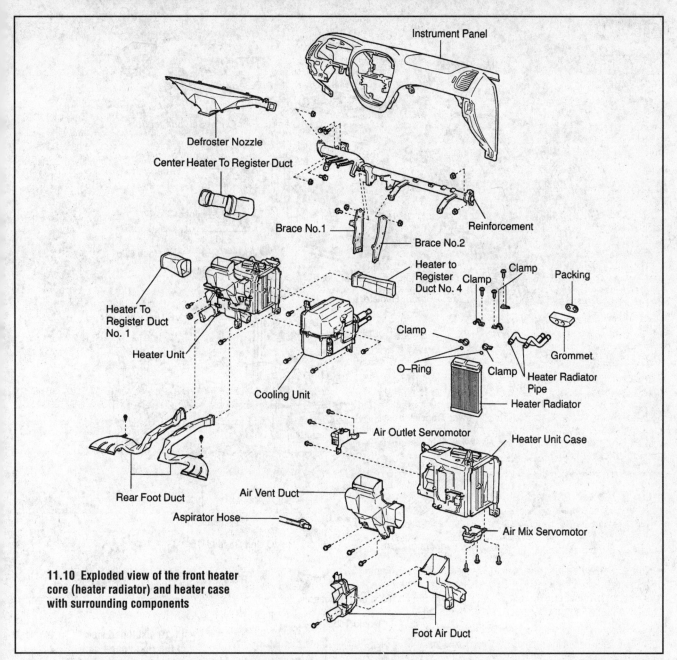

11.10 Exploded view of the front heater core (heater radiator) and heater case with surrounding components

Instrument Panel

Defroster Nozzle

Center Heater To Register Duct

Brace No.1

Reinforcement

Brace No.2

Heater to Register Duct No. 4

Clamp

Clamp

Packing

Clamp

Heater To Register Duct No. 1

Heater Unit

Clamp

O-Ring

Clamp

Grommet

Heater Radiator Pipe

Heater Radiator

Cooling Unit

Air Outlet Servomotor

Heater Unit Case

Rear Foot Duct

Air Vent Duct

Aspirator Hose

Air Mix Servomotor

Foot Air Duct

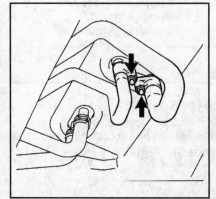

11.14 Remove the bolts and separate the liquid tube and the suction tube from the A/C unit case

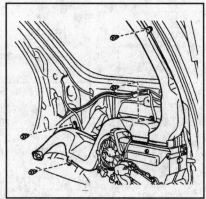

11.16 The air ducts are retained by clips

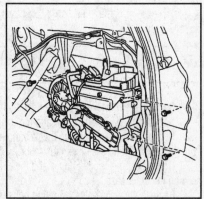

11.17 Location of the A/C unit mounting bolts

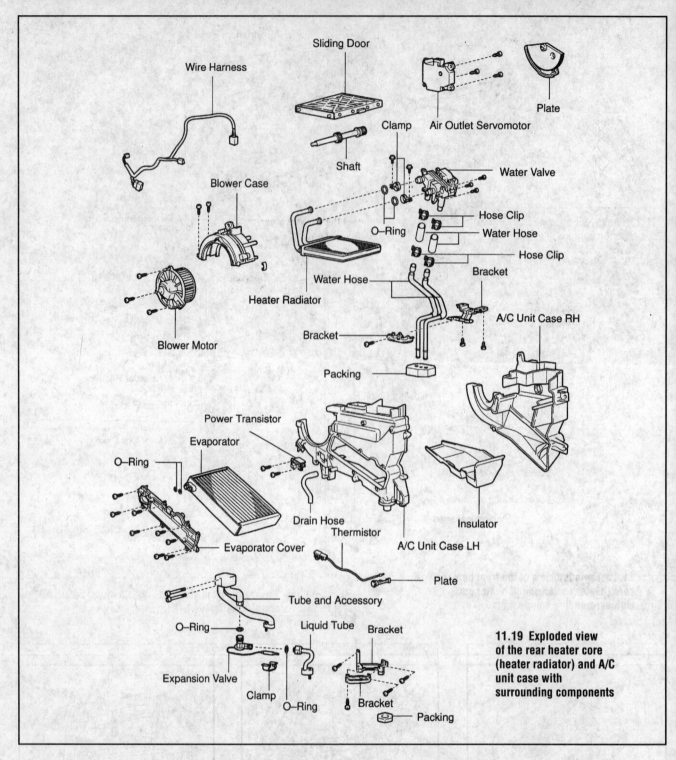

11.19 Exploded view of the rear heater core (heater radiator) and A/C unit case with surrounding components

18 Remove the blower motor (see Section 10).

19 Remove the power transistor and the thermistor (see illustration).

20 Remove the screws and the bracket and separate the coolant pipe from the rear A/C unit case (see illustration).

21 Disconnect the coolant hoses and the coolant pipes (see illustration) on the outer case.

22 Remove the water valve mounting bolts (see illustration) and separate the water valve from the A/C unit case.

23 Remove the expansion valve (see illustrations).

24 Remove the evaporator (see illustration).

25 Remove the air outlet servomotor (see illustration).

26 Separate the rear A/C unit case halves (see illustration) and remove the heater core (heater radiator) from the A/C unit case.

27 Installation is the reverse order of removal.

ALL MODELS

28 Refill the cooling system (see Chapter 1), reconnect the battery and run the engine. Check for leaks and proper system operation.

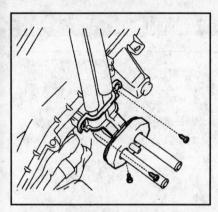

11.20 Remove the screws and separate the coolant pipe from the case

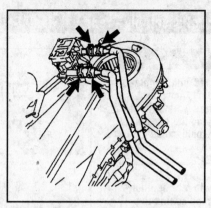

11.21 Disconnect the coolant pipes from the water valve

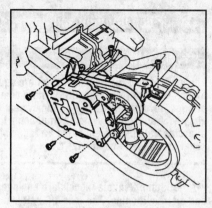

11.22 Location of the water valve mounting bolts

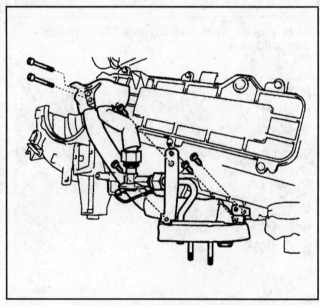

11.23a Remove the bracket bolts to access the expansion valve

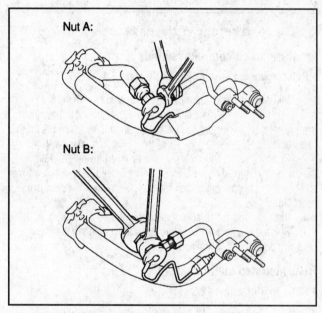

Nut A:

Nut B:

11.23b First remove the heat sensing tube then the expansion valve using two wrenches

11.24 Remove the mounting screws and the evaporator from the A/C unit case

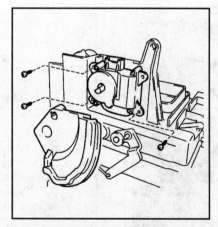

11.25 Remove the air outlet servomotor mounting screws and carefully pry the component from the case

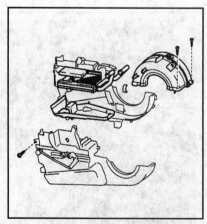

11.26 Remove the bolts and separate the case halves

12 Air conditioning and heater control assembly - removal, installation and adjustment

✳✳ WARNING:

The models covered by this manual are equipped with Supplemental Restraint systems (SRS), more commonly known as airbags. Always disable the airbag system before working in the vicinity of any airbag system components to avoid the possibility of accidental deployment of the airbag, which could cause personal injury (see Chapter 12).

➥Note: Some Sequoia models are equipped with rear heating and air conditioning. The rear A/C unit is mounted behind the rear quarter trim panel. The rear A/C unit houses the heater core (heater radiator) and the evaporator (see Section 11).

REMOVAL AND INSTALLATION

1   Disconnect the negative cable from the battery.

Front-mounted control assembly

▶ Refer to illustrations 12.3a, 12.3b and 12.4

2   Remove the center cluster integration panel (see Chapter 11).
3   Push the control assembly in (see illustration), toward the instrument panel until the locking tabs release, then pull the assembly down (see illustration).
4   Pull the control assembly out slightly (see illustration). Remove the cable(s) by spreading the claw of the cable clamp with a screwdriver and carefully prying it out of the assembly. Disconnect the electrical connectors.
5   Installation is the reverse of the removal procedure.
6   Run the engine and check for proper functioning of the heater (and air conditioning, if equipped).

Rear-mounted control assembly

▶ Refer to illustration 12.7

7   Using a screwdriver, release the two clips on each side of the heater control assembly (see illustration).
8   Pull the control assembly out slightly and disconnect the connector.

12.3a Push the control assembly toward the instrument panel until the tabs release . . .

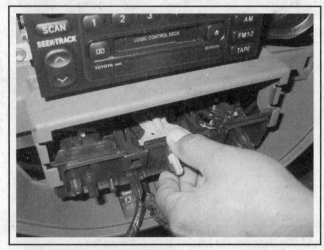

12.3b . . . then pull the assembly down to access the control cables

12.4 Remove the control cables and the connectors at the rear of the control assembly

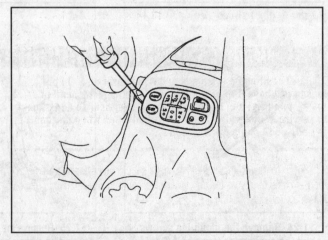

12.7 Release the clips, one on each side of the heater control assembly

9 Remove the heater control assembly from the console.
10 Installation is the reverse of the removal procedure.

CABLE ADJUSTMENT

Refer to illustrations 12.12, 12.13 and 12.14

11 With the cables attached at the control end and the control assembly installed in the dash, adjust the cables at their ends. The controls should be set to: FACE (center recirculation) and MAX COOL (counterclockwise) positions.

12 To adjust the mode control cable, pull the control linkage to the FACE position (down), connect the cable and lock the clamp in position (see illustration).

13 To adjust the air mix control cable, set the air mix damper to COOL, install the cable and lock the clamp while applying slight pressure on the outer cable linkage (see illustration).

14 To adjust the water valve control cable, pull the control linkage to the COOL position, connect the control cable and lock the clamp in position (see illustration).

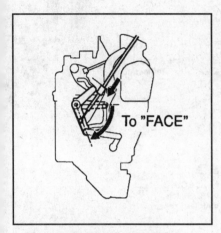

12.12 Pull the air inlet control linkage down to the FACE position and lock the cable clamp

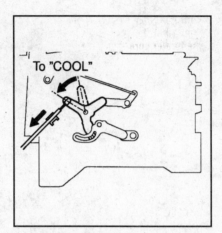

12.13 Lock the cable clamp while gently pushing the outer cable linkage to the COOL position

12.14 Pull the air inlet control linkage to the COOL position and lock the cable clamp

13 Air conditioning and heating system - check and maintenance

▶ **Refer to illustrations 13.1a and 13.1b**

❄❄ **WARNING:**

The air conditioning system is under high pressure. Do not loosen any hose fittings or remove any components until after the system has been discharged by a dealer service department or service station. Always wear eye protection when disconnecting air conditioning system fittings.

1 The following maintenance checks should be performed on a regular basis to ensure the air conditioner continues to operate at peak efficiency (see illustrations).

   a) Check the compressor drivebelt. If it's worn or deteriorated, replace it (see Chapter 1).

   b) Check the drivebelt tension and, if necessary, adjust it (see Chapter 1).

   c) Check the system hoses. Look for cracks, bubbles, hard spots and deterioration. Inspect the hoses and all fittings for oil bubbles and seepage. If there's any evidence of wear, damage or leaks, replace the hose(s).

   d) Inspect the condenser fins for leaves, bugs and other debris. Use a "fin comb" or compressed air to clean the condenser.

   e) Make sure the system has the correct refrigerant charge.

2 It's a good idea to operate the system for about 10 minutes at least once a month, particularly during the winter. Long term non-use can cause hardening, and subsequent failure, of the seals.

3 Because of the complexity of the air conditioning system and the special equipment necessary to service it, in-depth troubleshooting and repairs are not included in this manual. However, simple checks and component replacement procedures are provided in this Chapter.

4 The most common cause of poor cooling is simply a low system refrigerant charge. If a noticeable drop in cool air output occurs, the following quick check will help you determine if the refrigerant level is low.

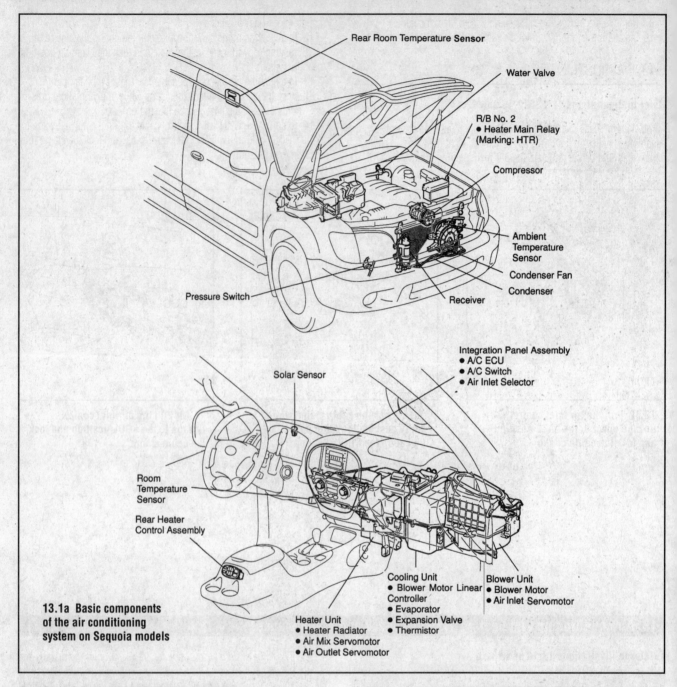

Rear Room Temperature **Sensor**

Water Valve

R/B No. 2
● Heater Main Relay
(Marking: HTR)

Compressor

Ambient
Temperature
Sensor

Condenser Fan

Condenser

Receiver

Pressure Switch

Solar Sensor

Integration Panel Assembly
● A/C ECU
● A/C Switch
● Air Inlet Selector

Room
Temperature
Sensor

Rear Heater
Control Assembly

**13.1a Basic components
of the air conditioning
system on Sequoia models**

Heater Unit
● Heater Radiator
● Air Mix Servomotor
● Air Outlet Servomotor

Cooling Unit
● Blower Motor Linear
Controller
● Evaporator
● Expansion Valve
● Thermistor

Blower Unit
● Blower Motor
● Air Inlet Servomotor

CHECKING THE REFRIGERANT CHARGE

▶ **Refer to illustration 13.8**

5   Warm the engine up to normal operating temperature.

6   Place the air conditioning temperature selector at the coldest setting and the blower at the highest setting. Open the doors (to make sure the air conditioning system doesn't cycle off as soon as it cools the passenger compartment).

7   With the compressor engaged, the clutch will make an audible click and the center of the clutch will rotate. If the compressor discharge line feels warm and the compressor inlet pipe feels cool, the system is properly charged.

8   Place a thermometer in the dashboard vent nearest the evaporator (see illustration) and then operate the system until the indicated

temperature is around 40 to 45 degrees F. If the ambient (outside) air temperature is very high, say 110 degrees F, the duct air temperature may be as high as 60 degrees F, but generally the air conditioning is 30 to 40 degrees F cooler than the ambient air.

➡**Note: Humidity of the ambient air also affects the cooling capacity of the system. Higher ambient humidity lowers the effectiveness of the air conditioning system.**

ADDING REFRIGERANT

▶ **Refer to illustrations 13.9 and 13.12**

9   Buy an automotive charging kit (see illustration) at an auto parts store. A charging kit includes a 14-ounce can of refrigerant, a tap valve and a short section of hose that can be attached between the tap valve

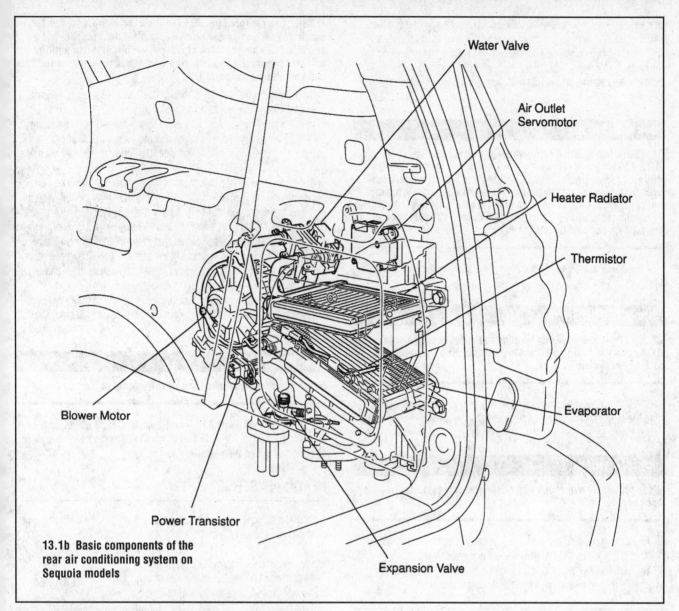

Water Valve

Air Outlet Servomotor

Heater Radiator

Thermistor

Evaporator

Blower Motor

Power Transistor

13.1b Basic components of the rear air conditioning system on Sequoia models

Expansion Valve

13.8 Check the temperature of the output air in the center register with a thermometer - it should be approximately 35 to 40-degrees F below the ambient air temperature

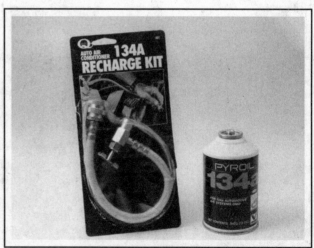

13.9 A basic charging kit is available at most auto parts stores - it must say R-134a and so must the cans of refrigerant you buy

and the system low side service valve. Because one can of refrigerant may not be sufficient to bring the system charge up to the proper level, it's a good idea to buy an additional can. Make sure that one of the cans contains red refrigerant dye. If the system is leaking, the red dye will leak out with the refrigerant and help you pinpoint the location of the leak.

✳✳ CAUTION:

There are two types of refrigerant used in automotive systems; R-12 - which has been widely used on earlier models and the more environmentally-friendly R-134a used in all models covered by this manual. These two refrigerants (and their appropriate refrigerant oils) are not compatible and must never be mixed or components will be damaged. Use only R-134a refrigerant in the models covered by this manual.

10 Hook up the charging kit by following the manufacturer's instructions.

✳✳ WARNING:

DO NOT hook the charging kit hose to the system high side! The fittings on the charging kit are designed to fit only on the low side of the system.

11 Back off the valve handle on the charging kit and screw the kit onto the refrigerant can, making sure first that the O-ring or rubber seal inside the threaded portion of the kit is in place.

✳✳ WARNING:

Wear protective eyewear when dealing with pressurized refrigerant cans.

12 Remove the dust cap from the low-side charging connection and attach the quick-connect fitting on the kit hose (see illustration).
13 Warm up the engine and turn on the air conditioner. Keep the charging kit hose away from the fan and other moving parts.

13.12 Add R-134a refrigerant to the low-side port only - the procedure will go faster if you wrap the can with a warm, wet towel to prevent icing

➡ **Note: The charging process requires the compressor to be running. Your compressor may cycle off if the pressure is low due to a low charge. If the clutch cycles off, you can pull the low-pressure cycling switch plug and attach a jumper wire. This will keep the compressor ON.**

14 Turn the valve handle on the kit until the stem pierces the can, then back the handle out to release the refrigerant. You should be able to hear the rush of gas. Add refrigerant to the low side of the system until both the receiver-drier surface and the evaporator inlet pipe feel about the same temperature. Allow stabilization time between each addition.
15 If you have an accurate thermometer, place it in the center air conditioning vent (see illustration 13.8) and then note the temperature of the air coming out of the vent. A fully-charged system which is working correctly should cool down to about 40 degrees F. Generally, an air conditioning system will put out air that is 30 to 40 degrees F cooler than the ambient air. For example, if the ambient (outside) air temperature is very high (over 100 degrees F), the temperature of air coming out of the registers should be 60 to 70 degrees F.
16 When the can is empty, turn the valve handle to the closed position and release the connection from the low-side port. Replace the dust cap.

✳✳ WARNING:

Never add more than two cans of refrigerant to the system.

17 Remove the charging kit from the can and store the kit for future use with the piercing valve in the UP position, to prevent inadvertently piercing the can on the next use.

HEATING SYSTEMS

18 If the carpet under the heater core is damp, or if antifreeze vapor or steam is coming through the vents, the heater core is leaking. Remove it (see Section 11) and install a new unit (most radiator shops will not repair a leaking heater core).
19 If the air coming out of the heater vents isn't hot, the problem could stem from any of the following causes:
   a) *The thermostat is stuck open, preventing the engine coolant from warming up enough to carry heat to the heater core. Replace the thermostat (see Section 3).*
   b) *There is a blockage in the system, preventing the flow of coolant through the heater core. Feel both heater hoses at the firewall. They should be hot. If one of them is cold, there is an obstruction in one of the hoses or in the heater core, or the heater control valve is shut. Detach the hoses and back flush the heater core with a water hose. If the heater core is clear but circulation is impeded, remove the two hoses and flush them out with a water hose.*
   c) *If flushing fails to remove the blockage from the heater core, the core must be replaced (see Section 11).*

ELIMINATING AIR CONDITIONING ODORS

▶ **Refer to illustration 13.24**

20 Unpleasant odors that often develop in air conditioning systems are caused by the growth of a fungus, usually on the surface of the evaporator core. The warm, humid environment there is a perfect breed-

ing ground for mildew to develop.

21 The evaporator core on most vehicles is difficult to access, and factory dealerships have a lengthy, expensive process for eliminating the fungus by opening up the evaporator case and using a powerful disinfectant and rinse on the core until the fungus is gone. You can service your own system at home, but it takes something much stronger than basic household germ-killers or deodorizers.

22 Aerosol disinfectants for automotive air conditioning systems are available in most auto parts stores, but remember when shopping for them that the most effective treatments are also the most expensive. The basic procedure for using these sprays is to start by running the system in the RECIRC mode for ten minutes with the blower on its highest speed. Use the highest heat mode to dry out the system and keep the compressor from engaging by disconnecting the wiring connector at the compressor (see Section 15).

23 Make sure that the disinfectant can comes with a long spray hose. Work the nozzle through the insulating foam surrounding the evaporator outlet pipe so that it protrudes inside the evaporator housing, and then spray according to the manufacturer's recommendations. Try to cover the whole surface of the evaporator core, by aiming the spray up, down and sideways. Follow the manufacturer's recommendations for the length of spray and waiting time between applications.

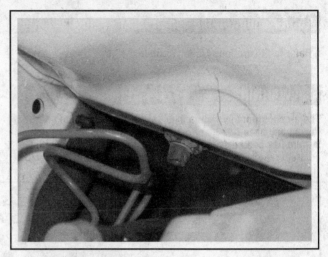

13.24 Location of the evaporator drain at the firewall

24 Once the evaporator has been cleaned, the best way to prevent the mildew from coming back again is to make sure your evaporator housing drain tube is clear (see illustration).

14 Air conditioning receiver/drier - removal and installation

▶ **Refer to illustration 14.3**

❄❄ WARNING:

The air conditioning system is under high pressure. Do not loosen any hose fittings or remove any components until the system has been discharged. Air conditioning refrigerant should be properly discharged into an EPA-approved recovery/recycling unit by a dealer service department or an automotive air conditioning repair facility. Always wear eye protection when disconnecting air conditioning system fittings.

1 Have the refrigerant discharged and recovered by an air conditioning technician.

2 Remove the radiator grille (see Chapter 11).

3 Disconnect the refrigerant lines from the receiver/drier and cap the open fittings to prevent entry of moisture (see illustration).

4 Loosen the pinch bolt or bracket mounting bolts, (whichever is easier to get a wrench on) and remove the receiver/drier.

5 Installation is the reverse of removal.

6 Have the system evacuated, charged and leak tested by the shop

14.3 After the system has been discharged, remove the bolts, detach the refrigerant lines from the top of the receiver/drier and cap them

that discharged it. If the receiver was replaced, have them add about 0.71 ounces of refrigerant oil to the system. Use only the refrigerant oil compatible with the refrigerant of your system (R-134a).

15 Air conditioning compressor - removal and installation

♦ **Refer to illustrations 15.5a, 15.5b and 15.6**

❊❊ **WARNING:**

The air conditioning system is under high pressure. Do not loosen any hose fittings or remove any components until the system has been discharged. Air conditioning refrigerant should be properly discharged into an EPA-approved recovery/recycling unit by a dealer service department or an automotive air conditioning repair facility. Always wear eye protection when disconnecting air conditioning system fittings.

1   Have the refrigerant discharged by an automotive air conditioning technician.
2   Remove the under-vehicle splash shield (see Chapter 1, illustration 8.6).

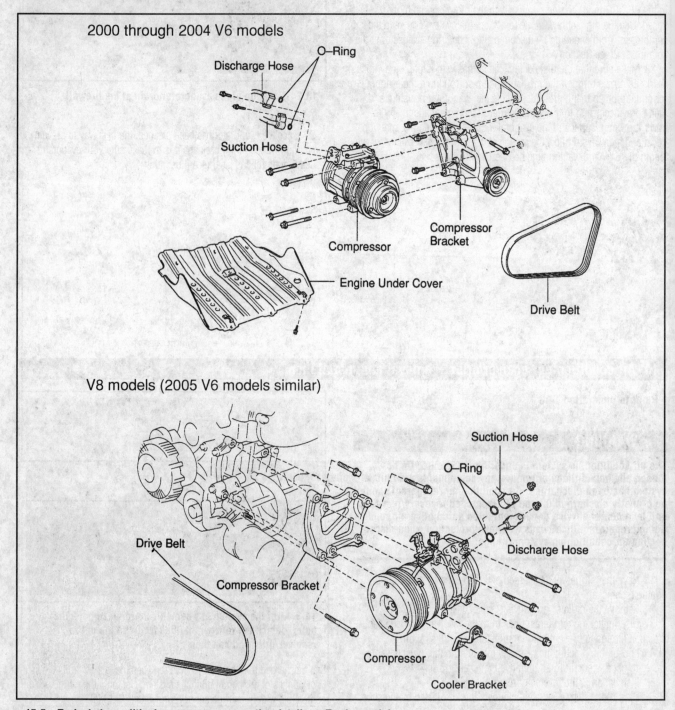

2000 through 2004 V6 models

Discharge Hose
O–Ring
Suction Hose
Compressor
Compressor Bracket
Engine Under Cover
Drive Belt

V8 models (2005 V6 models similar)

Suction Hose
O–Ring
Discharge Hose
Drive Belt
Compressor Bracket
Compressor
Cooler Bracket

15.5a Typical air conditioning compressor mounting details on Tundra models

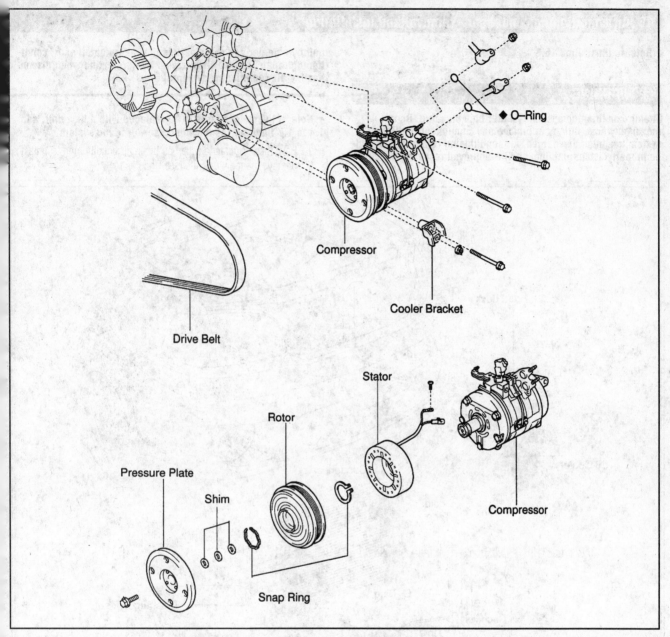

15.5b Typical air conditioning compressor mounting details on Sequoia models

3   Remove the drivebelt from the compressor (see Chapter 1).

4   Detach the electrical connector and disconnect the refrigerant lines.

5   Unbolt the compressor (see illustrations).

6   On V8 models, remove the cooler bracket (see illustration) and lift the compressor from the vehicle.

7   If a new or rebuilt compressor is being installed, follow the directions supplied with the compressor regarding the proper level of oil prior to installation.

8   Installation is the reverse of removal. Replace any O-rings with new ones specifically made for the type of refrigerant in your system and lubricate them with refrigerant oil, also designed specifically for your system (R-134a).

9   Have the system evacuated, recharged and leak tested by the shop that discharged it.

15.6 Remove the bolts and the cooler bracket from the bottom of the air conditioning compressor

16 Air conditioning condenser - removal and installation

▶ Refer to illustration 16.8

❄ **WARNING:**

The air conditioning system is under high pressure. Do not loosen any hose fittings or remove any components until the system has been discharged. Air conditioning refrigerant should be properly discharged into an EPA-approved recovery/recycling unit by a dealer service department or an automotive air conditioning repair facility. Always wear eye protection when disconnecting air conditioning system fittings.

➡**Note:** If the condenser is to be replaced with a new unit, add 1.4 to 1.7 fluid ounces of compressor oil to the system.

1   Have the refrigerant discharged by an air conditioning technician.
2   Remove the radiator grille (see Chapter 11).

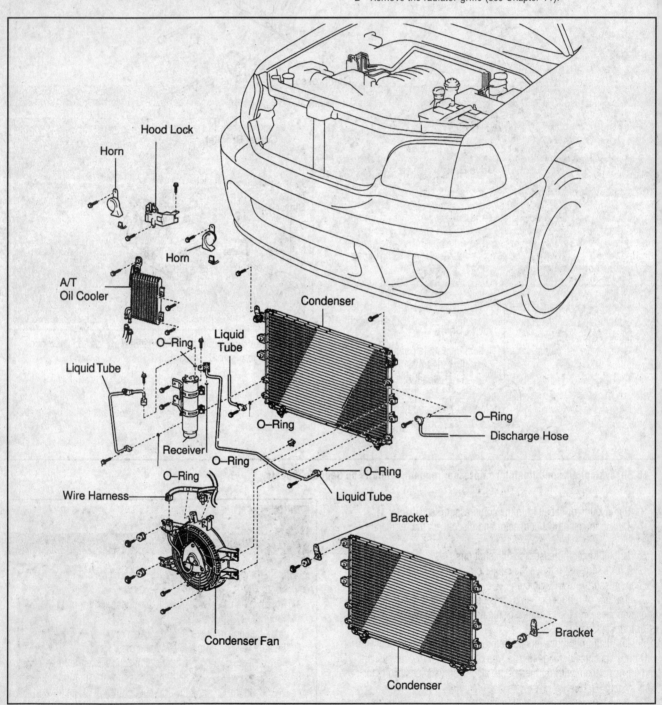

16.8 Typical condenser fan and condenser mounting details - Sequoia models shown

3   Disconnect the electrical connector, remove the condenser fan mounting bolts (see illustration 16.8) and lift it from the front of the vehicle.

4   On V8 models, remove the transmission oil cooler (see Chapter 7B).

5   Remove the horns (see Chapter 12).

6   Remove the receiver/drier (see Section 14).

7   Remove the hood lock brace and center support brace (see Chapter 11). Disconnect the condenser inlet and outlet fittings. Cap the open fittings immediately to keep moisture and contamination out of the system.

8   Remove the condenser mounting bolts, pull the condenser forward and lift it out (see illustration).

9   Install the condenser, brackets and bolts, making sure the rubber cushions fit on the mounting points properly.

10 Reconnect the refrigerant lines, using new O-rings where needed. If a new condenser has been installed, add approximately 1.4 to 1.7 ounces (50 cc) of new refrigerant oil of the correct type (R-134a).

11 Reinstall the remaining parts in the reverse order of removal.

12 Have the system evacuated, charged and leak tested by the shop that discharged it.

17 Air conditioning pressure cycling switch - removal and installation

▶ **Refer to illustration 17.2**

> ※ **WARNING:**
>
> The air conditioning system is under high pressure. Do not loosen any hose fittings or remove any components until the system has been discharged. Air conditioning refrigerant should be properly discharged into an EPA-approved recovery/recycling unit by a dealer service department or an automotive air conditioning repair facility. Always wear eye protection when disconnecting air conditioning system fittings.

1   Have the refrigerant discharged by an air conditioning technician.

2   Unplug the electrical connector from the pressure cycling switch (see illustration).

3   Unscrew the pressure cycling switch. Be sure to hold the fitting on the line with a wrench to prevent deforming the pressure line.

4   Lubricate the switch O-ring with clean refrigerant oil of the correct type.

5   Screw the new switch onto the threads until hand tight, and then tighten to the torque listed in this Chapter's Specifications.

17.2 When removing the pressure cycling switch, use two wrenches to prevent twisting the lines

6   Reconnect the electrical connector.

7   Have the system evacuated, charged and leak tested by the shop that discharged it.

18 Oil cooler - removal and installation

▶ **Refer to illustrations 18.4a, 18.4b and 18.4c**

> ※ **WARNING:**
>
> Wait until the engine is completely cool before beginning this procedure.

1   Drain the engine oil and remove the oil filter (see Chapter 1).

2   Drain the coolant (see Chapter 1) into a container.

3   Disconnect the oil cooler hoses from the oil cooler. Be sure to position an oil drain pan directly below the cooler lines to catch any residual oil.

4   Remove the union bolt (see illustrations) and separate the oil cooler from the oil filter adapter.

5   Installation is the reverse of removal.

6   Be sure to install a new O-ring and plate washer.

7   Torque the union bolt to the specifications listed in this Chapter.

8   Install a new oil filter. Refill the engine with oil and the cooling system with the proper type and concentration of antifreeze (see Chapter 1) and check for coolant leakage and proper gauge function.

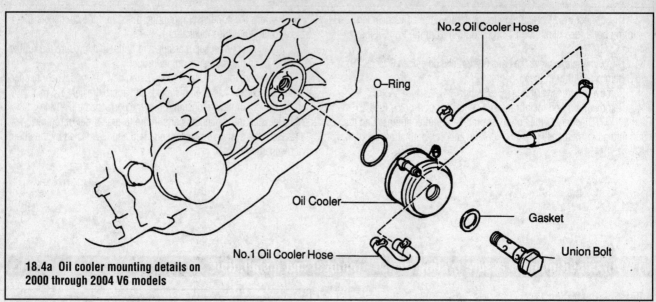

18.4a Oil cooler mounting details on 2000 through 2004 V6 models

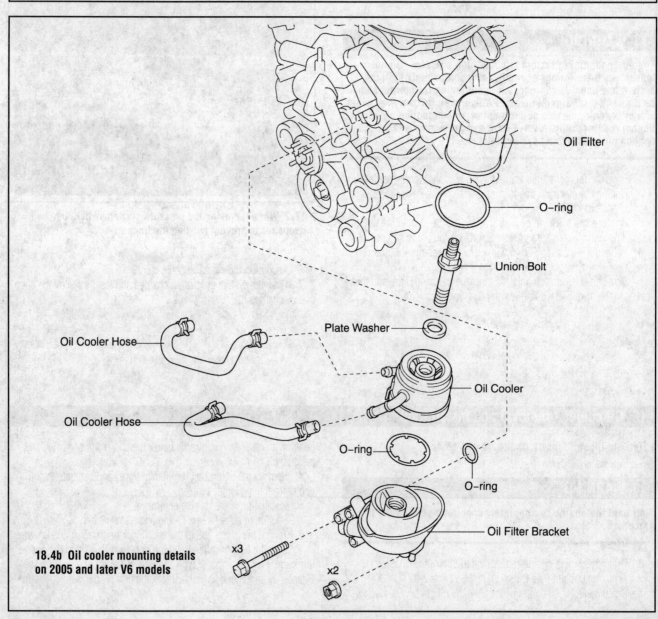

18.4b Oil cooler mounting details on 2005 and later V6 models

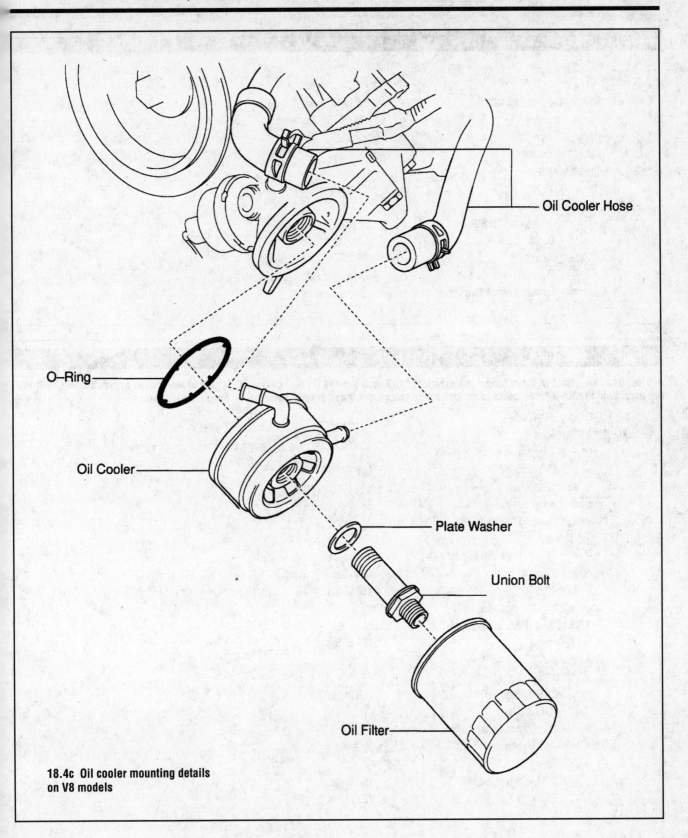

**18.4c Oil cooler mounting details
on V8 models**

Specifications

General

Radiator cap pressure rating	8.6 to 14.9 psi
Thermostat rating	176 to 183-degrees F
Refrigerant type	R-134a
Refrigerant capacity	
Tundra models	
2000 through 2003	20 to 23 ounces
2004 on	
Access cab models	20 to 23 ounces
Double cab models	17 to 20 ounces
Sequoia models	
Front A/C systems only	25 to 28 ounces
Front and rear A/C systems	35 to 38 ounces

Torque specifications   Ft-lbs (unless otherwise indicated)

→ Note: One foot-pound (ft-lb) of torque is equivalent to 12 inch-pounds (in-lbs) of torque. Torque values below approximately 15 ft-lbs are expressed in inch-pounds, since most foot-pound torque wrenches are not accurate at these smaller values.

Oil cooler union bolt	
V6 models	
2000 through 2004	44
2005 and later	50
V8 models	51
Pressure cycling switch	84 in-lbs
Thermostat housing mounting nuts	
V6 models	
2000 through 2004 (nuts)	168 in-lbs
2005 and later	80 in-lbs
V8 models	168 in-lbs
Water inlet housing mounting bolts	
V6 models (2005 and later only)	80 in-lbs
V8 models	156 in-lbs
Water pump-to-block bolts	
V6 models	
2000 through 2004	168 in-lbs
2005 and later	
Longer bolts	17
Shorter bolts	80 in-lbs
V8 models	
Nut and two stud bolts	156 in-lbs
Five bolts	180 in-lbs

Section

Reference to other Chapters

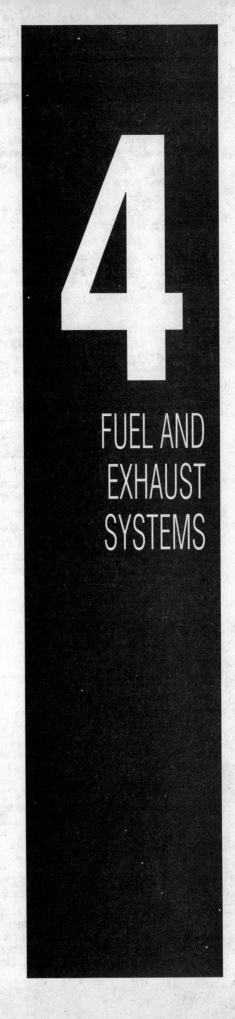

4

FUEL AND EXHAUST SYSTEMS

1   General information

All models are equipped with a Sequential Electronic Fuel Injection system. The fuel system consists of the following components:

Air filter, housing and air intake duct between housing and throttle body
Circuit opening relay
EFI main relay
Electric fuel pump and fuel level sending unit assembly (located in the fuel tank)
Fuel pressure pulsation damper (part of banjo fitting that connects fuel delivery pipe to left fuel rail)
Fuel pressure regulator (on the right fuel rail)
Fuel pump relay (V8 engine)
Fuel rail and fuel injector assembly
Fuel tank

Throttle body assembly, which includes throttle valve, Throttle Position (TP) sensor and either Idle Air Control (IAC) valve (V6 engine) or throttle control motor (V8 engine) (for information on the TP sensor, IAC valve and throttle control motor, see Chapter 6)

SEQUENTIAL FUEL INJECTION (SFI) SYSTEM

Intake air is drawn through the air filter housing and air intake duct into the throttle body. Inside the throttle body, the throttle valve controls the amount of air passing into the air intake plenum and then into the cylinders. The intake port for each cylinder is equipped with its own injector. The Sequential Fuel Injection (SFI) system injects fuel into the intake ports in the same sequence as the firing order of the engine. The injectors are controlled by the Powertrain Control Module (PCM). The PCM constantly monitors the operating conditions of the engine - temperature, speed, load, etc. - and then delivers the optimal amount of fuel for these conditions. The amount of fuel delivered by each injector is determined by its "on time," which the PCM can vary so quickly that two successive injectors can deliver a different amount of fuel. In other words, the PCM responds instantly to any changes in the operating conditions of the engine.

FUEL PUMP AND LINES

The electric fuel pump is located inside the fuel tank. To remove the fuel pump, lower the tank, unbolt the pump flange from the tank, then remove the fuel pump through the hole in the top of the tank.

On V8 models, the fuel pump circuit is controlled by the circuit opening relay and by the fuel pump relay, both of which in turn are controlled by the PCM. When the engine is cranked, current flows from the ignition switch to the starter relay coil and to the PCM, which also receives a signal from the Crankshaft Position (CKP) sensor. When the PCM receives these two signals (from the ignition switch and from the CKP sensor), the transistor inside the PCM that controls the circuit opening relay coil allows current to flow to the circuit opening relay, which closes and allows current to flow to the fuel pump. As long as the transistor inside the PCM that controls the circuit opening relay continues to receive the CKP sensor signal, it continues to supply current to the circuit opening relay, which continues to supply current to the fuel pump. The fuel pump speed is determined by the operating condition of the engine. When the engine is being cranked, the transistor inside the PCM that controls current to the fuel pump relay coil is switched off, so the fuel pump relay closes and battery voltage is applied directly to the fuel pump, which operates at high speed. After the engine has started, the transistor inside the PCM that operates the fuel pump relay is on, so the fuel pump relay opens and battery voltage is supplied to the fuel pump through the fuel pump resistor.

The fuel pump circuit on V6 models is similar, but the fuel pump only runs at one speed. When the engine is cranked, current flows from the ignition switch to the starter relay coil, the starter relay switches on and current flows through the starter relay to the PCM. When the engine is turning, the Crankshaft Position (CKP) sensor also sends a signal to the PCM. When the PCM receives these two signals, it switches on a transistor that controls the circuit opening relay coil, the circuit opening relay switches on and current is supplied to the fuel pump. As long as the PCM continues to receive a starter signal and a CKP sensor signal, the transistor continues to supply current to the circuit opening relay, and the fuel pump keeps running. (Note that there is no fuel pump relay on these models.)

Fuel is circulated from the fuel tank to the fuel injection system through a metal line running along the underside of the vehicle. An electric fuel pump and fuel level sending unit is located inside the fuel tank. The fuel pressure regulator, which is mounted on the fuel rail assembly, routes all excess fuel back to the fuel tank through a separate return line.

EXHAUST SYSTEM

The exhaust system consists of the exhaust manifolds, the exhaust pipe(s), the catalytic converters, the muffler(s) and the tail pipe(s). For information on servicing the exhaust system, refer to Section 16. The catalytic converters are emission control devices added to the exhaust system to reduce pollutants. For more information regarding the catalytic converters, refer to Chapter 6.

2   Fuel pressure relief procedure

♦ **Refer to illustration 2.3**

✳✳ WARNING:

Gasoline is extremely flammable, so take extra precautions when you work on any part of the fuel system. Don't smoke or allow open flames or bare light bulbs near the work area, and don't work in a garage where a gas-type appliance (such as a water heater or a clothes dryer) is present. Since gasoline is carcinogenic, wear latex gloves when there's a possibility of being exposed to fuel, and, if you spill any fuel on your skin, rinse it off immediately with soap and water. Mop up any spills immediately and do not store fuel-soaked rags where they could ignite. The fuel system is under constant pressure, so, if any fuel lines are to be disconnected, the fuel pressure in the system must be relieved first. When you perform any kind of work on the fuel system, wear safety glasses and have a Class B type fire extinguisher on hand.

1   Remove the fuel filler cap to relieve any pressure inside the fuel tank.

2   Start the engine.

3   On V8 models, remove the fuel pump relay or, on V6 models, remove the circuit opening relay, from the fuse and relay box located on the left side of the engine compartment (see illustration).

➡**Note: The illustration accompanying this step depicts a typical fuse and relay box in a V8 engine compartment. To locate the fuel pump relay or circuit opening relay in your vehicle's fuse and relay box, refer to the fuse and relay guide on the underside of the fuse and relay box lid, or in your owner's manual.**

4   Allow the engine to run until it stops. Turn the ignition switch to OFF and disconnect the cable from the negative terminal of the battery before working on the fuel system.

5   The fuel system pressure is now relieved. When you're finished working on the fuel system, install the circuit opening relay and connect the negative cable to the battery.

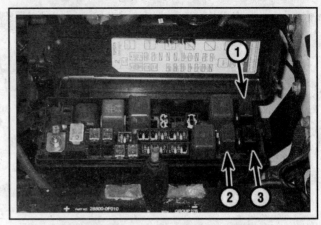

2.3 To relieve fuel pressure, disable the electric fuel pump by removing the fuel pump relay on V8 models, or by removing the circuit opening relay on V6 models; the relays are located in the fuse and relay box on the left side of the engine compartment (typical)

1	Circuit opening relay	3	EFI relay
2	Fuel pump relay		

3   Fuel pump/fuel pressure - check

❊❊ WARNING:

Gasoline is extremely flammable, so take extra precautions when you work on any part of the fuel system. See the Warning in Section 2.

GENERAL CHECKS

1   If you suspect insufficient fuel delivery check the following items first:

   a) *Check the battery and make sure that it's fully charged (see Chapter 5).*

   b) *Check the fuel filter (see Chapter 1) for obstructions.*

   c) *Inspect all fuel lines. Verify that the problem is not simply a leak in a line.*

2   Verify that the fuel pump actually runs. Remove the fuel filler cap and then have an assistant turn the ignition switch to ON while you listen carefully for the sound of the fuel pump operating. You should hear a brief whirring noise (for about two seconds) as the pump comes on and pressurizes the system. If the fuel pump makes no sound, check the fuel pump electrical circuit (see the wiring diagrams at the end of Chapter 12). If the fuel pump runs, but a fuel system problem persists, proceed to the fuel pump pressure check.

FUEL PUMP PRESSURE CHECK

Refer to illustrations 3.4, 3.5a, 3.5b and 3.5c

➡**Note 1: Before proceeding, make sure that you have a fuel pressure gauge capable of measuring fuel pressure up to 60 psi and a special adapter bolt of the correct size for connecting the fuel pressure gauge to the fuel injection system (see Step 2).**

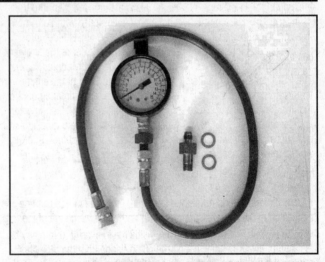

3.4 This typical fuel pressure testing setup includes a fuel pressure gauge capable of measuring the operating pressure range of fuel injection systems and a special banjo-type adapter bolt with a Schrader valve on top to allow you to easily connect the pressure gauge hose to it

➡**Note 2: On V8 engines a quick check of fuel pump operation can be made by observing the screw in the center of the fuel pulsation damper (located at the rear of the left fuel rail). When the engine is cranked over, the screw will pop up a little if the fuel pump is working.**

3   Relieve the fuel system pressure (see Section 2).

V8 models and 2000 through 2004 V6 models

4   Besides a fuel pressure gauge capable of reading fuel pressure up to 60 psi, you'll need a special adapter fitting to hook up the fuel pressure gauge to the fuel rail (see illustration). The fitting looks like a banjo

3.5a To check the fuel pressure, remove this banjo bolt from the front of the left fuel rail, remove the sealing washers above and below the banjo fitting . . .

3.5b . . . install the adapter bolt with the sealing washers placed above and below the banjo fitting . . .

3.5c . . . then hook up the fuel pressure gauge to the Schrader valve on top of the adapter (V8 engine shown, 2000 through 2004 V6 engines similar)

bolt with a Schrader valve on top. This fitting is available from specialty tool manufacturers and from some auto parts stores.

5   Remove the banjo bolt (see illustration) from the front of the left fuel rail and install the special banjo bolt/Schrader valve (see illustration). Then connect the fuel pressure gauge hose to the adapter (see illustration).

6   Turn all the accessories Off and switch the ignition key On. The fuel pump should run for about two seconds. Verify that it's running and then note the reading on the gauge. After the pump stops running, the pressure should hold steady. After five minutes it should not drop below the minimum listed in this Chapter's Specifications.

7   Start the engine and let it idle at normal operating temperature. Check the gauge and compare the fuel pressure with the value listed in this Chapter's Specifications. If the pressure is higher than specified, turn the engine off and depressurize the fuel system (see Section 2). Detach fuel return line from the fuel pressure regulator and blow through the return line to check it for an obstruction. If there is no obstruction, replace the fuel pressure regulator.

8   If the pressure is lower than specified, pinch the fuel return line (connected to the fuel pressure regulator) and watch the gauge; if the pressure increases, replace the fuel pressure regulator. If it does not increase, replace the fuel filter (see Chapter 1) and check the pressure again. If it is still low, the fuel pump is probably faulty.

V6 engine only

9   With the engine running, detach the vacuum hose from the fuel pressure regulator and check the gauge. Compare the reading on the gauge with the value listed in this Chapter's Specifications. If the pressure didn't rise when the vacuum hose was disconnected, check for vacuum at the hose. If there is no vacuum, check the hose for an obstruction or a break. If vacuum is present, replace the fuel pressure regulator.

All engines

10 After the testing is complete, relieve the fuel pressure (see Section 2), remove the fuel pressure gauge, install the crossover pipe-to-fuel rail banjo bolt and sealing washers, then tighten the banjo bolt to the torque listed in this Chapter's Specifications.

2005 and later V6 models

Refer to illustrations 3.12 and 3.14

11 Besides a fuel pressure gauge capable of reading fuel pressure up to 60 psi, you'll need a T-fitting to connect the pressure gauge between the fuel supply line and the fuel rail, and some hose clamps to secure the gauge hose and the fuel supply hose to the T-fitting.

12 After relieving the fuel pressure (see Section 2), locate the fuel supply hose quick-connect fitting at the rear end of the left fuel rail tube, just in front of the fuel pulsation damper (see illustration). (If you're unfamiliar with this type of fitting, refer to Section 4.)

13 Disconnect the fuel supply hose quick-connect fitting.

14 Using the T-fitting, connect the fuel pressure gauge between the fuel supply hose and the fuel rail (see illustration).

15 Start the engine and measure the fuel pressure at idle. Compare your reading to the fuel pressure range listed in this Chapter's Specifications.

16 If the fuel pressure is out of range, check the fuel pump (see Sections 7 and 8), the fuel pressure regulator (see Section 14) and the fuel

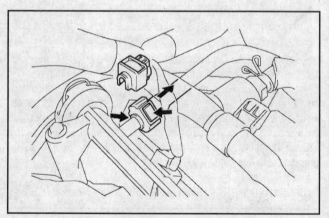

3.12 On 2005 and later V6 models, locate the fuel supply hose quick-connect fitting at the rear end of the left fuel rail tube, remove the clamp, depress the release button and pull two halves of the connector apart. Have some rags handy to mop up the fuel that will spill out

injectors (see Section 15).

17 Stop the engine, wait five minutes, then verify that the fuel pressure is still at least at the level listed in this Chapter's Specifications.

18 Relieve the system fuel pressure (see Section 2).

19 Loosen the hose clamp and remove the T-fitting and fuel pressure gauge.

20 Reconnect the fuel supply hose quick-connect fitting and install the clamp.

FUEL PUMP ELECTRICAL CIRCUIT CHECK

21 If the pump does not turn on (makes no sound) with the ignition switch in the ON position, check the IGN fuse (located on the fuse panel underneath the left end of the dash) and the EFI fuse (located in the engine compartment fuse and relay box). Also check the circuit opening relay, EFI relay and, on V8 models, the fuel pump relay.

➡**Note: These relays are located in the engine compartment fuse/relay box (see illustration 2.3).**

22 If the relays are good and the fuel pump does not operate, check the fuel pump circuit (see the wiring diagrams at the end of Chapter 12).

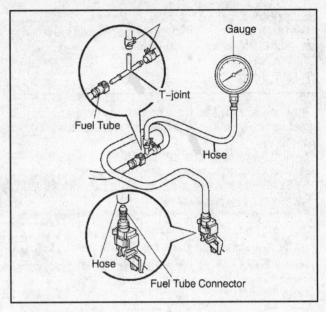

3.14 Fuel pressure connection details (2005 and later V6 models)

---

4 Fuel lines and fittings - general information

◆ **Refer to illustrations 4.2a, 4.2b and 4.2c**

❄❄ WARNING:

Gasoline is extremely flammable, so take extra precautions when you work on any part of the fuel system. See the Warning in Section 2.

1 Always relieve the fuel pressure before servicing fuel lines or fittings (see Section 2).

2 The fuel supply and return lines extend from the fuel tank to the engine compartment (see illustrations). The lines are secured to the underbody with clips. Anytime you raise the vehicle for underbody service, inspect the lines underneath the vehicle for leaks, kinks and dents.

3 If evidence of dirt is found in the system or fuel filter during

4.2a At the fuel tank, the fuel supply hose (lower arrow) and the fuel return hose (upper arrow) are connected to metal lines which are routed underneath the vehicle . . .

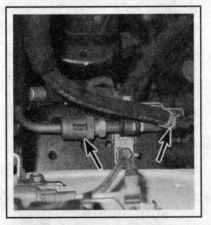

4.2b . . . to the left side of the engine compartment, where the fuel supply hose is screwed into a threaded fitting (left arrow) and the return hose is clamped to the return line (right arrow) . . .

4.2c . . . and then to the left-side fuel rail, where the fuel supply hose (left arrow) is permanently attached to a metal line connected to the fuel rail by the fuel pulsation damper (upper arrow); the return hose (right arrow) is clamped to a metal line connected by another banjo fitting to the fuel pressure regulator (not visible) on the rear end of the right fuel rail (V8 engine shown, V6 similar)

disassembly, disconnect the fuel supply line and blow it out with compressed air. And be sure to inspect the fuel strainer on the fuel pump (see Section 8) for damage and deterioration.

STEEL TUBING

4   If replacement of a fuel line or emission line is called for, use welded steel tubing meeting the manufacturer's specifications or its equivalent.

5   Don't use copper or aluminum tubing to replace steel tubing. These materials cannot withstand normal vehicle vibration.

6   Because fuel lines used on fuel-injected vehicles are under high pressure, they require special consideration.

7   Some fuel lines have threaded fittings with O-rings. Any time the fittings are loosened to service or replace components:

a) *Hold the stationary fitting of the fuel line with one wrench while loosening the tube nut with another. This will prevent the line from twisting.*

b) *Check all O-rings for cuts, cracks and deterioration. Replace any that appear hardened, worn or damaged.*

c) *If the lines are replaced, always use original equipment parts, or parts that meet the original equipment standards specified in this Section.*

FLEXIBLE HOSE

❋❋ CAUTION:

Use only original equipment replacement hoses or their equivalent. Others may fail from the high pressures of this system.

---

8   Don't route fuel hose within four inches of any part of the exhaust system or within ten inches of the catalytic converter. Metal lines and rubber hoses must never be allowed to chafe against the frame. A minimum of 1/4-inch clearance must be maintained around a hose to prevent contact with the frame.

9   Some models may be equipped with nylon fuel line and quick-connect fittings at the fuel filter and/or fuel pump. The quick-connect fittings cannot be serviced separately. Do not attempt to service these types of fuel lines in the event the retainer tabs or the line becomes damaged. Replace the entire fuel line as an assembly.

REPLACEMENT

10   If a fuel line or fuel hose is damaged, replace it with factory replacement parts. Do not substitute line or hose of inferior quality; it might not be suitable for, and it might fail from, the operating pressure of this system.

11   Relieve the fuel pressure.

12   Remove all clamps and/or clips attaching the lines to the vehicle body. Pay close attention to all clips; they not only secure the fuel lines and hoses, they also route them correctly. The hoses and lines must be reattached to their respective clips when reassembled.

13   On fuel hoses that are clamped onto the metal fuel lines, loosen the clamp and pull the hose off the fitting. If a hose sticks to the metal line, twisting it back and forth will help to loosen it. Before installing old clamps, make sure that they're still tight. Replace any worn clamps.

Quick-connect fittings

▶ **Refer to illustrations 4.14a, 4.14b and 4.14c**

14   Rotate the outer part of the fitting in the direction of the arrow until the tabs of the outer fitting are disengaged from the "windows" of the inner fitting and then pull the metal line and the hose apart (see illustrations).

15   Inspect the O-ring inside the inner fitting. If the O-ring is damaged or worn, replace it.

16   To reconnect a quick-connect fitting, push the metal line into the hose until it stops at the raised ridge on the line. Then rotate the outer part of the fitting in the opposite direction of the arrow until the tabs on the outer fitting seat into the windows in the inner part of the fitting.

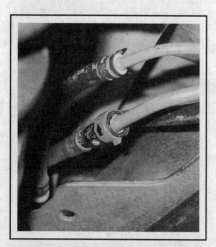
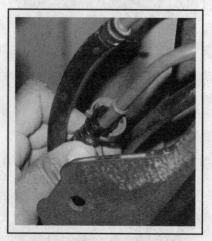
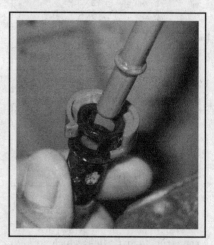

4.14a   To disconnect a quick-connect fitting, rotate the outer part of the fitting in the direction of the arrow . . .

4.14b   . . . until the tabs of the outer fitting are disengaged . . .

4.14c   . . . from the "windows" of the inner fitting, then pull the hose off the line; be sure to inspect the O-ring inside the inner fitting

2005 and later V6 models

▶ **Refer to illustrations 4.17 and 4.18**

17 These models utilize a slightly different style of quick-connect fitting to connect the fuel supply and return hoses to the fuel rail. First, remove the protective clamp from the connector (see illustration).

18 Disconnect the quick-connect fitting (see illustration).

19 To reconnect the fitting, push the male side into the female side until you hear/feel an audible click. To verify that the fitting is secure, try to pull the two halves of the fitting apart.

20 To install the protective clamp on the quick-connect fitting, push it onto the hose, then push it sideways until it clicks into place.

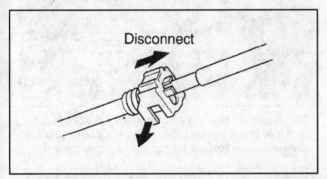

4.17 To remove the clamp from the quick-connect fitting, push it sideways, then pull it off (2005 and later V6 models)

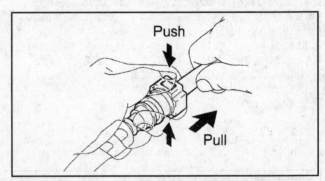

4.18 To disconnect the quick-connect fitting, depress the release buttons and pull the two halves of the fitting apart (2005 and later V6 models)

5 Fuel tank - removal and installation

❋❋ **WARNING:**

Gasoline is extremely flammable, so take extra precautions when you work on any part of the fuel system. See the Warning in Section 2.

REMOVAL

▶ **Refer to illustrations 5.4, 5.7, 5.9a, 5.9b, 5.10a and 5.10b**

1 Relieve the fuel system pressure (see Section 2) and remove the fuel tank cap.

2 Disconnect the cable from the negative terminal of the battery.

3 Raise the rear of the vehicle and place it securely on jackstands.

4 On 2000 through 2004 models, loosen the hose clamps that secure the fuel filler neck hose and the EVAP hose to the fuel filler neck pipe (see illustration), then disconnect both hoses. On 2005 models, the EVAP hose is connected to the fuel filler neck pipe with a quick-connect fitting. If you're unfamiliar with this type of fitting, refer to Section 4. On 2005 models, unscrew the fuel tank cap and remove the four bolts that secure the fuel filler neck pipe shield and remove the shield.

5 On 2005 models, remove the EVAP canister assembly (see Chapter 6).

6 On 2005 models, remove the nut that secures the speed sensor wire harness to the fuel tank and detach the harness from the tank.

7 Disconnect the fuel supply hose, the fuel return hose and the EVAP hose from the metal fuel supply line, fuel return hose and EVAP line (see illustration). If any of these hoses employs a quick-connect fitting to connect it to the metal line, refer to Section 4 for help with disconnecting this type of fitting.

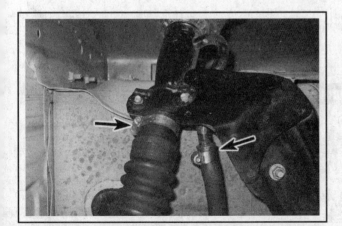

5.4 Loosen these hose clamps and disconnect the fuel filler neck hose and the fuel vapor hose (2000 through 2004 model shown)

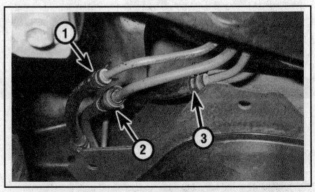

5.7 Before lowering the fuel tank, disconnect these hoses from the metal lines (2000 through 2004 models)

1 *Fuel return hose*
2 *Fuel supply hose (for instructions regarding quick-connect fittings, see Section 4)*
3 *EVAP hose*

8   On 2005 and later models, remove the two nuts that secure the fuel filler neck pipe bracket to the fuel tank and detach the bracket from the tank.

9   Place a transmission jack or a floor jack under the fuel tank assembly (see illustration). If you're using a floor jack, put a piece of plywood between the jack head and the tank assembly to protect the tank. Raise the jack just enough to take the weight of the tank off the fuel tank bands, then remove the bolts that secure the fuel tank bands (see illustration).

10   Carefully lower the tank just far enough to detach the fuel pump/ fuel level sending unit wiring harness clip and to disconnect the fuel pump/fuel level sending unit electrical connector (see illustrations).

11   Carefully lower the tank and remove it.

12   Installation is the reverse of removal.

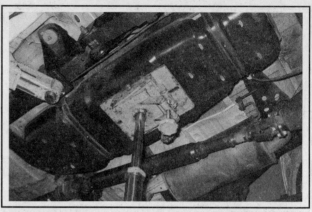

5.9a Support the tank with a transmission jack (shown) or use a floor jack; if you use a floor jack, put a piece of plywood between the jack head and the fuel tank to protect the tank

5.9b To detach the fuel tank, remove these tank band bolts

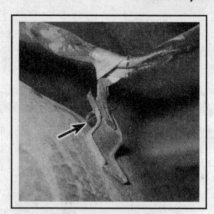

5.10a Detach the fuel pump/fuel level sending unit harness clip from this bracket on the fuel tank . . .

5.10b . . . then disconnect the fuel pump/fuel level sending unit electrical connector (2000 through 2004 model shown)

6   Fuel tank cleaning and repair - general information

1   The fuel tanks installed in the vehicles covered by this manual are not repairable. If the fuel tank becomes damaged, it must be replaced.

2   Cleaning the fuel tank (due to fuel contamination) should be performed by a professional with the proper training to carry out this critical and potentially dangerous work. Even after cleaning and flushing, explosive fumes may remain inside the fuel tank.

3   If the fuel tank is removed from the vehicle, it should not be placed in an area where sparks or open flames could ignite the fumes coming out of the tank. Be especially careful inside a garage where a gas-type appliance is located.

7   Fuel pump/fuel level sending unit assembly - removal and installation

▶ **Refer to illustration 7.3a and 7.3b**

1   Relieve the system fuel pressure (see Section 2).

2   Remove the fuel tank (see Section 5).

3   On 2000 through 2004 models, disconnect the fuel supply and return hoses from the fuel pump/fuel level sending unit, then remove the bolts that attach the fuel pump/fuel level sending unit flange to the fuel tank (see illustration). On 2005 and later models, disconnect the electrical connector from the fuel pump/fuel level sending unit, then remove the two tube joint U-clips (see illustration) and pull up on the fuel supply and return line fittings to disconnect them from the pump. Then use a large pair of pliers or a hammer and punch to loosen the threaded fuel pump retaining ring.

4   Carefully withdraw the fuel pump/fuel level sending unit assembly from the fuel tank. Make sure you don't damage the fuel pump filter or bend the sending unit float arm.

5   If you want to replace either the fuel pump or the fuel level sending unit, you'll have to separate the two components (see Section 8).

6   While the pump is removed, inspect the pump filter (see Section 8). Make sure that it's not clogged or damaged. If the filter is dirty, try washing it in clean solvent. If it's still clogged, replace it.

7   Installation is the reverse of removal. Be sure to tighten the fuel pump/fuel level sending unit flange bolts securely.

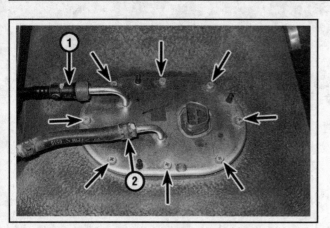

7.3a To remove the fuel pump/fuel level sending unit assembly, disconnect the fuel supply hose (1) and the fuel return hose (2), then remove all eight flange bolts

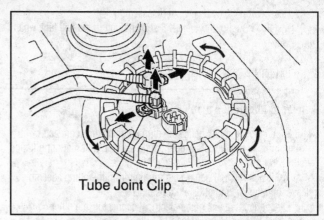

7.3b To disconnect the fuel supply and return line fittings on 2005 models, pull out the tube joint U-clips, then pull the tube fittings straight up. To detach the fuel pump from the fuel tank, unscrew the retaining ring

8   Fuel pump/fuel level sending unit - replacement

1   Remove the fuel tank (see Section 5) and remove the fuel pump/fuel level sending unit from the tank (see Section 7).

2000 THROUGH 2004 MODELS

▶ **Refer to illustrations 8.2, 8.4 and 8.7**

2   Remove the gasket from the fuel pump/fuel sending unit flange (see illustration).

3   Pry the lower end of the fuel pump loose from its retaining bracket.

4   Remove the rubber cushion from the lower end of the fuel pump and remove the clip that secures the filter to the pump (see illustration).

5   Remove the fuel pump filter and inspect it for contamination. If it's dirty, try washing it in clean solvent; use a soft brush to work loose silt and sludge. If the filter cannot be adequately cleaned, replace it.

6   If you're only replacing the fuel pump filter, install the new filter, the clip and the rubber cushion, push the lower end of the pump back into the slot at the end of the retaining bracket, then install the fuel pump/fuel level sending unit assembly in the fuel tank (see Section 7).

7   If you're replacing the fuel pump or the fuel level sending unit, remove the hose clamp at the upper end of the pump (see illustration) and then disconnect the pump from the hose.

8.2 Remove the old rubber gasket from the fuel pump/fuel level sending unit flange

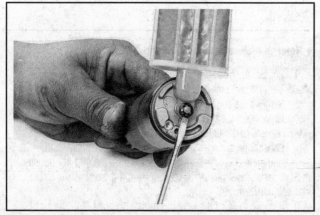

8.4 To disconnect the fuel filter from the fuel pump, remove this clip

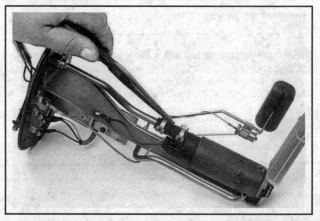

8.7 To disconnect the fuel pump from the fuel level sending unit assembly, remove the upper hose clamp and then pull off the hose

8   Reassembly is the reverse of disassembly.

9   Install the fuel pump/fuel level sending unit in the fuel tank when you're done.

2005 AND LATER MODELS

▶ **Refer to illustration 8.10**

10   Disconnect the fuel gauge sending unit electrical connector from the underside of the fuel pump/fuel gauge sending unit mounting flange (see illustration).

11   To detach the fuel gauge sending unit from the pump assembly, release the lock tab and slide off the sender.

12   Disconnect the fuel pump electrical connector (the other connector) from the underside of the fuel pump/fuel gauge sending unit mounting flange.

13   Detach the suction filter hose from the fuel sub-tank.

14   Disengage the two plastic tubes from the tube guides on top of the fuel suction plate.

15   Wrap the tip of a screwdriver with tape and disengage the three snap-claws that secure the fuel suction plate to the fuel sub-tank, then separate the fuel suction plate and fuel suction filter as a single assembly from the fuel sub-tank.

16   Using the screwdriver again, disengage the five snap-claws from the fuel suction filter and remove the suction filter from the fuel pump. If the suction filter is dirty, try cleaning it with solvent and an old toothbrush. If you're unable to clean the suction filter, replace it.

17   Disconnect the fuel pump harness electrical connector from the fuel pump.

18   Remove the O-ring from the fuel filter or fuel pump (it might be stuck to either component) and discard it.

19   If you're going to replace the main fuel filter, the component that's an integral part of the mounting pump/sending unit mounting flange, you'll have to detach the fuel suction plate. To do so, remove the two E-rings from the fuel suction plate, then slide off the suction plate and remove the two long springs.

20   Installation is the reverse of removal. Be sure to use a new O-ring between the pump and the main filter.

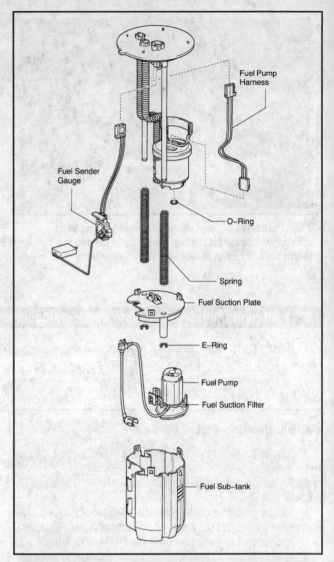

8.10 An exploded view of the fuel pump/fuel gauge sending unit used on 2005 and later models

9   Air filter housing - removal and installation

ALL MODELS EXCEPT 2005 AND LATER V6

▶ **Refer to illustrations 9.1 and 9.3**

1   Loosen the clamp (see illustration) that secures the air intake duct to the air filter housing and pull the air inlet duct away from the housing.

2   Remove the two screws from the Mass Air Flow (MAF) sensor unit on top of the air cleaner assembly, then remove the MAF sensor and set it aside.

3   Remove the three air filter housing mounting bolts (see illustration), lift up on the air cleaner housing and carefully work the air intake snorkel free of its grommet in the right fenderwell. (The air intake snorkel is riveted to the air cleaner housing assembly; so it must be removed with the air cleaner.)

2005 AND LATER V6 MODELS

▶ **Refer to illustrations 9.4, 9.6 and 9.9**

4   Loosen the hose clamp (see illustration) and disconnect the air intake duct from the air filter housing.

5   Remove the engine cover (see illustration 4.12 in Chapter 2A).

6   Disconnect the electrical connector from the Mass Air Flow (MAF) sensor (see illustration).

7   Loosen the spring-type clamp and disconnect the Positive Crankcase Ventilation (PCV) fresh air inlet hose from the air filter housing.

8   Loosen the screw on the hose clamp that secures the intake manifold to the throttle body.

9   Remove the air filter housing mounting bolts, pull the air filter housing forward, disconnect the vacuum hose from the back of the housing (see illustration) and remove the air filter housing.

10   Installation is the reverse of removal.

9.1 Loosen this hose clamp (lower right arrow) and detach the air intake duct from the air cleaner housing; to detach the MAF sensor from the air cleaner housing, remove the two screws

9.3 To detach the air cleaner housing, remove these three bolts (arrows), lift up the housing and carefully work the air intake snorkel out from its grommet in the right fenderwell; do not try to detach the air intake snorkel from the housing - it's attached to the housing by a rivet (center arrow) (housing cover and filter element removed for clarity)

9.4 Loosen the hose clamp that secures the air intake duct to the air filter housing and disconnect the duct from the housing (2005 and later V6 engine)

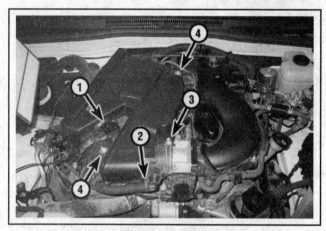

9.6 To detach the air filter housing from a 2005 and later V6 engine:

1 Disconnect the electrical connector from the MAF sensor
2 Loosen the spring clamp and disconnect the PCV fresh air inlet hose
3 Loosen the screw on the hose clamp that secures the air filter housing to the throttle body
4 Remove the two air filter housing mounting bolts

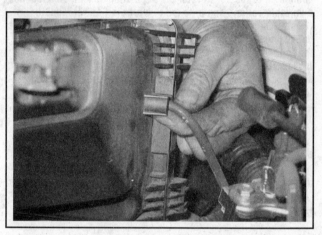

9.9 After removing the two filter housing mounting bolts, pull the housing forward slightly and disconnect this vacuum hose from the back of the housing

10 Accelerator cable - removal, installation and adjustment

➡Note: This procedure applies to 2000 through 2004 V6 models and to 2000 through 2002 V8 models. 2000 through 2004 V6 models are equipped with a conventional throttle body with a cable-actuated throttle plate. 2000 through 2002 V8 models are equipped with an electronic throttle body unit that uses an accelerator cable, whose only purpose is to actuate an accelerator pedal position sensor located on the throttle body; the throttle plate is actually rotated by a throttle motor, also mounted on the throttle body. This procedure does NOT apply to 2003 and later V8 models or to 2005 and later V6 models, all of which are equipped with electronic throttle bodies that do not use an accelerator cable.

REMOVAL AND INSTALLATION

▶ **Refer to illustrations 10.1, 10.2, 10.3, 10.4 and 10.5**

1 Loosen the accelerator cable locknut (see illustration) and slide the accelerator cable out of the cable bracket.

2 Pull the accelerator cable forward to create some slack and disengage the accelerator cable end plug from its slot in the throttle cam (see illustration).

3 Disengage the accelerator cable from the cable bracket on the air intake plenum (see illustration).

4 Using a flashlight under the dash, locate the rear end of the accelerator cable at the top of the accelerator pedal. Disengage the accelerator cable from the accelerator pedal arm (see illustration).

5 Detach the accelerator cable housing flange from the firewall (see illustration) and pull the cable through the opening in the firewall.

6 Installation is the reverse of removal. After installing the accelerator cable, adjust it as follows.

ADJUSTMENT

➡ **Note: Cable adjustment is especially important on V8 models because the cable is not directly linked to the throttle valve. It's connected to the accelerator pedal position sensor, which sends a variable voltage signal to the Powertrain Control Module (PCM), which opens and closes the throttle valve via the throttle control motor. If the cable is too tight, it prevents the accelerator pedal position sensor from closing fully, which will send an inaccurate voltage signal to the PCM, which will send the wrong signal to the throttle control motor, resulting in an incorrect throttle valve angle. See Chapter 6 for more information on the accelerator pedal position sensor and the throttle control motor.**

7 Lift up on the cable to remove any slack.

8 Turn the adjusting nut until it's about 1/8-inch from the cable bracket.

9 Tighten the locknut and check cable deflection between the throttle lever and the cable casing. Deflection should be about 3/8-inch.

10 If the deflection is too little or too much, loosen the locknut and turn the adjusting nut until the deflection is correct.

11 After you have adjusted the accelerator cable, have an assistant help you verify that the throttle cam opens all the way when you depress the accelerator pedal to the floor and that it returns to the fully closed

10.1 Holding the adjusting nut with a back-up wrench (right wrench), back off the locknut (left wrench) and disengage the accelerator cable from the cable bracket

10.2 Pull the accelerator cable forward to put some slack in it and disengage the cable end plug from its slot in the throttle cam

position when you release the accelerator. And make sure that the cable operates smoothly, without binding or sticking.

10.3 To disengage the accelerator cable from this bracket, simply pull it down (toward the air intake plenum)

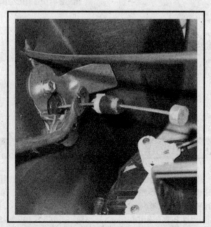

10.4 Disengage the other end of the accelerator cable from its slot at the upper end of the accelerator pedal

10.5 To detach the accelerator cable flange from the firewall, remove these two bolts

11 Sequential Fuel Injection (SFI) system - general information

The Sequential Fuel Injection (SFI) system consists of three sub-systems: air intake, electronic control and fuel delivery. The SFI system uses a Powertrain Control Module (PCM) and various information sensors to calculate the correct air/fuel ratio under all operating conditions. The SFI system and the emission control systems are closely linked. For more information on the emission control systems, refer to Chapter 6.

AIR INTAKE SYSTEM

The air intake system consists of the air cleaner, the air intake duct, the throttle body, the idle control system and the intake manifold. V6 models are equipped with a three-piece intake manifold and V8 models are equipped with a two-piece intake manifold (see Chapter 2A [V6 models] or 2B [V8 models] for the removal and installation procedures for the intake manifold).

The throttle body is a single-barrel, side-draft design. The throttle body is heated by engine coolant to prevent icing in cold weather. A Throttle Position (TP) sensor is attached to the throttle shaft to monitor changes in the throttle opening. 2000 through 2002 V6 models use a conventional, cable-actuated throttle body unit. The electronic throttle body unit employed on 2000 through 2002 V8 and on 2003 and 2004 V6 engines also uses an accelerator cable, but the cable itself doesn't actuate the throttle plate directly.

On these engines, the throttle body is equipped with an accelerator pedal position sensor and a throttle control motor. When the accelerator pedal is depressed, the accelerator cable rotates the accelerator pedal position sensor, which sends a variable voltage signal to the powertrain Control Module (PCM), which calculates the correct throttle opening based on inputs from the accelerator pedal position sensor and other sensors. The PCM then directs the throttle control motor to open the throttle valve the correct amount. The throttle bodies employed on 2003 and later V8 models and on 2005 V6 models are fully electronic (there is no accelerator cable). The accelerator pedal position sensor is located at, and is an integral component of, the accelerator pedal. For more information on the accelerator pedal position sensor and the throttle control motor, refer to Chapter 6.

When the engine is idling, the air/fuel ratio is controlled by the idle air control system. The idle air control system consists of the PCM, the Idle Air Control (IAC) valve (V6 engines) or throttle control motor (V8 engines) and various sensors such as the Engine Coolant Temperature (ECT) sensor, the Throttle Position (TP) sensor and the Mass Air Flow (MAF) sensor. The IAC valve or throttle control motor is controlled by the PCM in accordance with the running conditions of the engine (air conditioning on, power steering demand, cold or warm temperature, etc.). On 2000 through 2002 V6 engines, the IAC valve regulates the amount of airflow bypassing the throttle plate; on V8 engines and on 2003 and later V6 engines, the throttle control motor opens the throttle valve slightly to increase the idle as necessary.

ELECTRONIC CONTROL SYSTEM

For information on the electronic control system, the Powertrain Control Module (PCM) and sensors, refer to Chapter 6.

FUEL DELIVERY SYSTEM

The fuel delivery system consists of the fuel pump, the fuel filter, the fuel supply line connecting the fuel tank to the fuel rail, the fuel pulsation damper, the fuel rail and injectors, the fuel pressure regulator and the fuel return line between the fuel rail and the fuel tank.

The electric fuel pump is located inside the fuel tank. Fuel is drawn through a filter into the pump, flows through the fuel line and fuel filter and then to the fuel rail, from which it's sprayed by the injectors into the intake ports. The fuel pulsation damper, which is located at the connection between the fuel supply line and the fuel rail, dampens the pressure pulses from the fuel pump. The fuel pressure regulator, which is located at the connection between the fuel rail and the fuel return line, maintains a constant fuel pressure to the injectors.

Each injector consists of a solenoid coil, a pintle valve and the housing. When current is applied to the solenoid by the PCM, the pintle valve raises off its seat and the pressurized fuel inside the housing squirts out the nozzle. The amount of fuel injected is determined by the length of time that the pintle valve is open, which is determined by the length of time during which current is supplied to the solenoid. Because the injector on-time determines the air-fuel mixture ratio, injector timing must be very precise.

12 Fuel injection system - check

▶ Refer to illustrations 12.7 and 12.8

✳✳ WARNING:

Gasoline is extremely flammable, so take extra precautions when you work on any part of the fuel system. See the Warning in Section 2.

1 Inspect all system electrical connectors, especially the ground connections. Loose connectors and poor grounds are a common cause of many engine control system problems.

2 Verify that the battery is fully charged. The Powertrain Control Module (PCM) and sensors don't operate correctly without adequate supply voltage.

3 Inspect the air filter element (see Chapter 1). A dirty or partially blocked filter reduces performance and economy.

4 Check fuel pump operation (see Section 3). If the fuel pump fuse is blown, replace it and note whether it blows again. If it does, look for a short in the wiring harness to the fuel pump (see the wiring diagrams in Chapter 12).

5 Inspect all vacuum hoses connected to the intake manifold for damage, deterioration and leakage.

6 Remove the air intake duct from the throttle body and look for dirt, carbon, varnish, or other residue inside the throttle body bore, particularly around the throttle plate. If it's dirty, refer to Chapter 6 and troubleshoot the PCV system for the cause of the excessive residue

7 With the engine running, place an automotive stethoscope against

each injector, one at a time, and listen for a clicking sound that indicates operation (see illustration). If you don't have a stethoscope, you can place the tip of a long screwdriver against the injector and listen through the handle.

8   With the engine off and the fuel injector electrical connectors

disconnected, measure the resistance of each injector with an ohmmeter (see illustration). Refer to the Specifications at the end of this Chapter for the correct resistance.

9   Refer to Chapter 6 for other system checks.

12.7 Use a stethoscope to listen for the clicking sound that indicates that each injector is working correctly; the clicking sound should rise and fall with changes in engine speed

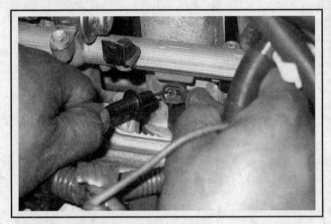

12.8 Using an ohmmeter, measure the resistance across the terminals of each injector with an ohmmeter

13   Throttle body - inspection, removal and installation

INSPECTION

▶ **Refer to illustrations 13.2a and 13.2b**

1   Verify that the throttle linkage operates smoothly. If it doesn't, check

the routing and adjustment of the accelerator cable (see Section 10).

2   Remove the air intake duct from the throttle body and inspect the area around and behind the throttle plate for carbon and residue build-up. If it's dirty, clean it with aerosol carburetor cleaner and a tooth brush. Make sure that the can specifically states that it is safe with oxygen sensor systems and catalytic converters (see illustrations).

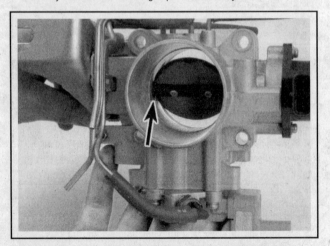

13.2a The area inside the throttle body behind the throttle plate suffers from sludge build-up because of vapors vented into the throttle body by the PCV hose from the crankcase (throttle body removed for clarity)

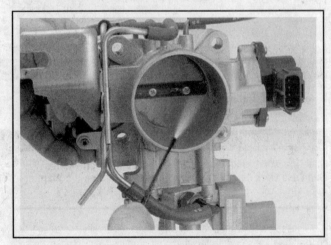

13.2b With the engine off, use aerosol carburetor cleaner (make sure it's approved for use with catalytic converters and oxygen sensors), a toothbrush and a rag to clean the throttle body; be sure to open the throttle plate and clean behind it

REMOVAL AND INSTALLATION

※※ WARNING:

Wait until the engine is completely cool before beginning this procedure.

2000 through 2002 V6 models

▶ **Refer to illustration 13.3**

3   Disconnect the air intake duct from the throttle body (see illustration).
4   Disconnect the accelerator cable (see Section 10).
5   If the vehicle is equipped with an automatic transmission, disconnect the throttle valve cable (see Chapter 7B).

6   Disconnect the cruise control cable, if equipped.
7   Disconnect the Throttle Position (TP) sensor electrical connector (see Chapter 6).
8   Disconnect the Idle Air Control (IAC) electrical connector (see Chapter 6).
9   Clearly label and disconnect all vacuum hoses.
10   Clamp off the coolant bypass hoses attached to the throttle body. Loosen the hose clamps that secure the hoses to the throttle body, then disconnect the hoses. Be prepared for some coolant spillage.
11   Disconnect the air assist hose from the throttle body.
12   Remove the two nuts and two bolts that secure the throttle body to the intake plenum.
13   Remove the throttle body gasket.
14   Installation is the reverse of removal. Be sure to tighten the throttle body bolts and nuts to the torque listed in this Chapter's Specifications. Check the coolant level, adding as necessary (see Chapter 1).

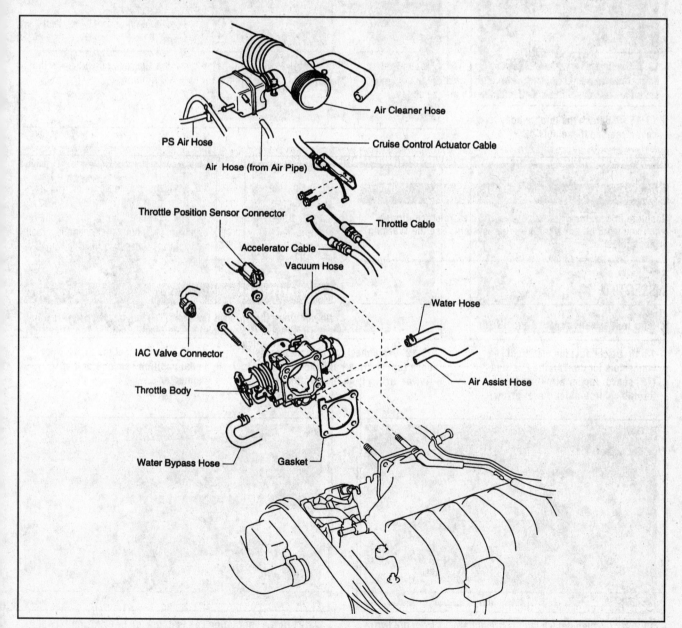

13.3 An exploded view of the throttle body assembly (2000 through 2002 V6 models)

2000 through 2002 V8 models

▶ **Refer to illustrations 13.15, 13.16, 13.18, 13.19, 13.20a, 13.20b, 13.21, 13.22, 13.23 and 13.24**

15 Remove the throttle body cover nuts (see illustration) and remove the cover.

16 Remove the air intake duct (see illustration).

17 Disconnect the accelerator cable from the throttle lever cam (see Section 10).

18 Disconnect the Throttle Position (TP) sensor and the throttle control motor electrical connectors (see illustration).

19 Clamp off the upper coolant bypass hose attached to the throttle body, then disconnect the PCV hose and the coolant bypass hose (see illustration).

20 Disconnect the accelerator pedal position sensor electrical connector and detach the electrical harness from the two clamps on the wire harness brackets (see illustrations).

21 Inspect the condition of the ventilation cap and remove it if necessary (see illustration).

13.15 To detach the throttle body cover, remove these nuts

13.16 To remove the air intake duct on a V8 model, disconnect, remove or loosen the following parts:

1 Power steering air hose
2 Vacuum hose
3 Air inlet hose for EVAP system
4 PCV hose
5 Air conditioning hose clamp bolt
6 Air intake duct-to-air filter housing clamp
7 Air intake duct-to-throttle body clamp

13.18 Disconnect the electrical connectors for the Throttle Position (TP) sensor (upper arrow) and the throttle control motor (lower arrow)

13.19 Disconnect the coolant bypass hose (upper arrow) and the PCV hose (lower arrow) from the throttle body

13.20a Disconnect the accelerator pedal position sensor electrical connector . . .

13.20b . . . then detach the electrical harness from the two clamps on top of the throttle body; to release each clamp, simply pry them open

13.21 If the ventilation cap is damaged or worn, replace it

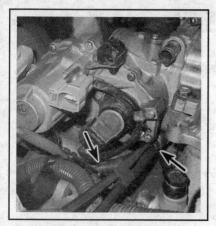

13.22 Clamp off the coolant bypass hose in the area indicated by the left arrow, then disconnect it (right arrow) from the pipe on the underside of the throttle body

13.23 Remove the throttle body nuts and bolts

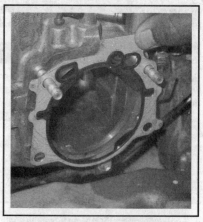

13.24 Remove and discard the old throttle body gasket; do NOT reuse this gasket or you will run the risk of an air leak

22 Clamp off the coolant bypass hose and disconnect it from the underside of the throttle body (see illustration).

23 Remove the two nuts and two bolts that secure the throttle body to the intake plenum (see illustration).

24 Remove the throttle body gasket (see illustration).

25 Installation is the reverse of removal. Be sure to tighten the throttle body bolts and nuts to the torque listed in this Chapter's Specifications. Check the coolant level, adding as necessary (see Chapter 1).

2003 and 2004 V6 models

26 Drain the engine coolant (see Chapter 1), or pinch off the coolant hoses to the throttle body.

27 Loosen the hose clamps and remove the air intake duct.

28 Disconnect the accelerator cable from the throttle body (see Section 10).

29 Disconnect the electrical connectors from the accelerator pedal position sensor, the Throttle Position (TP) sensor and the throttle control motor.

30 Disconnect the air hose from the throttle body.

31 Disconnect the two coolant bypass hoses from the throttle body.

32 Remove the two throttle body mounting nuts and the two mounting bolts.

33 Remove the throttle body assembly.

34 Remove the old throttle body gasket.

35 Installation is the reverse of removal. Be sure to use a new throttle body gasket and tighten the mounting bolts and nuts to the torque listed in this Chapter's Specifications.

2003 through 2005 V8 models and 2005 and later V6 models

▶ Refer to illustration 13.40

➡Note: All of these models are equipped with fully electronic throttle bodies, i.e. they have no accelerator cable. The throttle body unit employed on the 2003 and 2004 V8 models is different from the 2005 unit, which is similar in design to the 2005 and later V6 unit. However, the procedure for removing and installing all three is similar.

36 On V6 models, remove the engine cover (see illustration 4.12 in Chapter 2A). On V8 engines, remove the throttle body cover (see illus-

tration 13.15).

37 Drain the engine coolant (see Chapter 1), or pinch off the coolant hoses to the throttle body.

38 On V6 models, remove the air filter housing (see Section 9).

39 On V8 models, disconnect the air intake duct (see illustration 13.16).

40 Disconnect the electrical connector from the throttle body (see illustrations).

41 Loosen the hose clamps and disconnect both water bypass hoses from the throttle body.

42 Remove the throttle body mounting bolts and/or nut(s).

43 Remove the throttle body.

44 Remove and discard the old throttle body gasket.

45 Installation is the reverse of removal. Be sure to use a new gasket and tighten the throttle body mounting bolts and/or nut(s) to the torque listed in this Chapter's Specifications.

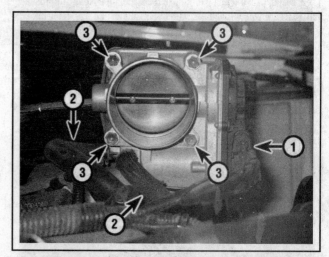

13.40 Throttle body details (2005 V6 throttle body shown, 2003 through 2005 V8 throttle body identical except for mounting fasteners):

1. Throttle motor/throttle position sensor electrical connector
2. Water bypass hoses
3. Throttle body mounting bolts (V6, shown) or three bolts and a nut (V8, not shown)

14 Fuel pressure regulator - removal and installation

✽✽ WARNING:

Gasoline is extremely flammable, so take extra precautions when you work on any part of the fuel system. See the Warning in Section 2.

1 Relieve the fuel system pressure (see Section 2).
2 Disconnect the negative battery cable.

2000 THROUGH 2004 V6 MODELS

3 Remove the air intake duct (see illustration 13.3).
4 Remove the air intake plenum (upper intake manifold) and the intake air connector assembly (middle intake manifold) (see Chapter 2A).
5 Disconnect the vacuum hose from the pipe on top of the fuel pressure regulator (see illustration 15.6).
6 Loosen the hose clamp and disconnect the fuel return hose from the regulator.
7 Remove the two bolts that secure the fuel pressure regulator to the fuel rail, then remove the regulator.
8 Remove and discard the old fuel pressure regulator O-ring. Install a new one before reinstalling the regulator.
9 Installation is the reverse of removal. Be sure to tighten the fuel pressure regulator bolts to the torque listed in this Chapter's Specifications.

2005 AND LATER V6 MODELS

▶ **Refer to illustration 14.13**

10 Remove the engine cover (see illustration 4.12 in Chapter 2A).
11 Remove the air filter housing (see Section 9).
12 Disconnect the fuel return hose (see "fuel pipe No. 2" in illustration 15.37). If you're unfamiliar with quick-connect fittings, refer to Section 4.
13 Disconnect the vacuum hose from the fuel pressure regulator (see illustration).

14 Remove the fuel pressure regulator mounting bolts and remove the pressure regulator. Remove and discard the old pressure regulator O-ring.
15 Installation is the reverse of removal. Be sure to use a new O-ring and tighten the pressure regulator mounting bolts to the torque listed in this Chapter's Specifications.

V8 MODELS

▶ **Refer to illustrations 14.16 and 14.19**

16 Disconnect the vacuum hose from the fuel pressure regulator (see illustration).
17 Loosen the fuel return hose clamp and then disconnect the fuel return hose from the regulator.
18 Remove the two bolts that secure the fuel pressure regulator to the fuel rail and then remove the regulator.
19 Remove and discard the old fuel pressure regulator O-ring (see illustration). Install a new one before reinstalling the regulator.
20 Installation is the reverse of removal. Be sure to tighten the fuel pressure regulator bolts to the torque listed in this Chapter's Specifications.

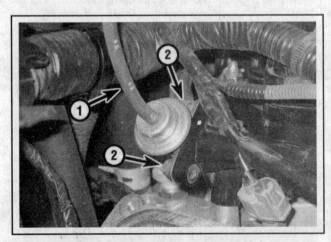

14.13 To detach the fuel pressure regulator from the fuel rail on a 2005 V6 model, disconnect the vacuum hose (1) and remove the two pressure regulator mounting bolts (2)

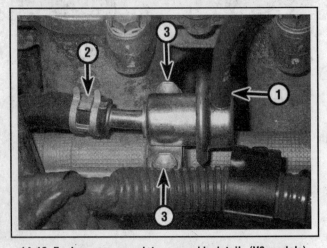

14.16 Fuel pressure regulator assembly details (V8 models)

1   Vacuum hose
2   Fuel return hose clamp
3   Fuel pressure regulator mounting bolts

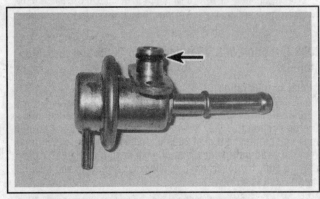

14.19 Remove and discard the old fuel pressure regulator O-ring

15 Fuel rail and injectors - removal and installation

ALL MODELS

✳✳ WARNING:

Gasoline is extremely flammable, so take extra precautions when you work on any part of the fuel system. See the Warning in Section 2.

1 Relieve the system fuel pressure (see Section 2).
2 Detach the cable from the negative terminal of the battery.
3 Remove the air intake duct (see Section 13).

2000 THROUGH 2004 V6 MODELS

♦ **Refer to illustration 15.6**

4 Remove the air intake plenum and the intake air connector (see Chapter 2A).
5 Remove the fuel pressure regulator (see Section 14).

6 Remove the banjo bolt and sealing washers, then disconnect the fuel inlet pipe from the rear end of the left fuel rail (see illustration). Discard the old sealing washers.
7 Remove the two banjo bolts and the sealing washers, then remove the fuel rail crossover pipe from the front end of the fuel rail assembly. Discard the old sealing washers.
8 Disconnect the fuel injector electrical connectors.
9 Remove the two retaining bolts from each fuel rail.
10 Carefully pull up on each fuel rail to disengage the injectors from their bores in the intake manifold.
11 Remove the two spacers from each side (the spacers are the cylindrical pieces that the fuel rail bolts go through). Inspect the spacers for damage. If they're worn or damaged, replace them.
12 Remove the injectors from the fuel rails (see illustration 15.29).
13 Remove the two O-rings and the insulator grommet from each injector (see illustration 15.30). Discard these pieces and replace them with new O-rings and a new grommet (see illustration 15.31).
14 Installation is the reverse of removal. Be sure to tighten the fuel rail retaining bolts to the torque listed in this Chapter's Specifications.
15 When you're done, start the engine and check for leaks at the banjo fittings for the fuel supply line and for the fuel rail crossover pipe.

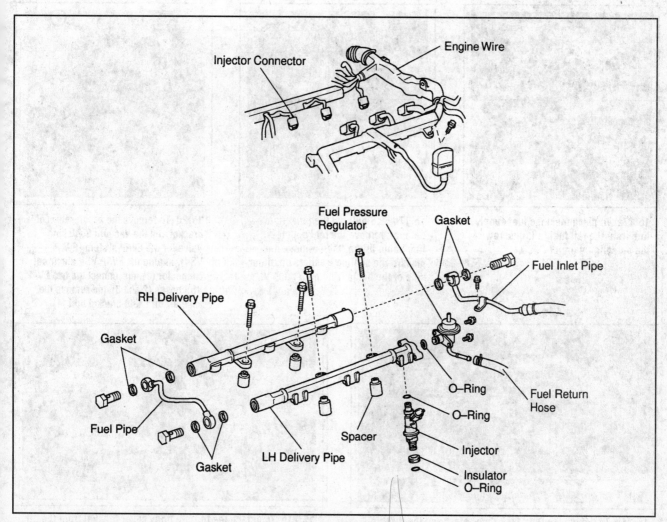

15.6 An exploded view of the fuel rail assembly (2000 through 2004 V6 models)

V8 MODELS

♦ Refer to illustrations 15.17a, 15.17b, 15.20, 15.21a, 15.21b, 15.22a, 15.22b, 15.25, 15.26, 15.27a, 15.27b, 15.29, 15.30 and 15.31

➡Note: The photos accompanying this subsection depict a typical 2000 through 2004 model. However, aside from a crossover pipe that is routed over the top of (instead of underneath) the intake manifold, the 2005 fuel rail assembly is basically identical to what you see here.

16 Remove the throttle body cover and the air intake duct (see Section 13).

17 Unscrew the pulsation damper, remove the fuel supply line banjo fitting, then remove and discard the old sealing washers (see illustrations).

18 Disconnect the accelerator cable from the throttle body (see Section 10), then remove the accelerator cable bracket retaining nuts. Also disengage the accelerator cable from its bracket on the left side of the intake manifold (see illustration 10.3) and set it aside.

19 Disconnect the PCV hose from the PCV valve (see Chapter 6).

20 Disconnect the electrical connector and the hoses from the Vacuum Switching Valve for the EVAP system (EVAP VSV), remove the EVAP VSV retaining bolt, then remove the accelerator cable bracket and the EVAP VSV from the intake manifold (see illustration).

21 Remove the Data Link Connector No. 1 (DLC1) from the throttle body cover bracket, remove the throttle body cover bracket bolt and the bracket (see illustrations).

22 Disconnect the fuel injector/ignition coil wiring harness clamps from the engine lifting hook and from the EVAP VSV bracket on the left side of the intake manifold, and from the two brackets bolted to the right fuel rail (see illustrations).

23 Disconnect the electrical connectors from the four ignition coils and from the four fuel injectors on each cylinder bank, then set the wiring harnesses aside.

24 Disconnect the vacuum hose from the fuel pressure regulator (see Section 14).

25 Remove the bolt (see illustration) that secures the fuel return pipe bracket to the left fuel rail.

26 Remove the left and right banjo bolts (see illustration) that connect the fuel rail crossover pipe to the front ends of the fuel rails, then remove the crossover pipe. Be sure to discard the old banjo bolt sealing washers.

27 Remove the two nuts that attach each fuel rail (see illustrations).

28 Remove the two fuel rails and eight fuel injectors by pulling straight up on each fuel rail while wiggling the injectors free of their

15.17a To disconnect the fuel supply line from the left fuel rail, unscrew the pulsation damper . . .

15.17b . . . pull out the pulsation damper, remove the fuel supply line banjo fitting, then remove and discard the sealing washers on either side of the banjo fitting

15.20 To remove the accelerator cable bracket and the Vacuum Switching Valve for the EVAP system (EVAP VSV), unplug the EVAP VSV electrical connector (1), disconnect the two EVAP VSV hoses (2 and 3) and remove the accelerator cable bracket bolt (4)

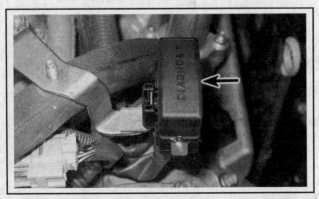

15.21a To detach the Data Link Connector from the throttle body cover bracket, simply pull it straight up

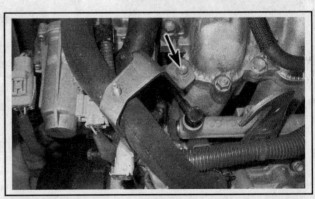

15.21b To detach the throttle body cover bracket from the air intake plenum, remove this bolt

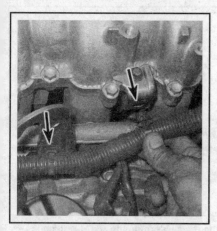

15.22a Detach the left fuel injector/ ignition coil wiring harness clips from the engine lifting hook and from the EVAP VSV bracket

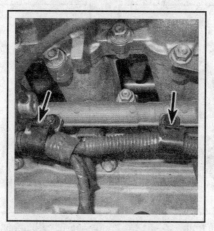

15.22b Detach the right fuel injector/ ignition coil wiring harness clamps from the brackets bolted to the right fuel rail

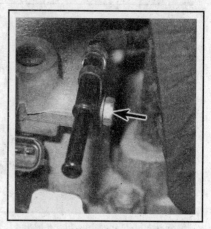

15.25 Remove the bolt that attaches the fuel return pipe bracket to the left fuel rail

bores in the intake manifold.

29 Remove each injector from the fuel rail (see illustration).

30 Remove the old O-ring and grommets from each injector (see illustration).

31 Install a new O-ring and grommets on each injector (see illustration). Coat the O-rings and grommets with a little clean engine oil to help them slide onto the injectors more easily.

32 Insert the injectors into the fuel rails, then push the injectors back into their bores in the intake manifold until they're fully seated. Coat the outer surfaces of the injector O-rings and grommets with a little clean engine oil to facilitate pushing the injectors back into the fuel rail and into their bores in the intake manifold. Install the fuel rail retaining nuts and tighten them to the torque listed in this Chapter's Specifications.

33 Using new sealing washers, install the fuel rail crossover pipe and tighten the banjo bolts to the torque listed in this Chapter's Specifications.

34 When installing the fuel supply line banjo bolt, be sure to use new sealing washers and then tighten the supply line banjo bolt to the torque listed in this Chapter's Specifications.

35 Installation is otherwise the reverse of removal.

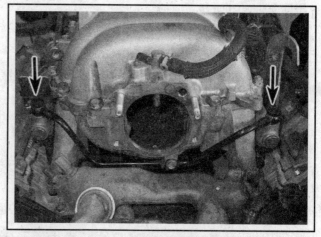

15.26 Remove the left and right fuel rail crossover pipe banjo bolts and then remove the crossover pipe (2000 through 2004 model shown; crossover is routed over the top of intake manifold on 2005 models)

15.27a To detach the left fuel rail from the intake manifold, remove these two nuts

15.27b To detach the right fuel rail from the intake manifold, remove these two nuts

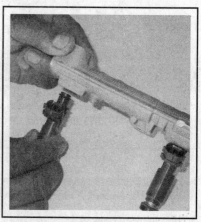

15.29 To work an injector out of its bore in the fuel rail, pull on it while wiggling it from side-to-side at the same time

15.30 Remove the old grommets and O-ring from each injector

15.31 When installing the new grommets and O-ring on each injector, this is how they should look when correctly installed; use a little clean engine oil on the new parts to help them slide into place more easily, and then coat them with a little oil to help them slide into the injector bores in the fuel rail and in the intake manifold

2005 AND LATER V6 MODELS

▶ **Refer to illustration 15.37**

36 Remove the upper intake manifold (see *Intake manifold - removal and installation* in Chapter 2A).

37 Disconnect the quick-connect fittings for fuel pipe No. 1 (the fuel supply hose) and fuel pipe No. 2 (the fuel return hose) (see illustration). If you're unfamiliar with quick-connect fittings, refer to Section 4.

38 Disconnect the vacuum hose from the fuel pressure regulator.

39 Disconnect the six electrical connectors from the fuel injectors.

40 Remove the fuel rail mounting bolts and remove the fuel rail and injectors as a single assembly.

41 Remove the injectors from the fuel rail (see illustration 15.29).

42 Remove and discard the O-ring and insulator from each injector (see illustration 15.30).

43 Install a new O-ring and insulator on each injector (see illustration 15.31).

44 Installation is the reverse of removal.

45 When you're done, start the engine and check for leaks at the quick-connect fittings and at the injectors.

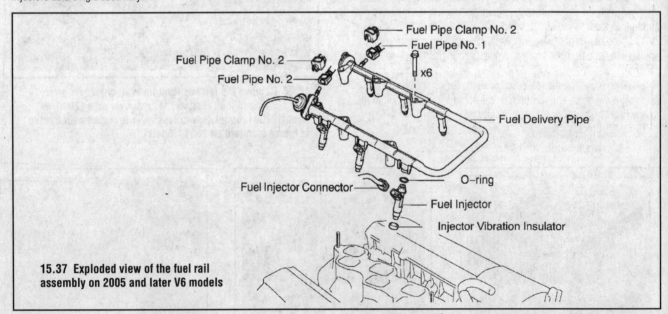

15.37 Exploded view of the fuel rail assembly on 2005 and later V6 models

Fuel Pipe Clamp No. 2
Fuel Pipe No. 1
Fuel Pipe Clamp No. 2
Fuel Pipe No. 2
x6
Fuel Delivery Pipe
Fuel Injector Connector
O-ring
Fuel Injector
Injector Vibration Insulator

16 Fuel pressure pulsation damper (2005 and later V6 models) - replacement

▶ **Refer to illustration 16.3**

➡**Note: The fuel pressure pulsation damper is located at the rear end of the left fuel rail tube.**

1 Remove the engine cover (see illustration 4.12 in Chapter 2A).

2 Relieve the fuel system pressure (see Section 2).

3 Remove the clip from the pulsation damper (see illustration), then pull the pulsation damper out of the fuel rail.

4   Remove and discard the old pulsation damper O-ring.
5   Installation is the reverse of removal. Be sure to use a new
O-ring. Apply a light coat of oil to the new O-ring before installing it on
the pulsation damper.

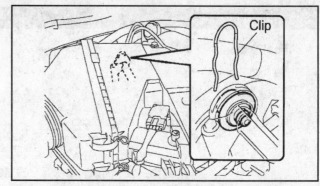

16.3 To remove the fuel pressure pulsation damper from
the fuel rail, remove this clip, then pull out the pulsation
damper. Be sure to use a NEW O-ring when installing the
pulsation damper

17 Exhaust system servicing - general information

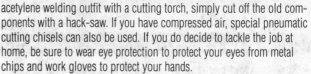

▶ Refer to illustrations 17.1a and 17.1b

✳✳ WARNING:

**Inspection and repair of exhaust system components should be
done only after the system components have cooled completely.**

1   The exhaust system consists of the exhaust manifolds, the catalytic converters, the muffler, the tailpipe and all connecting pipes, brackets, hangers and clamps. The exhaust system is attached to the body
with mounting brackets and rubber hangers (see illustration). If any of
these parts are damaged or deteriorated, excessive noise and vibration
will be transmitted to the body.
2   Conducting regular inspections of the exhaust system will keep it
safe and quiet. Look for any damaged or bent parts, open seams, holes,
loose connections, excessive corrosion or other defects which could
allow exhaust fumes to enter the vehicle. Deteriorated exhaust system
components should not be repaired - they should be replaced with new
parts.
3   If the exhaust system components are extremely corroded or
rusted together, they will probably have to be cut from the exhaust system. The convenient way to accomplish this is to have a muffler repair
shop remove the corroded sections with a cutting torch. If, however,
you want to save money by doing it yourself and you don't have an oxy/

acetylene welding outfit with a cutting torch, simply cut off the old components with a hack-saw. If you have compressed air, special pneumatic
cutting chisels can also be used. If you do decide to tackle the job at
home, be sure to wear eye protection to protect your eyes from metal
chips and work gloves to protect your hands.
4   Here are some simple guidelines to apply when repairing the
exhaust system:

a) *Work from the back to the front when removing exhaust system
    components.*
b) *Apply penetrating oil to the exhaust system component fasteners
    to make them easier to remove.*
c) *Use new gaskets, hangers and clamps when installing exhaust
    system components.*
d) *Apply anti-seize compound to the threads of all exhaust system
    fasteners during reassembly. Be sure to allow sufficient clearance
    between newly installed parts and all points on the underbody to
    avoid overheating the floor pan and possibly damaging the inte-
    rior carpet and insulation. Pay particularly close attention to the
    catalytic converter and its heat shield.*

✳✳ WARNING:

**The catalytic converter operates at very high temperatures and
takes a long time to cool. Wait until it's completely cool before
attempting to remove the converter. Failure to do so could result
in serious burns.**

17.1a A typical rubber exhaust hanger

17.1b Another typical rubber exhaust hanger

Specifications

Fuel system

Fuel system pressure
    V6 engines
        2000 through 2004 (at idle)
            Vacuum hose disconnected

from fuel pressure regulator	38 to 44 psi
Vacuum hose connected to fuel pressure regulator	33 to 38 psi
2005 and later (at idle)	41 to 42 psi
V8 engine (at idle)	38 to 44 psi
Residual fuel system pressure (five minutes after engine turned off)	21 psi minimum
Injector resistance (approximate)	
2000 through 2004 V6 and all V8 engines	13.4 to 14.2 ohms
2005 and later V6 engine	11.5 to 12.5 ohms

Torque specifications   Ft-lbs (unless otherwise indicated)

➡ **Note: One foot-pound (ft-lb) of torque is equivalent to 12 inch-pounds (in-lbs) of torque. Torque values below approximately 15 ft-lbs are expressed in inch-pounds, since most foot-pound torque wrenches are not accurate at these smaller values.**

Fuel supply pipe-to-fuel rail banjo bolt/pulsation damper	
V6 engines	
2000 through 2004 (banjo bolt)	25
2005 and later	Not applicable (uses a quick-connect fitting)
V8 engine (pulsation damper)	29
Fuel pressure regulator bolts	
V6 engines	
2000 through 2004	71 in-lbs
2005 and later	80 in-lbs
V8 engine	66 in-lbs
Fuel rail bolts	
V6 engines	
2000 through 2004	120 in-lbs
2005 and later	132 in-lbs
V8 engine	66 in-lbs
V6 engines	
2000 through 2004	25
2005 and later	Not applicable (no crossover pipe)
V8 engine	29
Throttle body mounting bolts/nuts	
V6 engines	
2000 through 2004	156 in-lbs
2005 and later	108 in-lbs
V8 engine	156 in-lbs

Section

Reference to other Chapters

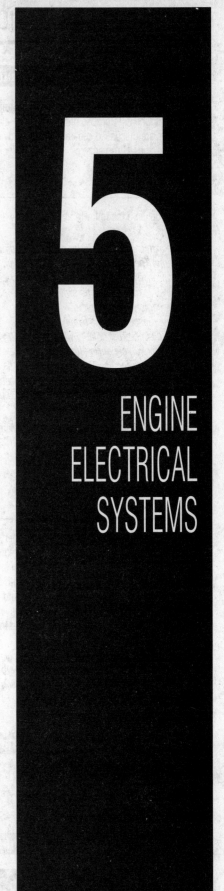

5

ENGINE ELECTRICAL SYSTEMS

1   General information, precautions and battery disconnection

The engine electrical systems include all ignition, charging and starting components. Because of their engine-related functions, these components are discussed separately from body electrical devices such as the lights, the instruments, etc. (which are included in Chapter 12).

PRECAUTIONS

Always observe the following precautions when working on the electrical system:

a) *Be extremely careful when servicing engine electrical components. They are easily damaged if checked, connected or handled improperly.*

b) *Never leave the ignition switched on for long periods of time when the engine is not running.*

c) *Never disconnect the battery cables while the engine is running.*

d) *Maintain correct polarity when connecting battery cables from another vehicle during jump starting - see the "Booster battery (jump) starting" section at the front of this manual.*

e) *Always disconnect the negative battery cable from the battery before working on the electrical system, but read the following battery disconnection procedure first.*

It's also a good idea to review the safety-related information regarding the engine electrical systems located in the "Safety first!" section at the front of this manual, before beginning any operation included in this Chapter.

BATTERY DISCONNECTION

Several systems on the vehicle require battery power to be available at all times, either to ensure their continued operation (such as the radio, alarm system, power door locks, windows, etc.) or to maintain control unit memories (such as that in the engine management system's Powertrain Control Module [PCM]) which would be lost if the battery were to be disconnected. Therefore, whenever the battery is to be disconnected, first note the following to ensure that there are no unforeseen consequences of this action:

a) *The engine management system's ECM will lose the information stored in its memory when the battery is disconnected. This includes idling and operating values, any fault codes detected and*

system monitors required for emissions testing. Whenever the battery is disconnected, the computer may require a certain period of time to "re-learn" the operating values.

b) *On any vehicle with power door locks, it is a wise precaution to remove the key from the ignition and to keep it with you, so that it does not get locked inside if the power door locks should engage accidentally when the battery is reconnected!*

Devices known as "memory-savers" can be used to avoid some of the above problems. Precise details vary according to the device used. Typically, it is plugged into the cigarette lighter and is connected by its own wires to a spare battery; the vehicle's own battery is then disconnected from the electrical system, leaving the "memory-saver" to pass sufficient current to maintain audio unit security codes and ECM memory values, and also to run permanently live circuits such as the clock and radio memory, all the while isolating the battery in the event of a short-circuit occurring while work is carried out.

❋❋ WARNING 1:

Some of these devices allow a considerable amount of current to pass, which can mean that many of the vehicle's systems are still operational when the main battery is disconnected. If a "memory-saver" is used, ensure that the circuit concerned is actually "dead" before carrying out any work on it!

❋❋ WARNING 2:

If work is to be performed around any of the airbag system components, the battery must be disconnected and no "memory saver" can be used. If a memory-saver device is used, power will be supplied to the airbag and personal injury may result if the airbag is accidentally deployed.

The battery on all vehicles is located in the front left corner of the engine compartment. To disconnect the battery for service procedures requiring power to be cut from the vehicle, loosen the negative cable clamp nut and detach the negative cable from the negative battery post. Isolate the cable end to prevent it from accidentally coming into contact with the battery post.

2   Battery - emergency jump starting

Refer to the *Booster battery (jump) starting* procedure at the front of this manual.

3   Battery - check and replacement

❋❋ WARNING:

Hydrogen gas is produced by the battery, so keep open flames and lighted cigarettes away from it at all times. Always wear eye protection when working around a battery. Rinse off spilled electrolyte immediately with large amounts of water.

CHECK

▶ **Refer to illustrations 3.1a, 3.1b and 3.1c**

1   A battery cannot be accurately tested until it is at or near a fully charged state. Disconnect the negative battery cable from the battery

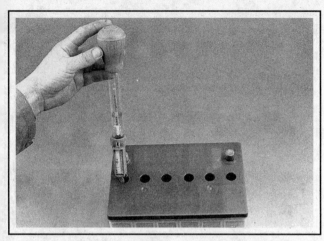

3.1a Use a battery hydrometer to draw electrolyte from the battery cell - this hydrometer is equipped with a thermometer to make temperature corrections

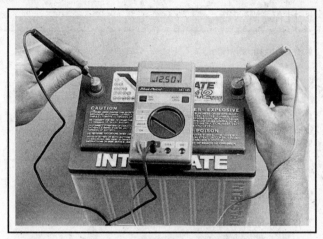

3.1b To test the open circuit voltage of the battery, connect the black probe of the voltmeter to the negative terminal and the red probe to the positive terminal of the battery - a fully charged battery should indicate 12.6 volts depending on the outside air temperature

and perform the following tests:

a) **Battery state of charge test** - Visually inspect the indicator eye (if equipped) on the top of the battery. If the indicator eye is dark in color, charge the battery as described in Chapter 1. If the battery is equipped with removable caps, check the battery electrolyte. The electrolyte level should be above the upper edge of the plates. If the level is low, add distilled water. DO NOT OVERFILL. The excess electrolyte may spill over during periods of heavy charging. Test the specific gravity of the electrolyte using a hydrometer (see illustration). Remove the caps and extract a sample of the electrolyte and observe the float inside the barrel of the hydrometer. Follow the instructions from the tool manufacturer and determine the specific gravity of the electrolyte for each cell. A fully charged battery will indicate approximately 1.270 (green zone) at 68-degrees F (20-degrees C). If the specific gravity of the electrolyte is low (red zone), charge the battery as described in Chapter 1.

b) **Open circuit voltage test** - Using a digital voltmeter, perform an open circuit voltage test (see illustration). Connect the negative probe of the voltmeter to the negative battery post and the positive probe to the positive battery post. The battery voltage should be greater than 12.5 volts. If the battery is less than the specified voltage, charge the battery before proceeding to the next test. Do not proceed with the battery load test until the battery is fully charged.

c) **Battery load test** - An accurate check of the battery condition can only be performed with a load tester (available at most auto parts stores). This test evaluates the ability of the battery to operate the starter and other accessories during periods of heavy amperage draw (load). Connect a battery load testing tool to the battery terminals (see illustration). Load test the battery according to the tool manufacturer's instructions. This tool increases the load demand (amperage draw) on the battery. Maintain the load on the battery for 15 seconds and observe that the battery voltage does not drop below 9.6 volts. If the battery condition is weak or defective, the tool will indicate this condition immediately.

➡Note: Cold temperatures will cause the minimum voltage reading to drop slightly. Follow the chart given in the tool manufacturer's instructions to compensate for cold climates. Minimum load voltage for freezing temperatures (32 degrees F/0-degrees C) should be approximately 9.1 volts.

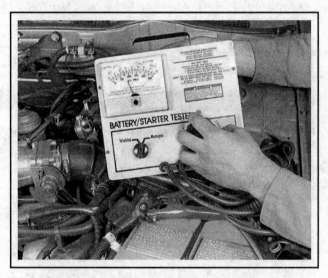

3.1c Some battery load testers are equipped with an ammeter which enables the battery load to be precisely dialed in, as shown - less expensive testers have a load switch and a voltmeter only

d) **Battery drain test** - This test will indicate whether there's a constant drain on the vehicle's electrical system that can cause the battery to discharge. Make sure all accessories are turned Off. If the vehicle has an underhood light, verify it's working properly, then disconnect it. Connect one lead of a digital ammeter to the disconnected negative battery cable clamp and the other lead to the negative battery post. A drain of approximately 100 milliamps or less is considered normal (due to the Powertrain Control Module, digital clocks, digital radios and other components which normally cause a key-off battery drain). An excessive drain (approximately 500 milliamps or more) will cause the battery to discharge. The problem circuit or component can be located by removing the fuses, one at a time, until the excessive drain stops and normal drain is indicated on the meter.

REPLACEMENT

♦ **Refer to illustration 3.4**

✷✷ CAUTION:

Always disconnect the negative cable first and hook it up last or the battery may be shorted by the tool being used to loosen the cable clamps.

2   Loosen the cable clamp nut and remove the negative battery cable from the negative battery post. Isolate the cable end to prevent it from accidentally coming into contact with the battery post.

3   Loosen the cable clamp nut and remove the positive battery cable from the positive battery post.

4   Remove the battery hold-down clamp (see illustration).

5   Lift out the battery. Be careful - it's heavy.

➡**Note: Battery straps and handlers are available at most auto parts stores for a reasonable price. They make it easier to remove and carry the battery.**

6   While the battery is out, inspect the battery tray for corrosion. If corrosion exists, clean the deposits with a mixture of baking soda and water to prevent further corrosion. Flush the area with plenty of clean water and dry thoroughly.

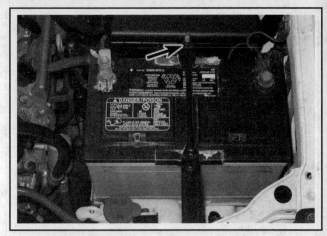

3.4 To remove the battery hold-down clamp, remove the nut and the bolt

7   If you are replacing the battery, make sure you replace it with a battery with the identical dimensions, amperage rating, cold cranking rating, etc.

8   When installing the battery, tighten the bolt and nut securely. Do not over-tighten the nut.

9   The remainder of installation is the reverse of removal.

4   Battery cables - replacement

♦ **Refer to illustrations 4.4a, 4.4b and 4.4c**

1   Periodically inspect the entire length of each battery cable for damage, cracked or burned insulation and corrosion. Poor battery cable connections can cause starting problems and decreased engine performance.

2   Check the cable-to-terminal connections at the ends of the cables for cracks, loose wire strands and corrosion. The presence of white, fluffy deposits under the insulation at the cable terminal connection is a sign that the cable is corroded and should be replaced. Check the terminals for distortion, missing mounting bolts and corrosion.

3   When removing the cables, always disconnect the negative cable from the negative battery post first and hook it up last, or the battery could be accidentally shorted by the tool used to loosen the cable clamps. Even if you're only replacing the positive cable, be sure to disconnect the negative cable first (see Chapter 1 for further information regarding battery cable maintenance).

4   Disconnect the old cables from the battery first, then disconnect them from the opposite end. To disconnect the positive cable from the starter solenoid (see illustration) on a V8 model, you'll have to remove the intake manifold (see Chapter 2B). The smaller ground cable is connected to the vehicle body next to the underhood fuse/relay box (see illustration) and the larger ground cable is attached to the left side of the engine block, near the flywheel/driveplate (see illustration). Note the routing of each cable to ensure correct installation.

4.4a The positive battery cable is connected to the terminal on the starter solenoid by this nut; on V8 models (shown here), the starter is located in the valley between the cylinder heads, and can only be accessed by removing the intake manifold

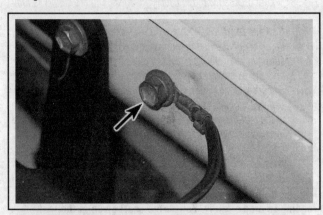

4.4b The smaller ground cable is attached to the body by this bolt near the engine compartment fuse/relay box (V8 model shown, V6 models similar)

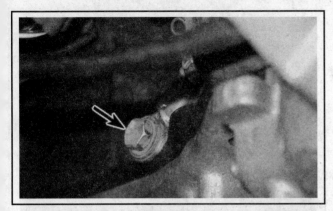

4.4c The larger ground cable is connected to the left side of the engine block by this bolt near the flywheel/driveplate (V8 model shown, V6 models similar)

5   If you are replacing either or both of the battery cables, take them with you when buying new cables. It is vitally important that you replace the cables with identical parts. Cables have a couple of easily identifiable characteristics: Positive cables are usually red and larger in cross-section; ground cables are usually black and smaller in cross-section.

6   Clean the threads of the starter solenoid or ground connection with a wire brush to remove rust and corrosion. Apply a light coat of battery terminal corrosion inhibitor or petroleum jelly to the threads to prevent future corrosion.

7   Attach the cable to the terminal and tighten the mounting nut/bolt securely.

8   Before connecting a new cable to the battery, make sure that it reaches the battery post without having to be stretched.

9   After installing the cables, connect the negative cable to the negative battery post.

5   Ignition system - general information and precautions

1   All models are equipped with a Direct Ignition System (DIS) which has no distributor. The DIS system includes the Powertrain Control Module (PCM), the Camshaft Position (CMP) sensor, the igniter(s), the ignition coils, the spark plug wires (on V6 engines only) and the spark plugs. For more information on the PCM and the CMP sensor, refer to Chapter 6.

2   On V6 models, the igniter for the ignition coils is mounted on the right side of the engine compartment. There are three ignition coils, mounted on the right valve cover, directly over the spark plugs. The DIS system on V6 models is a "waste spark" design. Each time the igniter fires an ignition coil, the coil fires two spark plugs: the one on top of which it's mounted, and the spark plug for its companion cylinder, to which it's connected by a spark plug wire. The firing order on V6 models is 1-2-3-4-5-6 and the interval between each power stroke is 120 degrees. So, for example, when the ignition coil for the No. 1 cylinder fires the No. 1 spark plug, it also fires the spark plug for the No. 4 cylinder, which is also at Top Dead Center (TDC), but which is on its exhaust stroke, not its power stroke. Hence the spark is "wasted." (The other two cylinder pairs are Nos. 2 and 5 and Nos. 3 and 6.)

3   V8 models have eight coils, mounted on the valve covers, directly over the spark plugs. There are no spark plug wires. And each ignition coil has an integral igniter. These models don't use a waste spark system, because they don't need to. Each coil fires one plug.

ALL MODELS

4   When working on the ignition system, take the following precautions:

a)   *Do not keep the ignition switch on for more than 10 seconds if the engine will not start.*

b)   *Always connect a tachometer in accordance with the manufacturer's instructions. Some tachometers may be incompatible with this ignition system. Consult an auto parts counterperson before buying a tachometer for use with this vehicle.*

c)   *Never allow the ignition coil terminals to touch ground. Grounding the coil could result in damage to the igniter and/or the ignition coil.*

d)   *Do not disconnect the battery when the engine is running.*

e)   *Make sure that the igniter is correctly grounded.*

6   Ignition system - check

V6 MODELS

❊❊ WARNING:

Because of the high voltage generated by the ignition system, extreme care should be taken whenever an operation is performed involving ignition components. This not only includes the igniter, coil and spark plug wires, but related components such as plug connectors, tachometer and other test equipment as well.

1   If the engine turns over but won't start on a V6 model, disconnect the spark plug wire from one of the three spark plugs on the left cylinder head and attach it to a calibrated ignition tester, which is available at most auto parts stores.

2   Connect the clip on the tester to a bolt or metal bracket on the engine.

3   Relieve the fuel system pressure (see Chapter 4). Keep the fuel system disabled while testing the ignition system.

4   Crank the engine and watch the end of the tester to see if bright blue, well-defined sparks occur.

5   If sparks occur, sufficient voltage is reaching the two companion spark plugs (Nos. 1 and 4, Nos. 2 and 5, or Nos. 3 and 6) to fire them. Repeat this test at each of the other two spark plug wires to verify that all three ignition coils are functioning.

6   If there is no spark, or only an intermittent spark, verify that there is battery voltage to the ignition coils (see the wiring diagrams at the end of Chapter 12).

7   If there is battery voltage to the coils, inspect the plug wires to the three spark plugs on the left cylinder bank (see Chapter 1).

8   If the three spark plug wires are okay, remove and then check

the spark plugs for fouling (see Chapter 1). Don't just inspect the three spark plugs which are connected to the coils by plug wires; inspect the other three plugs as well. If necessary, install new plugs.

9   If the spark plugs are good, check the coil resistance (see Section 7).

10 If the ignition system checks out, the CMP sensor or the PCM might be defective (see Chapter 6).

V8 MODELS

▶ **Refer to illustration 6.11**

➡**Note: Repeat the following test for each igniter/ignition coil and spark plug assembly.**

11 If the engine turns over but won't start on a V8 model, disconnect each igniter/ignition coil assembly (see Section 7) and hook it up to a calibrated ignition tester (see illustration). Make sure that the tester is designed for a direct ignition system.

12 Connect the clip on the tester to a bolt or metal bracket on the engine.

13 Relieve the fuel system pressure (see Chapter 4). Keep the fuel system disabled while testing the ignition system.

14 Crank the engine and watch the end of the tester to see if bright blue, well-defined sparks occur.

15 If sparks occur, sufficient voltage is reaching the spark plug. Repeat this test at each of the other igniter/ignition coils and verify that all eight igniter/ignition coils are functioning.

16 If there is no spark, or only an intermittent spark, verify that there is battery voltage to the ignition coil which fires that plug (see wiring diagrams at the end of Chapter 12).

17 If there is voltage to the igniter/ignition coil, remove and then check the spark plug for fouling (see Chapter 1). If necessary, install a new plug.

6.11 To use a calibrated ignition tester, simply connect it to the igniter/ignition coil assembly, clip the tester to a convenient ground and then, with the fuel system pressure relieved and the fuel system disabled, crank the engine; if there's enough power to fire the plug, sparks will be visible between the electrode tip and the tester body

18 If the spark plug is good, the coil might be bad, but there is no way to check the igniter/ignition coil units on V8 models, except to substitute a known good unit and note whether that solves the problem. Once purchased, electrical components cannot be returned, so keep this in mind if you decide to try this strategy. Your other option is to have the ignition coil(s) checked out by a dealer service department.

19 If the ignition system checks out, the CMP sensor or the PCM might be defective (see Chapter 6).

7   Ignition coils - check and replacement

CHECK (V6 MODELS ONLY)

▶ **Refer to illustrations 7.1 and 7.2**

➡**Note 1: The igniter/ignition coil assemblies on V8 models cannot be tested. Even at a dealership, these units are tested by substituting a known good unit; if substituting a known good unit produces a healthy spark, the bad unit is replaced.**

➡**Note 2: The following checks should be made with the engine cold. If the engine is hot, the resistance will be greater.**

1   Check the primary resistance of the ignition coil. With the ignition key turned to OFF, disconnect the electrical harness connector from each coil. Connect an ohmmeter across the coil primary terminals (see illustration) and compare your measurement to the resistance listed in this Chapter's Specifications. If the primary resistance is incorrect, replace the coil.

7.1 To check the primary resistance of a V6 ignition coil, measure the resistance across the primary terminals with an ohmmeter

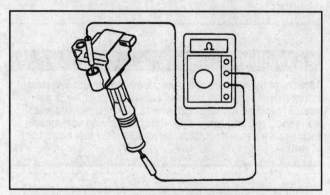

7.2 To check the secondary resistance of a V6 ignition coil, remove the spark plug wire and then measure the resistance across the high-tension terminals with an ohmmeter

2 Check the secondary resistance of each ignition coil. With the ignition key turned to OFF, label and detach the spark plug wires from each coil. Connect an ohmmeter across the two secondary terminals of each coil (see illustration) and compare your measurement to the resistance listed in this Chapter's Specifications. If the secondary resistance is incorrect, replace the coil.

REPLACEMENT

3 Disconnect the negative cable from the battery.

2000 through 2004 V6 models

♦ **Refer to illustration 7.5**

4 Remove the air intake duct (see "Throttle body - removal and installation" in Chapter 4).

5 Disconnect the ignition coil electrical connector(s) (see illustration). If you're going to replace more than one coil, label the connectors so that they don't get mixed up.

6 Label and detach the spark plug wire(s) (see Chapter 1). Pay close attention to the routing of the spark plug wires to ensure that they're correctly routed when re-installed.

7 Remove the ignition coil mounting bolt(s) and separate the coil(s) from the valve cover by pulling straight up.

8 Installation is the reverse of removal.

2005 and later V6 models

♦ **Refer to illustration 7.12**

9 Remove the engine cover (see illustration 4.12 in Chapter 2A).

10 Remove the air filter housing (see Chapter 4).

11 If you're removing a coil from the left cylinder bank, remove the upper intake manifold (see Chapter 2A).

12 Disconnect the electrical connector from the ignition coil (see illustration).

13 Remove the ignition coil mounting bolt and remove the coil.

14 Installation is the reverse of removal.

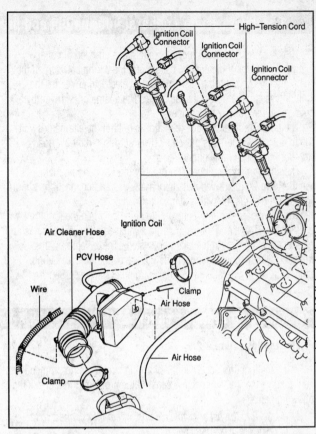

7.5 An exploded view of the ignition coils on 2000 through 2004 V6 models

V8 models

♦ **Refer to illustration 7.15**

15 Disconnect the electrical connector(s) from the igniter/ignition coil(s) (see illustration). If you're going to replace more than one coil, label the connectors so that they don't get mixed up.

16 Remove the bolt(s) securing the igniter/ignition coil(s) and remove the coil(s) by pulling straight up.

17 Installation is the reverse of the removal procedure.

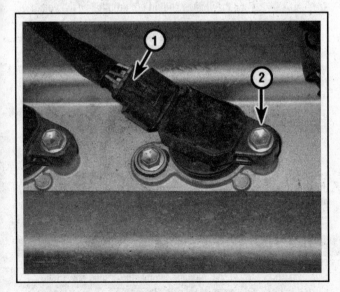

7.12 To remove an ignition coil, depress the release tab (1) and disconnect the electrical connector, then remove the coil mounting bolt (2)

7.15 To remove an igniter/ignition coil, disconnect the electrical connector from the coil, then remove the coil retaining bolt

8 Charging system - general information and precautions

The charging system includes the alternator, an internal voltage regulator, a charge indicator, the battery and the wiring between all the components. The charging system supplies electrical power for the ignition system, the lights, the radio, etc. The alternator is driven by a drivebelt at the front of the engine.

The purpose of the voltage regulator is to limit the alternator's voltage to a preset value. This prevents power surges, circuit overloads, etc., during peak voltage output.

The charging system doesn't ordinarily require periodic maintenance. However, the drivebelt, battery and wires and connections should be inspected at the intervals outlined in Chapter 1.

The dashboard warning light should come on when the ignition key is turned to START, and then should go off immediately. If it remains on, there is a malfunction in the charging system. These vehicles are also equipped with a voltage gauge. If the voltage gauge indicates abnormally high or low voltage, check the charging system (see Section 9).

Be very careful when making electrical circuit connections to a vehicle equipped with an alternator and note the following:

a) *When reconnecting wires to the alternator from the battery, be sure to note the polarity.*
b) *Before using arc welding equipment to repair any part of the vehicle, disconnect the wires from the alternator and the battery terminals.*
c) *Never start the engine with a battery charger connected.*
d) *Always disconnect both battery leads before using a battery charger.*
e) *The alternator is driven by an engine drivebelt which could cause serious injury if your hand, hair or clothes become entangled in it with the engine running.*
f) *Because the alternator is connected directly to the battery, it could arc or cause a fire if overloaded or shorted out.*

9 Charging system - check

▶ **Refer to illustration 9.2**

1 If a malfunction occurs in the charging circuit, do not immediately assume that the alternator is causing the problem. First, check the following items:

a) *Make sure the battery cable clamps, where they connect to the battery, are clean and tight.*
b) *Test the condition of the battery (see Section 3). If it does not pass all the tests, replace it with a new battery.*
c) *Check the external alternator wiring and connections.*
d) *Check the drivebelt condition and tension (see Chapter 1).*
e) *Check the alternator mounting bolts for tightness.*
f) *Run the engine and check the alternator for abnormal noise.*
g) *Check the charge light on the dash. It should illuminate when the ignition key is turned ON (engine not running). If it does not, check the circuit from the alternator to the charge light on the dash.*
h) *Check all the fuses that are in series with the charging system circuit. The location of these fuses may vary from year and model but the designations are generally the same. Refer to the wiring schematics at the end of Chapter 12 for additional information.*

2 With the ignition key off, check the battery voltage with no accessories operating (see illustration). It should be approximately 12.6 volts. It may be slightly higher if the engine has been operating within the last hour.

3 Start the engine and check the battery voltage again. It should now be greater than the voltage recorded in Step 2, but not more than 14.5 volts. Turn On all the vehicle accessories (air conditioning, rear window defogger, blower motor, etc.) and increase the engine

9.2 Connect a voltmeter to the battery terminals and check the battery voltage with the engine Off and again with the engine running

speed to 2,000 rpm - the voltage should not drop below the voltage recorded in Step 2.

4 If the indicated voltage is greater than the specified charging voltage, replace the voltage regulator.

➡**Note: It is recommended to replace the alternator/voltage regulator as a complete unit, using either a rebuilt or new alternator.**

5 If the indicated voltage reading is less than the specified charging voltage, the alternator is probably defective. Have the charging system checked at a dealer service department or other properly equipped repair facility.

➡**Note: Many auto parts stores will bench test an alternator off the vehicle. Refer to your local auto parts store regarding their policy; many will perform this service free of charge.**

10 Alternator - removal and installation

REMOVAL

▶ **Refer to illustrations 10.8a, 10.8b, 10.8c and 10.10**

1 Disconnect the cable from the negative battery terminal.

2 On 2005 and later V6 models, remove the engine cover (see illustration 4.12 in Chapter 2A).

3 Remove the accessory drivebelt or serpentine drivebelt (see Chapter 1).

4 On 2005 and later V6 models, remove the battery (see Section 3).

5   On 2005 and later V6 models, remove the bolt that secures the alternator wiring harness bracket to the battery tray.

6   Raise the front of the vehicle and place it securely on jackstands. If the vehicle is equipped with an engine under-cover, remove it (see illustration 8.6 in Chapter 1 for an example of a typical engine under-cover).

7   On 2005 and later V6 models, you might wish to unbolt the air conditioning compressor (see Chapter 3) to give yourself more room to work.

✳✳ WARNING:

Don't disconnect the refrigerant lines from the compressor.

8   Disconnect the electrical connectors from the alternator (see illustrations).

9   On 2005 and later V6 models, remove the two bolts that secure

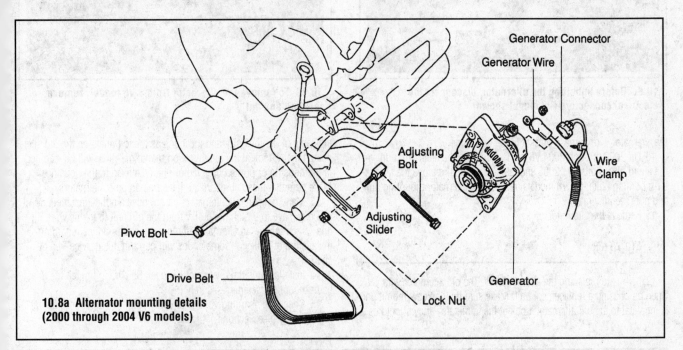

10.8a Alternator mounting details (2000 through 2004 V6 models)

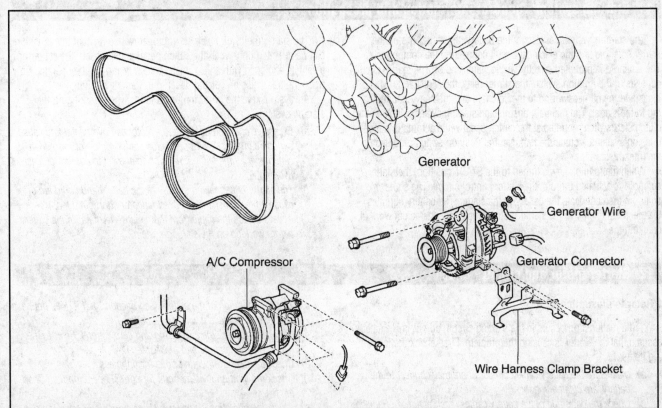

10.8b Alternator mounting details (2005 and later V6 models)

10.8c Before unbolting the alternator, disconnect the electrical connectors (V8 model shown)

10.10 To remove the alternator from a V8 model, remove this bolt and nut

the alternator wire harness bracket to the alternator.

10 On 2000 through 2004 V6 models, remove the alternator adjustment and pivot bolts. On 2005 and later V6 models, remove the alternator mounting bolts. On V8 models, remove the alternator mounting bolt and nut (see illustration).

11 Remove the alternator.

INSTALLATION

12 If you are replacing the alternator, take the old alternator with you when purchasing a replacement unit. Make sure that the new/rebuilt unit is identical to the old alternator. Look at the terminals - they should be the same in number, size and locations as the terminals on the old alternator. Finally, look at the identification markings - they will be stamped in the housing or printed on a tag or plaque affixed to the housing. Make sure that these numbers are the same on both alternators.

13 Many new/rebuilt alternators do not have a pulley installed, so you may have to switch the pulley from the old unit to the new/rebuilt one. When buying an alternator, find out the shop's policy regarding installation of pulleys - some shops will perform this service free of charge.

14 Installation is the reverse of removal. Be sure to tighten the alternator mounting fasteners securely.

15 Check the charging voltage to verify that the alternator is operating correctly (see Section 9).

11 Starting system - general information and precautions

The starting system consists of the battery, the starter motor, the starter solenoid and the electrical circuit connecting the components. The solenoid is mounted directly on the starter motor. The starter circuit consists of the ignition switch, the starter relay, the MAIN and AM2 fuses and the harness wiring to the solenoid and the heavy gauge wiring to the starter. The manual transmission starting systems include the clutch start switch mounted at the clutch pedal while automatic transmission systems include the Park/Neutral Position switch mounted on the transmission.

When the ignition key is turned to the START position, the starter solenoid is actuated through the starter control circuit. The starter solenoid then connects the battery to the starter. The battery supplies the electrical energy to the starter motor, which does the actual work of cranking the engine.

The starter motor on a vehicle equipped with a manual transmission can be operated only when the clutch pedal is depressed; the starter on a vehicle equipped with an automatic transmission can be operated only when the transmission selector lever is in Park or Neutral.

Always observe the following precautions when working on the starting system:

a) *Excessive cranking of the starter motor can overheat it and cause serious damage. Never operate the starter motor for more than 15 seconds at a time without pausing to allow it to cool for at least two minutes.*

b) *The starter is connected directly to the battery and could arc or cause a fire if mishandled, overloaded or short circuited.*

c) *Always detach the cable from the negative terminal of the battery before working on the starting system.*

12 Starter motor and circuit - check

♦ **Refer to illustration 12.3**

1 If a malfunction occurs in the starting circuit, do not immediately assume that the starter is causing the problem. First, check the following items:

a) *Make sure the battery cable clamps, where they connect to the battery, are clean and tight.*

b) *Check the condition of the battery cables (see Section 4). Replace any defective battery cables with new parts.*

c) *Test the condition of the battery (see Section 3). If it does not pass all the tests, replace it with a new battery.*

d) *Check the starter solenoid wiring and connections. Refer to the wiring diagrams at the end of Chapter 12.*

e) *Check the starter mounting bolts for tightness.*

f) *Check the ignition switch circuit for correct operation (see Chapter 12).*

g) *Check the operation of the Park/Neutral Position switch (automatic transmission) or clutch start switch (manual transmission).*

Make sure the shift lever is in PARK or NEUTRAL (automatic transmission) or the clutch pedal is pressed (manual transmission). Refer to Chapter 7 for the Park/Neutral Position switch check and adjustment procedure. Refer to Chapter 12 wiring diagrams, if necessary, when performing circuit checks. These systems must operate correctly to provide battery voltage to the ignition solenoid.

h) *Check the operation of the starter relay. The starter relay is located in the fuse/relay box inside the engine compartment. Refer to Chapter 12 for the testing procedure.*

2   If the starter does not actuate when the ignition switch is turned to the start position, check for battery voltage to the solenoid. This will determine if the solenoid is receiving the correct voltage signal from the ignition switch. Connect a test light or voltmeter to the starter solenoid positive terminal and while an assistant turns the ignition switch to the start position. If voltage is not available, refer to the wiring diagrams in Chapter 12 and check all the fuses and relays in the starting system. If voltage is available but the starter motor does not operate, remove the starter (see Section 13) and bench test it (see Step 4).

3   If the starter turns over slowly, check the starter cranking voltage and the current draw from the battery. This test must be performed with

12.3 To use an inductive ammeter, simply hold the ammeter over the positive or negative cable (whichever is easier in terms of clearance)

the starter assembly on the engine. Crank the engine over (for 10 seconds or less) and observe the battery voltage. It should not drop below 8.0 volts on manual transmission models or 8.5 volts on automatic transmission models. Also, observe the current draw using an ammeter (see illustration). It should not exceed 400 amps or drop below 250 amps.

✳✳ CAUTION:

The battery cables may be excessively heated because of the large amount of amperage being drawn from the battery. Discontinue the testing until the starting system has cooled down.

If the starter motor cranking amp values are not within the correct range, replace it with a new unit. There are several conditions that may affect the starter cranking potential. The battery must be in good condition and the battery cold-cranking rating must not be under-rated for the particular application. Be sure to check the battery specifications carefully. The battery terminals and cables must be clean and not corroded. Also, in cases of extreme cold temperatures, make sure the battery and/or engine block is warmed before performing the tests.

4   If the starter is receiving voltage but does not activate, remove and check the starter/solenoid assembly on the bench. Most likely the solenoid is defective. In some rare cases, the engine may be seized so be sure to try and rotate the crankshaft pulley (see Chapter 2A or 2B) before proceeding. With the starter/solenoid assembly mounted in a vise on the bench, install one jumper cable from the negative battery terminal to the body of the starter. Install the other jumper cable from the positive battery terminal to the B+ terminal on the starter. Install a starter switch and apply battery voltage to the solenoid S terminal (for 10 seconds or less) and see if the solenoid plunger, shift lever and overrunning clutch extends and rotates the pinion drive. If the pinion drive extends but does not rotate, the solenoid is operating but the starter motor is defective. If there is no movement but the solenoid clicks, the solenoid and/or the starter motor is defective. If the solenoid plunger extends and rotates the pinion drive, the starter/solenoid assembly is working properly.

13 Starter motor - removal and installation

▸ **Refer to illustrations 13.4a, 13.4b, 13.4c and 13.5**

➡**Note: On 2000 through 2004 V6 models, the starter motor is located on the right rear side of the engine block. On 2005 and later V6 models, the starter motor is located on the left rear side of the block. On V8 engines, the starter is located on top of the engine, between the cylinder heads, underneath the upper intake manifold.**

1   Disconnect the cable from the negative battery terminal.

2   On V6 models, raise the front of the vehicle and place it securely on jackstands. Then remove the engine under-cover.

3   On V8 models, remove the upper intake manifold (see Chapter 2B).

4   Disconnect the electrical connectors from the starter motor/solenoid assembly (see illustrations).

5   Remove the starter motor mounting bolts and remove the starter motor (see illustration).

6   Remove the starter motor assembly from the engine compartment.

7   Installation is the reverse of removal. Be sure to tighten the starter motor mounting bolts securely.

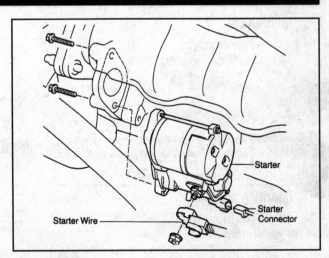

13.4a Starter mounting details (2000 through 2004 V6 models)

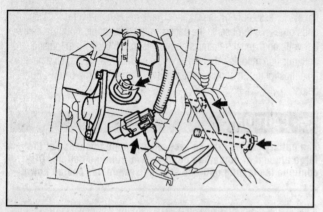

13.4b Starter motor mounting details (2005 and later V6 models)

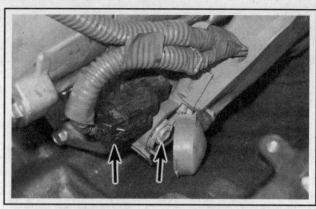

13.4c Before removing the starter assembly, disconnect the two electrical connections (V8 model shown)

13.5 To detach the starter motor assembly on a V8 model, remove the wiring protector bolt (lower arrow) and the two starter bolts (upper bolt indicates right starter bolt; left bolt not visible in this photo)

Specifications

Ignition coil

V6 models		
Primary resistance		
Cold		0.67 to 1.05 ohms
Hot		0.85 to 1.23 ohms
Secondary resistance		
Cold		9.3 to 16.0 k-ohms
Hot		11.7 to 18.8 k-ohms
V8 models		N/A

Charging system

Charging voltage		13.2 to 14.8 volts
Standard amperage		
No load		10 amps or less
With load		30 amps or more

6

EMISSIONS AND ENGINE CONTROL SYSTEMS

1   General information

▶ **Refer to illustrations 1.6a and 1.6b**

To prevent pollution of the atmosphere from incompletely burned and evaporating gases, and to maintain good driveability and fuel economy, a number of emission control systems are incorporated. They include the:

Acoustic Control Induction System (ACIS)
(2005 and later V6 models)
Catalytic converter
Electronic Throttle Control System-intelligent (ETCS-i)
Evaporative Emissions Control (EVAP) system
On-Board Diagnostics (OBD-II) system
Positive Crankcase Ventilation (PCV) system
Sequential Electronic Fuel Injection (SFI) system
Variable Valve Timing-intelligent (VVT-i) system

The Sections in this Chapter include general descriptions, checking procedures within the scope of the home mechanic and component replacement procedures (when possible) for each of the systems listed above.

Before assuming that an emissions control system is malfunctioning, check the fuel and ignition systems carefully. The diagnosis of some emission control devices requires specialized tools, equipment and training. If checking and servicing become too difficult or if a procedure is beyond your ability, consult a dealer service department or other repair shop. Remember, the most frequent cause of emissions problems is simply a loose or broken wire or vacuum hose, so always check the hose and wiring connections first.

This doesn't mean, however, that emissions control systems are particularly difficult to maintain and repair. You can quickly and easily perform many checks and do most of the regular maintenance at home with common tune-up and hand tools.

➡**Note: Because of a Federally mandated warranty which covers the emissions control system components, check with your dealer about warranty coverage before working on any emissions-related systems. Once the warranty has expired, you may wish to perform some of the component checks and/or replacement procedures in this Chapter to save money.**

Pay close attention to any special precautions outlined in this Chapter. It should be noted that the illustrations of the various systems may not exactly match the system installed on your vehicle because of changes made by the manufacturer during production or from year-to-year.

A Vehicle Emissions Control Information (VECI) label is attached to the underside of the hood (see illustration). This label contains important emissions specifications and adjustment information. Another label, the Vacuum Hose Routing Diagram (see illustration), provides a vacuum hose schematic with emissions components identified. When servicing the engine or emissions systems, the VECI label and the vacuum hose routing diagram in your particular vehicle should always be checked.

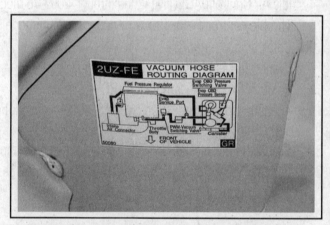

1.6a The Vehicle Emission Control Information (VECI) label contains such essential information as the types of emission control systems installed on the engine and the idle speed and ignition timing specifications

1.6b Vacuum hose routing diagram (V8 model shown, V6 models similar)

2   On Board Diagnostic (OBD) system and trouble codes

SCAN TOOL INFORMATION

▶ **Refer to illustrations 2.1 and 2.2**

1   Hand-held scanners are the most powerful and versatile tools for analyzing engine management systems used on later model vehicles (see illustration). Early model scanners handle codes and some diagnostics for many systems. Each brand scan tool must be examined carefully to match the year, make and model of the vehicle you are working on. Often, interchangeable cartridges are available to access the particular manufacturer (Ford, GM, Chrysler, Toyota etc.). Some manufacturers will specify by continent (Asia, Europe, USA, etc.).

➡**Note: An aftermarket generic scanner should work with any model covered by this manual. Before purchasing a generic scan tool, contact the manufacturer of the scanner you're planning to buy and verify that it will work properly with the OBD-II system you want to scan. If necessary, of course, you can always have the codes extracted by a dealer service department or an independent repair shop with a professional scan tool.**

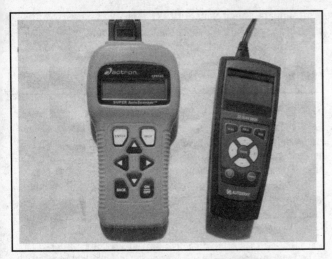

2.1 Scanners like these from Actron and AutoXray are powerful diagnostic aids - they can tell you just about anything you want to know about your engine management system

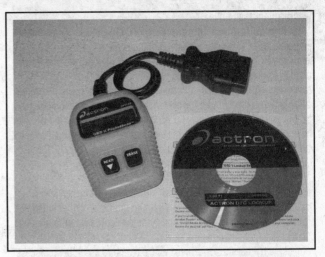

2.2 Simple code readers, like the Actron PocketScan, are an economical way to extract trouble codes when the CHECK ENGINE light comes on

2   With the arrival of the Federally mandated emission control system (OBD-II), a specially designed scanner has been developed. Several tool manufacturers have released OBD-II scan tools and code readers for the home mechanic (see illustration).

OBD SYSTEM GENERAL DESCRIPTION

3   All models are equipped with the second generation OBD-II system. This system consists of an on-board computer known as the Powertrain Control Module (PCM), and information sensors, which monitor various functions of the engine and send data to the PCM. This system incorporates a series of diagnostic monitors that detect and identify fuel injection and emissions control systems faults and store the information in the computer memory. This updated system also tests sensors and output actuators, diagnoses drive cycles, freezes data and clears codes.

4   This powerful diagnostic computer must be accessed using the new OBD-II scan tool and 16 pin Data Link Connector (DLC) located under the driver's dash area. The PCM is the "brain" of the electronically controlled fuel and emissions system. It receives data from a number of sensors and other electronic components (switches, relays, etc.). Based on the information it receives, the PCM generates output signals to control various relays, solenoids (i.e. fuel injectors) and other actuators. The PCM is specifically calibrated to optimize the emissions, fuel economy and driveability of the vehicle.

5   It isn't a good idea to attempt diagnosis or replacement of the PCM or emission control components at home while the vehicle is under warranty. Because of a Federally mandated warranty which covers the emissions system components and because any owner-induced damage to the PCM, the sensors and/or the control devices may void this warranty, take the vehicle to a dealer service department if the PCM or a system component malfunctions.

INFORMATION SENSORS

6   **Accelerator pedal position sensor** - On 2000 through 2002 V8 models and on 2003 and 2004 models, the accelerator pedal position sensor is located on the throttle body. This sensor, which is a variable potentiometer, produces an electrical output that varies with the angle of the throttle linkage, which the PCM uses to calculate the position of the accelerator pedal in order to calculate the correct position for the throttle plate inside the throttle body bore. The "accelerator cable" simply provides input for the accelerator pedal position sensor; it doesn't actually open and close the throttle plate. Instead, the PCM uses data from the accelerator pedal position sensor, along with information from the Throttle Position (TP) sensor and other sensors, to operate the throttle control motor (also located on the throttle body), which opens and closes the throttle plate.

On 2003 and later V8 models, and on 2005 and later V6 models, the accelerator pedal position sensor is located at, and is an integral component of, the accelerator pedal assembly. On these models, there is no accelerator cable at all. This accelerator pedal position sensor is also a variable potentiometer, but on these models it uses the actual position of the accelerator pedal as its input. Again, the PCM uses this data to calculate the correct position for the throttle plate, and directs the throttle motor (still inside the throttle body) to open and close the throttle plate accordingly.

7   **Air/Fuel Ratio Sensor** - Toyota also refers to the upstream oxygen sensors as air/fuel ratio sensors. But they're essentially the same devices; they're installed upstream in relation to the first catalytic converters and work the same way as oxygen sensors. The air/fuel ratio sensors, or upstream oxygen sensors, are located on the exhaust manifolds. See *oxygen sensor* for more information about the oxygen sensors.

8   **Camshaft Position (CMP) sensor** - The camshaft sensor produces a signal in which the PCM uses to identify number 1 cylinder and to time the sequential fuel injection. On 2000 through 2004 V6 models, the CMP sensor is located on the front of the right cylinder head. On V8 models, it's located on the front of the left cylinder head.

On 2005 and later V6 models, there are two CMP sensors. Toyota refers to the CMP sensors on these models as Variable Valve Timing-intelligent (VVT-i) sensors. The VVT-i sensors are located on the front inner walls of the cylinder heads, adjacent to the timing rotor installed on the front of each VVT-i controller. For more information about the VVT-i sensors, refer to Variable Valve Timing-intelligent (VVT-i) system - description and component replacement.

9   **Crankshaft Position (CKP) sensor** - The crankshaft sensor provides information on crankshaft position and the engine speed signal to the PCM. The CKP sensor is located on the front of all engines,

near the crankshaft pulley.

10 **Engine Coolant Temperature (ECT) sensor** - The coolant temperature (ECT) sensor monitors engine coolant temperature and sends the PCM a voltage signal that affects PCM control of the fuel mixture, ignition timing, and EGR operation. On 2000 through 2004 V6 models, the ECT sensor is located at the front of the engine, in the valley between the cylinder heads. On 2005 and later V6 models, it's located on the coolant bypass housing, which is located at the back of the engine (the coolant bypass housing is the casting that spans the valley between the two cylinder heads). On V8 models, the ECT sensor located at the front of the engine, below the throttle body.

11 **Knock sensors** - The knock sensors monitor engine "knock" (pre-ignition or detonation) and then signals the PCM when knock occurs, so that the PCM can retard ignition timing accordingly. There are two knock sensors on all engines: one on the inner wall of the left cylinder head, and another on the inner wall of the right head. On all engines, you'll have to remove the intake manifold assembly to access the knock sensors.

12 **Mass Air Flow (MAF) sensor** - The MAF sensor measures the mass of the intake air by monitoring the volume and weight of the air passing over a hot-wire element. The MAF sensor is located on top of the air filter housing.

13 **Oxygen sensors** - The oxygen sensors generate a voltage signal that varies in accordance with the difference between the oxygen content of the exhaust gases and the oxygen in the surrounding air. The downstream oxygen sensor monitors the content of the exhaust gases as they exit the downstream catalytic converter. This information is used by the PCM to predict catalyst deterioration and/or failure.

On 2000 through 2004 V6 models, the upstream sensor is located on the front exhaust pipe, just below the exhaust manifold flange and just ahead of the upstream catalyst. On 2005 and later V6 models, there are two upstream oxygen sensors and they too are located on the front exhaust pipes, below the exhaust manifold flange and ahead of the downstream catalysts. On 2000 through 2004 V6 models, there is one downstream, or post-converter, oxygen sensor. On 2000 through 2004 California models it's located on the center exhaust pipe, right behind the downstream catalyst. On 2000 through 2004 Federal models it's also located on the center exhaust pipe, but these models do not have a downstream catalyst. 2005 and later V6 models have two downstream oxygen sensors, one on each front exhaust pipe, just ahead of the downstream catalysts.

All V8 models have two upstream and two downstream oxygen sensors. The upstream sensors are located on the exhaust manifolds, just upstream from the flange. The downstream sensors are located right behind the downstream catalysts.

14 **Throttle Position (TP) Sensor** - The TP sensor, which is located on the throttle body, produces a variable voltage signal in proportion to the opening angle of the throttle valve. This signal, which is monitored by the PCM, enables the PCM to calculate throttle position. On V8 models and on 2003 and later V6 models, the TP sensor is one of the two principal sensors employed by the PCM to control the throttle control motor (the accelerator pedal position sensor is the other one).

15 **Vapor pressure sensor** - The vapor pressure sensor, which is a component of the evaporative emission control (EVAP) system, is located on the EVAP canister. The canister is located on the left side of the engine compartment (2000 through 2002 models) or underneath the vehicle, near the fuel tank (2003 and later models). The vapor pressure sensor monitors vapor pressure in the fuel tank. The PCM uses this information to turn a vacuum switching valve (VSV) on and off (see vapor pressure sensor switching valve below).

16 **Vehicle Speed Sensor (VSS)** - The vehicle speed sensor, which is located on the rear of the transmission, provides information to the PCM to indicate vehicle speed.

OUTPUT ACTUATORS

17 **EFI main relay** - The EFI main relay activates power to the fuel pump relay (or circuit opening relay). It is activated by the ignition switch and supplies battery power to the PCM and the EFI system when the ignition switch is in the START or RUN position. Refer to Chapter 12 or your owner's manual for more information on relay location.

18 **EVAP canister Vacuum Switching Valve (VSV), or canister purge valve** - The EVAP canister VSV, which is located on the intake manifold, is a solenoid valve operated by the PCM to purge the EVAP canister and route fuel vapors to the intake manifold for combustion. (This valve is also referred to as the canister purge valve.)

19 **Fuel injectors** - The PCM opens the fuel injectors sequentially (in firing order sequence). The PCM also controls the "pulse width," the interval of time during which each injector is open. The pulse width of an injector (measured in milliseconds) determines the amount of fuel delivered. For more information on the fuel delivery system and the fuel injectors, including injector replacement, refer to Chapter 4.

20 **Ignition coils** - The ignition coils are triggered by the PCM. Refer to Chapter 5 for more information on the ignition coils.

21 **Idle air control (IAC) valve (2000 through 2002 V6 models)** - The IAC valve, which is mounted on the bottom of the throttle body, allows a certain amount of air to bypass the throttle plate when the throttle valve is closed or at idle position. The IAC valve opening is controlled by the PCM. Vehicles equipped with electronic throttle bodies (all V8 models and 2003 and later V6 models) don't use an IAC valve; the idle speed is controlled by the throttle control motor, which is in turn controlled by the Powertrain Control Module (PCM).

22 **Throttle control motor (all V8 and 2003 and later V6 models)** - The throttle control motor, which is mounted on the throttle body, opens and closes the throttle plate. The PCM, which uses inputs from the accelerator pedal position sensor, the throttle position sensor and other sensors to calculate the correct throttle angle for the conditions, controls the throttle control motor, which in turn controls the throttle plate.

23 **Vapor pressure sensor vacuum switching valve (VSV)** - The vapor pressure sensor VSV, which is located on the EVAP canister, is controlled by the PCM. Normally, the vapor pressure sensor VSV routes vapors from the fuel tank to the vapor pressure sensor. When the vapor pressure sensor detects high vapor pressure in the fuel tank, it signals the PCM, which energizes the vapor pressure sensor VSV, which re-routes fuel tank vapors to the EVAP canister.

OBTAINING OBD-II SYSTEM TROUBLE CODES

▶ **Refer to illustration 2.25**

24 The PCM will illuminate the CHECK ENGINE light (also called the Malfunction Indicator Light) on the dash if it recognizes a component fault for two consecutive drive cycles. It will continue to set the light until the PCM does not detect any malfunction for three or more consecutive drive cycles.

25 The diagnostic codes for the OBD-II system can be extracted from the PCM by plugging a generic OBD-II scan tool (see illustrations 2.1 and 2.2) into the PCM's data link connector (see illustration), which is located under the left side of the dash.

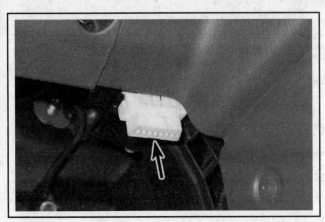

2.25 The 16-pin Data Link Connector (DLC) is located under the left side of the dash

26 Plug the scan tool into the 16-pin data link connector (DLC), and follow the instructions included with the scan tool to extract all the diagnostic codes.

CLEARING DIAGNOSTIC TROUBLE CODES

27 After the system has been repaired, the codes must be cleared from the PCM memory using a scan tool. Do not attempt to clear the codes by disconnecting battery power. If battery power is disconnected from the PCM, the PCM will lose the current engine operating parameters and driveability will suffer until the PCM "re-learns" the optimum operating parameters after driving the vehicle for a period of time.

28 Always clear the codes from the PCM before starting the engine after a new electronic emission control component is installed. The PCM stores the operating parameters of each sensor and may set a trouble code if a new sensor is allowed to operate before the parameters from the old sensor have been erased.

DIAGNOSTIC TROUBLE CODES

Trouble code	Code identification
P0010	Camshaft position A actuator circuit (Bank 1)
P0011	Camshaft position A, system performance or too much advance (Bank 1)
P0012	Camshaft position A, timing over-retarded (Bank 1)
P0016	Crankshaft position/camshaft position correlation (Bank 1, sensor A)
P0018	Crankshaft position/camshaft position correlation (Bank 2, sensor A)
P0020	Camshaft position A actuator circuit (Bank 2)
P0021	Camshaft position A, timing over-advanced/system performance (Bank 1)
P0022	Camshaft position A, timing over-retarded (Bank 1)
P0031	Oxygen sensor heater control circuit, low voltage (Bank 1, sensor 1)
P0032	Oxygen sensor heater control circuit, high voltage (Bank 1, sensor 1)
P0037	Oxygen sensor heater control circuit, low voltage (Bank 1, sensor 2)
P0038	Oxygen sensor heater control circuit, high voltage (Bank 1, sensor 2)
P0043	Oxygen sensor heater control circuit, low voltage (Bank 1, sensor 3)
P0044	Oxygen sensor heater control circuit, high voltage (Bank 1, sensor 3)
P0051	Oxygen sensor heater control circuit, low voltage (Bank 2, sensor 1)
P0052	Oxygen sensor heater control circuit, high voltage (Bank 2, sensor 1)
P0057	Oxygen sensor heater control circuit, low voltage (Bank 2, sensor 2)
P0058	Oxygen sensor heater control circuit, high voltage (Bank 2, sensor 2)
P0100	Mass Air Flow (MAF) sensor or circuit fault
P0101	Mass Air Flow (MAF) sensor range or performance problem

DIAGNOSTIC TROUBLE CODES (CONTINUED)

Trouble code	Code identification
P0102	Mass Air Flow (MAF) sensor circuit, low input voltage
P0103	Mass Air Flow (MAF) sensor circuit, high input voltage
P0110	Intake Air Temperature (IAT) sensor (in MAF sensor) or circuit fault
P0112	Intake Air Temperature (IAT) sensor circuit, low input voltage
P0113	Intake Air Temperature (IAT) sensor circuit, high input voltage
P0115	Engine Coolant Temperature (ECT) sensor or circuit fault
P0116	Engine Coolant Temperature (ECT) sensor range or performance problem
P0117	Engine Coolant Temperature (ECT) sensor, low input voltage
P0118	Engine Coolant Temperature (ECT) sensor, high input voltage
P0120	Throttle/Accelerator Pedal Position sensor A or circuit fault
P0121	Throttle/Accelerator Pedal Position sensor A, range/performance problem
P0122	Throttle/Accelerator Pedal Position sensor A circuit, low voltage input
P0123	Throttle/Accelerator Pedal Position sensor A circuit, high voltage input
P0125	Insufficient coolant temperature for closed-loop fuel control
P0128	Coolant temperature below thermostat regulating temperature
P0130	Pre-converter oxygen sensor circuit fault (Bank 1, sensor 1)
P0133	Pre-converter oxygen sensor circuit, slow response (Bank 1, sensor 1)
P0135	Pre-converter oxygen sensor heater or circuit fault (Bank 1, sensor 1)
P0136	Post-converter oxygen sensor circuit fault (Bank 1, sensor 2)
P0137	Post-converter oxygen sensor circuit, low voltage (Bank 1, sensor 2)
P0138	Post-converter oxygen sensor circuit, high voltage (Bank 1, sensor 2)
P0141	Post-converter oxygen sensor heater or circuit fault (bank 1, sensor 2)
P0150	Pre-converter oxygen sensor circuit malfunction (bank 2, sensor 1)
P0153	Pre-converter oxygen sensor circuit slow response (bank 2, sensor 1)
P0155	Pre-converter oxygen sensor heater circuit malfunction (bank 2, sensor 1)
P0156	Post-converter oxygen sensor circuit malfunction (Bank 2, sensor 2)
P0157	Post-converter oxygen sensor circuit, low voltage (Bank 2, sensor 2)
P0158	Post-converter oxygen sensor circuit, high voltage (Bank 2, sensor 2)
P0161	Post-converter oxygen sensor heater circuit problem (Bank 2, sensor 2)
P0171	Fuel injection system too lean (Bank 1)

DIAGNOSTIC TROUBLE CODES (CONTINUED)

Trouble code	Code identification
P0172	Fuel injection system too rich (Bank 1)
P0174	Fuel injection system lean (Bank 2)
P0175	Fuel injection system rich (Bank 2)
P0220	Throttle/Accelerator Pedal Position sensor B circuit
P0222	Throttle/Accelerator Pedal Position sensor B circuit, low voltage input
P0223	Throttle/Accelerator Pedal Position sensor B circuit, high voltage input
P0230	Fuel pump primary circuit
P0300	Random/multiple cylinder misfire detected
P0301	Cylinder No. 1 misfire detected
P0302	Cylinder No. 2 misfire detected
P0303	Cylinder No. 3 misfire detected
P0304	Cylinder No. 4 misfire detected
P0305	Cylinder No. 5 misfire detected
P0306	Cylinder No. 6 misfire detected
P0307	Cylinder No. 7 misfire detected
P0308	Cylinder No. 8 misfire detected
P0325	Knock Sensor No. 1 or circuit fault
P0327	Knock Sensor No. 1 circuit, low voltage (Bank 1 or single sensor)
P0328	Knock Sensor No. 1 circuit, high voltage (Bank 1 or single sensor)
P0330	Knock sensor no. 2 or circuit fault
P0331	Knock Sensor No. 2 circuit, low voltage (Bank 2)
P0332	Knock Sensor No. 2 circuit, high voltage (Bank 2)
P0335	Crankshaft Position (CKP) sensor A circuit fault
P0339	Crankshaft Position (CKP) sensor A circuit intermittent
P0340	Camshaft Position (CMP) sensor A circuit (Bank 1 or single sensor)
P0345	Camshaft Position (CMP) sensor A circuit (Bank 2)
P0346	Camshaft Position (CMP) sensor A circuit range or performance problem
P0351	Ignition coil A primary/secondary circuit
P0352	Ignition coil B primary/secondary circuit
P0353	Ignition coil C primary/secondary circuit

DIAGNOSTIC TROUBLE CODES (CONTINUED)

Trouble code	Code identification
P0354	Ignition coil D primary/secondary circuit
P0355	Ignition coil E primary/secondary circuit
P0356	Ignition coil F primary/secondary circuit
P0357	Ignition coil G primary/secondary circuit
P0358	Ignition coil H primary/secondary circuit
P0412	Air injection system air switching valve malfunction
P0418	Air injection system air pump malfunction
P0420	Catalyst system efficiency below threshold (Bank 1)
P0430	Catalyst system efficiency below threshold (Bank 2)
P043E	Evaporative Emission Control (EVAP) system reference orifice clogged
P043F	Evaporative Emission Control (EVAP) system reference orifice high flow
P0440	Evaporative Emission Control (EVAP) system malfunction
P0441	Evaporative Emission Control (EVAP) system incorrect purge flow
P0446	EVAP canister vent control valve circuit fault
P0450	EVAP system pressure sensor (fuel tank pressure sensor)
P0451	EVAP system pressure sensor range or performance problem
P0452	EVAP system pressure sensor, low voltage input
P0453	EVAP system pressure sensor, high voltage input
P0455	Evaporative Emission Control (EVAP) system, big leak detected
P0456	Evaporative Emission Control (EVAP) system, very small leak detected
P0500	Vehicle Speed Sensor (VSS) A malfunction
P0503	Vehicle Speed Sensor (VSS) A intermittent, erratic or and/or high voltage
P0504	Brake switch A/B correlation
P0505	Idle air control system
P0560	System voltage
P0604	Internal Control Module Random Access Memory (RAM) error
P0606	Powertrain Control Module (PCM) processor
P0607	Powertrain Control Module (PCM) performance
P0617	Starter relay circuit, high voltage
P0630	Vehicle Identification Number (VIN) not programmed or mismatched PCM

DIAGNOSTIC TROUBLE CODES (CONTINUED)

Trouble code	Code identification
P0657	Actuator supply voltage circuit open
P0705	Transmission Range (TR) sensor circuit malfunction (PRNDL input)
P0710	Transmission fluid temperature sensor A circuit malfunction
P0711	Transmission fluid temperature sensor A performance
P0712	Transmission fluid temperature sensor A circuit, low input voltage
P0713	Transmission fluid temperature sensor A circuit, high input voltage
P0717	Input/turbine speed sensor A circuit, no signal
P0722	Output speed sensor circuit, no signal
P0724	Brake switch B circuit, high voltage
P0748	Pressure control solenoid A electrical (shift solenoid valve SL1)
P0751	Shift solenoid A performance (shift solenoid valve S1)
P0756	Shift solenoid B performance (shift solenoid valve S2)
P0771	Shift solenoid E performance (shift solenoid valve SR)
P0776	Pressure control solenoid B performance (shift solenoid valve SL2)
P0778	Pressure control solenoid B electrical (shift solenoid valve SL2)
P0781	1-2 shift (1-2 shift valve)
P0973	Shift solenoid A control circuit, low voltage (shift solenoid valve S1)
P0974	Shift solenoid A control circuit, high voltage (shift solenoid valve S1)
P0976	Shift solenoid B control circuit, low voltage (shift solenoid valve S2)
P0977	Shift solenoid B control circuit, high voltage (shift solenoid valve S2)
P0985	Shift solenoid E control circuit, low voltage (shift solenoid valve SR)
P0986	Shift solenoid E control circuit, high voltage (shift solenoid valve SR)

3 Powertrain Control Module (PCM) - removal and installation

◆ Refer to illustration 3.4

✳✳ WARNING:

The models covered by this manual are equipped with Supplemental Restraint systems (SRS), more commonly known as airbags. Always disable the airbag system before working in the vicinity of any airbag system components to avoid the possibility of accidental deployment of the airbag, which could cause personal injury (see Chapter 12).

✳✳ CAUTION:

To avoid electrostatic discharge damage to the PCM, handle the PCM only by its case. Do not touch the electrical terminals during removal and installation. If available, ground yourself to the vehicle with an anti-static ground strap, available at computer supply stores.

➡Note: The Powertrain Control Module (PCM) is located inside the dash, in front of the glove box.

1 Disconnect the cable from the negative battery terminal.
2 Remove the glovebox (see Chapter 11).
3 Disconnect the electrical connectors from the PCM.

✳✳ CAUTION:

The ignition switch must be turned OFF when pulling out or plugging in the electrical connectors to prevent damage to the PCM.

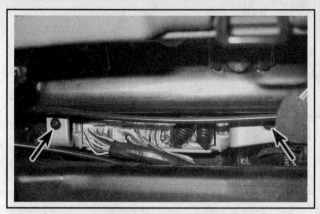

3.4 To remove the PCM, unplug the electrical connectors and then remove these two mounting bolts

4 Remove the PCM retaining bolts (see illustration) and remove the PCM.

✳✳ CAUTION:

Avoid any static electricity damage to the computer by grounding yourself to the body before touching the PCM and using a special anti-static pad to store the PCM on once it is removed.

5 Installation is the reverse of removal.

4 Accelerator pedal position sensor - replacement

2000 THROUGH 2002 V8 MODELS AND 2003 AND 2004 V6 MODELS

◆ Refer to illustrations 4.2, 4.3 and 4.5

1 Disconnect the negative battery cable.
2 Disconnect the electrical connector from the accelerator pedal position sensor (see illustration).
3 Remove the three sensor retaining screws (see illustration) and remove the sensor.
4 Before installing the accelerator pedal position sensor, make sure that the throttle plate is closed.
5 Install the accelerator pedal position sensor so that it's about 20 degrees to the left of its actual installed position (see illustration).
6 Gradually rotate the accelerator pedal position sensor clockwise until it touches the throttle valve shaft, then tighten the three screws securely.
7 Installation is otherwise the reverse of removal.

4.2 Disconnect the electrical connector from the accelerator pedal position sensor (V8 model shown, V6 similar)

4.3 To detach the accelerator pedal position sensor, remove these three screws (V8 model shown, V6 similar)

4.5 When installing the accelerator pedal position sensor, make sure that the throttle plate is closed, install the sensor so that it's about 20 degrees to the left of its actual installed position, then gradually rotate it clockwise until it touches the throttle valve shaft (V8 model shown, V6 similar)

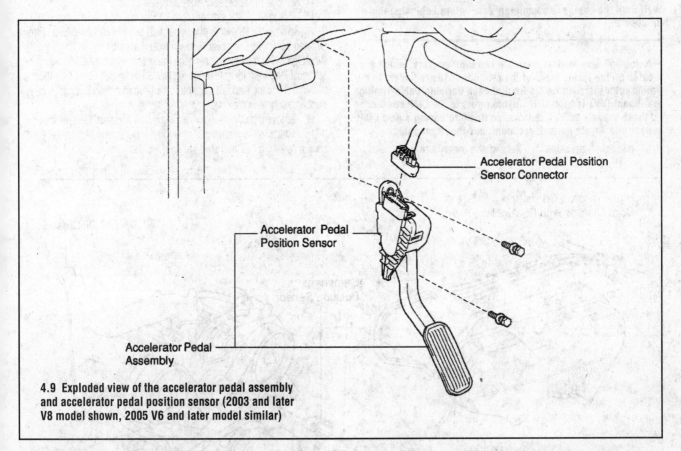

Accelerator Pedal Position Sensor Connector

Accelerator Pedal Position Sensor

Accelerator Pedal Assembly

4.9 Exploded view of the accelerator pedal assembly and accelerator pedal position sensor (2003 and later V8 model shown, 2005 V6 and later model similar)

2003 AND LATER V8 MODELS AND 2005 AND LATER V6 MODELS

▶ Refer to illustration 4.9

➡ Note: On these models the accelerator pedal position sensor is an integral component of the accelerator pedal assembly. To replace the accelerator pedal position sensor you must replace the entire accelerator pedal assembly.

8 Disconnect the cable from the negative battery terminal.
9 Working under the dash with a flashlight, disconnect the electrical connector from the accelerator pedal position sensor (see illustration).
10 Remove the two accelerator pedal assembly bolts and remove the accelerator pedal assembly.
11 Installation is the reverse of removal.

5 Camshaft Position (CMP) sensor - replacement

2000 THROUGH 2004 V6 MODELS

▶ Refer to illustrations 5.3

1 Disconnect the negative battery cable.

2 Remove the upper timing belt cover (see "Timing belt and sprockets - removal and installation" in Chapter 2A).

3 Unplug the CMP sensor electrical connector (see illustration).

4 Remove the CMP sensor retaining bolt and remove the CMP sensor.

5 Installation is the reverse of removal. Be sure to tighten the CMP sensor retaining bolt securely.

2005 AND LATER V6 MODELS

Refer to illustration 5.10

※※ WARNING:

Wait until the engine is completely cool before beginning this procedure.

➡**Note: On these models there are two CMP sensors, which are located on the inside walls of the cylinder heads, adjacent to the timing rotor installed on the front of each Variable Valve Timing-intelligent (VVT-i) controller. Toyota refers to the CMP sensors on these models as VVT sensors, so if you're buying a new CMP sensor at a Toyota parts department, use that terminology.**

6 Disconnect the cable from the negative battery terminal.

7 Remove the engine cover (see illustration 4.12 in Chapter 2A).

8 Remove the air intake duct and the air filter housing (see Chapter 4).

9 If you're going to remove the CMP/VVT sensor from the left cylinder head, drain the engine coolant (see Chapter 1) and disconnect the two coolant bypass hoses from the throttle body (see illustration 13.40 in Chapter 4).

➡**Note: As an alternative to draining the coolant, the hoses can be clamped off using locking pliers.**

10 Disconnect the electrical connector from the CMP/VVT sensor (see illustration).

11 Remove the CMP/VVT sensor mounting bolt and remove the sensor.

12 Installation is the reverse of removal.

V8 MODELS

▶ Refer to illustrations 5.15, 5.16 and 5.17

13 Disconnect the negative battery cable.

14 Remove the accessory drivebelt (see Chapter 1).

15 Unplug the CMP sensor electrical connector (see illustration).

16 Remove the left No. 3 (the upper left) timing belt cover (see "Timing belt and sprockets - removal, inspection and installation" in Chapter 2B). When removing the cover, disengage the grommet for the CMP sensor electrical lead from its hole in the left No. 3 timing belt cover (see illustration) and slide the lead out the slot in the grommet, then pull the lead through the hole in the cover.

17 Remove the CMP sensor retaining bolt and stud (see illustration).

18 Installation is the reverse of removal. Be sure to tighten the CMP sensor retaining bolt and stud securely.

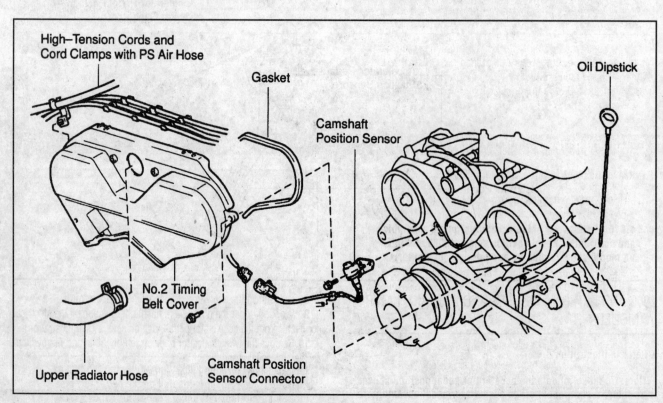

5.3 Camshaft Position (CMP) sensor installation details (V6 models)

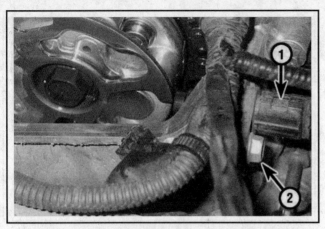

5.10 To remove the CMP/VVT sensor from a 2005 and later V6 engine, disconnect the electrical connector (1) and remove the sensor mounting bolt (2) (CMP/VVT sensor for right cylinder head shown, sensor for left head identical)

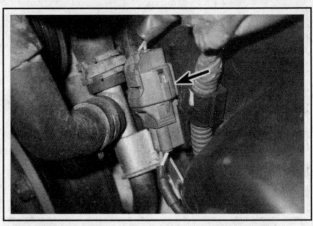

5.15 Disconnect the electrical connector for the CMP sensor (V8 models)

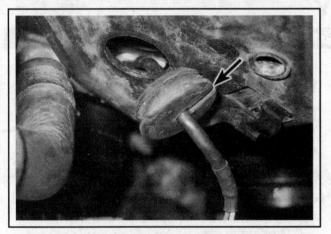

5.16 When removing the left No. 3 timing belt cover from a V8, pull the grommet for the CMP sensor electrical lead out of its hole in the cover, slide the lead out the slot (arrow) in the grommet, then thread the lead through the hole in the cover

5.17 To remove the CMP sensor on V8 models, remove the sensor retaining bolt and stud

6 Crankshaft Position (CKP) sensor - replacement

▶ Refer to illustrations 6.5a, 6.5b and 6.5c

1   Disconnect the negative battery cable.
2   Remove the engine under-cover (see "Engine oil and filter change" in Chapter 1).
3   On 2005 and later V6 models, remove the alternator (see Chapter 5).
4   On 2005 and later V6 models, remove the bolt that secures the air conditioning suction line at the front of the engine, disconnect the electrical connector from the air conditioning compressor and unbolt the compressor (see Chapter 3). Do NOT disconnect the air conditioning hoses from the compressor.
5   Disconnect the CKP sensor electrical connector (see illustrations).
6   Remove the CKP sensor retaining bolt.
7   Installation is the reverse of removal. Be sure to tighten the CKP sensor retaining bolt securely.

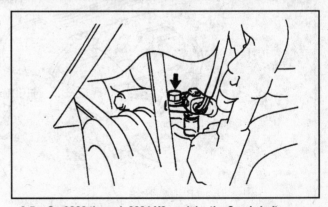

6.5a On 2000 through 2004 V6 models, the Crankshaft Position (CKP) sensor is located on the upper left side of the oil pump; to remove it, unplug the electrical connector and remove the sensor retaining bolt

6.5b On 2005 and later V6 models, the Crankshaft Position (CKP) sensor is located on the lower left side of the timing chain cover. To remove the CKP sensor, disconnect the electrical connector and remove the sensor mounting bolt

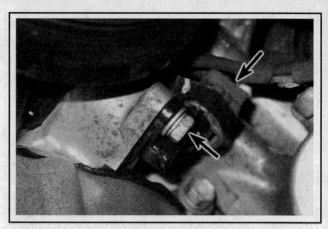

6.5c On V8 models, the Crankshaft Position (CKP) sensor is located on the left underside of the oil pump; to remove it, unplug the electrical connector and remove the sensor retaining bolt

7   Engine Coolant Temperature (ECT) sensor - replacement

✳✳ WARNING:

Wait until the engine has cooled completely before beginning this procedure.

✳✳ CAUTION:

Handle the Engine Coolant Temperature (ECT) sensor with care. Damage to the ECT sensor will affect the operation of the entire fuel injection system.

1   Drain the engine coolant until the coolant level is below the sensor (see Chapter 1).

2000 THROUGH 2004 V6 MODELS

▶ Refer to illustrations 7.4 and 7.6

2   Remove the No. 2 (the upper) timing belt cover (see "Timing belt and sprockets - removal and installation" in Chapter 2A).
3   Remove the fuel rail crossover pipe that connects the two fuel rails (see "Fuel rail and injectors - removal and installation" in Chapter 4).
4   Disconnect the electrical connector from the ECT sensor (see illustration).

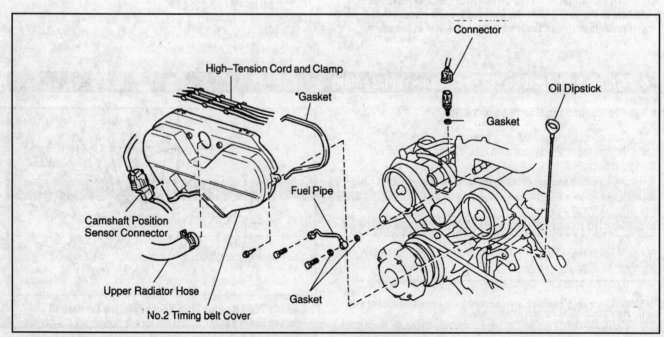

7.4 On V6 models, the Engine Coolant Temperature (ECT) sensor is located in the valley between the cylinder heads

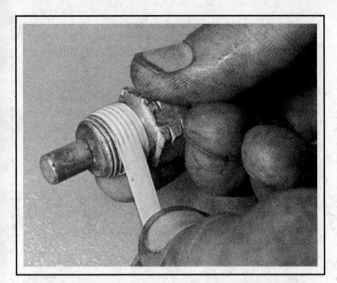

7.6 Wrap the threads of the ECT sensor with Teflon tape to prevent coolant leakage

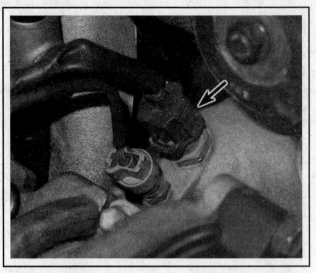

7.15 On V8 models, the Engine Coolant Temperature (ECT) sensor is located on the right end of the coolant crossover pipe under the throttle body

5   To remove the ECT sensor, unscrew it with a deep socket.

6   Before installing the new ECT sensor, wrap the threads of the sensor with Teflon tape to prevent coolant leakage (see illustration).

7   Installation is otherwise the reverse of removal. Be sure to use a new washer and tighten the ECT sensor to the torque listed in this Chapter's Specifications. Refill the cooling system (see Chapter 1).

2005 AND LATER V6 MODELS

8   Remove the engine cover (see illustration 4.12 in Chapter 2A).

9   The ECT sensor is located at the back of the engine, on the coolant bypass housing. The coolant bypass housing is the casting that spans the valley between the two cylinder heads.

10   Disconnect the electrical connector from the ECT sensor.

11   Unscrew the ECT sensor from the coolant bypass housing with a deep socket.

12   Before installing the new ECT sensor, wrap the threads of the sensor with Teflon tape to prevent coolant leakage (see illustration 7.6).

13   Installation is otherwise the reverse of removal. Be sure to use a new washer and tighten the ECT sensor to the torque listed in this Chapter's Specifications.

V8 MODELS

▶ **Refer to illustration 7.15**

14   Remove the throttle body cover and the air intake duct (see "Throttle body - removal and installation" in Chapter 4).

15   Disconnect the electrical connector from the ECT sensor (see illustration).

16   Unscrew and remove the ECT sensor. Discard the old washer.

17   Before installing the new ECT sensor, wrap the threads of the sensor with Teflon tape to prevent coolant leakage (see illustration 7.6). Installation is otherwise the reverse of removal. Be sure to use a new washer and tighten the ECT sensor to the torque listed in this Chapter's Specifications. Refill the cooling system (see Chapter 1).

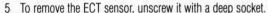

8   Knock sensor - replacement

※※ WARNING:

Wait for the engine to cool completely before performing this procedure.

➡**Note: The knock sensors are located on the inner walls of the cylinder heads on all models.**

2000 THROUGH 2004 V6 MODELS

▶ **Refer to illustrations 8.2 and 8.4**

1   Drain the engine coolant (see Chapter 1), then remove all three parts of the intake manifold assembly (air intake plenum, intake air connector and intake manifold) (see "Intake manifold - removal and installation" in Chapter 2A).

2   Remove the coolant bypass pipe (see illustration). Have some rags ready to mop up the slight coolant loss that will occur when you disconnect the bypass pipe.

3   Disconnect the electrical connector from the knock sensor.

4   Unscrew the knock sensor with a deep socket (see illustration).

5   Installation is the reverse of removal. Be sure to tighten the knock sensor to the torque listed in this Chapter's Specifications. Refill the cooling system (see Chapter 1).

2005 AND LATER V6 MODELS

▶ **Refer to illustrations 8.8 and 8.9**

6   Drain the engine coolant (see Chapter 1).

7   Remove the right cylinder head (see Chapter 2A).

8   Disconnect the heater coolant inlet hose from the coolant outlet pipe, then detach the four wiring harness clips from the outlet pipe, remove the three outlet pipe retaining bolts (see illustration) and remove the outlet pipe assembly.

9   Disconnect the electrical connectors from the knock sensors (see illustration).

10   Carefully note how the knock sensors are oriented, then remove the sensor retaining bolts and remove the sensors.

11   When installing the knock sensors, make sure that the centerline of each sensor is positioned within 15 degrees of the direction in which it was facing before removal (see illustration 8.9).

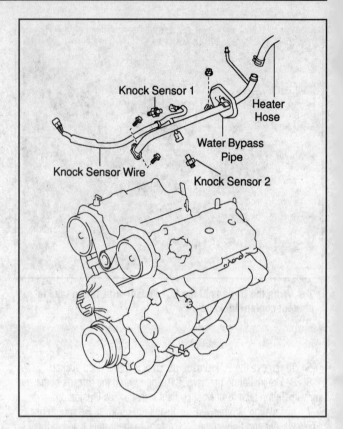

8.2   Installation details for the knock sensors, which are screwed into the inner walls of the cylinder heads (V6 models)

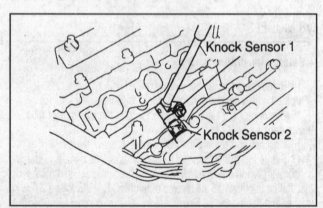

8.4   On V6 models, unscrew the knock sensors with a deep socket

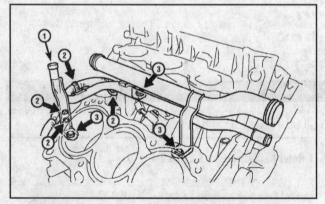

8.8   To remove the heater coolant outlet pipe, disconnect the heater inlet hose from the small pipe at the rear of the engine (1), then detach the four knock sensor wiring harness clips (2) and remove the three coolant outlet pipe mounting bolts (3)

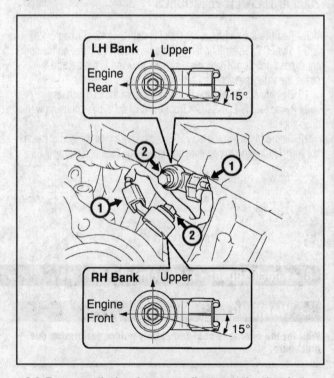

8.9   To remove the knock sensors, disconnect the electrical connectors (1), then remove the sensor retaining bolts (2). When installing the knock sensors, make sure that the centerline of each sensor is positioned within 15 degrees of the direction in which it was facing before removal

12 When the knock sensors are correctly positioned, tighten the sensor retaining bolts to the torque listed in this Chapter's Specifications.

13 The remainder of installation is the reverse of removal.

V8 MODELS

▶ **Refer to illustration 8.15**

14 Drain the engine coolant (see Chapter 1), then remove the upper and lower intake manifolds (see "Intake manifold - removal and installation" in Chapter 2A).

15 Disconnect the knock sensor electrical connector (see illustration).

16 Unscrew the knock sensor with a deep socket.

17 Installation is the reverse of removal. Be sure to tighten the knock sensor to the torque listed in this Chapter's Specifications. Refill the cooling system (see Chapter 1).

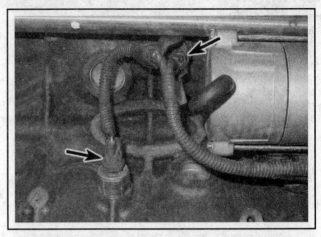

8.15 Knock sensor electrical connectors and knock sensors (V8 models)

9 Mass Air Flow (MAF) sensor - replacement

Refer to illustration 9.2a and 9.2b

1 Make sure the ignition key is in the Off position.

2 Disconnect the MAF sensor electrical connector (see illustrations).

3 Remove the MAF sensor retaining screws and remove the sensor.

4 Installation is the reverse of removal.

9.2a To detach the Mass Air Flow (MAF) sensor, unplug the electrical connector and remove the two MAF sensor retaining screws (V8 model shown, 2000 through 2004 V6 models similar)

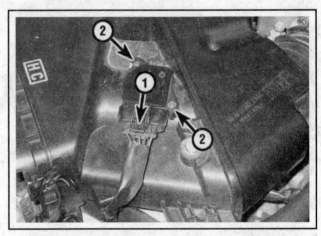

9.2b To detach the Mass Air Flow (MAF) sensor from the air filter housing on a 2005 and later V6 model, depress this release tab (1) and pull off the electrical connector, then remove the sensor retaining screws (2)

10 Oxygen sensor and Air/Fuel (A/F) ratio sensor - general information and replacement

GENERAL INFORMATION

1 Use special care when servicing an oxygen sensor or air/fuel ratio sensor:

a) *Oxygen sensors and air/fuel ratio sensors have a permanently attached pigtail and electrical connector which cannot be removed from the sensor. Damage or removal of the pigtail or electrical connector will ruin the sensor.*

b) *Grease, dirt and other contaminants should be kept away from the electrical connector and the louvered end of the sensor.*

c) *Do not use cleaning solvents of any kind on an oxygen sensor or air/fuel ratio sensor.*

d) *Do not drop or roughly handle an oxygen sensor or air/fuel ratio sensor.*

e) *Be sure to install the silicone boot in the correct position to prevent the boot from melting and to allow the sensor to operate properly.*

REPLACEMENT

➡**Note: Because it is installed in an exhaust manifold or exhaust pipe, which contract when cool, an oxygen sensor or air/fuel ratio sensor might be very difficult to loosen when the engine is cold. Rather than risk damage to the sensor, start and run the engine for a minute or two, then shut it off. Be careful not to burn yourself during the following procedure.**

2   Make sure the ignition key is in the Off position.
3   Raise the vehicle and place it securely on jackstands.

Upstream oxygen sensor or air/fuel ratio sensor

V6 models

4   On 2000 through 2004 models, the upstream oxygen sensor or air/fuel ratio sensor is located on the front end of the exhaust pipe, just behind the exhaust manifold flange and just ahead of the upstream cata-lyst (see illustration 16.6a). On 2005 and later V6 models, there are two upstream oxygen sensors and they're located on the front exhaust pipes, below the exhaust manifold flange and ahead of the downstream cata-lysts (see illustration 16.6b). Disconnect the sensor electrical connector, remove the sensor mounting nuts, then remove the sensor and gasket. Discard the old mounting nuts and the old gasket.

5   Apply anti-seize compound to the threads of the sensor mounting studs to facilitate future removal. Be sure to use a new gasket and new mounting nuts. Installation is otherwise the reverse of removal.

V8 models

⧫ **Refer to illustrations 10.6a, 10.6b and 10.6c**

6   The upstream oxygen sensor is located on the rear end of the exhaust manifold, just ahead of the flange for the exhaust pipe (see illustration). Disconnect the sensor electrical connector and unscrew the sensor from the exhaust manifold with a special oxygen sensor socket (see illustrations).

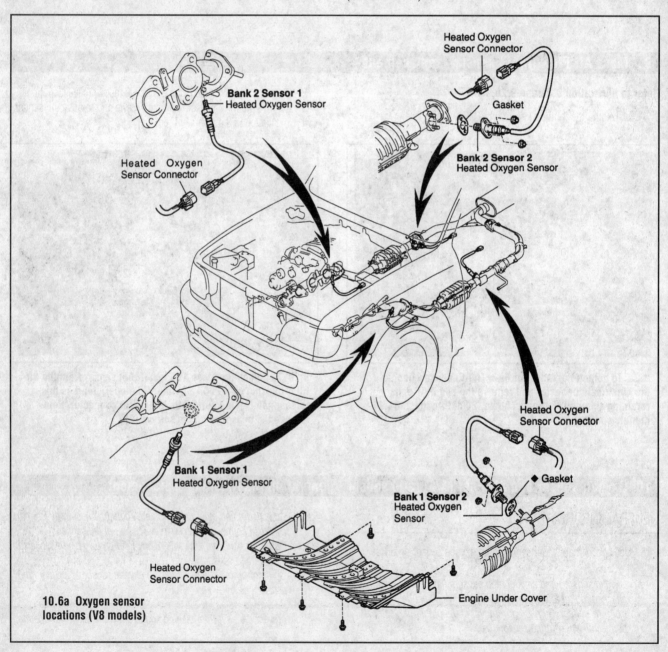

10.6a Oxygen sensor locations (V8 models)

7   Apply anti-seize compound to the threads of the sensor to facilitate future removal.

➡**Note: Most new sensors already have anti-seize compound applied to the threads. Installation is otherwise the reverse of removal.**

Downstream oxygen sensor

V6 models

8   On 2000 through 2004 V6 models, there is one downstream, or post-converter, oxygen sensor (see illustration 16.6a). On 2000 through 2004 California models it's located on the center exhaust pipe, right behind the downstream catalyst. On 2000 through 2004 Federal models it's also located on the center exhaust pipe, but these models do not have a downstream catalyst. 2005 and later V6 models have two downstream oxygen sensors, one on each front exhaust pipe, just ahead of the downstream catalysts (see illustration 16.6b). Disconnect the oxygen sensor electrical connector, then remove the sensor mounting nuts, the heat shield (if equipped), the sensor and the sensor gasket. Discard the old mounting nuts and the old gasket.

V8 models

▶ **Refer to illustrations 10.9a and 10.9b**

9   On V8 models, the downstream oxygen sensor is located behind the upstream (the front) converter in each exhaust pipe (see illustration 10.6a). Disconnect the oxygen sensor electrical connector, then remove the sensor mounting nuts, the heat shield (if equipped), the sensor and the sensor gasket (see illustrations). Discard the old mounting nuts and the old gasket.

All models

10   Apply anti-seize compound to the threads of the sensor mounting studs to facilitate future removal. Be sure to use a new gasket and new mounting nuts. Installation is otherwise the reverse of removal.

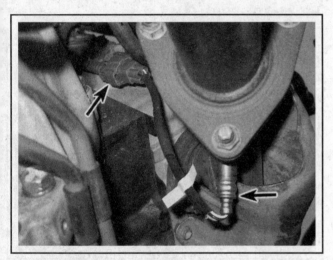

10.6b A typical upstream oxygen sensor on a V8 model, as seen from underneath the vehicle; to remove an upstream sensor, unplug the electrical connector . . .

10.6c . . . and unscrew the sensor with a special oxygen sensor socket (available from specialty tool suppliers and most auto parts stores)

10.9a A typical downstream oxygen sensor on a V8 model; to remove it, unplug the electrical connector . . .

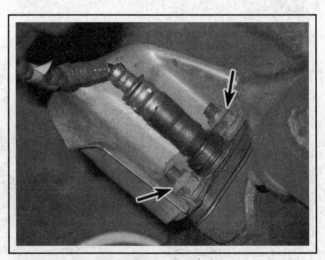

10.9b . . . remove the sensor mounting nuts, heat shield, the sensor and the gasket (discard the old mounting nuts and the old gasket)

11 Throttle Position (TP) sensor - replacement

➡Note: These procedures do NOT apply to 2003 and later V6 or 2003 and later V8 models. On these models, the Throttle Position (TP) sensor is not removable. If the TP sensor is defective you must replace the throttle body assembly.

1 Make sure the ignition key is in the Off position.

2000 THROUGH 2002 V6 MODELS

2 The Throttle Position (TP) sensor is located on the throttle body on the end of the throttle shaft.

3 Disconnect the electrical connector from the TP sensor.

4 Remove the two TP sensor mounting screws and remove the TP sensor.

5 Installation is the reverse of removal.

2000 THROUGH 2002 V8 MODELS

❊❊ CAUTION:

Toyota specifies the use of a Toyota hand-held tester, or a suitable OBD-II scan tool capable of reading the opening percentage of the throttle plate, for adjusting the TP sensor, so don't remove the TP sensor until you know whether your scan tool can do the job. If it can't, this procedure is best left to a qualified technician with the right tools.

Removal

▶ Refer to illustration 11.7

6 The Throttle Position (TP) sensor is located on the throttle body on the end of the throttle shaft.

7 Disconnect the electrical connector from the TP sensor (see illustration).

8 Remove the two TP sensor mounting screws and remove the TP sensor.

Installation and adjustment

▶ Refer to illustrations 11.10a, 11.10b and 11.10c

9 Remove the air intake duct (see "Throttle body - removal and installation" in Chapter 4) and verify that the throttle plate is closed.

10 When installing the TP sensor, make sure that the tangs on the two arms of the throttle shaft engage the slots in the backside of the TP sensor (see illustrations). Install the TP sensor rotated about 15 degrees to the right of its actual installed position (see illustration), rotate the TP sensor counterclockwise until it touches the throttle shaft and temporarily tighten the two TP sensor retaining screws.

11 Hook up a suitable scan tool and turn the ignition switch to ON. While reading the value of the throttle valve opening percentage, turn the throttle position sensor slowly to the left and the right. Set the TP sensor at the center of the standard value (14.4 to 16 percent), then tighten the screws securely.

12 Installation is otherwise the reverse of removal.

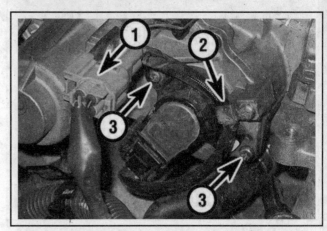

11.7 To remove the Throttle Position (TP) sensor on a V8 model, unplug the electrical connector (1), disengage the wiring harness from the clip (2), then remove the TP sensor mounting screws (3)

11.10a When installing the TP sensor on V8 models, make sure that the tangs on the end of these two arms . . .

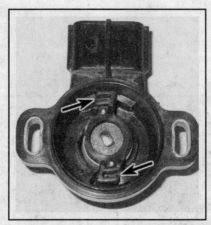

11.10b . . . are engaged with these slots in the backside of the TP sensor

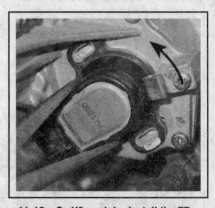

11.10c On V8 models, install the TP sensor at about 15 degrees to the right of its actual installed position, rotate it counterclockwise until it touches the throttle shaft, then install the sensor mounting screws and tighten them enough to hold the sensor in position

12 Transmission Range (TR) sensor - replacement

Even though the Society of Automotive Engineers (SAE) has recommended since 1996 that all manufacturers refer to the Park/Neutral Position (PNP) switch as the Transmission Range (TR) sensor, Toyota continues to refer to the TR sensor as the PNP switch. To avoid confusion, we have therefore included the TR sensor/PNP switch in Chapter 7B.

13 Vehicle Speed Sensor (VSS) - replacement

VEHICLE SPEED SENSOR (VSS) (ALL MODELS)

▶ **Refer to illustration 13.1**

1   The Vehicle Speed Sensor (VSS) (see illustration) is located on the left side of the transmission extension housing on 2WD models, or on the left side of the transfer case on 4WD models.
2   Disconnect the electrical connector from the VSS.
3   Remove the VSS retaining bolt, then remove the VSS from the transmission or transfer case.
4   Remove and discard the old VSS O-ring.
5   Coat the new O-ring with clean gear lube (manual transmissions) or ATF (automatic transmissions).
6   Installation is the reverse of removal.

OVERDRIVE (O/D) DIRECT CLUTCH SPEED SENSOR (AUTOMATIC MODELS ONLY)

▶ **Refer to illustration 13.7**

7   The O/D direct clutch speed sensor (see illustration) is located on the left side of the transmission.
8   Disconnect the electrical connector from the O/D direct clutch speed sensor.
9   Remove the sensor retaining bolt and remove the sensor.
10 Remove and discard the old sensor O-ring.
11 Coat the new sensor O-ring with clean ATF.
12 Installation is the reverse of removal.

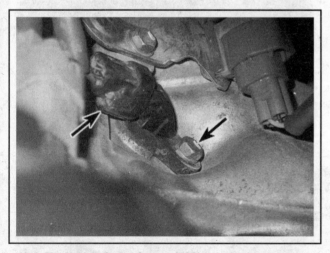

13.1 The Vehicle Speed Sensor (VSS) is located on the left side of the transfer case (4WD) or on the left side of the extension housing (2WD); to remove it, simply unplug the electrical connector and remove the sensor retaining bolt

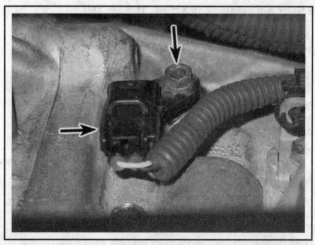

13.7 The O/D direct clutch speed sensor (automatics only) is located on the left side of the transmission, near the front of the transmission housing; to remove it, simply unplug the electrical connector and remove the sensor retaining bolt

14 Idle Air Control (IAC) valve (2000 through 2002 V6 models) - replacement

▶ Refer to illustration 14.2

1   Remove the throttle body (see Chapter 4).

2   Remove the Idle Air Control (IAC) valve mounting screws (see illustration) and detach the IAC valve from the throttle body. Remove and discard the old IAC valve gasket.

3   Installation is the reverse of removal. Be sure to use a new gasket and tighten the IAC valve screws securely.

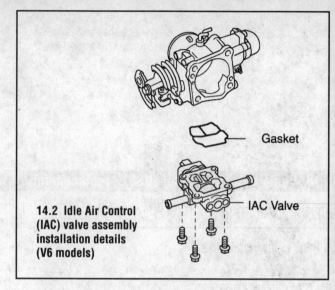

14.2 Idle Air Control (IAC) valve assembly installation details (V6 models)

15 Throttle control motor (2000 through 2002 V8 models) - removal and installation

REMOVAL

▶ Refer to illustrations 15.3, 15.4, 15.5 and 15.6

1   Make sure the ignition key is in the Off position.

2   Remove the throttle position (TP) sensor (see Section 11).

✳✳ CAUTION:

Before removing the TP sensor from the throttle control motor, be sure to read the Caution in Section 11 regarding adjustment of the TP sensor.

3   Unplug the electrical connector from the throttle control motor (see illustration).

4   Remove the six throttle motor assembly retaining screws (see illustration), then remove the throttle control motor cover and the throttle motor as a single assembly.

5   Remove the three throttle motor retaining screws (see illustration), carefully pry the grommet for the electrical harness from the throttle motor cover, then remove the throttle motor from the cover.

6   Remove the wave washer (see illustration).

INSTALLATION

▶ Refer to illustrations 15.8 and 15.12

7   Install the waver washer into the throttle control motor cover.

8   Install the throttle control motor into the cover. Make sure that the holes in the throttle control motor are aligned with the positioning pins of the cover (see illustration).

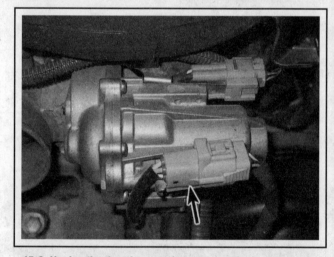

15.3 Unplug the throttle control motor electrical connector

9   Install the electrical harness grommet in the throttle control motor cover.

10 Tighten the three throttle control motor screws securely.

11 Lightly grease the gear teeth of the throttle control motor.

12 Install the washer on the throttle control motor driven gear (see illustration).

13 Align the pin hole of the cover with the positioning pin of the throttle body.

14 Install the throttle motor and cover assembly and tighten the screws securely.

15 Install and adjust the TP sensor (see Section 11).

15.4 To detach the throttle control motor cover from the throttle body, remove these six screws

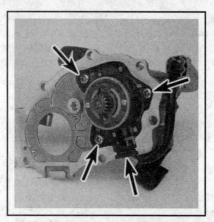

15.5 To separate the throttle control motor from the cover, remove these three screws and pry out the grommet for the electrical harness

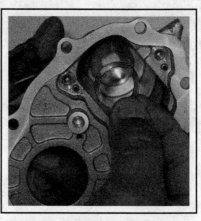

15.6 After removing the throttle control motor from the cover, remove this wave washer

15.8 When installing the throttle control motor in the cover, align the holes in the motor with the cover positioning pins, then push the grommet into position

15.12 Before installing the throttle control motor and cover, be sure to install this washer on the throttle control motor driven gear; when installing the cover, make sure that the hole in the cover is aligned with the positioning pin on the throttle body

16 Catalytic converter

➡Note: Because of the Federally mandated extended warranty which covers emissions-related components such as the catalytic converter, check with a dealer service department before replacing the converter at your own expense.

GENERAL DESCRIPTION

1   The catalytic converter is an emission control device added to the exhaust system to reduce pollutants from the exhaust gas stream. There are two types of converters: The oxidation catalyst reduces the levels of hydrocarbon (HC) and carbon monoxide (CO) by adding oxygen to the exhaust stream. The reduction catalyst lowers the levels of oxides of nitrogen (NOx) by removing oxygen from the exhaust gases. These two types of catalysts are combined into a three-way catalyst that reduces all three pollutants.

CHECK

2   The equipment for testing a catalytic converter is expensive. If you suspect that the converter on your vehicle is malfunctioning, take it to a dealer or authorized emissions inspection facility for diagnosis and repair.

3   Whenever the vehicle is raised for servicing underbody components, inspect the converter for leaks, corrosion, dents and other damage. Inspect the welds/flange bolts that attach the front and rear ends of the converter to the exhaust system. If damage is discovered, the

converter should be replaced.

4   Although catalytic converters don't break too often, they can become plugged. The easiest way to check for a restricted converter is to use a vacuum gauge to diagnose the effect of a blocked exhaust on intake vacuum.

a) Connect a vacuum gauge to an intake manifold vacuum source (see Chapter 2C).
b) Warm the engine to operating temperature, place the transaxle in Park (automatic) or Neutral (manual) and apply the parking brake.
c) Note and record the vacuum reading at idle.
d) Quickly open the throttle to near full throttle and release it shut. Note and record the vacuum reading.
e) Perform the test three more times, recording the reading after each test.

f) If the reading after the fourth test is more than one in-Hg lower than the reading recorded at idle, the exhaust system may be restricted (the catalytic converter could be plugged or an exhaust pipe or muffler could be restricted).

REPLACEMENT

▶ Refer to illustrations 16.6a, 16.6b and 16.6c

5   Be sure to spray the nuts on the exhaust flange studs with penetrant before removing them from the catalytic converter.

6   Remove the nuts and separate the catalytic converter from the exhaust system (see illustrations).

7   Installation is the reverse of removal.

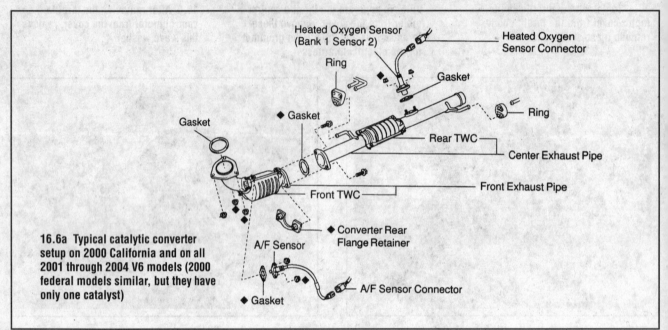

16.6a Typical catalytic converter setup on 2000 California and on all 2001 through 2004 V6 models (2000 federal models similar, but they have only one catalyst)

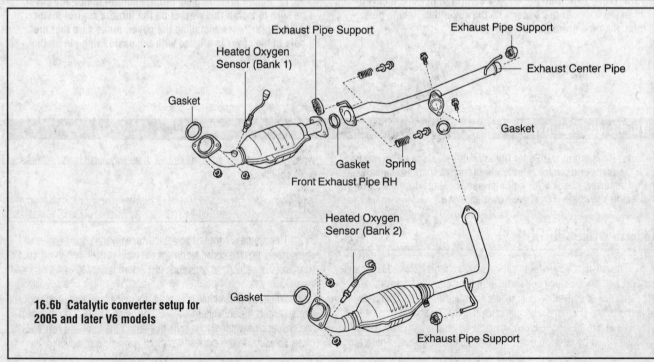

16.6b Catalytic converter setup for 2005 and later V6 models

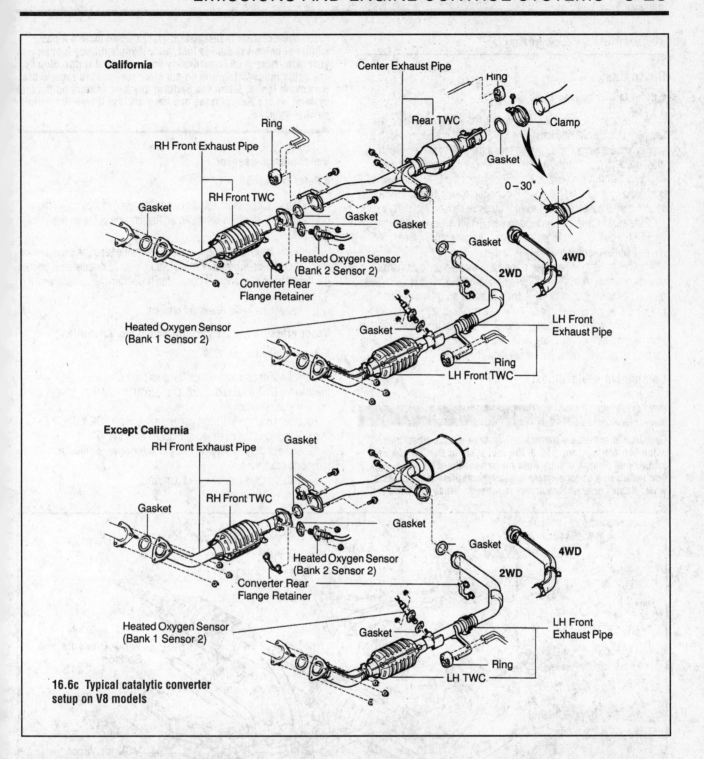

16.6c Typical catalytic converter setup on V8 models

17 Evaporative emission control (EVAP) system

GENERAL DESCRIPTION (ALL MODELS)

1   The fuel evaporative emission control (EVAP) system absorbs fuel vapors (unburned hydrocarbons) and, during engine operation, releases them into the intake manifold from which they're drawn into the intake ports where they mix with the incoming air-fuel mixture.

2   When the engine is off, gasoline in the fuel tank, and residual fuel in the other components such as the intake manifold, warms up and evaporates, producing unburned hydrocarbon fuel vapors that would waste gas and pollute the air if released into the atmosphere. In an EVAP system, these vapors are routed from the fuel tank, throttle body and intake manifold through a system of hoses to the EVAP system's charcoal canister, where they're stored until the vehicle is operated again. The charcoal canister is able to store these vapors because it's filled with activated charcoal, a substance that can absorb many times its mass in hydrocarbon vapors.

2000 THROUGH 2002 MODELS

Description

▶ **Refer to illustrations 17.3a and 17.3b**

3   The charcoal canister is mounted on the left side of the engine compartment (see illustrations). The EVAP system is equipped with a vapor pressure sensor, mounted on top of the EVAP canister, which monitors system pressure. A vapor pressure sensor Vacuum Switching Valve (VSV), which is also located on top of the EVAP canister, is controlled by the PCM. Normally, the vapor pressure sensor VSV routes vapors from the fuel tank to the vapor pressure sensor. When the vapor pressure sensor detects high vapor pressure in the fuel tank, it signals the PCM, which energizes the vapor pressure sensor VSV, which re-routes fuel tank vapors to the EVAP canister.

4   After the engine has been started and warmed up to its normal operating temperature, the EVAP Vacuum Switching Valve (EVAP VSV) opens (this valve is commonly referred to as the purge valve). The EVAP VSV, which is mounted on the intake manifold, allows manifold vacuum to draw fuel vapors out of the EVAP canister and into the intake manifold, where they are mixed with intake air before being burned with the air/fuel mixture inside the combustion chambers.

Component replacement

✳✳ WARNING:

Gasoline is extremely flammable, so take extra precautions when you work on any part of the fuel system. Don't smoke or allow open flames or bare light bulbs near the work area, and don't work in a garage where a gas-type appliance (such as a water heater or a clothes dryer) is present. Since gasoline is carcinogenic, wear fuel-resistant gloves when there's a possibility of being exposed to fuel, and, if you spill any fuel on your skin, rinse it off immediately with soap and water. Mop up any spills immediately and do not store fuel-soaked rags where they could ignite. When you perform any kind of work on the fuel system, wear safety glasses and have a Class B type fire extinguisher on hand.

Vapor pressure sensor

▶ **Refer to illustration 17.5**

5   The vapor pressure sensor (see illustration) is located on top of the EVAP canister, which is located on the left side of the engine compartment.

6   Disconnect the electrical connector from the vapor pressure sensor.

7   Pull the vapor pressure sensor straight up to disengage it from its mounting bracket and disconnect the hose from the underside of the sensor.

8   Installation is the reverse of removal.

Vapor pressure sensor Vacuum Switching Valve (VSV)

▶ **Refer to illustration 17.9**

9   The vapor pressure sensor Vacuum Switching Valve (VSV) (see illustration) is also located on top of the EVAP canister, in front of the vapor pressure sensor.

10   Disconnect the electrical connector from the VSV.

11   Disconnect the three vacuum hoses from the VSV.

12   Remove the VSV retaining screw and remove the VSV from its mounting bracket.

13   Installation is the reverse of removal.

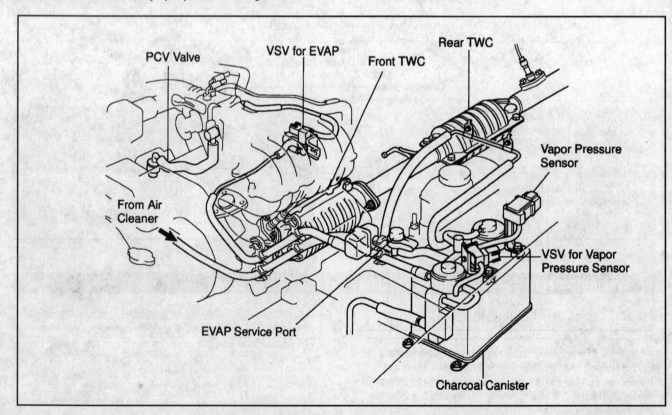

17.3a EVAP system component location, EVAP hose routing and PCV system component location (2000 through 2002 V6 models)

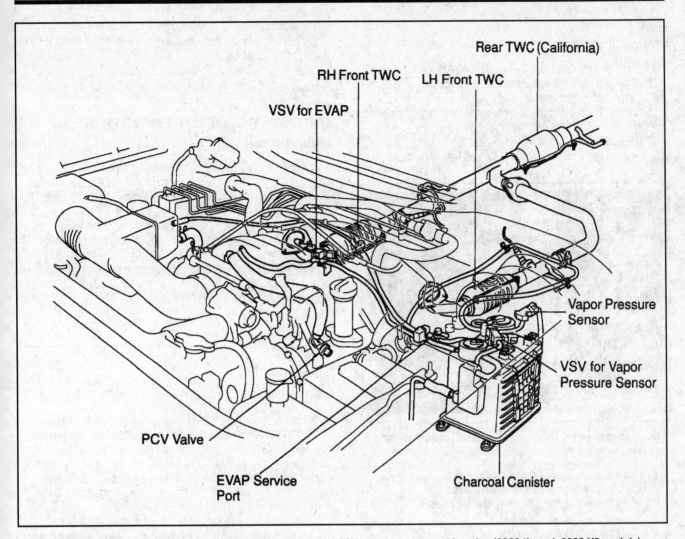

17.3b EVAP system component location, EVAP hose routing and PCV system component location (2000 through 2002 V8 models)

17.5 The vapor pressure sensor is located on top of the EVAP canister; to remove it, simply unplug the electrical connector, pull the sensor straight up to disengage it from its mounting bracket, then disconnect the vapor hose from the underside of the sensor

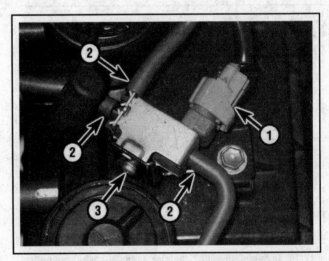

17.9 The vapor pressure sensor Vacuum Switching Valve (VSV) is also located on top of the EVAP canister; to remove it, unplug the electrical connector (1), disconnect the three vapor hoses (2), then remove the retaining screw (3) that attaches the sensor to its mounting bracket

Charcoal canister

▶ **Refer to illustrations 17.15 and 17.19**

14 Disconnect the cable from the negative battery terminal.

15 Unplug the electrical connectors from the vapor pressure sensor and the vapor pressure sensor VSV (see illustration).

16 Clearly label and disconnect the two hoses that connect the vapor pressure sensor VSV to the EVAP canister.

17 Remove the bolts that attach the brackets for the vapor pressure sensor and the vapor pressure sensor VSV, then remove the vapor pres-

sure sensor and VSV and set them aside.

18 Clearly label and disconnect the four hoses that connect the EVAP canister to the EVAP system.

19 Remove the three EVAP canister retaining bolts (see illustration) and remove the canister from the engine compartment.

20 Installation is the reverse of removal.

EVAP Vacuum Switching Valve (EVAP VSV)/purge valve

▶ **Refer to illustration 17.21**

21 The EVAP VSV (commonly referred to as the purge valve), is located at the rear of the intake manifold on V6 models (see illustration 17.3) and on the left side of the intake manifold on V8 models (see illustration).

22 Disconnect the electrical connector from the purge valve.

23 Loosen the spring-type hose clamp and disconnect the EVAP hoses from the purge valve.

24 Remove the EVAP VSV mounting bolt and remove the purge valve.

25 Installation is the reverse of removal.

2003 AND LATER MODELS

Description

▶ **Refer to illustrations 17.26a through 17.26e**

26 The EVAP system (see illustrations) consists of the charcoal canister, the canister closed valve, the EVAP system Vacuum Switching Valve (EVAP VSV) (more commonly referred to as the purge valve), the vapor pressure sensor, the air filter and the refueling valve. All components of the system except the purge valve are located underneath the vehicle. The purge valve is located in the engine compartment, on the intake manifold.

27 The vapor pressure sensor, which is mounted on top of the fuel tank and is connected to the Powertrain Control Module, monitors the pressure inside the fuel tank. When the vapor pressure sensor detects

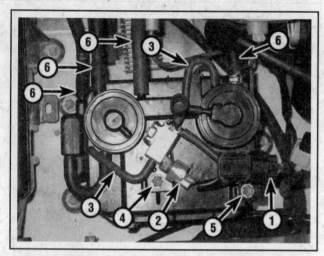

17.15 To remove the EVAP canister, unplug, disconnect or remove the following:

1 *Vapor pressure sensor electrical connector*
2 *Vapor pressure sensor VSV electrical connector*
3 *Vapor pressure sensor VSV-to-EVAP canister hoses*
4 *Vapor pressure sensor VSV bracket bolt*
5 *Vapor pressure sensor bracket bolt*
6 *EVAP canister-to-EVAP system hoses*

17.19 To detach the EVAP canister, remove these three bolts

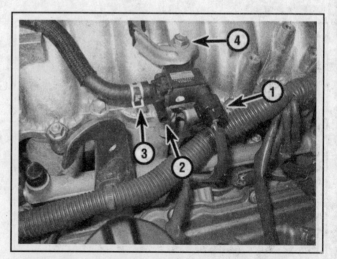

17.21 To remove the EVAP VSV from the intake manifold on a 2000 through 2002 V8 model, disconnect the electrical connector (1), disconnect the purge hose coming from the charcoal canister (2) and the hose going to the intake manifold (3), then remove the EVAP VSV mounting bolt (4) (V6 models similar, except that the EVAP VSV is located at the rear of the intake manifold)

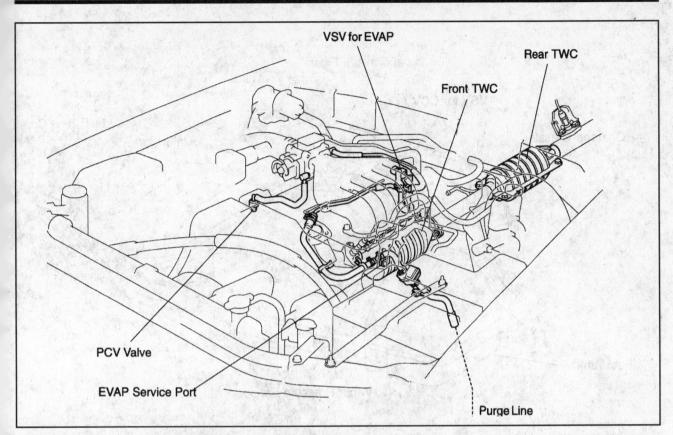

17.26a Engine compartment EVAP system component location, EVAP hose routing and PCV system component location (2003 and 2004 V6 models)

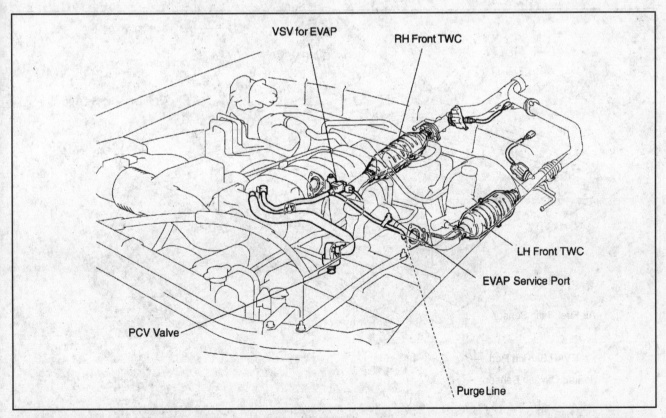

17.26b Engine compartment EVAP system component location, EVAP hose routing and PCV system component location (2003 and 2004 V8 models)

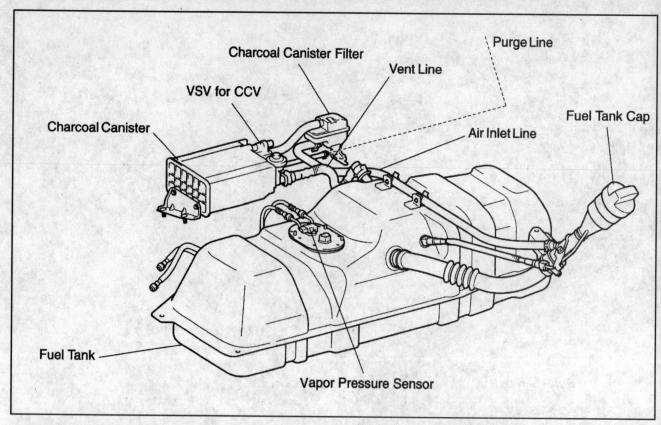

17.26c Under-vehicle EVAP system component location and EVAP hose routing (all 2003 and 2004 models)

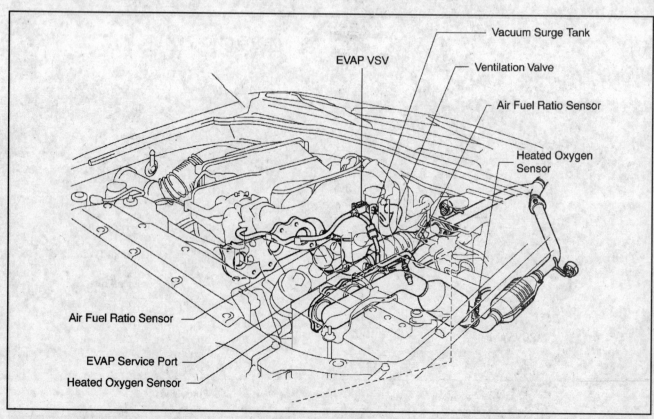

17.26d Engine compartment EVAP system component location, EVAP hose routing and PCV system component location (2005 and later V6 models)

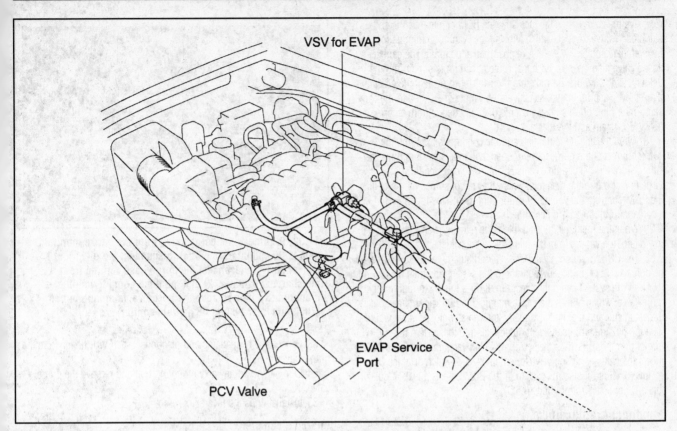

17.26e Engine compartment EVAP system component location, EVAP hose routing and PCV system component location (2005 and later V8 models)

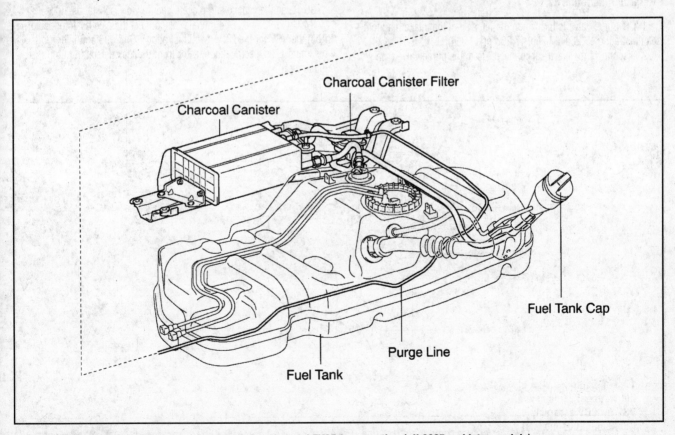

17.26f Under-vehicle EVAP system component location and EVAP hose routing (all 2005 and later models)

excessive pressure inside the fuel tank, the PCM commands the canister closed valve to open, which allows these vapors to migrate to the charcoal canister, where they're absorbed and stored. The canister closed valve also opens a hose that routes fresh (outside) air into the fuel tank as the vapors are routed to the canister to prevent the creation of a relative vacuum inside the tank as the vapors are vented to the canister. An air filter on this hose prevents dust and debris from entering the EVAP system and the fuel tank.

28 When the engine is running and the conditions are right, the PCM commands the purge valve to open, and intake vacuum pulls the vapors stored in the charcoal canister out of the canister, through a purge line and into the manifold, where they're consumed by the engine. The PCM controls the volume of vapors drawn into the manifold in accordance with the driving conditions.

29 When the fuel tank is being filled, the refueling valve controls the flow rate of the vapors from the fuel tank to the canister.

30 The PCM monitors the EVAP system regularly for leaks by using the purge valve to induce a vacuum into the system and then measuring how well the system can hold a vacuum. It does this by turning on the purge valve, which allows the intake manifold to pull vapor gases out of the canister and fuel tank. Normally, the canister closed valve would open the outside air hose to vent outside air into the system during normal purging. But when the system is being vacuum tested, the PCM directs the canister closed valve to remain closed, which produces the (relative) vacuum inside the system. If the PCM detects a leak, it outputs a Diagnostic Trouble Code (DTC).

Component replacement

Purge valve

▶ **Refer to illustration 17.32**

31 The purge valve is located on the left side of the intake manifold (see illustration 17.26a, 17.26b, 17.26d or 17.26e).

32 Disconnect the electrical connector from the purge valve (see illustration).

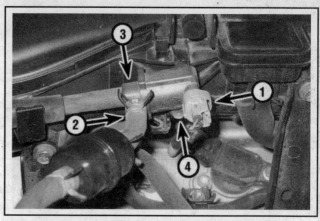

17.32 To remove the purge valve from the intake manifold on a 2005 and later V6 model, disconnect the electrical connector (1), disconnect the purge hose coming from the charcoal canister (2) and the hose going to the intake manifold (3), then remove the purge valve mounting bolt (4) (2003 and later V6 and V8 models similar)

33 Loosen the hose clamp and disconnect the EVAP hoses from the purge valve.

34 Remove the purge valve mounting bolt and remove the purge valve.

35 Installation is the reverse of removal.

Charcoal canister

▶ **Refer to illustration 17.37a and 17.37b**

36 Raise the vehicle and place it securely on jackstands.

37 On 2003 and 2004 models, disconnect the electrical connector from the VSV for the CCV (see illustration). On 2005 and later models, disconnect the electrical connector from the leak detection pump (see illustration).

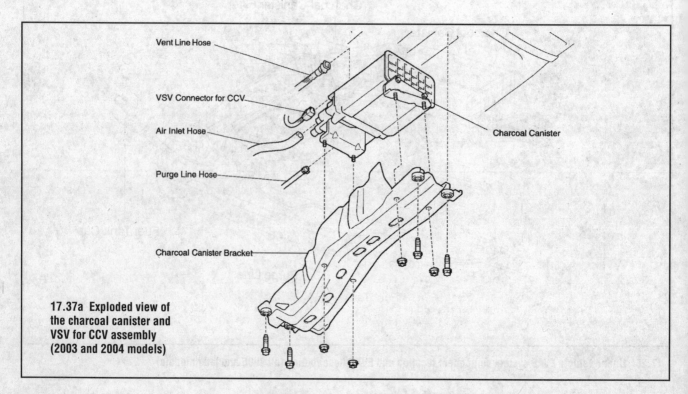

Vent Line Hose

VSV Connector for CCV

Air Inlet Hose

Purge Line Hose

Charcoal Canister

Charcoal Canister Bracket

17.37a Exploded view of the charcoal canister and VSV for CCV assembly (2003 and 2004 models)

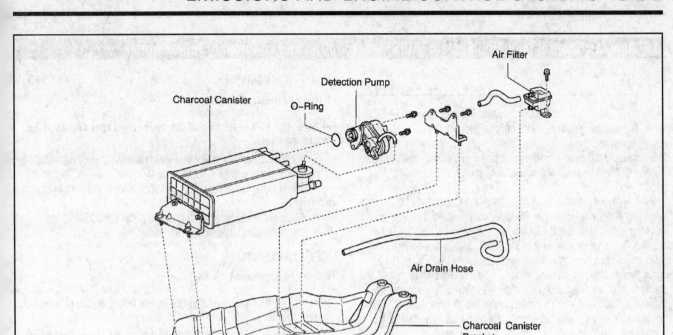

17.37b Exploded view of the charcoal canister, leak detection pump and air filter assembly (2005 and later models)

38 Remove the charcoal canister mounting bracket bolts and nuts and lower the bracket and canister as a single assembly.

39 Separate the canister from its mounting bracket.

40 Installation is the reverse of removal.

Canister Closed Valve (CCV), Vacuum Switching Valve for CCV (VSV for CCV) and air filter (2003 and 2004 models)

41 Remove the charcoal canister (see Steps 36 through 39), if necessary.

42 Refer to illustration 17.26c and 17.37a and use them as a guide for removal and installation.

43 Pay particular attention to which hoses are connected to which components. Label them if necessary.

Leak detection pump and air filter (2005 and later models)

44 Remove the charcoal canister (see Steps 36 through 39).

45 Refer to illustration 17.26f and 17.37b and use them as a guide for removal and installation.

46 Pay particular attention to which hoses are connected to which components. Label them if necessary.

18 Positive Crankcase Ventilation (PCV) system

DESCRIPTION

1 The Positive Crankcase Ventilation (PCV) system reduces hydrocarbon emissions by scavenging crankcase vapors. It does this by circulating fresh air from the air cleaner through the crankcase, where it mixes with blow-by gases, before being drawn through a PCV valve into the intake manifold.

2 The PCV system consists of the PCV valve and two hoses, one of which connects the air cleaner to the crankcase and the other (containing the PCV valve) which connects the crankcase to the throttle body.

3 To maintain idle quality, the PCV valve restricts the flow when the intake manifold vacuum is high. If abnormal operating conditions (such as piston ring problems) arise, the system is designed to allow excessive amounts of blow-by gases to flow back through the crankcase vent tube into the air cleaner to be consumed by normal combustion.

CHECK

4 To check and/or replace the PCV valve, refer to Chapter 1. To find the PCV valve on:

2000 through 2002 V6 models (see illustration 17.3a)
2000 through 2002 V8 models (see illustration 17.3b)
2003 and 2004 V6 models (see illustration 17.26a)
2003 and 2004 V8 models (see illustration 17.26b)
2005 and later V6 models (see illustration 17.26d)
2005 and later V8 models (see illustration 17.26e)

19 Acoustic Control Induction System (ACIS) - description and component replacement

DESCRIPTION

1   The PCM-controlled Acoustic Control Induction System (ACIS) varies the effective length of the intake manifold runners in response to engine speed and the angle of the throttle plate inside the throttle body. This capability increases efficiency and power at low and high speeds. Slightly different versions of the ACIS are used on 2005 and later V6 and V8 models.

2   The ACIS consists of a PCM-controlled Vacuum Switching Valve (VSV), an actuator and an intake air control valve. The VSV is a PCM controlled-device that controls the intake vacuum applied to the actuator. The actuator is a vacuum diaphragm that uses a pushrod and bellcrank to open and close the intake air control valve. The intake air control valve, which is an integral part of the intake manifold, opens and closes to alter the effective length of the intake manifold runners in two stages. On 2005 and later V6 models, there is one large intake air control valve for all six intake runners. On 2005 and later V8 models, there is one air control valve for each intake manifold runner.

3   At low-to-medium speeds, the PCM activates the VSV, sending vacuum to the actuator diaphragm. The actuator closes the intake air control valve, lengthening the length of the intake manifold and improving intake efficiency.

4   At higher speeds, the PCM deactivates the VSV, cutting vacuum to the actuator diaphragm. The actuator opens the intake air control valve, shortening the length of the intake manifold and improving engine power.

COMPONENT REPLACEMENT

➡Note: The actuator and the intake air control valve are integral components of the intake manifold. If either component fails, replace the intake manifold (see Chapter 2A or 2B). The VSV is the only component that you can replace at home.

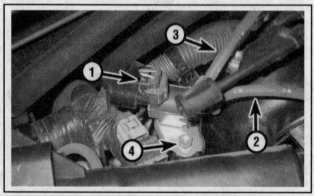

19.6 To remove the Acoustic Control Induction System Vacuum Switching Valve (ACIS VSV) from the intake manifold on a 2005 and later V6 model:

1   Disconnect the electrical connector
2   Disconnect the vacuum hose that connects the VSV to its manifold vacuum source
3   Disconnect the vacuum hose that directs manifold vacuum to the intake air control valve actuator
4   Remove the VSV mounting bolt

VSV (V6 models)

▶ Refer to illustration 19.6

➡Note: The VSV is located at the upper right rear corner of the intake manifold.

5   Remove the engine cover (see illustration 4.12 in Chapter 2A).
6   Disconnect the electrical connector from the VSV (see illustration).
7   Clearly label both vacuum hoses, then disconnect both hoses from the VSV.
8   Remove the VSV mounting bolt and remove the VSV.
9   Installation is the reverse of removal.

VSV (V8 models)

▶ Refer to illustration 19.10

➡Note: The VSV is located at the upper left rear corner of the intake manifold.

10 Disconnect the electrical connector from the VSV (see illustration).
11 Clearly label both vacuum hoses, then disconnect both hoses from the VSV.
12 Remove the VSV mounting bolt and remove the VSV.
13 Installation is the reverse of removal.

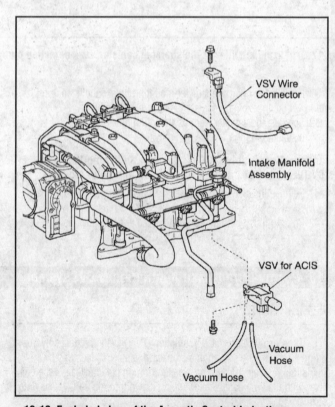

19.10 Exploded view of the Acoustic Control Induction System Vacuum Switching Valve (ACIS VSV) on a 2005 and later V8 model

20 Secondary air injection system - description and component replacement

DESCRIPTION

1 An air injection system is used on 2005 V8 models. Air injection helps the upstream catalytic converters reach their effective operating temperature more rapidly during warm-ups. The system consists of the PCM, an electronic air injection control driver, an electric air pump, an air pressure sensor, an electric air switching valve, two Vacuum Switching Valves (VSVs) and two vacuum-type air switching valves. These components are connected by an array of plastic and metal plumbing and wiring harnesses.

2 Using inputs from the Engine Coolant Temperature (ECT) sensor and Intake Air Temperature (IAT) sensor, the PCM determines whether the engine is cold, or already warmed up. If the engine is already warmed up, the PCM doesn't turn on the air injection system. If the engine is cold, or not sufficiently warmed up, the PCM calculates how much air is needed to warm up the catalysts. To do so, it uses inputs from the ECT, IAT and Mass Air Flow (MAF) sensors to calculate how long the air injection system should be turned on to bring the upstream catalysts up to their operating temperature.

3 When the PCM activates the air injection system, it turns on the two VSVs and it turns on the air injection control driver, which turns on the air injection pump and the electric air switching valve. The two VSVs send vacuum to the vacuum-type air switching valves, which allows the air pump to pump air through the electric air switching valve, through the two vacuum-type air switching valves, and then through lines to the exhaust manifolds.

4 Using input from the air pressure sensor, the PCM monitors the air injection system and uses this data to control the air injection control driver, which in turn controls the electric air switching valve.

5 When air is pumped into the exhaust manifolds, it helps to burn up any residual unburned fuel vapors, which are always present in the rich mixture used during warm-ups. Without this additional air, the engine exhaust would still warm up, but not as quickly, because the unburned fuel actually cools the exhaust temperature. But when the oxygen in the extra air combines with and promotes the burning of this unburned fuel, it heats up the exhaust much more quickly, and the upstream catalysts reach their operating temperature more quickly too.

COMPONENT REPLACEMENT

6 Because it is used only during cold-start warm-ups, the air injection system should be trouble-free for years. But if a component of the system fails, you can replace any of the critical components yourself. However, be aware that replacing some of these components - the air pressure sensor, the electric air switching valve and the air pump - is a little more difficult because you'll have to remove the intake manifold to access them.

Air injection control driver

▶ **Refer to illustration 20.7**

➡ **Note: The air injection control driver is located on the left side of the engine compartment, near the rear end of the left cylinder head.**

7 Disconnect the two electrical connectors from the air injection control driver (see illustration).

8 Remove the two air injection control driver mounting bolts and remove the driver.

9 Installation is the reverse of removal.

Vacuum Switching Valves (VSVs)

▶ **Refer to illustration 20.12**

➡ **Note: The VSVs are located on the right side of the intake manifold.**

10 Disconnect the electrical connectors from the VSVs.

11 Disconnect the vacuum hoses from the VSVs (see illustration).

12 Remove the VSV mounting bracket bolts and remove the VSV assembly.

13 Installation is the reverse of removal.

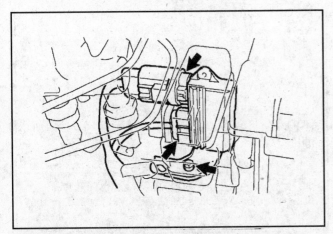

20.7 To remove the air injection control driver, disconnect the two electrical connectors and remove the two driver mounting bolts

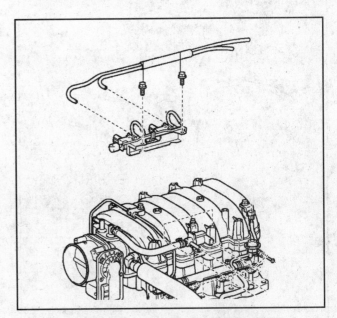

20.12 To remove the Vacuum Switching Valves (VSVs) from the intake manifold, disconnect the two electrical connectors, disconnect the two vacuum hoses and remove the two mounting bolts

Vacuum-type air switching valves

▶ Refer to illustration 20.14

➡**Note: The vacuum-type air switching valves are located on the back of the engine, where they're bolted to the rear coolant bypass housing. Toyota calls this component the rear water bypass joint (see illustration 5.10 in Chapter 2B).**

14 Remove the two nuts from each air tube flange at the exhaust manifolds (see illustration) and disconnect the two air tubes from the exhaust manifolds. Remove the two bolts from each air tube mounting flange at the air switching valves and disconnect the air tubes from the switching valves. Remove both air tubes and remove and discard the old air tube mounting flange gaskets.

15 Remove the four air switching valve mounting bolts, pull back the switching valve assembly far enough to disconnect the two vacuum hoses and remove the switching valve assembly.

16 Installation is the reverse of removal. Be sure to use new gaskets.

Air pressure sensor, electric air switching valve and air pump

17 Remove the intake manifold (see Chapter 2B).

Air pressure sensor

▶ Refer to illustration 20.19

18 Disconnect the electrical connector from the air pressure sensor
19 Remove the two pressure sensor mounting bolts (see illustration) and remove the pressure sensor from the air switching valve.
20 Installation is the reverse of removal.

Air switching valve

21 Disconnect the electrical connector from the air switching valve.

22 Clearly label the two hoses connected to the air switching valve, then disconnect them.
23 Remove the air switching valve mounting bolts (see illustration 20.19) and remove the air switching valve.
24 Installation is the reverse of removal.

Air pump only

25 If you're replacing the air pump, remove the air pressure sensor and the air switching valve (see Steps 18 through 23).
26 Disconnect the electrical connector from the pump.
27 Disconnect the hoses from the pump.
28 Remove the three pump mounting bolts, bushings and spacers (see illustration 20.19) and remove the pump from its mounting bracket.
29 Installation is the reverse of removal.

Air pressure sensor/air switching valve/air pump assembly

▶ Refer to illustration 20.33

30 If you're only removing the air pump to service something below it, you can remove the air pressure sensor, air switching valve and air pump as a single assembly.

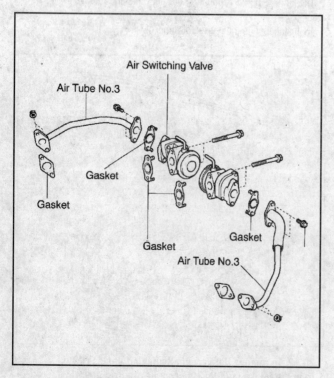

20.14 Exploded view of the vacuum-type air switching valves

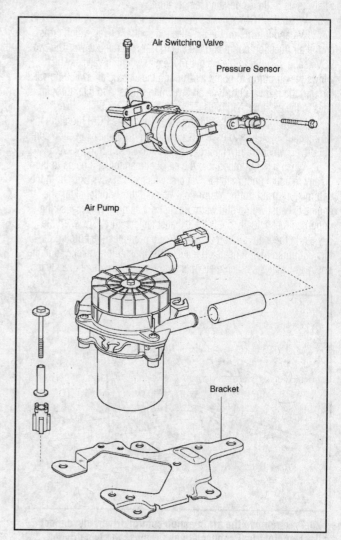

20.19 Exploded view of the air pressure sensor, electric air switching valve and air pump assembly

31 Disconnect the electrical connectors from the air pressure sensor, the air switching valve and the air pump.

32 Disconnect all hoses from the air pressure sensor, the air switching valve and the air pump.

33 Remove the air pump assembly mounting bracket bolts and nuts (see illustration) and remove the air pressure sensor, air switching valve and air pump as a single assembly.

34 Installation is the reverse of removal.

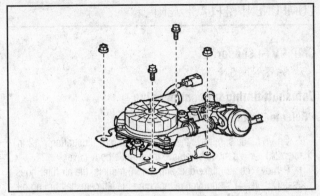

20.33 To detach the air pressure sensor/electric air switching valve/air pump assembly from the engine, remove these two bolts and two nuts

21 Variable Valve Timing-intelligent (VVT-i) system - description and component replacement

DESCRIPTION

1 The Variable Valve Timing (VVT-i) system varies intake camshaft timing within a range of 50 degrees (V6 models) or 44 degrees (V8 models) to produce valve timing that is optimized for the driving conditions. The VVT-i system achieves this by using engine oil pressure to advance or retard the controller on the front end of each intake camshaft.

2 The VVT-i system consists of two CMP/VVT-i sensors, the Powertrain Control Module (PCM), two VVT-i oil control valves and two controller (intake camshaft sprocket/actuator assemblies). The VVT-i systems on V6 and V8 models are virtually identical except for the design of the controllers.

3 On V6 models, each controller consists of a timing rotor (for the CMP/VVT-i sensor), a housing with an impeller-type vane inside it, a lock pin and the actual timing chain sprocket for the intake camshaft. The vane is fixed on the end of the camshaft. Oil pressure directed into the housing from the advance or retard side of the intake cam causes the vane to rotate in relation to the intake camshaft, advancing or retarding the valve timing. The higher the oil pressure (or flow) the more the actuator assembly will rotate, thereby advancing or retarding the camshaft. When the engine is turned off, the intake cam is in its most retarded position to ensure easy starting. When the engine is first started and no oil pressure has yet been applied to the controller housing, the lock pin locks the housing and vane together to prevent it from making a knocking sound. When oil pressure enters the housing and is applied to the lock pin spring, the spring is compressed and the lock pin retracts, allowing the housing to rotate in rotate in relation to the vane.

4 On V8 models, each controller consists of a housing, four vanes and a lock pin. Oil pressure directed into the housing from the advance or retard side of the intake cam causes the vane to rotate in relation to the intake camshaft, advancing or retarding the valve timing. The higher the oil pressure (or flow) the more the actuator assembly will rotate,

thereby advancing or retarding the camshaft. When the engine is turned off, the intake cam is in its most retarded position to ensure easy starting. When the engine is first started and no oil pressure has yet been applied to the controller housing, the lock pin locks the housing and vane together to prevent it from making a knocking sound. When oil pressure enters the housing and is applied to the lock pin spring, the spring is compressed and the lock pin retracts, allowing the housing to rotate in rotate in relation to the four vanes.

5 When oil is applied to the advance side of the vane(s), the actuator advances the camshaft in a clockwise direction. When oil is applied to the retard side of the vanes, the actuator rotates the camshaft counterclockwise back to 0 degrees advance, which is the normal position of the actuator during engine operation under no-load or idle conditions. The PCM can also send a signal to the timing oil control valve to stop oil flow to both (advance and retard) passages to hold camshaft advance in its current position. Under light engine loads, the VVT system retards the camshaft timing to decrease valve overlap and stabilize engine output. Under medium engine loads, the VVT system advances the camshaft timing to increase valve overlap, thereby increasing fuel economy and decreasing exhaust emissions. Under heavy engine loads at low RPM, the VVT system advances the camshaft timing to help close the intake valve faster, which improves low to midrange torque. Under heavy engine loads at high RPM, the VVT system retard the camshaft timing to slow the closing of the intake valve to improve engine horsepower.

6 The camshaft timing oil control valve is a PCM-controlled device that controls and directs the flow of oil to the advance or retard passages leading to the controller. There are two oil control valves, one for each intake camshaft. A spring-loaded spool valve inside the oil control valve directs oil pumped into the valve toward either the advance outlet port or the retard outlet port, depending on the engine speed, which determines oil pressure. When the engine is turned off, the spring is extended and the spool valve is in its most retarded state. Once the engine is started and oil pressure begins to rise with engine rpm, the spool is slowly pushed against its spring.

COMPONENT REPLACEMENT

CMP/VVT-i sensor

7   See Section 5.

Camshaft timing oil control valve

♦ **Refer to illustration 21.10**

8   On V6 models, remove the engine cover (see illustration 4.12 in Chapter 2A). On V8 models, remove the throttle body cover.

9   Remove the air intake duct and, on V6 models, the air filter housing (see Chapter 4). And if you're replacing the left camshaft timing oil control valve on a V6 model, remove the upper intake manifold (see Chapter 2A).

10 Disconnect the electrical connector from the camshaft timing oil control valve (see illustration).

11 Remove the camshaft timing oil control valve mounting bolt and remove the oil control valve.

12 Installation is the reverse of removal.

21.10 To remove the camshaft timing oil control valve (1), disconnect the electrical connector (2), and remove the mounting bolt (3)

Torque specifications   Ft-lbs (unless otherwise noted)

➡**Note: One foot-pound (ft-lb) of torque is equivalent to 12 inch-pounds (in-lbs) of torque. Torque values below approximately 15 ft-lbs are expressed in inch-pounds, since most foot-pound torque wrenches are not accurate at these smaller values.**

Engine Coolant Temperature (ECT) sensor	
2000 through 2004 V6 models	168 in-lbs
2005 and later V6 models and V8 models	180 in-lbs
Knock sensor	
V6 models	
2000 through 2004	29
2005 and later	180 in-lbs
V8 models	
2000 through 2004	33
2005 and later	180 in-lbs

Section

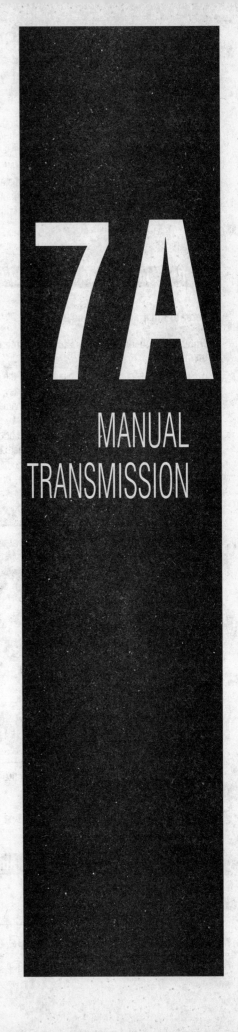

7A

MANUAL
TRANSMISSION

1 General information

All vehicles covered by this manual are equipped with a five-speed manual transmission or a four-speed automatic transmission. Information on the manual transmission and certain procedures common to both transmissions - such as oil seal replacement - can be found in this Part of Chapter 7. Service procedures for the automatic transmission can be found in Part B, and information about the transfer case used on 4WD models can be found in Part C of this Chapter.

Depending on the expense involved in having a transmission overhauled, it might be a better idea to consider replacing it with either a new or rebuilt unit. Your local dealer or transmission shop should be able to supply information concerning cost, availability and exchange policy. Regardless of how you decide to remedy a transmission problem, you can still save a lot of money by removing and installing the unit yourself.

2 Rear output shaft oil seal (2WD) - replacement

♦ **Refer to illustrations 2.4 and 2.5**

➡**Note: This procedure applies to both manual and automatic transmissions.**

1   Oil leaks frequently occur due to wear of the extension housing oil seal. Replacement of this seal is relatively easy, since the repair can usually be performed without removing the transmission from the vehicle.

2   The extension housing oil seal is located at the extreme rear of the transmission, where the driveshaft is attached. Raise the vehicle and support it securely on jackstands. If the seal is leaking, transmission lubricant will be built up on the front of the driveshaft and may be dripping from the rear of the transmission.

3   Remove the driveshaft (see Chapter 8).

4   Using a puller or seal removal tool, carefully remove the oil seal out of the rear of the transmission (see illustration). Do not damage the splines on the transmission output shaft.

5   Using a large section of pipe or a very large deep socket as a drift, install the new oil seal (see illustration). Drive it into the bore squarely and make sure it's completely seated.

6   Lubricate the splines of the transmission output shaft and the outside of the driveshaft slip yoke with light-weight grease, then install the driveshaft. Be careful not to damage the lip of the new seal.

7   Check the lubricant level in the transmission, adding as necessary (see Chapter 1).

2.4 Using a seal removal tool to remove the seal

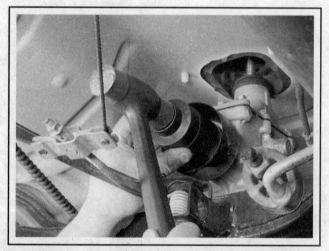

2.5 To install the new seal, tap it into place with a large socket and hammer

3 Transmission mount - check and replacement

CHECK

♦ **Refer to illustration 3.1**

1   Insert a large screwdriver or pry bar into the space between the

transmission and the crossmember and try to pry the transmission up slightly (see illustration).

2   The transmission should not move away from the insulator much. If there is any separation of the rubber, the mount is worn out.

3.1 To check the transmission mount, insert a prybar or large screwdriver between the mount rubber and the mount bracket, then try to pry the transmission up off its mount; if the transmission moves significantly, replace the mount

3.3 Transmission mount-to-crossmember retaining bolts (4WD shown, 2WD similar)

REPLACEMENT

▶ **Refer to illustration 3.3 and 3.4**

3   Support the transmission with a jack and remove the transmission mount-to-crossmember retaining bolts (see illustration).

4   Raise the transmission slightly with the jack and remove the mount-to-transmission fasteners (see illustration).

➡**Note: On 4WD models, the mount is attached the transfer case.**

5   Installation is the reverse of the removal procedure. Be sure to tighten the fasteners securely.

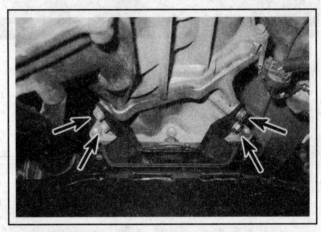

3.4 Remove the four mount-to-transmission fasteners

4   Shift lever - removal and installation

▶ **Refer to illustrations 4.2, 4.3a, 4.3b and 4.4**

1   Unscrew the shift lever knob(s). Remove the four shift lever boot retaining screws and remove the boot and retainer.

2   Remove the small dust cover from the shift lever cap (see illustration).

3   Press down on the shift lever cap and rotate it counterclockwise to remove the shift lever (see illustrations).

4.2 Remove the dust cover from the shift lever cap

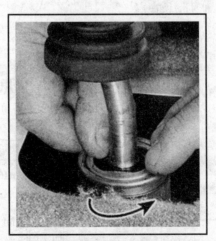

4.3a Push down on the shift lever cap and rotate it counterclockwise . . .

4.3b . . . then remove the shift lever

4   Installation is the reverse of removal. Before installing the shift lever, lubricate the friction surfaces with multi-purpose grease (see illustration).

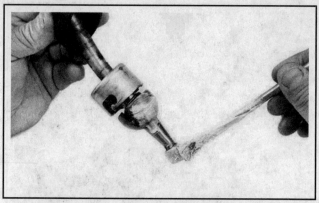

4.4 Lubricate the friction surfaces of the shift lever with multi-purpose grease before installing the shift lever

5   Back-up light switch - removal and installation

▶ **Refer to illustration 5.2**

1   Raise the vehicle and support it securely on jackstands.
2   Unplug the back-up light switch electrical connector (see illustration).
3   Unscrew the back-up light switch from the transmission case and remove the switch and gasket.
4   Install the switch and gasket in the transmission case and tighten it securely.
5   Plug in the electrical connector.
6   Lower the vehicle and check the operation of the back-up lights.

5.2 The back-up light switch is located on the right side of the transmission

6   Manual transmission - removal and installation

REMOVAL

1   Disconnect the cable from the negative terminal of the battery.
2   Remove the shift lever (see Section 4).

2WD models

▶ **Refer to illustrations 6.9, 6.12 and 6.13**

3   Raise the vehicle and support it securely on jackstands.
4   Disconnect the driveshaft (see Chapter 8).
5   Unbolt the clutch release cylinder and hydraulic line brackets (see Chapter 8).

❋❋ CAUTION:

Don't depress the clutch pedal with the release cylinder removed.

6   Disconnect the electrical connectors for the vehicle speed sensor and back-up light switch.

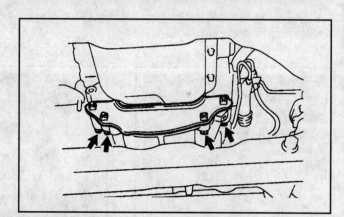

6.9 Remove the four bolts securing the flywheel inspection cover

7   Remove the front exhaust pipe(s) (see Chapter 4).
8   Remove the starter (see Chapter 5).
9   Remove the flywheel inspection cover (see illustration).
10 Support the engine by placing a floor jack and a block of wood

under the engine oil pan.

11 Support the transmission with a jack, preferably a special jack made for this purpose. Safety chains will help steady the transmission on the jack. If you don't have a transmission jack, support the transmission with a floor jack.

12 Remove the transmission mount-to-crossmember retaining bolts and the crossmember-to-frame mounting bolts and remove the crossmember (see illustration).

13 Remove the six transmission-to-engine bolts (see illustration).

14 Make a final check that all wires have been disconnected from the transmission.

15 Move the transmission and jack toward the rear of the vehicle until the transmission input shaft is clear of the clutch or clutch housing.

16 Inspect the clutch components. It's a good idea to install new clutch components whenever the transmission is removed (see Chapter 8).

4WD models

▶ **Refer to illustration 6.27**

➡**Note: The following procedure involves removing the transfer case with the transmission.**

17 On 4WD models with a manually-operated transfer case, remove the transfer case shift lever (see Chapter 7C).

18 Raise the vehicle and support it securely on jackstands.

19 Disconnect both the front and rear driveshafts (see Chapter 8).

20 Unplug the electrical connectors from the vehicle speed sensor, 4WD position switch and L4 position switch. Detach all wire harnesses from the transmission and crossmember.

21 Unbolt the clutch release cylinder and hydraulic line brackets (see Chapter 8).

✳✳ CAUTION:

Don't depress the clutch pedal with the release cylinder removed.

22 Remove the front exhaust pipe(s) (see Chapter 4).

23 Remove the starter (see Chapter 5).

24 Remove the flywheel inspection cover (see illustration 6.9).

25 Support the engine by placing a floor jack and a block of wood under the engine oil pan.

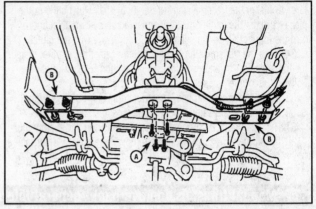

6.12 Remove the crossmember-to-transmission mount bolts (A), and the crossmember-to-frame bolts (B)

26 Support the transmission and transfer case with a jack, preferably a special jack made for this purpose. Safety chains will help steady the transmission on the jack.

27 Remove the four transfer case mount-to-crossmember retaining bolts (see Section 3) and the crossmember-to-frame mounting bolts and remove the crossmember (see illustration).

28 Remove the six transmission-to-engine bolts (see illustration 6.13).

29 Make a final check that all wires have been disconnected from the transmission and transfer case, then move the transmission and transfer case with the jack toward the rear of the vehicle until the transmission input shaft is clear of the clutch or clutch housing.

30 Lower the transmission and transfer case from the vehicle.

31 Inspect the clutch components. It's a good idea to install new clutch components whenever the transmission is removed (see Chapter 8).

INSTALLATION

32 Install the clutch components, if they were removed (see Chapter 8).

33 Apply a light coat of high-temperature grease to the splines and end of the input shaft. With the transmission secured to the jack, raise it into position behind the engine, and then carefully slide it forward, engaging the input shaft with the clutch plate hub. Do not use excessive

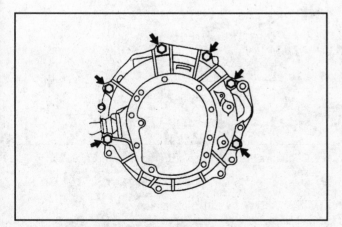

6.13 There are six transmission-to-engine bolts

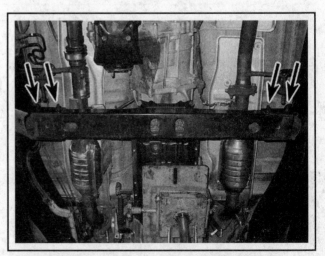

6.27 Remove the crossmember-to-frame mounting bolts

force to install the transmission - if the input shaft won't slide into place, readjust the angle of the transmission so that it's level and/or turn the input shaft so the splines engage properly with the clutch.

34 Once the transmission is fully seated against the engine, install the transmission-to-engine bolts (see illustration 6.13). Tighten the bolts to the torque listed in this Chapter's Specifications.

35 Install the crossmember. Tighten all nuts and bolts securely.

36 Bolt the transmission mount to the crossmember.

37 Remove the jacks supporting the transmission and the engine.

38 Install the various components removed previously. Refer to

Chapter 8 for driveshaft installation and clutch release cylinder installation procedures. Refer to Chapter 4 for help with reconnecting the exhaust system components.

39 Make a final check to verify all wires and hoses have been reconnected to the transmission and, if applicable, the transfer case.

40 If necessary, fill the transmission and or transfer case with lubricant to the proper level (see Chapter 1). Lower the vehicle.

41 Install the shift lever(s) and boot (see Section 4).

42 Connect the negative cable to the negative terminal of the battery. Road test the vehicle and check for leaks.

7 Manual transmission overhaul - general information

Overhauling a manual transmission is a difficult job for the do-it-yourselfer. It involves the disassembly and reassembly of many small parts. Numerous clearances must be precisely measured and, if necessary, changed with select fit spacers and snap-rings. As a result, if transmission problems arise, it can be removed and installed by a competent do-it-yourselfer, but overhaul should be left to a transmission repair shop. Rebuilt transmissions may be available - check with your dealer parts department or a local transmission repair shop. At any rate, the time and money involved in an overhaul is almost sure to exceed the cost of a rebuilt unit.

Nevertheless, it's not impossible for an inexperienced mechanic to rebuild a transmission if the special tools are available and the job is done in a deliberate step-by-step manner so nothing is overlooked.

The tools necessary for an overhaul include internal and external snap-ring pliers, a bearing puller, a slide hammer, a set of pin punches, a dial indicator and possibly a hydraulic press. In addition, a large, sturdy workbench and a vise or transmission stand will be required.

During disassembly of the transmission, make careful notes of how each piece comes off, where it fits in relation to other pieces and what holds it in place.

Before taking the transmission apart for repair, it will help if you have some idea what area of the transmission is malfunctioning.

Certain problems can be closely tied to specific areas in the transmission, which can make component examination and replacement easier. Refer to the *Troubleshooting* Section at the front of this manual for information regarding possible sources of trouble.

Specifications

General

Fluid type and capacity	See Chapter 1

Torque specifications

	Ft-lbs
Crossmember-to-frame bolts	53
Transmission-to-engine bolts	53

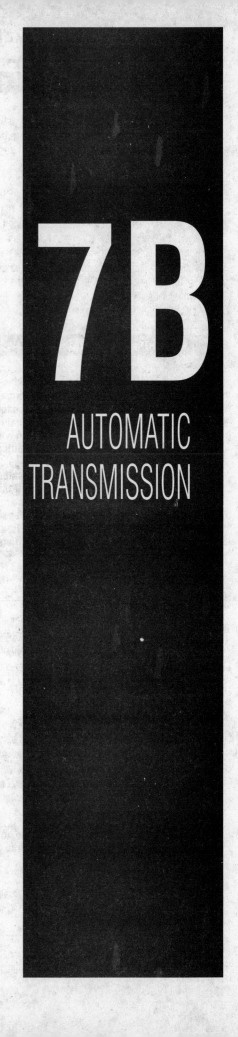

7B

AUTOMATIC TRANSMISSION

Section

Reference to other Chapters

1 General information

The 2000 through 2004 vehicles equipped with an automatic transmission use a four-speed (three speeds plus overdrive) unit, either a model A340E or a model A340F, depending on the engine and drivetrain (2WD or 4WD) combination. 2005 and later vehicles equipped with an automatic use a five speed unit designated as A750E (2WD) or A750F (4WD). All models are equipped with a lock-up torque converter, known as a Torque Converter Clutch (or TCC). The clutch provides a direct connection between the engine and the drive wheels for improved efficiency and fuel economy. A340E and A340F models are electronically controlled; upshifts and downshifts are initiated by a computer, which controls the solenoids mounted on the valve body.

Due to the complexity of the clutches and the hydraulic control system, and because of the special tools and expertise needed to overhaul an automatic transmission, diagnosis and repair of the transmission must be handled by a dealer service department or a transmission repair shop. The procedures in this Chapter are limited to general diagnosis, routine maintenance, adjustment and transmission removal and installation. However, even though the repair work must be done by a transmission specialist, you can save money by removing and installing the transmission yourself. You can also check and adjust the shift linkage or shift cable and, on V6 models, the Throttle Valve (TV) cable, replace the extension housing seal and check and replace the Park/Neutral position switch.

✳✳ CAUTION:

If a vehicle with an automatic transmission is disabled, do NOT tow it at speeds greater than 30 mph or distances over 50 miles.

2 Diagnosis - general

➡**Note: Automatic transmission malfunctions may be caused by five general conditions: poor engine performance, improper adjustments, hydraulic malfunctions, mechanical malfunctions or malfunctions in the computer or its signal network. Diagnosis of these problems should always begin with a check of the easily repaired items: fluid level and condition (Chapter 1), shift linkage adjustment and throttle linkage adjustment. Next, perform a road test to determine if the problem has been corrected or if more diagnosis is necessary. If the problem persists after the preliminary tests and corrections are completed, additional diagnosis should be done by a dealer service department or transmission repair shop. Refer to the Troubleshooting Section at the front of this manual for information on symptoms of transmission problems.**

PRELIMINARY CHECKS

1 Drive the vehicle to warm the transmission to normal operating temperature.

2 Check the fluid level as described in Chapter 1:

   a) *If the fluid level is unusually low, add enough fluid to bring the level within the designated area of the dipstick, then check for external leaks (see below).*

   b) *If the fluid level is abnormally high, drain off the excess, then check the drained fluid for contamination by coolant. The presence of engine coolant in the automatic transmission fluid indicates that a failure has occurred in the internal radiator walls that separate the coolant from the transmission fluid (see Chapter 3).*

   c) *If the fluid is foaming, drain it and refill the transmission, then check for coolant in the fluid or a high fluid level.*

3 Check the engine idle speed.

➡**Note: If the engine is malfunctioning, do not proceed with the preliminary checks until it has been repaired and runs normally.**

4 Inspect the shift linkage (see Section 4) or shift cable (see Section 5). Make sure it's properly adjusted and operates smoothly.

FLUID LEAK DIAGNOSIS

5 Most fluid leaks are easy to locate visually. Repair usually consists of replacing a seal or gasket. If a leak is difficult to find, the following procedure may help.

6 Identify the fluid. Make sure it's transmission fluid and not engine oil or brake fluid (automatic transmission fluid is a deep red color).

7 Try to pinpoint the source of the leak. Drive the vehicle several miles, then park it over a large sheet of cardboard. After a minute or two, you should be able to locate the leak by determining the source of the fluid dripping onto the cardboard.

8 Make a careful visual inspection of the suspected component and the area immediately around it. Pay particular attention to gasket mating surfaces. A mirror is often helpful for finding leaks in areas that are hard to see.

9 If the leak still cannot be found, clean the suspected area thoroughly with a degreaser or solvent, then dry it.

10 Drive the vehicle for several miles at normal operating temperature and varying speeds. After driving the vehicle, visually inspect the suspected component again.

11 Once the leak has been located, the cause must be determined before it can be properly repaired. If a gasket is replaced but the sealing flange is bent, the new gasket will not stop the leak. The bent flange must be straightened.

12 Before attempting to repair a leak, check to make sure the following conditions are corrected or they may cause another leak.

➡**Note: Some of the following conditions cannot be fixed without highly specialized tools and expertise. Such problems must be referred to a transmission repair shop or a dealer service department.**

Gasket leaks

13 Check the pan periodically. Make sure the bolts are tight, no bolts are missing, the gasket is in good condition and the pan is flat (dents in the pan may indicate damage to the valve body inside).

14 If the pan gasket is leaking, the fluid level or the fluid pressure may be too high, the vent may be plugged, the pan bolts may be too tight, the pan sealing flange may be warped, the sealing surface of the transmission housing may be damaged, the gasket may be damaged or the transmission casting may be cracked or porous. If sealant instead of gasket material has been used to form a seal between the pan and the transmission housing, it may be the wrong sealant.

Seal leaks

15 If a transmission seal is leaking, the fluid level or pressure may be too high, the vent may be plugged, the seal bore may be damaged, the seal itself may be damaged or improperly installed, the surface of the shaft protruding through the seal may be damaged or a loose bearing may be causing excessive shaft movement.

16 Make sure the dipstick tube seal is in good condition and the tube is properly seated. Periodically check the area around the speedometer gear or sensor for leakage. If transmission fluid is evident, check the O-ring for damage.

Case leaks

17 If the case itself appears to be leaking, the casting is porous and will have to be repaired or replaced.

18 Make sure the oil cooler hose fittings are tight and in good condition.

Fluid comes out vent pipe or fill tube

19 If this condition occurs, the transmission is overfilled, there is coolant in the fluid, the case is porous, the dipstick is incorrect, the vent is plugged or the drain back holes are plugged.

3 Throttle Valve (TV) cable - replacement and adjustment

➡ **Note: This procedure applies to V6 models only.**

REPLACEMENT

▶ **Refer to illustrations 3.5, 3.6, 3.7, 3.8, 3.9, 3.12, 3.13, 3.14a, 3.14b and 3.15**

1 Disconnect the TV cable from the throttle linkage (see Chapter 4).

2 Disengage the TV cable from the cable clamps in the engine compartment.

3 Remove the bolt that attaches the TV cable clamp to the torque converter housing.

4 Remove the oil pan (see Chapter 1).

5 Disconnect the connectors for the shift solenoid valves (see illustration).

6 Remove the oil strainer and oil pipes (see illustration).

7 Remove the valve body bolts (see illustration). Before removing the last bolt, support the valve body, or have an assistant support it for you, so that it doesn't fall when the last bolt is removed.

8 Lower the valve body and disengage the TV cable from the throttle cam (see illustration).

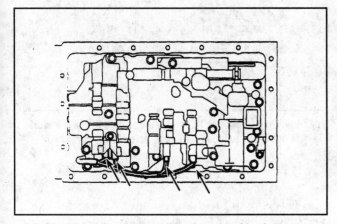

3.5 Disconnect the connectors for the solenoid valves

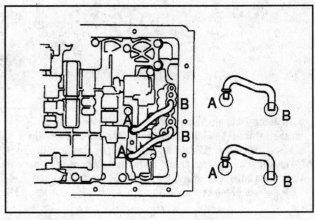

3.6 If equipped, remove the oil tubes by carefully prying them out

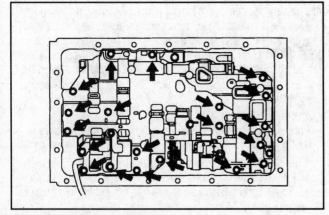

3.7 Valve body bolts (typical)

3.8 Lower the valve body and disengage the TV cable from the throttle cam

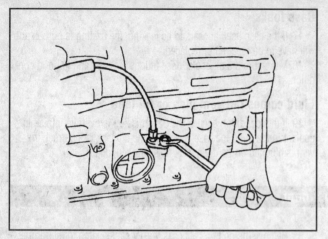

3.9 Remove the cable hold-down bolt from the transmission and pull out the cable

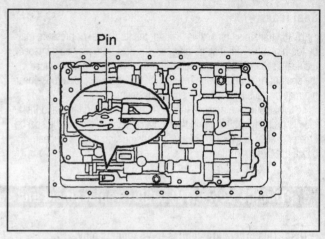

3.12 Before installing the valve body, make sure that the groove of the manual valve is aligned with the pin of the lever

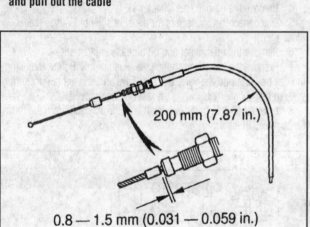

3.13 If the TV cable being installed is new, the cable stopper must be staked: bend the upper end of the cable to a radius of about 200 mm (7.87 inches), pull the inner cable until you feel a slight resistance, hold it there, and stake the stopper so that it's between 0.8 and 1.5 mm (0.031 to 0.059 inch) from the end of the cable sheath

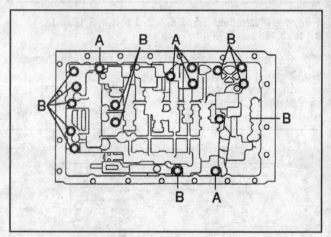

3.14a When tightening the bolts, make sure that you install a bolt of the correct length in each hole (2000 V6 model)

A   23 mm (0.91 inch)
B   32 mm (1.26 inch)

9   Remove the cable hold-down bolt from the transmission case (see illustration) and pull out the old TV cable.

10 Install the cable hold-down bolt and tighten it securely.

11 Connect the new TV cable to the cam.

12 Before installing the valve body, make sure that the groove of the manual valve is aligned with the pin of the lever (see illustration).

13 If the throttle cable being installed is a new cable, the cable stopper is not staked. First, bend the upper end of the cable to a radius of about 200 mm (7.87 inches). Pull the inner cable until you feel a slight resistance and hold it there. Stake the stopper so that it's between 0.8 and 1.5 mm (0.031 to 0.059 inch) from the end of the cable sheath (see illustration).

14 Install the valve body.

➡**Note: The valve body bolts are different lengths, be sure to install a bolt of correct length in each hole (see illustrations).**

15 Install the oil strainer (see illustration).

16 The remainder of installation is the reverse of removal.

17 Fill the transmission with new ATF and check the fluid level (see Chapter 1).

18 Check and, if necessary, adjust the TV cable (see Steps 17 through 20).

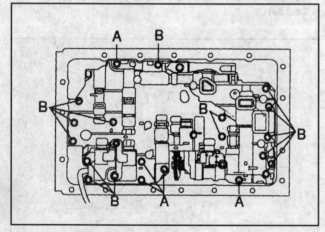

3.14b 2001 and later V6 models

A   23 mm (0.91 inch)
B   32 mm (1.26 inch)

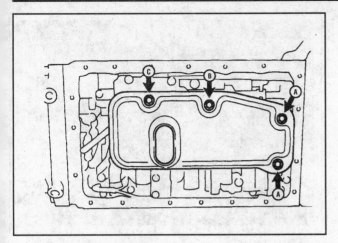

3.15 Oil strainer mounting bolts

> A   14 mm (0.55 inch)   C   23 mm (0.91 inch)
> B   20mm (0.79 inch)

ADJUSTMENT

♦ **Refer to illustration 3.21**

19 Verify that the accelerator pedal is fully released and the throttle valve is fully closed.

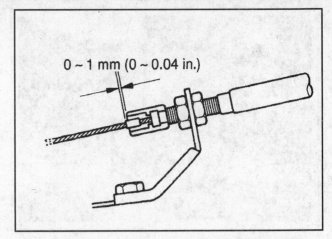

3.21 To check the adjustment of the TV cable, measure the distance between the cable sheath and the stopper on the cable

20 Verify that the TV cable is taut (the cable itself, not the cable sheath).

21 Measure the distance between the cable sheath (the outer cable) and the stopper on the TV cable (see illustration), then compare your measurement to the dimension shown.

22 If the TV cable is out of adjustment, adjust it by utilizing the adjustment nuts at the bracket.

4   Shift cable - removal, installation and adjustment

♦ **Refer to illustrations 4.2, 4.5, 4.6 and 4.7**

☀ WARNING:

These models are equipped with airbags. Always disable the airbag system before working in the vicinity of any airbag system component to avoid the possibility of accidental deployment of the airbag(s), which could cause personal injury (see Chapter 12).

1 Remove the upper and lower steering column covers and the knee bolster trim panel (see Chapter 11).

2 Detach the upper end of the shift cable from the steering column (see illustration).

3 Disengage the cable from the cable bracket (see illustration 4.2).

4 Raise the vehicle and place it securely on jackstands.

5 Remove the shift cable heat shield (see illustration).

6 Remove the nut that attaches the shift cable to the manual lever

4.2 Remove the wire retaining pin and detach the upper end of the shift cable from the shifter lever (A), then squeeze the tabs on the cable casing and detach the cable from the column (B)

4.5 Shift cable heat shield mounting nuts

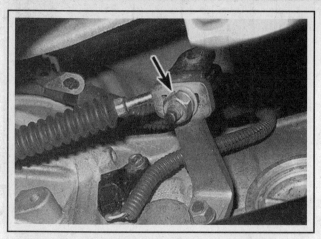

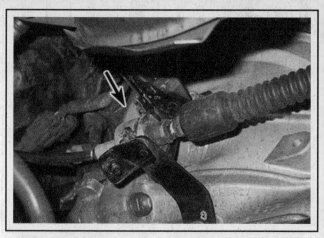

4.6 Remove the nut that attaches the shift cable to the shifter lever

4.7 Remove the retainer clip and disengage the cable from the cable bracket

and disconnect the shift cable from the manual lever (see illustration).

7   Remove the retainer clip and disengage the cable from the cable bracket (see illustration).

8   Remove or disconnect any cable retaining clamps or clips on the

firewall or the transmission.

9   Pull the cable through the firewall and remove it.

10 Installation is the reverse of removal.

5   Park/Neutral Position (PNP) switch - removal, installation and adjustment

▶ **Refer to illustration 5.3, 5.5, 5.6 and 5.7**

1   The Park/Neutral Position switch prevents the engine from starting in any gear other than Park or Neutral. If the engine starts in any position other than Park or Neutral, it's either out of adjustment or defective.

2   Raise the vehicle and place it securely on jackstands.

3   On 2000 through 2004 models, disconnect the oil cooler pipe (see illustration).

4   Unplug the electrical connector from the Park/Neutral Position switch.

5   Pry off the lock washer and remove the nut (see illustration).

6   Remove the Park/Neutral Position switch retaining bolt (see illustration).

7   Remove the switch.

8   Installation is the reverse of removal. Check and, if necessary, adjust the switch.

ADJUSTMENT

9   Raise the vehicle and place it securely on jackstands.

10 Loosen the switch retaining bolt (see illustration 5.6).

11 Align the groove and neutral basic line.

12 Holding the switch in this position, tighten the switch retaining bolt.

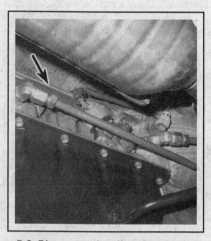

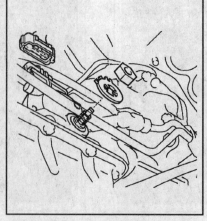

5.3 Disconnect the oil cooler pipe for access to the PNP switch

5.5 To remove the Park/Neutral Position switch, pry off the lock washer and remove the nut

5.10 Loosen the Park/Neutral Position switch mounting bolt

6   Transmission oil cooler - removal and installation

▶ **Refer to illustration 6.2**

1   Disconnect the negative battery cable from the battery.
2   Working through the front grille, disconnect the fluid lines (see illustration).
3   Unscrew the three mounting bolts and remove the cooler.
4   Installation is the reverse of removal. Tighten all cooler fasteners and cooler lines securely.
5   Check the automatic transmission fluid level and add some, if necessary (see Chapter 1).

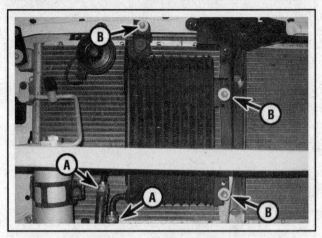

6.2 Disconnect the transmission lines (A) and remove the three mounting bolts (B) (front grille removed for clarity)

7   Automatic transmission - removal and installation

REMOVAL

▶ **Refer to illustrations 7.13a, 7.13b, 7.15a, 7.15b, 7.17, 7.21**

1   Disconnect the negative cable from the battery.
2   Remove the dipstick.
3   On six-cylinder models, disconnect the TV cable from the throttle body (see Section 3).
4   On 4WD models, remove the transfer shift lever, if equipped (see Chapter 7C).
5   Raise the vehicle and support it securely on jackstands.
6   Drain the transmission fluid (see Chapter 1), then reinstall the pan.
7   Remove any engine/transmission undercovers that are in the way.
8   Remove the dipstick tube bracket bolt(s) and remove the dipstick tube.
9   Disconnect the driveshaft(s) (see Chapter 8).
10 Remove the front and center exhaust pipes (see Chapter 4).
11 Unplug the electrical connectors from the vehicle speed sensor(s), the shift solenoids, the ATF temperature sensor, the Park/Neutral position switch and, on 4WD models, the transfer Neutral position switch, transfer L4 position switch and transfer indicator switch, then detach all wire harnesses from the transmission and set them aside.
12 Disconnect the shift cable from the manual lever (see Section 5).
13 Remove the transmission cooler line bracket bolts, detach all brackets and clamps and disconnect the two cooler lines from the transmission (see illustrations).
14 On six-cylinder models, remove the starter and disconnect the wiring harness from the transmission (see Chapter 5).

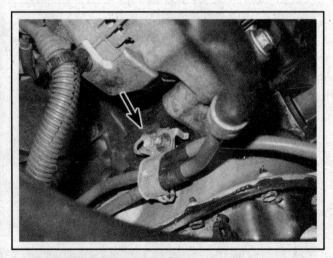

7.13a Detach the three transmission oil cooler line brackets . . .

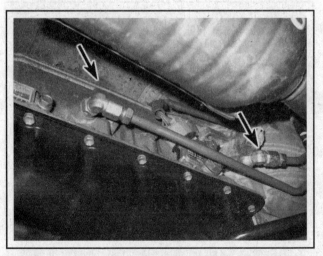

7.13b . . . and disconnect the lines from the transmission

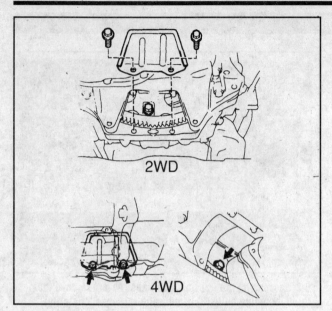

7.15a V8 model torque converter cover details

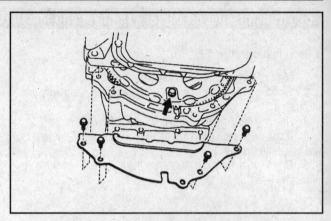

7.15b V6 model torque converter cover details

15 Remove the torque converter access cover (see illustrations).

16 Mark the torque converter and the driveplate with a scribe or chalk so they can be installed in the same position.

17 Remove the six driveplate-to-torque converter bolts (see illustration). Turn the crankshaft (in a clockwise direction only, viewed from the front) for access to each bolt.

18 Support the transmission with a jack - preferably a jack made for this purpose. Safety chains will help steady the transmission on the jack.

19 Support the engine with a jack. Use a block of wood under the oil pan to spread the load.

20 Remove the bolts securing the transmission mount to the frame crossmember or, on 4WD models, the transfer case mount-to-crossmember.

21 Raise the transmission enough to allow removal of the crossmember, then remove the nuts/bolts securing the crossmember to the frame side rails and remove the crossmember and the mount (see illustration).

22 Remove the transmission-to-engine bolts.

23 Slowly lower the jack until you can remove the upper bolts securing the transmission to the engine. Several long extensions may have to be used to reach the upper bolts.

24 Move the transmission to the rear to disengage it from the engine block dowel pins and make sure the torque converter is detached from the driveplate. Secure the torque converter to the transmission so it won't fall out during removal.

INSTALLATION

25 Prior to installation, make sure the torque converter hub is securely engaged in the front pump of the transmission. This can be confirmed by pushing in on the converter and turning it; if it isn't seated completely it will drop into place as this is done.

26 With the transmission secured to the jack, raise it into position. Be sure to keep it level so the torque converter doesn't slide out.

27 Turn the torque converter until the marks on the converter and driveplate are aligned.

28 Move the transmission forward carefully until the dowel pins engage with the holes in the bellhousing.

29 Install the transmission-to-engine bolts. Tighten them to the torque listed in this Chapter's Specifications.

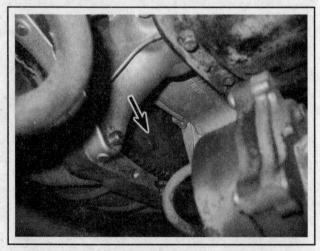

7.17 Remove the six driveplate-to-torque converter bolts by turning the crankshaft (in a clockwise direction only, viewed from the front) for access to each bolt

7.21 Remove the four transmission mount-to-crossmember bolts (two bolts not visible in picture)

30 Install the driveplate-to-torque converter bolts and tighten them to the torque listed in this Chapter's Specifications.

➡**Note: Install all of the bolts before tightening any of them.**

31 Install the crossmember and tighten the bolts and nuts securely.

32 Lower the transmission extension housing until the mount is seated on and aligned with the crossmember and tighten the fasteners securely.

33 The remainder of installation is the reverse of removal.

34 Remove all jacks supporting the transmission and engine and lower the vehicle.

35 Fill the transmission with the specified fluid (Chapter 1), run the engine and check for fluid leaks.

8 Automatic transmission overhaul - general information

In the event of a fault occurring, it will be necessary to establish whether the fault is electrical, mechanical or hydraulic in nature, before repair work can be contemplated. Diagnosis requires detailed knowledge of the transmission's operation and construction, as well as access to specialized test equipment, and so is deemed to be beyond the scope of this manual. It is therefore essential that problems with the automatic transmission are referred to a dealer service department or other qualified repair facility for assessment.

Note that a faulty transmission should not be removed before the vehicle has been assessed by a knowledgeable technician equipped with the proper tools, as troubleshooting must be performed with the transmission installed in the vehicle.

Specifications

General

Transmission fluid type and capacity	See Chapter 1

Torque specifications   Ft-lbs (unless otherwise indicated)

➡**Note: One foot-pound (ft-lb) of torque is equivalent to 12 inch-pounds (in-lbs) of torque. Torque values below approximately 15 ft-lbs are expressed in inch-pounds, since most foot-pound torque wrenches are not accurate at these smaller values.**

Flywheel-to-torque converter bolts	30
Transmission-to-engine bolts	53
Valve body bolts	84 in-lbs
Crossmember-to-frame bolts	53

Notes

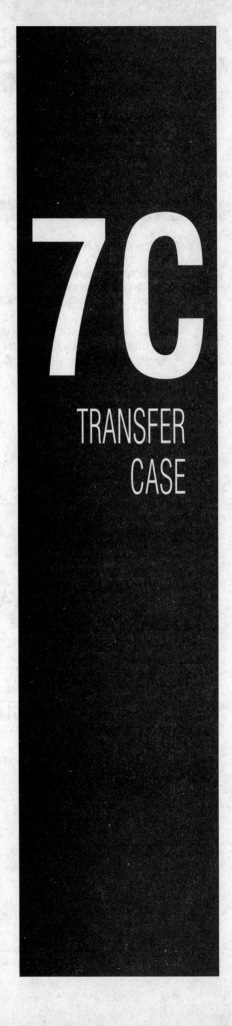

7C

TRANSFER CASE

Section

Reference to other Chapters

1 General information

Four-wheel drive (4WD) models are equipped with a transfer case mounted on the rear of the transmission. Drive is transmitted from the engine, through the transmission and the transfer case to the front and rear axles by driveshafts.

We don't recommend trying to rebuild a transfer case at home. It's difficult to overhaul without special tools, and rebuilt units are available for less than it would cost to rebuild your own. However, there are a number of components that you can check, adjust and/or replace - and those are the items covered in this Chapter.

2 Oil seals - replacement

◆ Refer to illustrations 2.3a, 2.3b, 2.4, 2.5, 2.7 and 2.8

➡Note: This procedure applies to both the front and rear seals. Although the accompanying photos show the transfer case out of the vehicle, it's not necessary to remove the transfer case to replace a seal.

1 Raise the vehicle and support it securely on jackstands.
2 If you're replacing the front seal, remove the front driveshaft; if you're replacing the rear seal, remove the rear driveshaft (see Chapter 8).
3 Unstake and remove the companion flange retaining nut (see illustrations).
4 Remove the companion flange (see illustration). If the flange is difficult to remove from the output shaft, use a puller.
5 Pry out the seal with a screwdriver or a seal removal tool (see illustration). Don't damage the seal bore.
6 Lubricate the new seal lip with multi-purpose grease.
7 Drive the seal into place with a large socket (see illustration).

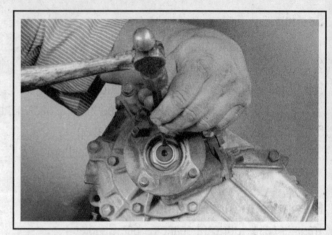

2.3a Unstake the companion flange retaining nut . . .

2.3b . . . then break the nut loose while holding the flange as shown

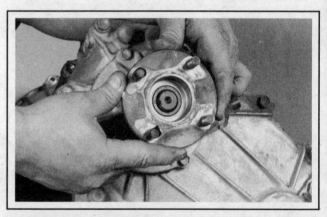

2.4 Pull the companion flange off the output shaft - it may be necessary to use a small puller

2.5 Pry out the seal with a screwdriver or a seal removal tool

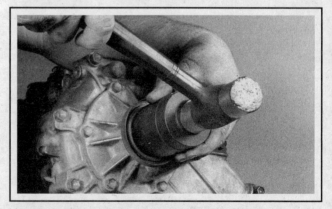

2.7 Drive the seal into place with seal installer tool or a large socket

The outside diameter of the socket should be slightly smaller than the outside diameter of the seal.

8   There's a smaller seal inside the companion flange (see illustration). If the seal needs to be replaced, pry it out and install a new one the same way you did the larger seal.

9   The remainder of installation is the reverse of removal. Be sure to tighten the companion flange nut to the torque listed in this Chapter's Specifications.

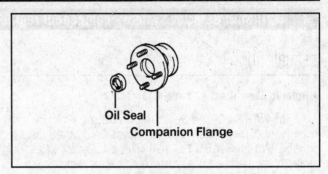

Oil Seal

Companion Flange

2.8 If necessary, replace the smaller inner seal

3   Shift lever - removal and installation

▶ **Refer to illustrations 3.3, 3.4a, 3.4b and 3.4c**

➡**Note: Although the accompanying photos show the transfer case out of the vehicle, it's not necessary to remove the transfer case for shifter removal.**

1   On manual transmission models with the 4WD Shift Selector System, place the transfer shift lever in the High position and turn the 2WD-4WD selector switch to the Off position. For models without the 4WD Shift Selector System, place the transfer case in the 4WD High position.

2   If equipped, remove the center console upper trim panel (see Chapter 11).

3   Remove the four shift lever boot screws, shift lever knob(s), shift lever boot retainer and shifter boot (see illustration).

4   Remove the small shift lever boot, remove the shift lever snapring and pull out the shift lever (see illustrations).

5   Installation is the reverse of removal. Lubricate the lower end of the shift lever with multi-purpose grease before installing it.

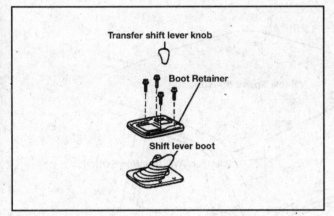

Transfer shift lever knob

Boot Retainer

Shift lever boot

3.3 Shifter boot details

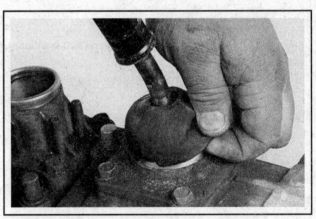

3.4a To detach the transfer shift lever from the transfer case, remove the small dust cover . . .

3.4b . . . remove the snap-ring with a pair of snap-ring pliers . . .

3.4c . . . and pull out the shift lever

4   4WD Shift Selector System - component replacement

DESCRIPTION

▶ **Refer to illustrations 4.1a and 4.1b**

1   The 4WD Shift Selector System (see illustrations) allows the driver to switch the settings of the front differential and transfer case by selecting 2WD, High 4WD or Low 4WD with a 2-4 selector switch located on the dashboard. The system consists of the Electronic Control Unit (ECU), ADD Actuator (see Chapter 8), dash mounted selector switch, the actuator assembly, the 4WD position switches No. 1 and No. 2 (Sequoia models), the L4 (Low 4WD) switch, the Neutral position switch, the vehicle speed sensor, and the 4WD indicator light (located on the instrument cluster, see Chapter 12).

REPLACEMENT

Selector switch

2   The 2-4 selector switch is part of the center cluster integration panel and is replaced as an assembly. Refer to Chapter 11 for panel removal.

No.1 Vehicle Speed Sensor

▶ **Refer to illustration 4.6**

3   Make sure the ignition key is in the Off position.
4   Raise the vehicle and support it securely on jackstands.
5   Disconnect the speed sensor electrical connector.
6   Remove the speed sensor mounting fasteners (see illustration).
7   Apply a small amount of gear oil to the new speed sensor O-ring seals.
8   Installation is the reverse of removal.

Transfer Position Switches

▶ **Refer to illustration 4.11**

9   Make sure the ignition key is in the Off position.
10   Raise the vehicle and support it securely on jackstands.
11   The Transfer 4WD position and Transfer L4 position switch are located on the actuator assembly (see illustration).
12   Disconnect the 4WD sensor or L4 sensor electrical connector.
13   Using the proper wrench, remove the sensor.
14   Using a new gasket, install the new sensor.
15   The remainder of installation is the reverse of removal.

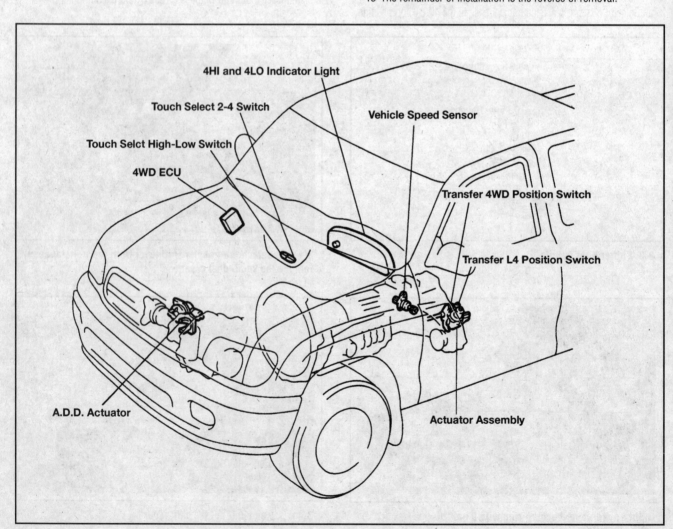

4.1a 4WD Shift Selector System - Tundra models

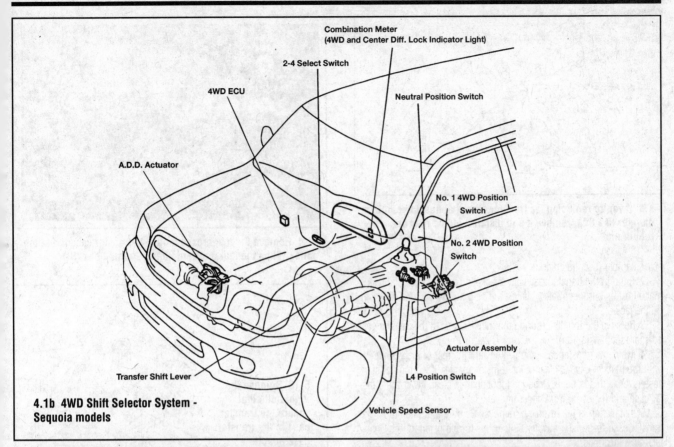

4.1b 4WD Shift Selector System - Sequoia models

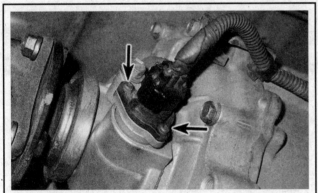

4.6 Unplug the electrical connector and remove the speed sensor mounting fasteners

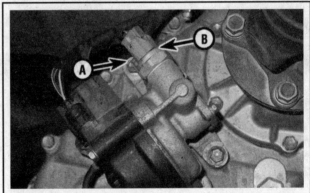

4.11 4WD Position switch (A) and the L4 position switch (B)

5 Transfer case - removal and installation

REMOVAL

▶ **Refer to illustrations 5.8, 5.11 and 5.12**

1   On manual transmission models with the 4WD Shift Selector System, place the transfer shift lever in the High position and turn the 2WD-4WD selector switch to the Off position. On automatic Tundra models place the switch in the On position. For models without the 4WD Shift Selector System, place the transfer case in the 4WD High position.

2   If applicable, remove the transfer case shift lever (see Section 3).

3   Detach the breather hose from the top of the transfer case.

4   Raise the vehicle and support it securely on jackstands.

5   Drain the transfer case lubricant (see Chapter 1).

6   Remove the front and rear driveshafts (see Chapter 8).

7   On V8 models, remove the right and left front exhaust pipes (see Chapter 4).

8   If equipped, remove the dynamic damper (see illustration).

9   Unplug all electrical connectors, such as vehicle speed sensor, 4WD position switch, neutral position switch and any other connector or wiring harnesses attached to the transfer case.

10 Support the transmission with a jack or jackstand. The transmission should remain supported at all times while the transfer case is out of the vehicle.

11 Remove the four crossmember-to-transfer case mount retaining bolts and the crossmember-to-frame mounting bolts, then remove the

5.8 If you're replacing the transfer case that has a dynamic damper like this, remove it and install it on the new or rebuilt unit

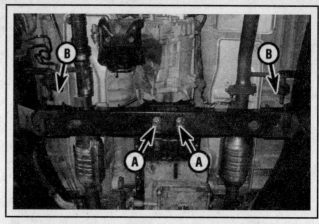

5.11 Remove the four crossmember-to-transfer case mounting bolts (A) and the four crossmember-to-frame bolts (B)

crossmember (see illustration).

12 Support the transfer case with a jack - preferably a special jack made for this purpose. Safety chains will help steady the transfer case on the jack.

13 Remove the 8 bolts securing the transfer case to the transfer case adapter (see illustration).

14 Move the transfer case and jack toward the rear of the vehicle until the transfer case is clear of the transfer adapter. Keep the transfer case level as this is done. Once the input shaft is clear, lower the transfer case and remove it from under the vehicle.

15 Installation is the reverse of removal. Be sure to tighten the transfer case-to-transfer adapter bolts to the torque listed in this Chapter's Specifications.

16 Fill the transfer case with the specified fluid (Chapter 1), drive the vehicle and check for fluid leaks.

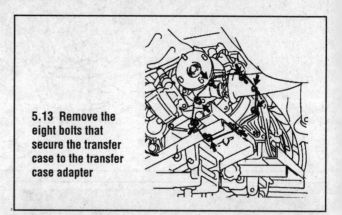

5.13 Remove the eight bolts that secure the transfer case to the transfer case adapter

6   Transfer case overhaul - general information

1   Overhauling a transfer case is a difficult job for the do-it-yourselfer. It involves the disassembly and reassembly of many small parts. Numerous clearances must be precisely measured and, if necessary, changed with select-fit spacers and snap-rings. As a result, if transfer case problems arise, it can be removed and installed by a competent do-it-yourselfer, but overhaul should be left to a transmission repair shop. Rebuilt transfer cases may be available - check with your dealer parts department and auto parts stores. At any rate, the time and money involved in an overhaul is almost sure to exceed the cost of a rebuilt unit.

2   Nevertheless, it's not impossible for an inexperienced mechanic to rebuild a transfer case if the special tools are available and the job is done in a deliberate step-by-step manner so nothing is overlooked.

3   The tools necessary for an overhaul include internal and external

snap-ring pliers, a bearing puller, a slide hammer, a set of pin punches, a dial indicator and possibly a hydraulic press. In addition, a large, sturdy workbench and a vise or transfer case stand will be required.

4   During disassembly of the transfer case, make careful notes of how each piece comes off, where it fits in relation to other pieces and what holds it in place. Noting how they are installed when you remove the parts will make it much easier to get the transfer case back together.

5   Before taking the transfer case apart for repair, it will help if you have some idea what area of the transfer case is malfunctioning. Certain problems can be closely tied to specific areas in the transfer case, which can make component examination and replacement easier. Refer to the *Troubleshooting* section at the front of this manual for information regarding possible sources of trouble.

Specifications

General

Transfer case - lubricant type	See Chapter 1

Torque specifications

	Ft-lbs
Companion flange nut (front and rear)	87
Transfer adapter-to-transfer case bolts	17

8

CLUTCH AND DRIVELINE

Section

1 General information

The Sections in this Chapter deal with the components from the rear of the engine to the rear wheels (except for the transmission and transfer case, which are dealt with in Chapter 7) and forward to the front wheels on four-wheel drive (4WD) models. In this Chapter, the components are grouped into three categories: clutch, driveshaft(s) and axle(s). Separate Sections within this Chapter cover checks and repair procedures for components in each of these three groups.

Since nearly all these procedures involve working under the vehicle, make sure it's safely supported on sturdy jackstands or a hoist where the vehicle can be safely raised and lowered.

2 Clutch - description and check

♦ **Refer to illustration 2.1**

1   All vehicles with a manual transmission have a single dry plate, diaphragm spring type clutch (see illustration). The clutch disc has a splined hub which allows it to slide along the splines of the transmission input shaft. The clutch and pressure plate are held in contact by spring pressure exerted by the diaphragm in the pressure plate.

2   The clutch release system is operated by hydraulic pressure. The hydraulic release system consists of the clutch pedal, a master cylinder and fluid reservoir, the hydraulic line, a release (or slave) cylinder which actuates the clutch release lever and the clutch release (or throwout) bearing.

3   When pressure is applied to the clutch pedal to release the clutch, hydraulic pressure is exerted against the outer end of the clutch release lever. As the lever pivots, the shaft fingers push against the release bearing. The bearing pushes against the fingers of the diaphragm spring of the pressure plate assembly, which in turn releases the clutch plate.

4   Terminology can be a problem when discussing the clutch components because common names are in some cases different from those used by the manufacturer. For example, the driven plate is also called the clutch plate or disc, the clutch release bearing is sometimes called a throwout bearing, the release cylinder is sometimes called the slave cylinder.

5   Other than to replace components with obvious damage, some preliminary checks should be performed to diagnose clutch problems.

a) The first check should be of the fluid level in the clutch master cylinder. If the fluid level is low, add fluid as necessary and inspect the hydraulic system for leaks. If the master cylinder reservoir is dry, bleed the system as described in Section 8 and recheck the clutch operation.

b) To check "clutch spin-down time," run the engine at normal idle speed with the transmission in Neutral (clutch pedal up - engaged). Disengage the clutch (pedal down), wait several seconds and shift the transmission into Reverse. No grinding noise should be heard. A grinding noise would most likely indicate a bad pressure plate or clutch disc.

c) To check for complete clutch release, run the engine (with the parking brake applied to prevent vehicle movement) and hold the clutch pedal approximately 1/2-inch from the floor. Shift the transmission between 1st gear and Reverse several times. If the shift is rough, component failure is indicated. Check the release cylinder pushrod travel. With the clutch pedal depressed completely, the release cylinder pushrod should extend substantially (you may have to remove an inspection plug to see the pushrod). If it doesn't, check the fluid level in the clutch master cylinder.

d) Visually inspect the pivot bushing at the top of the clutch pedal to make sure there's no binding or excessive play.

e) Crawl under the vehicle and make sure the clutch release lever is securely attached to the ball stud.

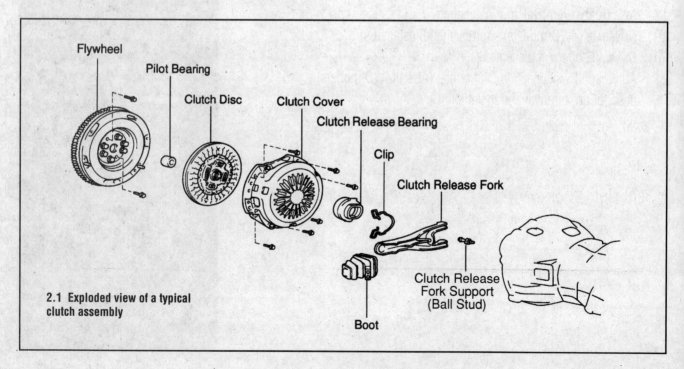

2.1 Exploded view of a typical clutch assembly

Flywheel
Pilot Bearing
Clutch Disc
Clutch Cover
Clutch Release Bearing
Clip
Clutch Release Fork
Clutch Release Fork Support (Ball Stud)
Boot

3 Clutch master cylinder - removal and installation

▶ **Refer to illustrations 3.2 and 3.4**

1 Draw out the hydraulic fluid from the clutch master cylinder reservoir with a syringe.

2 Disconnect the clutch fluid hydraulic line (see illustration); use a flare-nut wrench to protect the tube nut. Have rags handy, as some fluid will be lost as the line is removed.

※※ CAUTION:

Don't allow fluid to come into contact with the paint since it will damage the finish. Also have a plug ready and immediately plug the line to prevent leakage and fluid contamination.

3 Working inside the passenger compartment, remove the driver's side lower finish panel(s).

4 Disengage the return spring, remove the retaining clip and clevis pin (see illustration) and disconnect the clutch master cylinder pushrod from the clutch pedal.

5 Remove the master cylinder mounting nuts (see illustration 3.2) and detach the master cylinder from the firewall.

6 Installation is the reverse of removal. Be sure to tighten the master cylinder mounting nuts securely.

7 Fill the clutch master cylinder reservoir with the fluid specified in Chapter 1 and bleed the clutch system (see Section 8).

3.2 Loosen the hydraulic line retaining nut (A) with a flare nut wrench, then remove the two mounting nuts (B) (left nut visible, right nut not visible in this photo)

3.4 To disconnect the clutch master cylinder pushrod from the clutch pedal, pry off the return spring (A), remove the retaining clip (B) and push out the clevis pin

4 Clutch release cylinder - removal and installation

▶ **Refer to illustrations 4.2 and 4.3**

1 Raise the vehicle and support it securely on jackstands.

2 Disconnect the clutch fluid hydraulic line from the release cylinder (see illustration). Use a flare-nut wrench to protect the tube nut. Have rags handy, as some fluid will be lost as the line is removed.

※※ CAUTION:

Don't allow fluid to come into contact with the paint - it will damage the finish.

Also have a plug ready and immediately plug the line to prevent leakage.

4.2 Use a flare-nut wrench to protect the tube nut, while disconnecting the clutch fluid hydraulic line from the front of the release cylinder

3   Remove the two release cylinder mounting bolts (see illustration).

4   Detach the release cylinder.

5   Installation is the reverse of removal. Make sure the pushrod dust boot is in good condition and the pushrod is seated correctly in its pocket in the release lever. Tighten the release cylinder mounting bolts securely.

6   Fill the clutch fluid reservoir with the recommended fluid (see Chapter 1).

7   Bleed the clutch hydraulic system (see Section 8).

8   Remove the jackstands and lower the vehicle.

4.3 To detach the clutch release cylinder, remove these mounting bolts (upper arrows) (lower arrow indicates bleeder plug)

5   Clutch components - removal, inspection and installation

※※ WARNING:

Dust produced by clutch wear and deposited on clutch components is hazardous to your health. DO NOT blow it out with compressed air and DO NOT inhale it. DO NOT use gasoline or petroleum-based solvents to remove the dust. Brake system cleaner should be used to flush the dust into a drain pan. After the clutch components are wiped clean with a rag, dispose of the contaminated rags and cleaner in a covered, marked container.

REMOVAL

♦ **Refer to illustrations 5.4a, 5.4b and 5.5**

➡**Note: The following procedure is based on the assumption that the transmission is being removed and the engine remains in place. However, anytime the transmission or engine is removed, you should inspect the clutch assembly for wear. Unless the clutch components are new or in near-perfect condition, their relatively low cost - compared to the time and trouble it takes to get to them - warrants their replacement anytime the engine or transmission is removed.**

1   Raise the vehicle and support it securely on jackstands.

2   Remove the transmission (see Chapter 7, Part A). Support the engine while the transmission is out. An engine hoist should be used to support it from above. If you use a jack underneath the engine instead, make sure a piece of wood is positioned between the jack and oil pan to spread the load.

※※ CAUTION:

The pick-up for the oil pump is very close to the bottom of the oil pan. If the pan is bent or distorted in any way, engine oil starvation could occur.

3   The clutch release lever and release bearing can remain attached to the housing for the time being.

4   To support the clutch disc during removal, install an alignment tool through the clutch disc hub (see illustrations).

5   Carefully inspect the flywheel and pressure plate for indexing marks. The marks are usually an X, an O or a white letter. If they cannot be found, paint a mark so the pressure plate and the flywheel will be in the same alignment during installation (see illustration).

5.4a A clutch alignment tool like this one is available at most auto part stores

5.4b Insert the clutch alignment tool into the clutch disc splines prior to loosening the pressure plate bolts, or to center the clutch disc when reinstalling the pressure plate

5.5 If you're going to re-use the same pressure plate, mark its relationship to the flywheel

5.8 Inspect the surface of the flywheel for cracks, dark-colored areas (signs of overheating) and other obvious defects

6 Loosen the six pressure plate-to-flywheel bolts in 1/4-turn increments until they can be removed by hand. Work in a criss-cross pattern until all spring pressure is relieved, then hold the pressure plate securely and completely remove the bolts, followed by the pressure plate and clutch disc.

INSPECTION

♦ **Refer to illustrations 5.8, 5.10 and 5.12**

7 Ordinarily, when a problem occurs in the clutch, it can be attributed to wear of the clutch driven plate assembly (clutch disc). However, all components should be inspected at this time.

➟**Note: If the clutch components are contaminated with oil, there will be shiny, black glazed spots on the clutch disc lining, which will cause the clutch to slip. Replacing clutch components won't completely solve the problem - be sure to check the crankshaft rear oil seal and the transmission input shaft seal for leaks. If it looks like a seal is leaking, be sure to install a new one to avoid the same problem with the new clutch.**

8 Check the flywheel for cracks, heat checking, grooves and other obvious defects (see illustration). If the imperfections are slight, a machine shop can machine the surface flat and smooth, which is highly recommended regardless of the surface appearance. Refer to Chapter 2 for the flywheel removal and installation procedure.

9 Inspect the pilot bearing (see Section 7).

10 Check the lining on the clutch disc. There should be at least 1/32-inch of lining above the rivet heads. Check for loose rivets, distortion, cracks, broken springs and other obvious damage (see illustration). As mentioned above, ordinarily the clutch disc is routinely replaced, so if you're in doubt about its condition, replace it.

11 The release bearing should also be replaced along with the clutch disc (see Section 6).

12 Check the machined surfaces and the diaphragm spring fingers of the pressure plate (see illustration). If the surface is grooved or otherwise damaged, replace the pressure plate. Also check for obvious damage, distortion, cracks, etc. Light glazing can be removed with medium-grit emery cloth. If a new pressure plate is required, new and factory-rebuilt units are available.

5.10 Inspect the clutch plate lining, springs and splines for wear

INSTALLATION

♦ **Refer to illustration 5.16**

13 Before installation, clean the flywheel and pressure plate machined surfaces with lacquer thinner or acetone. It's important that no oil or grease is on these surfaces or the lining of the clutch disc. Handle the parts only with clean hands.

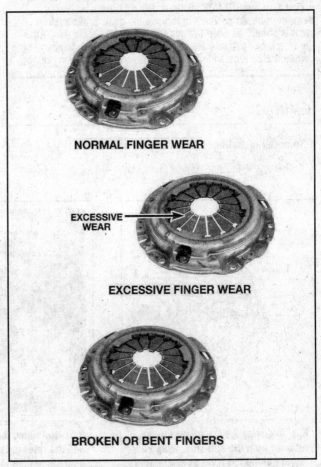

NORMAL FINGER WEAR

EXCESSIVE WEAR →

EXCESSIVE FINGER WEAR

BROKEN OR BENT FINGERS

5.12 Replace the pressure plate if excessive wear or damage is noted

14 Position the clutch disc and pressure plate against the flywheel with the clutch held in place with an alignment tool (see illustration 5.4b). Make sure it's installed properly (most replacement clutch plates will be marked "flywheel side" or something similar - if it's not marked, install the clutch disc with the damper springs toward the transmission).

15 Tighten the pressure plate-to-flywheel bolts only finger-tight, working around the pressure plate.

16 Center the clutch disc by ensuring the alignment tool extends through the splined hub and into the pilot bearing in the crankshaft. Wiggle the tool up, down or from side-to-side as needed to bottom the tool in the pilot bearing. Tighten the pressure plate-to-flywheel bolts a little at a time, working in the order shown (see illustration), to prevent distorting the cover. After all the bolts are snug, tighten them to the torque listed in this Chapter's Specifications. Remove the alignment tool.

17 Using high-temperature grease, lubricate the inner groove of the release bearing, the release lever contact areas and the transmission input shaft bearing retainer (see Section 6).

18 Install the clutch release bearing (see Section 6).

19 Install the transmission, release cylinder and all components removed previously.

5.16 Tighten the six pressure plate bolts in the order shown, a little at a time until they're snug, then tighten them gradually and uniformly to the torque listed in this Chapter's Specifications

6   Clutch release bearing - removal, inspection and installation

✳✳ WARNING:

Dust produced by clutch wear and deposited on clutch components is hazardous to your health. DO NOT blow it out with compressed air and DO NOT inhale it. DO NOT use gasoline or petroleum-based solvents to remove the dust. Brake system cleaner should be used to flush the dust into a drain pan. After the clutch components are wiped clean with a rag, dispose of the contaminated rags and cleaner in a covered, marked container.

REMOVAL

▶ **Refer to illustrations 6.3, 6.4a, 6.4b and 6.4c**

1 Remove the release cylinder (see Section 4).

2 Remove the transmission (see Chapter 7, Part A).

3 Disengage the release lever retainer from the ball stud (see illustration).

4 Note how the release lever fingers are engaged by the wire retainer on the bearing, then disengage the bearing from the lever and slide it off the input shaft (see illustration). Remove the release lever (see illustration). Inspect the release lever boot for cracks or tears. If it's worn or damaged, replace it. Inspect the wire retainer in the lever. If its damaged or distorted, remove it (see illustration) and discard it.

INSPECTION

▶ **Refer to illustration 6.5**

5 Hold the outer portion of the bearing and rotate the center while applying pressure (see illustration). If the bearing doesn't turn

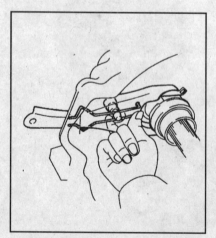

6.3 To disengage the release lever retainer from the ball stud, wrap your finger behind the lever as shown and pull toward you until the retainer "pops" off the ball stud

6.4a Note how the release bearing and the release lever fit together, then disengage them and slide off the bearing

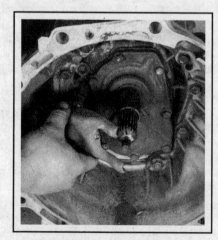

6.4b Remove the release lever

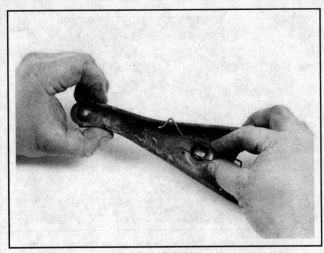

6.4c If the release lever retainer is damaged or distorted, remove and discard it - don't install the release lever with a weak or bent retainer

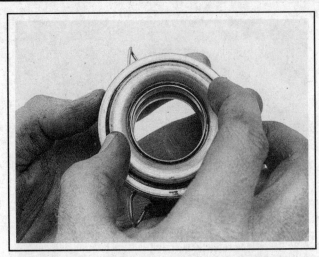

6.5 To check the condition of the release bearing, hold it by the outer race and rotate the inner race while applying pressure; the bearing should turn smoothly - if it doesn't, replace it

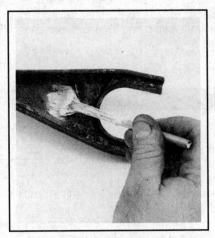

6.6a Lubricate the "pocket" for the ball stud in the backside of the release lever . . .

6.6b . . . the release lever "fingers" and the "pocket" for the release cylinder pushrod with high temperature grease

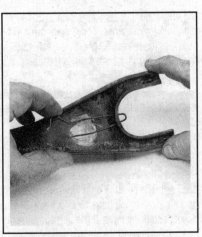

6.6c If you removed the old release lever retainer, install the new one

smoothly or if it's noisy, replace it with a new one. Wipe the bearing with a clean rag and inspect it for damage, wear and cracks. Don't immerse the bearing in solvent - it's sealed for life and to do so would ruin it.

INSTALLATION

▶ Refer to illustrations 6.6a, 6.6b, 6.6c, 6.6d, 6.8a and 6.8b

6 Lightly lubricate the clutch release lever at the indicated spots (see illustrations). If you removed the old retainer from the lever, install the new one (see illustration). Also lubricate the groove around the circumference of the release bearing that's engaged by the lever, and the bearing retainer around the input shaft with high-temperature grease (see illustration).

7 Attach the release bearing to the release lever. Make sure the bearing is properly engaged by the retainer clip.

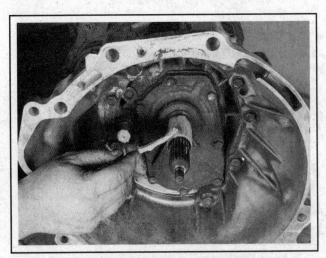

6.6d Lubricate the bearing retainer surrounding the input shaft with high temperature grease

6.8a Install the release lever and release bearing and push the release lever onto the ball stud (you should feel it snap into place when the ball stud pops through the wire retainer)

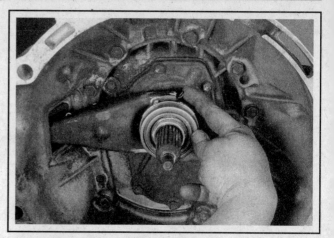

6.8b Slide the release bearing in and out on the retainer to verify that the release lever is locked onto the ball stud, the bearing slides freely and the two parts are properly engaged

8   Lubricate the clutch release lever ball stud with high-temperature grease, insert the release lever through the boot, slide the release bearing onto the input shaft splines and push the lever onto the ball stud until the lever retainer "pops" onto the stud (see illustration). Make sure that the release lever pivots freely and the release bearing slides freely on the input shaft splines (see illustration).

9   Apply a light coat of high-temperature grease to the face of the release bearing, where it contacts the pressure plate diaphragm fingers.
10   The remainder of installation is the reverse of the removal procedure. Tighten all transmission-to-engine bolts to the torque listed in the Chapter 7A Specifications.

7   Pilot bearing - inspection and replacement

◆ **Refer to illustrations 7.5 and 7.6**

1   The clutch pilot bearing is a needle roller type bearing which is pressed into the rear of the crankshaft. It's greased at the factory and does not require additional lubrication. Its primary purpose is to support the front of the transmission input shaft. The pilot bearing should be inspected whenever the clutch components are removed from the engine. Because of its inaccessibility, replace it with a new one if you have any doubt about its condition.

➡**Note: If the engine has been removed from the vehicle, disregard the following Steps which don't apply.**

2   Remove the transmission (see Chapter 7A).

3   Remove the clutch components (see Section 5).
4   Using a flashlight, inspect the bearing for excessive wear, scoring, dryness, roughness and any other obvious damage. If any of these conditions are noted, replace the bearing.
5   Removal can be accomplished with a special puller (see illustration), which is available at most auto parts stores.
6   To install the new bearing, lightly lubricate the outside surface with grease, then drive it into the recess with a bearing installer or a socket (see illustration). If the bearing is equipped with a seal in one end, the seal must face out.
7   Install the clutch components, transmission and all other components removed previously.

7.5 A small slide-hammer puller is handy for removing an old pilot bearing

7.6 Tap the bearing into place with a bearing installer or a socket that is slightly smaller than the outside diameter of the bearing

8 Clutch hydraulic system - bleeding

▶ **Refer to illustration 8.4**

1   The hydraulic system should be bled to remove all air whenever any part of the system has been removed or if the fluid level has been allowed to fall so low that air has been drawn into the master cylinder. The procedure is very similar to bleeding a brake system.

2   Fill the master cylinder with new brake fluid conforming to DOT 3 specifications.

❋❋ CAUTION:

Do not re-use any of the fluid coming from the system during the bleeding operation or use fluid which has been inside an open container for an extended period of time.

3   Raise the vehicle and support it securely on jackstands to gain access to the release cylinder, which is located on the left side of the clutch housing.

4   Remove the dust cap which fits over the bleeder valve and push a length of plastic hose over the valve (see illustration). Place the other end of the hose into a clear container with about two inches of brake fluid. The hose end must be in the fluid at the bottom of the container.

5   Have an assistant depress the clutch pedal and hold it. Open the bleeder valve on the release cylinder, allowing fluid to flow through the hose. Close the bleeder valve when your assistant signals the clutch pedal is at the bottom of its travel. Once closed, have your assistant release the pedal.

6   Continue this process until all air is evacuated from the system, indicated by a solid stream of fluid being ejected from the bleeder valve

8.4 The setup for bleeding the clutch hydraulic system is simple: a length of rubber hose between the bleeder plug and a small container with about two inches of brake fluid in it; make sure the hose is submerged in the fluid

each time with no air bubbles in the hose or container. Keep a close watch on the fluid level inside the clutch master cylinder reservoir - if the level drops too low, air will be sucked back into the system and the process will have to be started all over again.

7   Install the dust cap and lower the vehicle. Check carefully for proper operation before placing the vehicle in normal service.

9 Clutch start switch/clutch start cancel switch - replacement

CLUTCH START SWITCH

▶ **Refer to illustration 9.1**

Check

1   The clutch start switch is located on the clutch pedal bracket (see illustration).

2   Verify that the engine will not start when the clutch pedal is depressed all the way.

3   If the clutch start switch doesn't perform as described above, check switch continuity.

4   Unplug the electrical connector from the switch and verify that there is continuity between the two clutch start switch terminals when the pedal is depressed.

5   Verify that no continuity exists between the switch terminals when the pedal is released.

6   If the switch fails either of these continuity tests, replace it.

Replacement

7   Remove the driver's side lower finish panel (see Chapter 11).

8   Unplug the electrical connector from the switch, remove the switch locknut and remove the switch.

9   Installation is the reverse of removal. Adjust the switch.

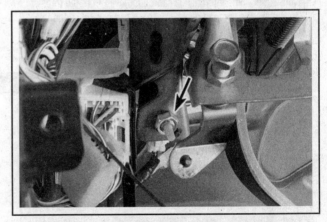

9.1 The clutch start switch (arrow) is located on a small bracket near the top of the clutch pedal; it's secured to the bracket by a pair of nuts, one above the bracket, one below, which are also used to adjust the position of the switch plunger in relation to its stopper on the pedal.

Adjustment

10 With the switch in the OFF position, loosen the switch locknut and screw the switch in or out so that the switch plunger protrudes between 0.20 and 0.31 inches past the locknut.

CLUTCH START CANCEL SWITCH (4WD MODELS)

▶ **Refer to illustration 9.15**

11 The clutch start switch on 4WD models can be turned off by the clutch start cancel switch so that the vehicle can be driven out of difficult situations by cranking the engine with the clutch engaged (pedal not depressed). The button for the clutch start cancel switch is located on the left end of the dash.

12 Remove the driver's side lower finish panel (see Chapter 11).

13 Remove the switch base from the instrument panel (see Chapter 11).

14 Unplug the electrical connectors.

15 Remove the cancel switch from the backside of the switch base (see illustration).

16 Installation is the reverse of removal.

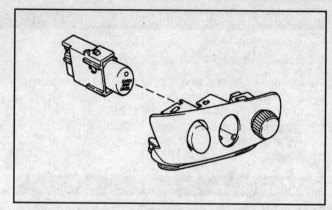

9.15 After removing the switch base from the instrument panel, remove the switch from the base

10 Driveshaft(s) and universal joints - general information

1 A driveshaft is a tube, or a pair of tubes, that transmits power between the transmission (or transfer case on 4WD models) and the differential. Universal joints are located at either end of the driveshaft; a third U-joint is employed just behind the center on two-piece driveshafts. The driveshaft is attached to the rear differential by a companion flange; on 4WD models, the front driveshaft is attached to the front differential the same way.

2 Driveshafts on 2WD models employ a splined yoke, known as a "slip yoke" or "sleeve yoke," at the front, which slips into the extension housing of the transmission. This arrangement allows the driveshaft to slide back-and-forth within the transmission during vehicle operation. An oil seal prevents leakage of fluid at this point and keeps dirt from entering the transmission. If leakage is evident at the front of the driveshaft, replace the oil seal (see Chapter 7, Part A).

3 On 4WD models, each driveshaft is attached to the transfer case by a companion flange. Once a front or rear driveshaft has been removed, either companion flange can be removed from the transfer case to replace the companion seal(s) (each companion flange uses two seals: one seal between the companion flange and the transfer case, the other, smaller, seal inside the companion flange itself). Refer to Chapter 7C for the transfer case seal replacement procedure.

4 On two-piece driveshafts, center bearings support the driveline. The center bearing is a ball-type bearing mounted in a rubber cushion attached to a frame crossmember. The bearing is pre-lubricated and sealed at the factory. On two-piece driveshafts, a sleeve yoke is employed in the rear driveshaft section.

5 The driveshaft assembly requires periodic lubrication. See Chapter 1 for the lubrication procedure and maintenance interval.

6 Since the driveshaft is a balanced unit, it's important that no undercoating, mud, etc. be allowed to stay on it. When the vehicle is raised for service it's a good idea to clean the driveshaft and inspect it for any obvious damage. Also, make sure the small weights used to originally balance the driveshaft are in place and securely attached. Whenever the driveshaft is removed it must be reinstalled in the same relative position to preserve the balance.

7 Problems with the driveshaft are usually indicated by a noise or vibration while driving the vehicle. A road test should verify if the problem is the driveshaft or another vehicle component. Refer to the *Troubleshooting* Section at the front of this manual. If you suspect trouble, inspect the driveline (see the next Section).

11 Driveline inspection

1 Raise the rear of the vehicle and support it securely on jackstands. Block the front wheels to keep the vehicle from rolling off the stands.

2 Crawl under the vehicle and visually inspect the driveshaft. Look for any dents or cracks in the tubing. If any are found, the driveshaft must be replaced.

3 Check for oil leakage at the front and rear of the driveshaft. Leakage where the driveshaft enters the transmission or transfer case indicates a defective transmission/transfer case seal (see Chapter 7). Leakage where the driveshaft joins the differential indicates a defective pinion seal (see Section 17).

4 While under the vehicle, have an assistant rotate a rear wheel so the driveshaft will rotate. As it does, make sure the universal joints are operating properly without binding, noise or looseness. Listen for any noise from the center bearing (if equipped), indicating it's worn or damaged. Also check the rubber portion of the center bearing for cracking or separation, which will necessitate replacement.

5 The universal joint can also be checked with the driveshaft motionless, by gripping your hands on either side of the joint and attempting to twist the joint. Any movement at all in the joint is a sign of considerable wear. Lifting up on the shaft will also indicate movement in the universal joints.

6 Finally, check the driveshaft mounting bolts at the ends to make sure they're tight.

7 On 4WD models, the above driveshaft checks should be repeated on all driveshafts. In addition, check for grease leakage around the sleeve yoke, indicating failure of the yoke seal.

8 Check for leakage where the driveshafts connect to the transfer case and front differential. Leakage indicates worn oil seals.

9 At the same time, check for looseness in the joints of the front driveaxles. Also check for grease or oil leakage from around the driveaxles by inspecting the rubber boots and both ends of each axle. Oil leakage at the differential junction indicates a defective side oil seal. Leakage at the wheel side indicates a defective front hub seal, while leakage at the boots means a damaged rubber boot. For servicing of these components, see the appropriate Sections.

12 Driveshaft - removal and installation

1   Raise the vehicle and support it securely on jackstands. Place the transmission in Neutral with the parking brake off. Block the front wheels to prevent the vehicle from rolling.

REMOVAL

▶ **Refer to illustrations 12.2, 12.3 and 12.4**

2   Using a scribe, a hammer and punch, or paint, make marks on the driveshaft and the differential flange in line with each other (see illustration). This is to make sure the driveshaft is reinstalled in the same position to preserve the balance.

3   Remove the bolts securing the flange yoke to the rear differential (see illustration). Turn the driveshaft (or wheels) as necessary to bring the bolts into the most accessible position.

4   On vehicles with a two-piece driveshaft, remove the center bearing protector and remove the bolts, nuts and washers from the center support bearing bracket (see illustration).

5   Lower the rear of the driveshaft. Slide the front of the driveshaft out of the transmission or transfer case or, if equipped with a companion flange, separate the flange at the transfer case.

6   On 2WD models, wrap a plastic bag over the transmission extension housing and hold it in place with a rubber band. This will prevent loss of fluid and protect against contamination while the driveshaft is out.

INSTALLATION

7   Remove the plastic bag from the transmission or transfer case and wipe the area clean. Inspect the oil seal carefully. Slide the front of the driveshaft into the transmission (2WD models) or bolt the U-joint flange yoke to the companion flange, installing the fasteners finger-tight (4WD models).

8   Raise the center bearing (if equipped) into place and screw the retaining bolts in a few turns. Raise the rear of the driveshaft into position, checking to be sure the marks are in alignment. If not, turn the rear wheels to match the pinion flange and the driveshaft.

9   Tighten all nuts to the torque listed in this Chapter's Specifications. Remove the jackstands and lower the vehicle.

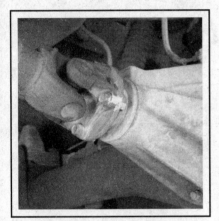

12.2 Mark the relationship of the driveshaft U-joint to the pinion flange

12.3 Using a backup wrench to hold each bolt, break loose all four nuts securing the flange yoke to the differential

12.4 To detach the center bearing on two-piece driveshafts, remove the two mounting bolts

13 Universal joints - replacement

▶ **Refer to illustration 13.2, 13.3a, 13.3b, 13.3c, 13.3d, 13.3e, 13.4 and 13.5**

➡**Note: A press or large vise will be required for this procedure. It may be a good idea to take the driveshaft to a repair or machine shop where the U-joints can be replaced for you, usually at a reasonable charge.**

1   Remove the driveshaft (see Section 12).
2   Place the driveshaft on a bench equipped with a vise.

❊❊ CAUTION:

Some models use driveshafts equipped with "double-cardan" type universal joints (see illustration). Do NOT attempt to disassemble a double-cardan U-joint. The Toyota OEM double-cardan U-joint used on vehicles covered by this manual is not rebuildable. If you determine that a double-cardan U-joint is worn out, replace the driveshaft assembly with a new or rebuilt unit.

13.2 If the driveshaft has a U-joint that looks like this, it's a "double-cardan" type (meaning it has two U-joints instead of one) and it can't be rebuilt; if a double-cardan U-joint is damaged or worn out, replace the driveshaft

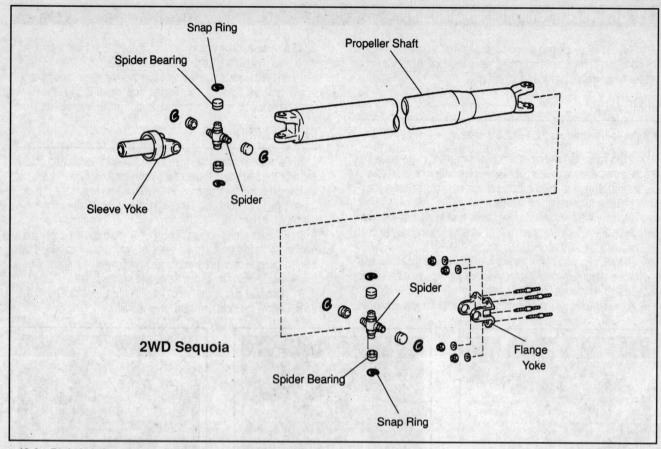

2WD Sequoia

13.3a **Exploded view of a one-piece driveshaft (2WD Sequoia)**

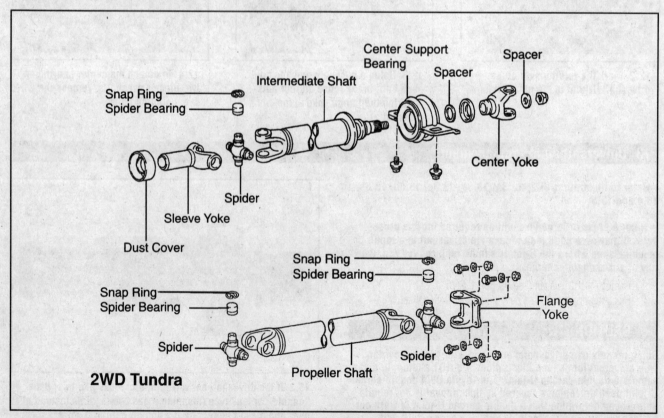

2WD Tundra

13.3b **Exploded view of a typical three-joint driveshaft assembly (2WD Tundra)**

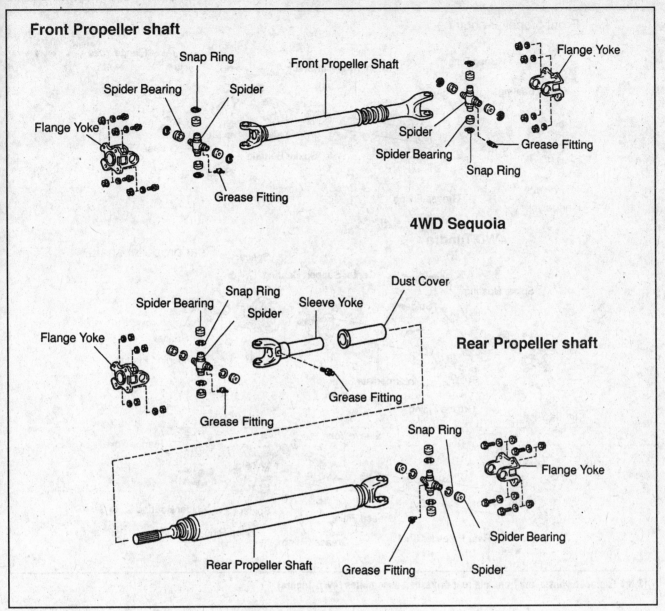

13.3c Exploded view of the front and rear driveshaft assemblies (4WD Sequoia)

3   Mark the shaft and yoke for proper reassembly, then remove the snap-rings from the U-joint (see illustrations).

➡Note: The driveshafts covered in this manual are equipped with U-joints that use either outer snap-rings or inner snap-rings to retain the bearing cap. Outer snap-rings can be removed using a pair of pliers while inner snap-rings require a punch and a hammer to dislodge the snap-ring from the groove on the bearing cap.

4   Place a piece of pipe or a large socket with the same inside diameter over one of the bearing cups. Position a socket which is of slightly smaller diameter than the cup on the opposite bearing cup (see illustration) and use the vise to force the cup out (inside the pipe or large socket), stopping just before it comes completely out of the yoke.

5   Use the vise or large pliers to work the cup the rest of the way out (see illustration).

6   Transfer the sockets to the other side and press the opposite bearing cup out in the same manner.

7   After the bearing cups have been removed, lift the U-joint from the yoke and thoroughly clean all dirt and debris from the yokes on both ends of the driveshaft. Be sure to remove any metal burrs from the yoke bores.

8   Pack the new U-joint bearing cups with grease, this will allow the needle bearings to be held in place while your installing the bearing cups. Ordinarily, specific instructions for lubrication will be included with the U-joint servicing kit and should be followed carefully.

9   Position the U-joint body in the yoke and partially install one bearing cup in the yoke. If the U-joint is equipped with a grease fitting, be sure it points in the same direction as the grease fitting on the opposite end of the driveshaft.

10 Start the U-joint body into the bearing cup and partially install the other cup. Align the U-joint body between the bearing cups and press the bearing cups into position, being careful not to damage the dust seals.

11 Install the snap-rings. If difficulty is encountered in seating the snap-rings, strike the driveshaft yoke sharply with a hammer. This will

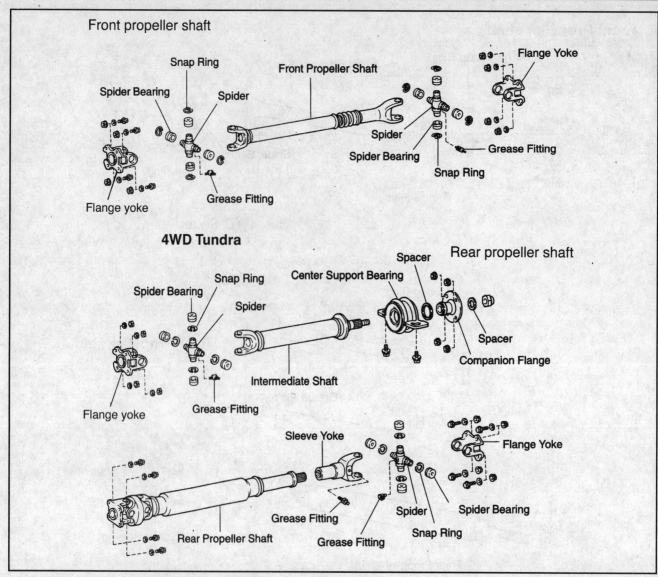

13.3d Exploded view of the front and rear driveshaft assemblies (4WD Tundra)

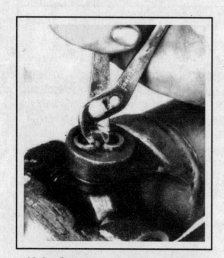

13.3e Outer type snap-rings can be removed with a small pair of pliers

13.4 To remove the U-joint from the driveshaft, use a vise as a press; the small socket will push the cross and bearing cup into the large socket

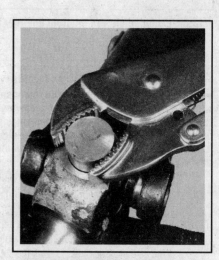

13.5 Grip the bearing cup with locking pliers and remove it from the yoke

spring the yoke ears slightly and allow the snap-rings to seat in the groove. This should also be done to center the U-joint after assembly.

➡**Note: If you still have difficulty seating the snap-rings, one of the small needle bearings may have become stuck between the bearing cap and the end of the spider. Disassemble and inspect the joint.**

12 Install the driveshaft (see Section 12).

13 If the U-joint is equipped with a grease fitting, lubricate it as described in Chapter 1.

14 Remove the jackstands and lower the vehicle.

14 Center bearing - removal and installation

▶ **Refer to illustrations 14.6a, 14.6b and 14.7**

1 Raise the vehicle and support it securely on jackstands.

2 Remove the driveshaft (see Section 13).

3 Clamp the driveshaft securely into a bench vise.

4 Separate the intermediate (front) part of the driveshaft from the rear part: On 2WD Tundra models with a outer snap-ring type driveshaft, mark the center U-joint yoke and driveshaft, then disassemble the center U-joint (see Section 13). On 4WD Tundra models, mark the relationship of the center U-joint flange yoke to the flange behind the center bearing, then unbolt the center U-joint from the flange.

5 Unstake the center yoke (outer snap-ring type) or flange (inner snap-ring type) retaining nut and remove it.

6 Mark the relationship of the intermediate shaft to the yoke or flange (see illustrations).

7 Remove the yoke or flange from the intermediate shaft (see illustration).

8 Remove the center bearing from the intermediate shaft.

➡**Note: This usually requires a puller or a punch and a hammer to separate the center bearing from the intermediate shaft.**

9 Holding the center bearing assembly in one hand, turn the bearing with the other hand and verify that it operates freely and smoothly. If it's stiff or noisy, replace it.

10 Installation is the reverse of removal. Be sure to tighten the yoke or flange retaining nut to the torque listed in this Chapter's Specifications, then stake it.

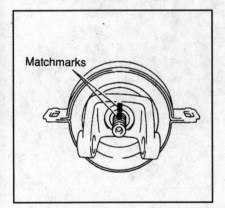

14.6a Mark the relationship of the intermediate shaft to the yoke . . .

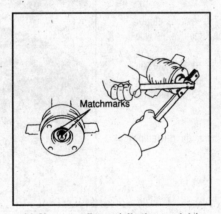

14.6b . . . or flange (all other models)

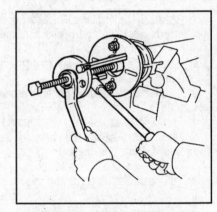

14.7 Remove the yoke or flange from the intermediate shaft with a puller

15 Axles - description and check

DESCRIPTION

1 The rear axle assembly is a hypoid, semi-floating type (the centerline of the pinion gear is below the centerline of the ring gear). When the vehicle goes around a corner, the differential allows the outer rear tire to turn more quickly than the inner tire. The axleshafts are splined to the differential side gears, so when the vehicle goes around a corner, the inner tire, which turns more slowly than the outer tire, turns its side gear more slowly than the outer tire turns its side gear. The differential pinion gears roll around the slower side gear, driving the outer side gear - and tire - more quickly. The differential is housed within a casting with a pressed steel cover, known as the "carrier." The steel axle tubes are pressed into and welded to the carrier.

2 A locking limited-slip rear axle is used on some models. This differential allows for normal operation until one wheel loses traction. A limited-slip unit is similar in design to a conventional differential, except for the addition of a pair of clutch "cones" which slow the rotation of the differential case when one wheel is on a firm surface and the other on a slippery one. The difference in wheel rotational speed produced by this condition applies additional force to the pinion gears and through the cone, which is splined to the axleshafts, equalizes the rotation speed of the axleshaft driving the wheel with traction.

3 On 4WD models, a fully independent front axle assembly is used. This consists of a differential and a pair of driveaxles. Each driveaxle has an inner and outer constant velocity (CV) joint.

CHECK

4   Often, a suspected "axle" problem lies elsewhere. Do a thorough check of other possible causes before assuming the axle is the problem.

5   The following noises are those commonly associated with axle diagnosis procedures:

a) *Road noise is often mistaken for mechanical faults. Driving the vehicle on different surfaces will show whether the road surface is the cause of the noise. Road noise will remain the same if the vehicle is under power or coasting.*

b) *Tire noise is sometimes mistaken for mechanical problems. Tires which are worn or low on pressure are particularly susceptible to emitting vibrations and noises. Tire noise will remain about the same during varying driving situations, where axle noise will change during coasting, acceleration, etc.*

c) *Engine and transmission noise can be deceiving because it will travel along the driveline. To isolate engine and transmission noises, make a note of the engine speed at which the noise is most pronounced. Stop the vehicle and place the transmission in Neutral and run the engine to the same speed. If the noise is the same, the axle is not at fault.*

6   Because of the special tools needed, overhauling the differential isn't cost effective for a do-it-yourselfer. The procedures included in this Chapter describe axleshaft removal and installation, axleshaft oil seal replacement, axleshaft bearing replacement and removal of the entire unit for repair or replacement. Any further work should be left to a dealer service department or other qualified repair shop.

➡**Note: If the rear axle must be replaced, refer to the identification code and manufacturer's code stamped on the front side of the right rear axle tube. This number contains information on the rear axle ratio, differential type, manufacturer and build date information, all of which are necessary to ensure that you get the right axle.**

16 Axleshaft, bearing and oil seals (rear) - removal and installation

▶ **Refer to illustrations 16.6, 16.7, 16.8 and 16.9**

1   Release the parking brake. Raise the rear of the vehicle, support it securely on jackstands and block the front wheels. Remove the wheel and, if equipped, the brake drum.

2   If the vehicle is equipped with rear disc brakes, remove the caliper and disc (see Chapter 9).

3   If the vehicle is equipped with ABS, remove the ABS sensor (see Chapter 9).

4   Remove the brake assembly (see Chapter 9).

5   Disconnect the parking brake cable and the hydraulic brake line to the wheel cylinder (see Chapter 9).

6   Remove the four backing plate mounting nuts (see illustration).

7   Pull the axleshaft out of the rear axle housing (see illustration).

8   Remove the O-ring from the rear axle housing (see illustration).

9   Remove the axleshaft inner oil seal from the axle housing with a seal removal tool or a big screwdriver (see illustration).

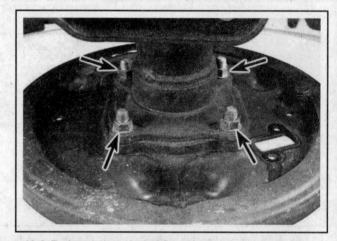

16.6 To detach the axleshaft from the rear axle housing, remove the four backing plate nuts

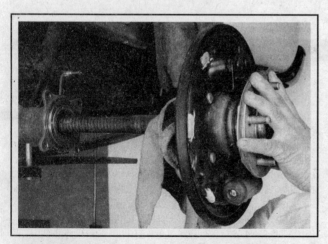

16.7 Extract the axleshaft very carefully from the axle housing, especially if you don't want to replace the axleshaft seal

16.8 Remove this O-ring from the rear axle housing; be sure to discard the old O-ring and install a new one before installing the axleshaft

10 Further disassembly of the axleshaft assembly requires special tools and a hydraulic press. If the axleshaft, bearing or outer oil seal needs to be replaced, take the axleshaft assembly to an automotive machine shop.

11 Drive a new axleshaft inner seal into the end of the axle tube with a seal installer or a big socket. Coat the lip of the seal with clean oil or multi-purpose grease.

12 Install a new axle housing O-ring. Apply a light coat of oil to the new O-ring.

13 Make sure the axleshaft is clean and there are no burrs or metal splinters on it. Deburr any surface irregularities so the axleshaft doesn't damage the seal during installation. Lightly coat the axleshaft with clean oil, then insert it into the axle housing. Make sure the splined inner end of the axleshaft doesn't damage the lip of the new axleshaft seal.

14 Installation is the reverse of removal. Tighten the four backing plate mounting nuts to the torque listed in this Chapter's Specifications.

16.9 Use a seal removal tool or a big screwdriver to pry out the old axleshaft seal; use a seal installer or a big socket to install the new seal

17 Differential pinion seal - replacement

▶ **Refer to illustrations 17.3, 17.4, 17.6, 17.8, 17.9 and 17.10**

➡**Note: This procedure applies to the rear pinion seal on all vehicles and the front pinion seal on 4WD models as well.**

1 Loosen the wheel lug nuts, raise the vehicle and support it securely on jackstands. Block the wheels at the opposite end to keep the vehicle from rolling off the stands. Remove the wheels (this will allow you to obtain a more accurate pinion shaft preload reading).

2 Disconnect the driveshaft from the differential (see Section 13) and fasten it out of the way.

3 Use an inch-pound torque wrench to check the torque required to rotate the pinion (see illustration). Record it for use later.

4 Scribe or punch alignment marks on the pinion shaft, nut and flange (see illustration).

5 Count the number of threads visible between the end of the nut and the end of the pinion shaft and jot it down for later use.

6 A special flange holding tool is the best way to keep the companion flange from moving while the pinion nut is loosened. If you're unable to obtain a flange holding tool, immobilize the flange by inserting a big screwdriver through one of the U-joint bolt holes in the flange

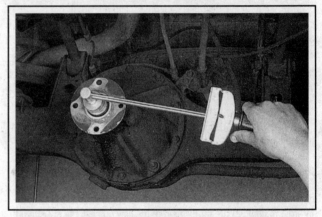

17.3 Use an inch-pound torque wrench to check the torque necessary to rotate the pinion shaft

and wedge it against a bracket (see illustration) or reinforcement rib on the differential carrier.

17.4 Mark the relative positions of the pinion, nut and flange before removing the nut

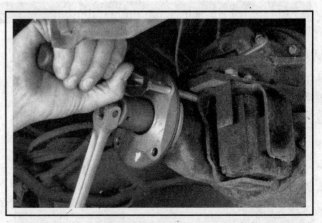

17.6 If you don't have a flange holding tool, lock the flange by jamming a large screwdriver through a bolt hole in the flange and wedge it underneath a bracket as shown, or under a reinforcement rib on the differential carrier

17.8 If you can't pull off the pinion flange by hand, remove it with a small puller

17.9 Pry out the old pinion seal with a seal removal tool or a big screwdriver or tap it out with a small punch

7   Remove the pinion nut.

8   Withdraw the companion flange. It may be necessary to use a puller to draw it out (see illustration). Do NOT attempt to pry behind the flange or hammer on the flange or the end of the pinion shaft.

9   Pry out the old seal (see illustration) and discard it.

10 Lubricate the lips of the new seal with high-temperature grease and tap it evenly into position with a seal installation tool or a large socket. Make sure it enters the housing squarely and is tapped in to its full depth (see illustration).

11 Align the mating marks made before disassembly and install the companion flange. If necessary, tighten the pinion nut to draw the flange into place. Do not try to hammer the flange into position.

12 Apply non-hardening sealant to the ends of the splines visible in the center of the flange so oil will be sealed in.

13 Install the washer (if equipped) and pinion nut. Tighten the nut carefully, until the original number of threads are exposed.

14 Measure the torque required to rotate the pinion and tighten the nut in small increments until it matches the figure recorded in Step 3. In order to compensate for the drag of the new oil seal, the nut should be tightened more until the rotational torque of the pinion slightly exceeds what was recorded earlier, but not by more than 5 in-lbs.

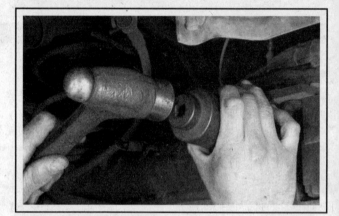

17.10 Lubricate the lips of the new pinion seal and seat it squarely in the bore, then drive it into the carrier with a seal driver or a large socket

15 Connect the driveshaft, install the wheels and lower the vehicle. Tighten the lug nuts to the torque listed in the Chapter 1 Specifications.

18 Axle (rear) - removal and installation

1   Loosen the rear wheel lug nuts, raise the rear of the vehicle and support it securely on jackstands placed under the frame (not under the axle). Block the front wheels to keep the vehicle from rolling off the stands. Remove the rear wheels.

2   Position a jack under the rear axle differential housing.

3   Disconnect the driveshaft from the differential (see Section 12). Fasten the driveshaft out of the way with a piece of wire from the underbody.

4   If you're working on a Tundra model, disconnect the load sensing proportioning and bypass valve (LSP & BV) height sensing spring from the axle (see Chapter 9).

5   Detach all brake hoses and/or lines from the axle housing, then plug them to prevent fluid leakage.

6   Remove the rear brake assemblies (see Chapter 9).

7   Disconnect the parking brake cables from the brake assemblies

and detach the cables from the rear axle housing (see Chapter 9).

8   Detach the vent hose from the axle housing and fasten it out of the way.

9   Disconnect the shock absorbers from the spring seats or from the axle brackets (see Chapter 10).

10 On Tundra models, disconnect the leaf spring U-bolts and remove the spacers, bumpers and spring seats (see Chapter 10).

11 On Sequoia models, disconnect the stabilizer bar, lateral control rod and upper and lower suspension arms (see Chapter 10).

12 Lower the jack under the differential, then remove the rear axle assembly from under the vehicle.

13 Installation is the reverse of removal. Be sure to tighten all suspension fasteners to the torque listed in the Chapter 10 Specifications.

14 Bleed the brakes (see Chapter 9).

19 Driveaxle (4WD models) - removal and installation

▶ **Refer to illustrations 19.2, 19.3a, 19.3b and 19.6**

1   Loosen the front wheel lug nuts, raise the front of the vehicle and support it securely on jackstands. Block the rear wheels to keep the vehicle from rolling off the stands. Remove the wheels. Drain the differential (see Chapter 1).

2   Remove the grease cap (see illustration).

3   Remove the cotter pin and nut lock, place a prybar or large screwdriver between the wheel studs to hold the driveaxle and break the driveaxle/hub nut loose with a large breaker bar, or have an assistant apply the brakes. Remove the nut.

4   Disconnect the lower control arm from the balljoint (see Chapter 10). If you're removing the left driveaxle, remove the left side shock absorber (see Chapter 10).

5   Knock the driveaxle loose from the steering knuckle with a brass drift and hammer. Do NOT use a steel punch or strike the end of the driveaxle with a steel hammer; a steel punch or hammer will damage the threads or the splines on the end of the driveaxle.

6   Swing the steering knuckle outward and pull the driveaxle assembly out of the steering knuckle, then detach the driveaxle from the differential. If you're removing the right driveaxle, tap the inner CV joint out of the differential with a hammer and a brass drift; if you're

19.2 Using a hammer and chisel, tap the grease cap from the hub

removing the left driveaxle, a slide hammer with a special hooked adapter (available at most auto parts stores) will be needed to pull the inner CV joint from the differential (see illustration).

7   Installation is the reverse of removal. Be sure to tighten the driveaxle/hub nut to the torque listed in this Chapter's Specifications, then install the nut lock and a new cotter pin. Tighten the lug nuts to the torque listed in the Chapter 1 Specifications. Tighten all suspension fasteners to the torque listed in the Chapter 10 Specifications.

19.3a Remove the cotter pin and the nut lock

19.3b Place a prybar between two of the wheel studs, then loosen the driveaxle/hub nut

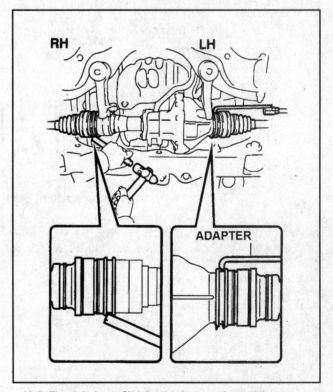

19.6 The right inner CV joint can be tapped out of the differential with a hammer and brass punch; the left side must be pulled out with a slide hammer equipped with a special hooked adapter

20 Driveaxle boot - replacement

▶ **Refer to illustration 20.1**

1   Remove the driveaxle (see Section 19) (see illustration).

DISASSEMBLY

▶ **Refer to illustrations 20.3, 20.5 and 20.6**

2   Mount the driveaxle in a vise with wood lined jaws (to prevent damage to the axleshaft). Check the CV joint for excessive play in the radial direction, which indicates worn parts. Check for smooth operation throughout the full range of motion for each CV joint. If a boot is torn, disassemble the joint, clean the components and inspect for damage due to loss of lubrication and possible contamination by foreign matter.

3   Using a pair of pliers, squeeze the retaining tabs of the large inboard joint boot clamp and slide it away from the CV joint (see illustration). Using a pair of side cutters cut the boot clamps. Old and worn boots can be cut off.

4   Mark the inboard joint to the shaft to ensure that they are reassembled properly.

5   Using a pair of snap-ring pliers, expand the snap-ring that connects the outboard joint shaft to the inboard joint and separate the shafts (see illustration).

6   If you haven't already cut them off, remove both boots. Wrap the splines on the inner end of the axleshaft with electrical or duct tape to protect the boots from the sharp edges of the splines (see illustration).

CHECK

7   Thoroughly clean all components, including the outer CV joint assembly, with solvent until the old CV joint grease is completely removed. Inspect all visible bearing surfaces for cracks, pitting, scoring and other signs of wear. If the inner CV joint is worn, you can buy a new inner CV joint and install it on the old axleshaft; if the outer CV joint is worn, you'll have to purchase a new outer CV joint and axleshaft (they're sold preassembled).

REASSEMBLY

▶ **Refer to illustrations 20.9 and 20.10**

8   Slide the clamps and boot(s) onto the axleshaft. Pack the outboard and inboard joint assemblies and boots with grease. Align the matchmarks and place the inboard joint on the shaft, then using a pair of snap-ring pliers, expand the snap-ring and seat the joint on the shaft.

9   Slide the boot into place, making sure both ends seat in their grooves. Adjust the length of the driveaxle to the dimension shown at the end of the Chapter (see illustration).

10 Equalize the pressure in the boot, then tighten and secure the boot clamps (see illustration).

11 Install the driveaxle assembly (see Section 15).

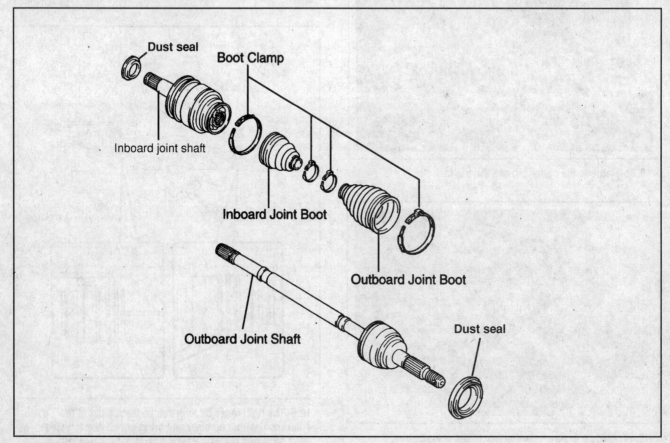

20.1 Exploded view of a typical driveaxle assembly

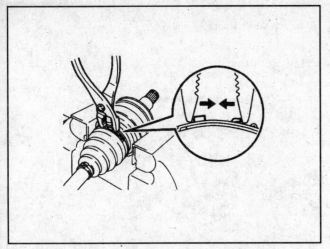

20.3 Squeeze the retaining tabs with a pair of pliers

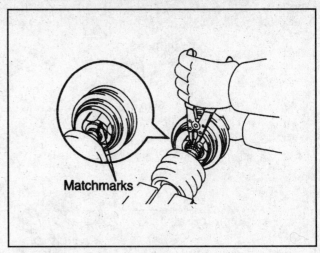

Matchmarks

20.5 Expand the snap-ring to separate the inboard CV joint from the shaft

20.6 Wrap the splined area of the axleshaft with tape to prevent damage to the boots when removing or installing them

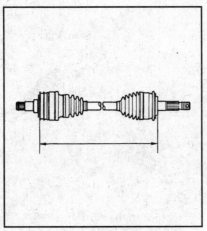

20.9 Set the driveaxle length to the dimension listed in this Chapter's Specifications before the boot clamps are tightened

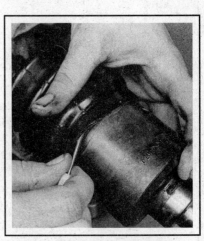

20.10 Equalize the pressure inside the boot by inserting a small, dull screwdriver between the boot and the outer race

21 Automatic Disconnecting Differential (ADD) (4WD models) - description, removal and installation

DESCRIPTION

1   The Automatic Disconnecting Differential (ADD) connects the power flow through the left axleshaft when 4WD mode is selected, and disconnects the power flow when 2WD mode is selected. Although ADD-equipped vehicles make selecting 2WD or 4WD more convenient (there are no locking hubs to deal with), they also increase wear on the CV joints and dust boots, as well as some of the axle and differential components, which rotate all the time, even in 2WD. If your vehicle is equipped with ADD, be sure to inspect the CV joints and boots regularly. If shifting into or out of 4WD becomes a problem, have the ADD system checked out by a dealer service department or other qualified repair shop that specializes in 4WD vehicles.

REMOVAL AND INSTALLATION

Differential carrier

♦ **Refer to illustrations 21.7 and 21.9**

2   Loosen the wheel lug nuts, raise the vehicle and support it securely on jackstands. Remove the wheels.

3   Remove the driveaxles (see Section 19).

4   Remove the engine under cover (see illustration 8.6 in Chapter 1).

5   Drain the lubricant from the differential (see Chapter 1).

6   Disconnect the driveshaft from the front differential (see Section 12) and support the front end of the driveshaft with a piece of wire.

7   Disconnect the breather hose, detach the fasteners that retain the

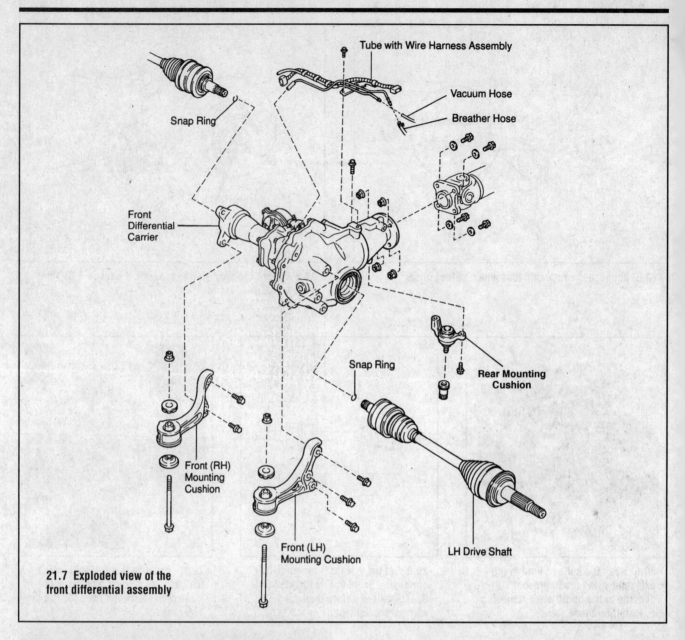

21.7 Exploded view of the front differential assembly

21.9 Differential mounting bolt locations

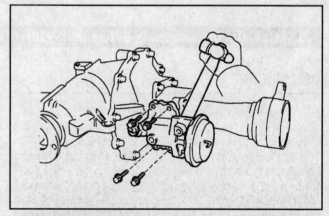

21.12 To detach the ADD actuator from the front differential, unplug the electrical connector and detach the vacuum hoses, then remove the four retaining bolts - if the actuator is stuck to the front differential, use a wooden hammer handle to pry it off as shown

vacuum tubing bracket to the differential, unplug the actuator electrical connector and detach the actuator vacuum hoses. Remove the tube and wire harness assembly from the differential (see illustration).

8   Support the differential with a transmission jack or a floor jack.

9   Remove the differential rear mounting nut and the two differential front mounting bolts (see illustration) and lower the differential.

10 Installation is the reverse of removal. Refill the differential with the proper lubricant (see Chapter 1) and tighten the lug nuts to the torque listed in the Chapter 1 Specifications.

ADD actuator

▶ **Refer to illustration 21.12**

11 Remove the four retaining bolts.

12 Remove the actuator (see illustration).

13 Installation is the reverse of removal. Before installing the actuator, remove the old RTV sealant from the mating surfaces of the differential and the actuator and apply a thin bead of new RTV sealant to those surfaces. Tighten the actuator bolts securely.

Specifications

General

Clutch pedal height	See Chapter 1
Clutch pedal freeplay	See Chapter 1
Clutch fluid type	See Chapter 1
Driveaxle length	20-5/8 inches

Torque specifications   Ft-lbs (unless otherwise indicated)

➡**Note: One foot-pound (ft-lb) of torque is equivalent to 12 inch-pounds (in-lbs) of torque. Torque values below approximately 15 ft-lbs are expressed in inch-pounds, since most foot-pound torque wrenches are not accurate at these smaller values.**

Clutch

Pressure plate-to-flywheel bolts	168 in-lbs

Driveshaft

Flange bolts/nuts	
Front (4WD)	54
Rear (2WD/4WD)	54
Center support bearing bolts	30
Center support bearing yoke or flange nut	
Step 1	133
Step 2	loosen one full turn
Step 3	60

Front driveaxle (4WD models)

Driveaxle hub nut	173

Rear axle

Brake backing plate nuts	
Tundra	51
Sequoia	90

Notes

Section

Reference to other Chapters

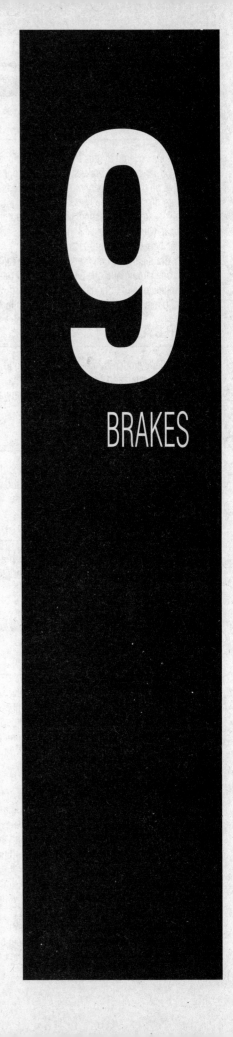

9

BRAKES

1 General information

GENERAL

All models covered by this manual are equipped with hydraulically operated, power-assisted brake systems. All front brakes are disc type. Rear brakes on the Tundra are drum type, while rear brakes on the Sequoia are disc type.

All brakes are self-adjusting. Disc brakes automatically compensate for pad wear, while drum brakes incorporate an adjustment mechanism which is activated as the brakes are applied.

The hydraulic system has separate circuits for the front and rear brakes. If one circuit fails, the other circuit will remain functional and a warning indicator will light up on the dashboard when a substantial amount of brake fluid is lost, showing that a failure has occurred.

MASTER CYLINDER

The master cylinder is located under the hood, mounted to the power brake booster, and can be identified by the large fluid reservoir on top. The master cylinder has separate primary and secondary piston assemblies for the front and rear circuits.

POWER BRAKE BOOSTER

The power brake booster uses engine manifold vacuum to provide assistance to the brakes. It is mounted on the firewall in the engine compartment, directly behind the master cylinder.

ANTI-LOCK BRAKE SYSTEM (ABS)

An optional Anti-Lock Brake System (ABS) prevents wheel lockup at all four wheels. Refer to Section 2 for a description of ABS operation.

PARKING BRAKE

The parking brake mechanically operates the rear brakes only. On Tundra models, the parking brake cables pull on a lever attached to the brake shoe assembly, causing the shoes to expand against the drum. Sequoia models use small brake shoes that expand against a drum surface integral with the rear brake discs.

PRECAUTIONS

There are some general precautions and warnings related to the brake system:

a) *Use only brake fluid conforming to DOT 3 specifications.*

b) *The brake pads and linings may contain fibers which are hazardous to your health if inhaled. Whenever you work on brake system components, clean all parts with brake system cleaner or denatured alcohol. Do not allow the fine dust to become airborne.*

c) *Safety should be paramount whenever any servicing of the brake components is performed. Do not use parts or fasteners which are not in perfect condition, and be sure all clearances and torque specifications are adhered to. If you are at all unsure about a certain procedure, seek professional advice. Upon completion of any brake system work, test the brakes carefully in a controlled area before driving the vehicle in traffic. If a problem is suspected in the brake system, don't drive the vehicle until it's fixed.*

2 Anti-lock Brake System (ABS) - general information

DESCRIPTION

1 The Anti-lock Brake System (ABS) is designed to maintain vehicle maneuverability, directional stability and optimum deceleration under severe braking conditions on most road surfaces. It does so by monitoring the rotational speed of the wheels and controlling the brake line pressure during braking. This prevents the wheels from locking up.

Electronic control unit (ECU)

2 The electronic control unit (ECU) for the Anti-lock Brake System is an integral part of the ABS actuator in the engine compartment.

3 The ECU monitors the rotation of each wheel with four wheel speed sensors; it processes this information and avoids wheel lockup by controlling the hydraulic line pressure accordingly. Here's how it works: When the brakes are applied too firmly during a "panic stop," hydraulic line pressure inside the brake lines builds to such a high level that it "locks up" the wheels, causing the vehicle to skid out of control. On a vehicle equipped with ABS, the ECU prevents the hydraulic pressure from reaching this dangerously high level by monitoring the rotational speed of the wheels. When a wheel begins to slow down, i.e. "lock up," in relation to the other wheels, the ECU energizes the solenoid (inside the actuator) controlling the hydraulic brake fluid cir-

cuit to that wheel. The energized solenoid opens the circuit, allowing some of the brake fluid into a reservoir, thereby lowering the pressure and preventing the wheel from locking up. As soon as the rotation speed of the wheel equals that of the other wheels, the solenoid closes and pressure begins to build again. This cycle of opening and closing the circuit occurs many times a second at each wheel. The ECU can regulate the pressure to a single wheel, or to two, three or all four wheels, simultaneously.

4 The ECU also monitors the ABS system for malfunctions. If the ECU detects a problem, the ABS warning light on the instrument cluster lights up and a diagnostic code is stored which, when retrieved by a service technician, will indicate the problem area or component. When the engine is started, the ABS warning light glows for about three seconds (indicating that the ECU is monitoring the system for faults), then goes out; if the ABS light remains on, there's a problem in the ABS system. Take the vehicle to a dealer service department or an authorized service facility.

Actuator

5 The actuator assembly, which is mounted inside the engine compartment, houses the solenoids which regulate brake fluid pressure in response to signals from the ECU.

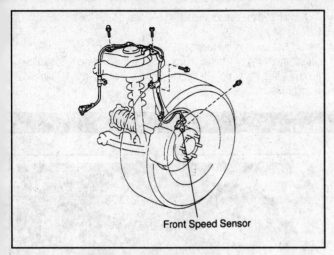

2.13 Front wheel speed sensor details

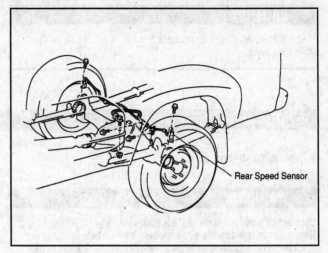

2.17a Rear wheel speed sensor details - Tundra

Speed sensors

6   Each wheel has its own speed sensor. The speed sensor is a small variable reluctance sensor (pick-up coil) which sends a variable voltage signal (actually, an alternating current sine wave output) to the ECU. The ECU converts this analog signal into a digital code which it compares to its "map" (program), then either ignores it (if the wheel is rotating at the same speed as the other wheels) or executes a command to open a solenoid for the circuit to that wheel (if the wheel is starting to slow down in relation to the other wheels).

Brake light switch

7   The brake light switch signals the control unit when the driver steps on the brake pedal. Without this signal the anti-lock system won't activate.

Diagnosis and repair

8   If the ABS warning light on the instrument cluster comes on and stays on, make sure the parking brake is released and there's no problem with the brake hydraulic system. If neither of these is the cause, the anti-lock system is probably malfunctioning. Although special test procedures are necessary to properly diagnose the system, the home mechanic can perform a few preliminary checks before taking the vehicle to a dealer service department.

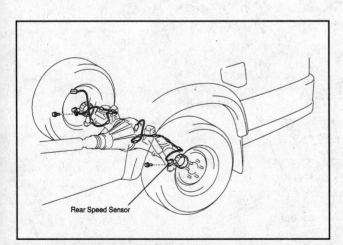

2.17b Rear wheel speed sensor details - Sequoia

a) Make sure the brakes, calipers and wheel cylinders are in good condition.
b) Inspect the electrical connectors at the ECU. Make sure they're clean and tight.
c) Check the fuses.
d) Follow the wiring harness to the speed sensor(s) and brake light switch and make sure all connections are clean and tight and the wiring isn't damaged.

9   If the above preliminary checks don't rectify the problem, the vehicle should be diagnosed by a dealer service department or other qualified repair shop.

WHEEL SPEED SENSOR - REMOVAL AND INSTALLATION

Front wheel speed sensor

♦ **Refer to illustration 2.13**

10   Loosen the wheel lug nuts, raise the front of the vehicle and support it securely on jackstands. Block the wheels at the opposite end.
11   Remove the inner fender liner (see Chapter 11).
12   Follow the sensor wiring harness up to the electrical connector, then unplug the connector.
13   Unbolt the wiring harness securing brackets (see illustration).
14   Remove the sensor mounting bolt and detach the sensor from the steering knuckle.
15   Installation is the reverse of removal. Be sure to tighten the sensor mounting bolt to the torque listed in this Chapter's Specifications. Tighten the wheel lug nuts to the torque listed in the Chapter 1 Specifications.

Rear wheel speed sensor

♦ **Refer to illustrations 2.17a and 2.17b**

16   Loosen the wheel lug nuts, raise the rear of the vehicle and support it securely on jackstands. Block the wheels at the opposite end.
17   Unplug the electrical connector from the wheel speed sensor (see illustrations). Remove the harness securing clips and brackets.

3 Disc brake pads - replacement

◗ Refer to illustration 3.2

1   Loosen the wheel lug nuts, raise the vehicle and support it securely on jackstands. Block the wheels at the opposite end. Remove the wheels.

2   Remove about two-thirds of the fluid from the master cylinder reservoir and discard it; as the pistons are pushed in to make room for the new pads, the fluid will be forced back into the reservoir. Position a drain pan under the brake assembly and clean the caliper and surrounding area with brake system cleaner (see illustration).

3   While the pads are removed, inspect the caliper for brake fluid leaks and ruptures in the piston boot(s). Replace the caliper if it's damaged or leaking (see Section 4). Also inspect the brake disc carefully (see Section 5). If machining is necessary, follow the information in that Section to remove the disc.

18 Remove the sensor mounting bolt and detach the sensor from the axle housing.

19 Installation is the reverse of the removal procedure. Tighten the sensor mounting bolt to the torque listed in this Chapter's Specifications. Tighten the wheel lug nuts to the torque listed in the Chapter 1 Specifications.

3.2 Wash the disc and caliper with brake system cleaner to remove the brake dust; DO NOT blow off the brake dust with compressed air

Front brake pads

◗ Refer to illustrations 3.4a through 3.4h

4   To replace the front brake pads, follow the accompanying photos, beginning with illustration 3.4a. Be sure to stay in order and read the caption under each illustration. Work on one brake assembly at a time so you'll have something to refer to if you get in trouble.

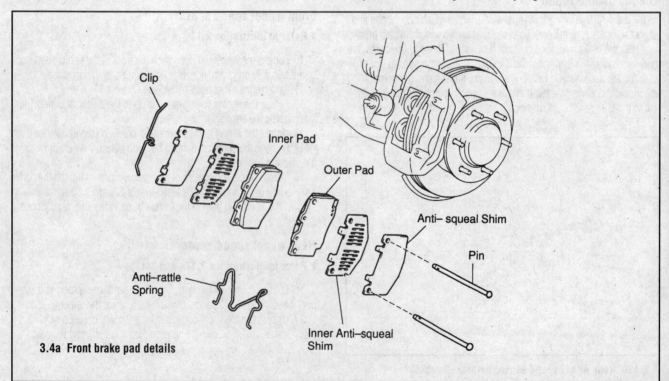

3.4a Front brake pad details

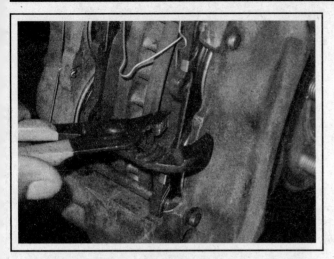

3.4b Squeeze the pads towards the caliper frame to free-up the pads and depress the pistons into their bores.

➡Note: Replace one pad at a time to prevent the pistons on the other side of the caliper from coming out

3.4c Detach the pin retaining clip

3.4d Pull out the pad retaining pins

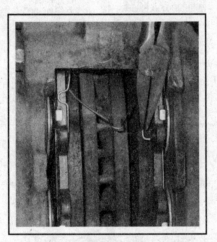

3.4e Remove the anti-rattle spring

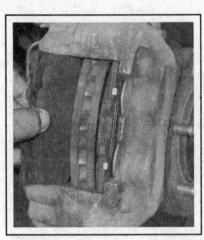

3.4f Remove the inner brake pad

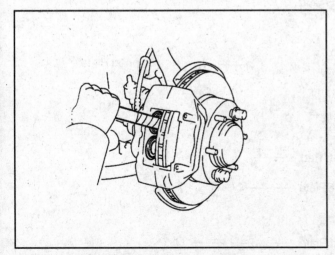

3.4g Push the inner pistons back into their bores to provide room for the new pad

3.4h Apply anti-squeal compound to the back of the new pads, then stick the anti-squeal shims to them. Install the new inner pad, then replace the outer pad the same way. Reinstall the pad retaining pins, anti-rattle spring and the pin retaining clip, then proceed to Step 13.

REAR BRAKE PADS (SEQUOIA MODELS)

▶ **Refer to illustrations 3.5 and 3.6**

5   Using a C-clamp, depress the caliper piston into its bore to make room for the new, thicker pads (see illustration).

6   Remove the caliper mounting bolts (see illustration), then hang the caliper with a piece of wire - DON'T let it hang by the brake hose.

7   Remove the brake pads from the caliper mounting bracket.

8   Remove the pad support plates from the caliper mounting bracket and check them for damage. If they aren't cracked and they fit tightly, they can be reused.

9   Apply anti-squeal compound to the back of the new pads, then stick the anti-squeal shims to them (see illustration 3.4h).

10 Check the bushings and boots in the caliper mounting bracket and replace them if necessary.

11 Install the brake pads in the caliper mounting bracket, then install the caliper over them.

12 Before installing the caliper mounting bolts, clean and check them for corrosion and damage. If significantly corroded or damaged, replace them. Lubricate the sliding surfaces of the bolts with high-temperature brake grease, then install and tighten them to the torque listed in this Chapter's Specifications.

FRONT OR REAR BRAKE PADS

13 Install the brake pads on the opposite wheel, then install the wheels and lower the vehicle. Tighten the lug nuts to the torque listed

3.5 To make room for the new pads, use a C-clamp to depress the piston into its bore before removing the caliper (typical)

in the Chapter 1 Specifications.

14 Add brake fluid to the reservoir until it's full (see Chapter 1). Pump the brakes several times to seat the pads against the disc, then check the fluid level again.

15 Check the operation of the brakes before driving the vehicle in traffic. Try to avoid heavy brake applications until the brakes have been applied lightly several times to seat the pads.

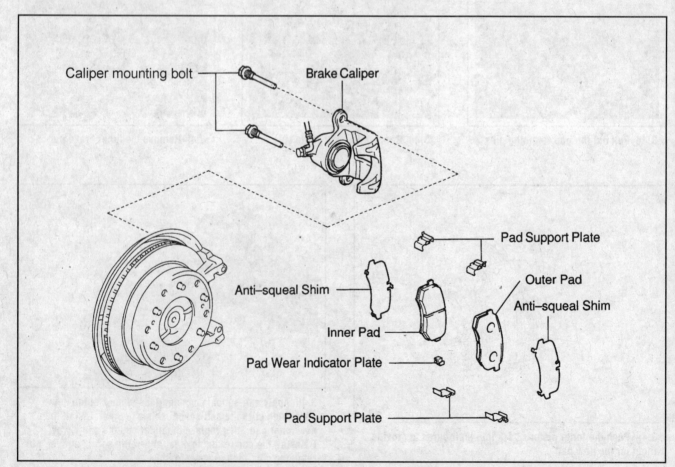

Caliper mounting bolt — Brake Caliper

Pad Support Plate

Anti-squeal Shim

Outer Pad

Anti-squeal Shim

Inner Pad

Pad Wear Indicator Plate

Pad Support Plate

3.6 Rear brake pad details (Sequoia models)

4 Disc brake caliper - removal and installation

➡Note: If caliper replacement is indicated (usually because of fluid leaks, a stuck piston or broken bleeder screw) explore your options. New and factory rebuilt calipers are available on an exchange basis.

1 Loosen the front wheel lug nuts, raise the front of the vehicle and support it securely on jackstands. Apply the parking brake. Remove the wheels. Position a drain pan under the brake assembly and clean the caliper and surrounding area with brake system cleaner (see illustration 3.2).

2 Remove about two-thirds of the fluid from the master cylinder reservoir and discard it; as the pistons are pushed in for clearance to allow the pads to be removed, the fluid will be forced back into the reservoir. Position a drain pan under the brake assembly and clean the caliper and surrounding area with brake system cleaner.

FRONT CALIPER

▶ Refer to illustration 4.3

3 If you're removing a front caliper, unscrew the tube nut fitting and detach the brake line from the caliper, then remove the caliper mounting bolts (see illustration).

➡Note: Use a flare-nut wrench, if available, to prevent rounding-off the corners of the fitting.

If the caliper can't be pulled off, the brake pads are hanging up on the ridge around the circumference of the disc; remove the brake pads (see Section 3).

REAR CALIPER

▶ Refer to illustrations 4.5a and 4.5b

4 Push the piston back into the bore with a C-clamp to provide

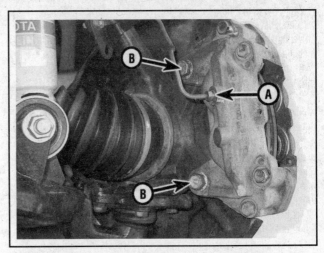

4.3 To remove a front caliper, unscrew the hydraulic line fitting (A) with a flare-nut wrench, then remove the caliper mounting bolts (B)

room for the new brake pads (see illustration 3.5). As the piston is depressed to the bottom of the caliper bore, the fluid in the master cylinder will rise. Make sure it doesn't overflow. If necessary, siphon off some of the fluid.

5 Remove the brake hose-to-caliper union bolt (see illustration).

➡Note: If you're removing the caliper for access to other components, leave the hose connected.

Discard the two sealing washers on each side of the banjo fitting; use new ones when you reattach the brake hose to the caliper. Plug the banjo fitting with a piece of rubber hose (see illustration).

6 Remove the caliper mounting bolts and detach the caliper from the mounting bracket.

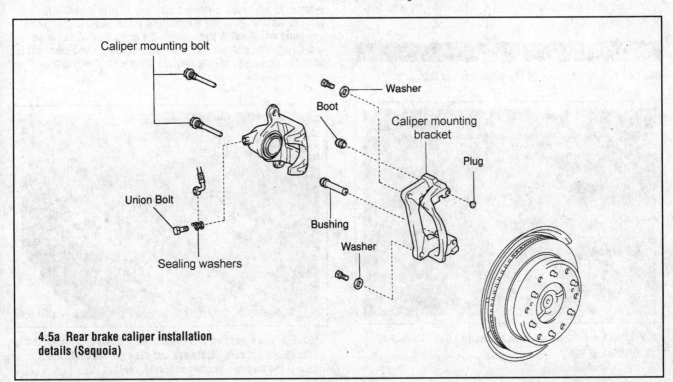

4.5a Rear brake caliper installation details (Sequoia)

FRONT OR REAR CALIPER

7   If you're planning to install the same caliper, clean the caliper with brake system cleaner. DO NOT use kerosene or petroleum-based solvents. Carefully inspect the caliper for leaks and damage. DO NOT install a caliper that is leaking or damaged.

8   Install the caliper and tighten the caliper mounting bolts to the torque listed in this Chapter's Specifications.

9   Connect the brake hose or line to the caliper; use new sealing washers if you're installing a rear caliper and tighten the union bolt to the torque listed in this Chapter's Specifications.

10 Bleed the brake system (see Section 11). If you capped or plugged the line or hose and not much fluid was lost, you'll probably only have to bleed the circuit to the caliper that was removed. If the brake fluid hose was not disconnected (caliper removed for access to other components), the brakes will not require bleeding.

11 Install the wheel and lug nuts, lower the vehicle and tighten the lug nuts to the torque listed in the Chapter 1 Specifications. Check brake operation carefully before placing the vehicle into service.

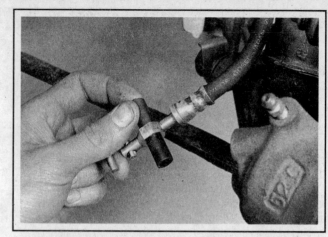

4.5b Using a piece of hose of the appropriate size, plug the banjo fitting to prevent brake fluid from dripping out of the hose and to prevent contaminants from entering the brake system

5   Brake disc - inspection, removal and installation

INSPECTION

▸ **Refer to illustrations 5.2, 5.3, 5.4a, 5.4b, 5.5a, 5.5b and 5.7**

1   Loosen the wheel lug nuts, raise the front of the vehicle and support it securely on jackstands. Apply the parking brake. Remove the wheel. Reinstall the lug nuts to hold the disc in place against the hub (washers may have to be used to allow the nuts to apply pressure to the disc).

2   Remove the brake caliper and hang it with a length of wire (don't disconnect the brake line or hose from the caliper) (see illustration). If you're removing a front caliper, unclip the brake line from the bracket on the steering knuckle.

✳✳ CAUTION:

Don't let the caliper hang by the brake hose or line.

If you're checking a rear disc, remove the brake pads (see Section 3).

3   Visually inspect the disc surface for score marks and other damage (see illustration). Light scratches and shallow grooves are normal after use and won't affect brake operation. Deep grooves require disc removal and refinishing by an automotive machine shop. Be sure to check both sides of the disc.

4   To check disc runout, place a dial indicator at a point about 1/2-inch from the outer edge of the disc (see illustration). Set the indicator to zero and turn the disc. The indicator reading should not exceed the allowable runout listed in this Chapter's Specifications. If it does, the disc should be refinished by an automotive machine shop.

➡**Note: To produce a smooth finish and ensure a perfectly smooth surface - thereby eliminating brake pedal pulsation or any other undesirable symptoms - the discs should be resurfaced regardless of the dial indicator reading. If you elect not to have the discs resurfaced, deglaze them with sandpaper or emery cloth (see illustration).**

5.2 Hang the caliper with a piece of wire - don't let it hang by the hose or brake line!

5.3 The brake pads on this vehicle were obviously neglected - they wore out completely and cut deep grooves into the disc; if the disc is worn this severely, replace it

5.4a Measure the brake disc runout with a dial indicator; if the reading exceeds the maximum allowable runout limit, the disc must be resurfaced or replaced

5.4b Using a swirling motion, remove the glaze from the disc surface with sandpaper or emery cloth

5.5a Measure the brake disc thickness at several points with a micrometer

5.5b The minimum allowable thickness dimension is cast into the back side of the disc

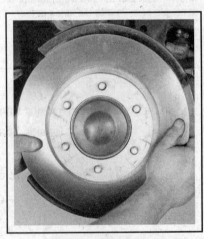

5.7 To detach the disc from the hub, simply pull it off

5   The disc must not be machined to a thickness less than the specified minimum thickness, which is cast into the disc. Measure disc thickness with a micrometer (see illustrations).

REMOVAL AND INSTALLATION

6   Remove the brake caliper, if not already done (see Section 4). If you're removing a rear disc, also remove the caliper mounting bracket.

7   Remove the lug nuts that were installed in Step 1 and pull off the disc (see illustration).

8   Installation is the reverse of removal. Be sure to tighten all brake fasteners to the torque values listed in this Chapter's Specifications. Tighten the wheel lug nuts to the torque listed in the Chapter 1 Specifications.

---

6   Drum brake shoes - replacement

⬦ **Refer to illustrations 6.2, 6.3, 6.4a through 6.4x, 6.5, 6.7a and 6.7b**

❊❊ WARNING:

Brake shoes must be replaced on both wheels at the same time - never replace the shoes on only one wheel. Also, brake system dust is hazardous to your health. DO NOT blow it out with compressed air and DO NOT inhale it. DO NOT use gasoline or solvent to remove the dust. Use brake system cleaner only.

1   Loosen the wheel lug nuts, raise the rear of the vehicle and support it securely on jackstands. Block the front wheels to keep the vehicle from rolling off the stands. Remove the rear wheels.

2   Remove the brake drum. If the drum is stuck because of corrosion between the axle flange and the wheel studs or brake drum, spray penetrating oil around the flange and studs and allow it to soak in. Tap around the studs and flange with a hammer to break the drum loose, then tap around the back edge of the drum to remove it. If the drum is still stuck, screw a couple of bolts of the proper diameter and thread pitch into the threaded holes in the drum and tighten them (see illus-

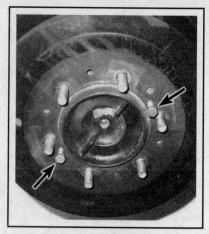

6.2 If the brake drum is "frozen" onto the wheel studs, it can usually be loosened by applying penetrant to the studs; if that doesn't work, thread a couple of bolts into the holes in the drum and tighten them until they push the drum off

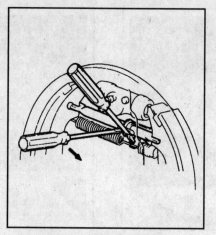

6.3 If the drum binds on the shoes, remove the adjusting hole plug, push the adjuster lever off the star wheel with a screwdriver and turn the star wheel with another screwdriver to retract the shoes

6.4a Wash the brake assembly with brake cleaner; DO NOT blow it out with compressed air!

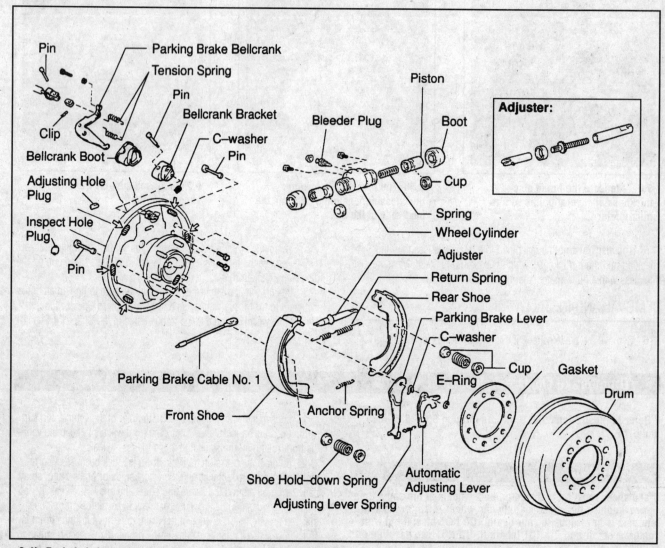

6.4b Exploded view of the drum brake assembly (Tundra models)

6.4c Disengage the return spring from the top of the rear shoe

6.4d Remove the hold-down spring from the rear shoe. To remove a hold-down spring, place the hold-down spring tool (available at most auto parts stores) over the hold-down spring, push down and give it a quarter-turn and release

6.4e Disengage the shoe and the anchor spring

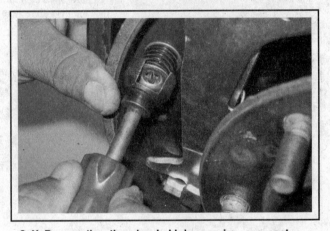

6.4f Remove the other shoe hold-down spring, cup, seat and pin

6.4g Disengage the adjusting lever spring from the adjusting lever

6.4h Remove the adjuster assembly

6.4i Pry off the E-ring that secures the automatic adjusting lever to the pin on the shoe

tration). When they come in contact with the axleshaft flange, they'll push the drum off.

3   If the drum is locked onto the shoes because of excessive drum wear, you'll have to retract the shoes (see illustration).

4   Clean the brake assembly with brake system cleaner before beginning work. Follow illustrations 6.4a through 6.4x for the inspec-

tion and replacement of the brake shoes. Be sure to stay in order and read the caption under each illustration.

5   Clean the brake drum and check it for score marks, deep grooves, hard spots (which will appear as small discolored areas) and cracks. If the drum is worn, scored or out-of-round, it can be resurfaced by an automotive machine shop.

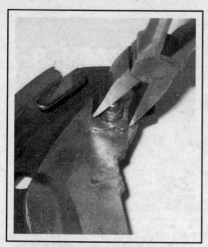

6.4j Remove the C-clip that secures the parking brake lever to the pin on the shoe

6.4k Drive out the pivot pin from the old brake shoe and drive it into the new shoe (if a new pin comes with the new shoe, use it instead)

6.4l After installing the parking brake lever on the new brake shoe, squeeze the ends of the new C-clip together

6.4m Install the automatic adjusting lever on the pivot pin, then secure it with a new E-ring

6.4n Install the adjusting lever spring

6.4o Lubricate the brake shoe contact points on the backing plate with high-temperature brake grease

6.4p Clean the adjuster screw and lubricate its moving parts with a light film of high-temperature brake grease

6.4q Attach the adjuster screw assembly

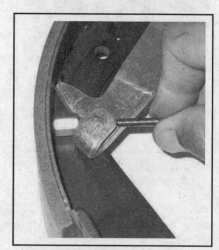

6.4r Attach the short parking brake cable to the lever on the shoe

6.4s Position the shoe/parking brake lever assembly on the backing plate and install the hold-down pin, spring, seat and retainer. Also attach the short parking brake cable to the bellcrank

6.4t Connect the anchor spring between the bottoms of the shoes . . .

6.4u . . . install the shoe on the backing plate, engaging the adjuster with the notch in the shoe, then install the hold-down spring and retainer

6.4v Connect the return spring between the tops of the two shoes; make sure the space between the short and long coils is positioned over the adjuster screw star wheel

6.4w Center the shoes on the backing plate, making sure the upper ends of the shoes engage the wheel cylinder pistons, and the adjuster assembly engages the slots in the shoes

6.4x This is how the assembled brake should look

6.5 The maximum diameter of the rear drum is cast into the inside of the drum

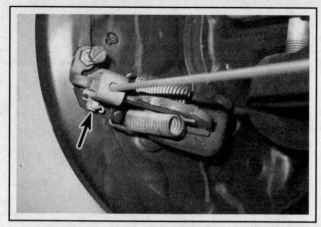

6.7a To disconnect the parking brake cable clevis from the bellcrank, remove the cotter pin and slide the clevis pin out

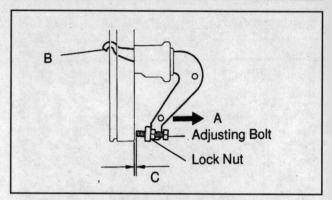

6.7b To adjust the parking brake cable bellcrank, pull the parking brake cable bellcrank in direction A until there is no clearance at B, loosen the bellcrank adjusting bolt locknut and turn the adjusting bolt in or out to bring dimension C within the range listed in this Chapter's Specifications

➡Note: Professionals recommend resurfacing the drums whenever a brake job is done. Resurfacing will eliminate the possibility of out-of-round drums. If the drums are worn so much they can't be resurfaced without exceeding the maximum allowable diameter (stamped into the drum) (see illustration), new ones will be required. At the very least, if you elect not to have the drums resurfaced, remove the glazing from the surface with sandpaper or emery cloth using a swirling motion.

6 Repeat Steps 2 through 5 for the other rear brake assembly.

7 Adjust the parking brake cable bellcrank. Disconnect the parking brake cable from the bellcrank (see illustration). Lightly pull the parking brake cable bellcrank in direction A until there is no clearance at B (see illustration), loosen the bellcrank adjusting bolt locknut and turn the adjusting bolt in or out to bring dimension C within the range listed in this Chapter's Specifications. Tighten the adjusting bolt locknut. Connect the parking brake cable clevis to the bellcrank with the clevis pin and install the clevis pin retainer clip. Hook the tension spring to the bellcrank lever. Repeat this procedure at the other rear brake.

8 On all models, move the parking brake lever back and forth a few times and verify that the adjuster screw turns. Repeat this procedure for the other brake.

9 Install the brake drums. Pump the brake several times, then turn the adjuster star wheel using a screwdriver inserted through the hole in the backing plate until the shoes slightly drag on the drums as the drums are turned. Now, back off the adjuster until the shoes don't drag on the drums. Install the plug in the brake backing plate.

10 Install the rear wheels, install the lug nuts, lower the vehicle and tighten the wheel lug nuts to the torque listed in the Chapter 1 Specifications.

11 Check the brake pedal position. If the brake pedal goes too close to the floor, further adjustment of the brakes is required. Back the vehicle up, making repeated stops, to actuate the self-adjusters, which work only when the vehicle is in reverse. Test the brakes for proper operation before driving in traffic.

7 Wheel cylinder - removal and installation

▸ Refer to illustration 7.2

➡Note: Never replace only one wheel cylinder. Always replace both of them at the same time.

1 Remove the brake drum and brake shoes (see Section 6).

2 Unscrew the brake line fitting from the rear of the wheel cylinder (see illustration). If available, use a flare-nut wrench to avoid rounding off the corners on the fitting. Don't pull the metal line out of the wheel cylinder - it could bend, making installation difficult.

3 Remove the two bolts securing the wheel cylinder to the brake backing plate.

4 Remove the wheel cylinder.

5 Plug the end of the brake line to prevent the loss of brake fluid and the entry of dirt.

6 Installation is the reverse of removal. Attach the brake line to the wheel cylinder before installing the mounting bolts and tighten the line fitting after the wheel cylinder mountings bolts have been tightened. If available, use a flare-nut wrench to tighten the line fitting.

7 Install the brake shoes and brake drum (see Section 6).

8 Bleed the brakes (see Section 11). Don't drive the vehicle in traffic until the operation of the brakes has been thoroughly tested.

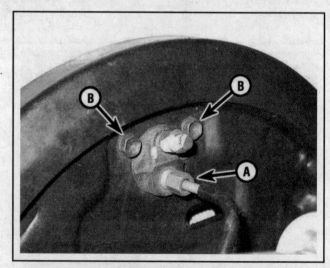

7.2 To remove the wheel cylinder, unscrew the brake line threaded fitting (A) with a flare-nut wrench, then remove the wheel cylinder mounting bolts (B)

8 Master cylinder - removal and installation

REMOVAL

▶ **Refer to illustration 8.2**

1 Place rags under the brake line fittings and prepare caps or plastic bags to cover the ends of the lines once they're disconnected.

✳✳ CAUTION:

Brake fluid will damage paint. Cover all painted surfaces and avoid spilling fluid during this procedure.

2 Unplug the brake fluid level warning switch electrical connector (see illustration). On Sequoia models, also unplug the electrical connectors for the master cylinder pressure sensors, which are threaded into the bottom of the cylinder.

3 Loosen the tube nuts at the ends of the brake lines where they enter the master cylinder. To prevent rounding off of the flats on these nuts, a flare-nut wrench, which wraps around the nut, should be used. Pull the brake lines away from the master cylinder slightly and plug the ends to prevent contamination. Be careful not to kink the hydraulic lines.

4 Remove the master cylinder mounting nuts (see illustration 8.2). Remove the master cylinder from the vehicle.

5 Remove the reservoir cap and discard any fluid remaining in the reservoir.

6 On Sequoia models, check the O-ring on the end of the master cylinder, replacing it if it is cracked or hardened.

INSTALLATION

▶ **Refer to illustrations 8.8 and 8.17**

✳✳ WARNING:

If you're replacing the master cylinder assembly or master cylinder pressure sensor(s) on a Sequoia model, the zero point calibration of the steering angle, master cylinder pressure, yaw rate and deceleration sensors for the Vehicle Skid Control (VSC) system must be performed. This will require taking the vehicle to a dealer service department or other qualified repair shop equipped with the special tool necessary to carry out this procedure. The brake hydraulic system will work as a conventional braking system, but the ABS, Traction Control and VSC system may not function properly.

➡Note: Before installing a new or rebuilt master cylinder on a Tundra model, check the clearance between the booster pushrod and the pocket in the master cylinder piston. If necessary, adjust the length of the power brake booster pushrod (see Section 12).

7 Bench bleed the master cylinder before installing it. Because it will be necessary to apply pressure to the master cylinder piston and, at the same time, control flow from the brake line outlets, it is recommended that the mater cylinder be mounted in a vise, with the jaws of the vise clamping on the mounting flange.

8 Attach a pair of bleeder tubes (available at most auto parts stores) to the outlet ports of the master cylinder (see illustration).

9 Fill the reservoir with brake fluid of the recommended type (see Chapter 1).

10 Slowly push the pistons into the master cylinder (a large Phillips screwdriver can be used for this) - air will be expelled from the pressure chambers and into the reservoir. Because the tubes are submerged in fluid, air can't be drawn back into the master cylinder when you release the pistons.

11 Repeat the procedure until no more air bubbles are present.

12 Remove the bleed tubes, one at a time, and install plugs in the open ports to prevent fluid leakage and air from entering. Install the reservoir cap.

13 Install the master cylinder over the studs on the power brake booster and tighten the nuts only finger-tight at this time. Don't forget to use a new gasket.

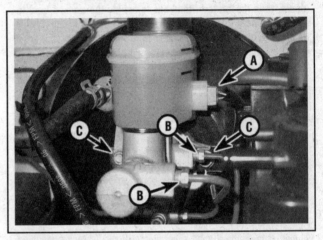

8.2 Master cylinder mounting details (Tundra shown; Sequoia models have two master cylinder pressure sensor electrical connectors on the underside of the cylinder, in addition to the items shown here)

A   *Fluid level sensor electrical connector*
B   *Brake line fittings*
C   *Mounting nuts*

8.8 The best way to bleed air from the master cylinder before installing it on the vehicle is with a pair of bleeder tubes that direct brake fluid into the reservoir during bleeding

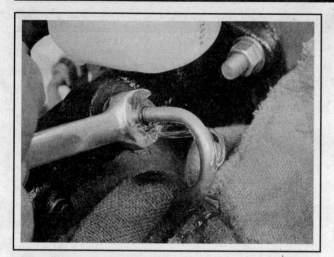

8.17 Have an assistant depress the brake pedal and hold it down, then loosen the fitting nut, allowing the air and fluid to escape; repeat this procedure on both fittings until the fluid is clear of air bubbles

14 Thread the brake line fittings into the master cylinder. Since the master cylinder is still a bit loose, it can be moved slightly so the fittings thread in easily. Don't strip the threads as the fittings are tightened.

15 Tighten the mounting nuts to the torque listed in this Chapter's Specifications. Tighten the brake line fittings securely.

16 Plug in the electrical connector to the fluid level warning switch. On Sequoia models, also plug in the electrical connectors to the master cylinder pressure sensors.

17 Fill the master cylinder reservoir with fluid, then bleed the master cylinder and the brake system (see Section 11). To bleed the master cylinder on the vehicle, have an assistant depress the brake pedal and hold it down. Loosen the fitting to allow air and fluid to escape (see illustration). Tighten the fitting, then allow your assistant to return the pedal to its rest position. Repeat this procedure on both fittings until the fluid is free of air bubbles. Check the operation of the brake system carefully before driving the vehicle.

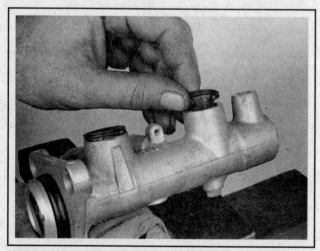

8.22 After the reservoir has been removed, pull the grommets from the master cylinder body; if they're hard, cracked or damaged, or have been leaking, replace them

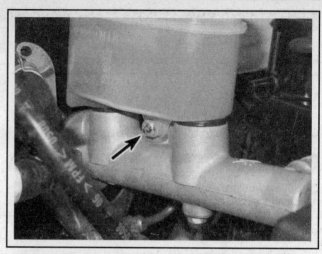

8.20 The master cylinder fluid reservoir is secured by a single screw

RESERVOIR/GROMMET REPLACEMENT

◆ **Refer to illustrations 8.20 and 8.22**

➡**Note:** The brake fluid reservoir can be replaced separately from the master cylinder body if it becomes damaged. If there is leakage between the reservoir and the master cylinder body, the grommets on the reservoir can be replaced.

18 Remove as much fluid as possible from the reservoir with a suction gun, large syringe or a poultry baster.

❋❋ **WARNING:**

If a poultry baster is used, never again use it for the preparation of food.

19 Place rags under the master cylinder to absorb any fluid that may spill out once the reservoir is detached from the master cylinder.

❋❋ **CAUTION:**

Brake fluid will damage paint. Cover all body parts and be careful not to spill fluid during this procedure.

20 Remove the screw that retains the reservoir to the master cylinder (see illustration).

21 Pull the reservoir out of the master cylinder body.

22 Pull the grommets out of the master cylinder (see illustration).

23 Lubricate the new grommets with clean brake fluid, then press them into place.

24 Push the reservoir into the grommets and secure it with the screw.

25 Refill the reservoir with the recommended brake fluid (see Chapter 1) and check for leaks.

26 Bleed the master cylinder (see illustration 8.17), followed by the remainder of the system (see Section 11).

9 Brake hoses and lines - check and replacement

1   About every six months, with the vehicle raised and placed securely on jackstands, the flexible hoses which connect the steel brake lines with the front and rear brake assemblies should be inspected for cracks, chafing of the outer cover, leaks, blisters and other damage. These are important and vulnerable parts of the brake system and inspection should be complete. A light and mirror will be needed for a thorough check. If a hose exhibits any of the above defects, replace it with a new one.

FLEXIBLE HOSES

▶ **Refer to illustrations 9.3 and 9.4**

2   Clean all dirt away from the ends of the hose.
3   Disconnect the brake line from the hose fitting (see illustration). Be careful not to bend the frame bracket or line. If necessary, soak the connections with penetrating oil.
4   Remove the U-clip from the female fitting at the bracket (see illustration) and remove the hose from the bracket.
5   If you're removing a rear brake hose on a Sequoia, disconnect the hose from the caliper, discarding the copper washers on either side of the fitting. Using new copper washers, attach the new brake hose to the caliper.
6   Pass the female fitting through the frame or frame bracket. With the least amount of twist in the hose, install the fitting in this position.
➡**Note: The weight of the vehicle must be on the suspension, so the vehicle should not be raised while positioning the hose.**
7   Install the U-clip in the female fitting at the frame bracket.

8   Attach the brake line to the hose fitting using a back-up wrench on the fitting. Tighten the tube nut securely.
9   Carefully check to make sure the suspension or steering components don't make contact with the hose. Have an assistant push down on the vehicle and also turn the steering wheel lock-to-lock during inspection.
10 Bleed the brake system as described in Section 11.

METAL BRAKE LINES

11 When replacing brake lines, be sure to use the correct parts. Don't use copper tubing for any brake system components. Purchase steel brake lines from a dealer parts department or auto parts store.
12 Prefabricated brake lines, with the ends already flared and fittings installed, are available at auto parts stores and dealer service departments. If necessary, carefully bend the line to the proper shape. A tube bender is recommended for this.

✳✳ CAUTION:

Don't crimp or damage the line.

13 When installing the new line make sure it's well supported in the brackets and has plenty of clearance between moving or hot components.
14 After installation, check the master cylinder fluid level and add fluid as necessary. Bleed the brake system as outlined in Section 11 and test the brakes carefully before placing the vehicle into normal operation.

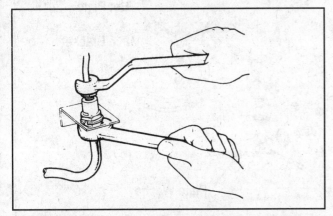

9.3 Hold the hose fitting with a wrench to prevent twisting the line, then loosen the tube nut with a flare-nut wrench to prevent rounding off the corners of the nut

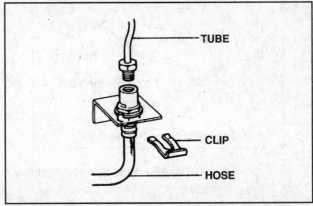

9.4 Once the tube nut has been completely loosened, pull off the clip with a pair of pliers or pry it off with a screwdriver

10 Load sensing proportioning and bypass valve - description, removal and installation

DESCRIPTION

▶ **Refer to illustration 10.1**

1   The center of gravity (cg) is higher on a truck than a car, and the majority of an unloaded truck's weight is up front. When the brakes are

applied, the combination of higher cg and forward weight bias tends to "pitch" the vehicle onto its front tires, simultaneously unloading the rear end. The front brakes therefore require more hydraulic pressure, and the rear brakes require less (or else the rear brakes will lock under heavy braking). This is also true of a car, which is equipped with a proportioning valve to "bias" braking power to the front brakes. But on a truck the forces involved are more extreme, and they change in proportion to the load in the bed; a loaded truck has more weight on the rear end than an unloaded truck. So, Tundra models are equipped with a load sensing proportioning and bypass valve (LSP & BV) which alters front/rear braking bias in accordance with the weight of the load (see illustration).

2 Because of the special gauge set needed to check and adjust the LSP & BV, servicing this unit at home is not recommended. If the LSP & BV malfunctions, you can replace it yourself, but you'll have to take the vehicle to a dealer service department or other qualified repair shop to have the unit adjusted. Also, to remove the rear axle, you must disconnect the LSP & BV height sensing rod from the axle. As long as the adjustment isn't changed, disconnecting and reattaching the height sensing rod should not affect the function of the LSP & BV.

REMOVAL AND INSTALLATION

▶ Refer to illustrations 10.4 and 10.8

❊❊ WARNING:

If replaced, the valve must be checked and, if necessary, adjusted at a dealer service department or other qualified repair shop equipped with the necessary gauges.

10.1 The load sensing proportioning and bypass valve (LSP & BV) is mounted to a bracket on the left frame rail, near the top of the left rear shock absorber

❊❊ CAUTION:

Brake fluid will damage paint. Cover all painted surfaces and avoid spilling fluid during this procedure.

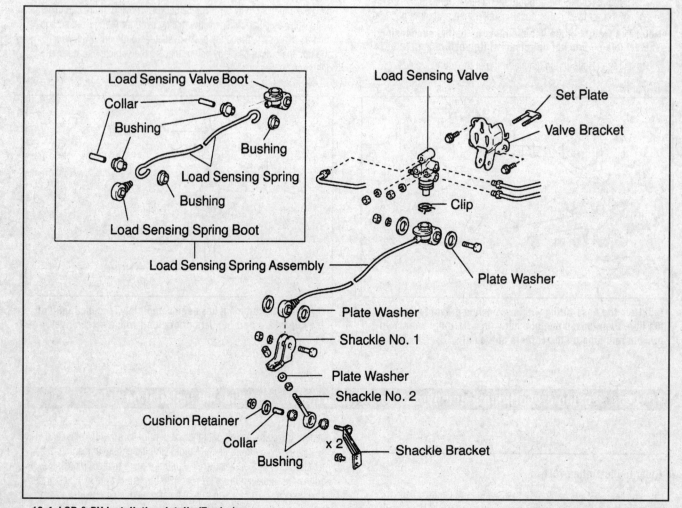

10.4 LSP & BV installation details (Tundra)

3   Raise the vehicle and support it securely on jackstands.

4   Disconnect the height sensing spring shackle No. 2 from the axle (see illustration).

5   Loosen the tube nuts that connect the brake lines to the LSP& BV valve. Use a flare-nut wrench to prevent rounding off of the flats on these nuts. Pull the brake lines away from the LSP & BV slightly and plug the ends of the lines to prevent contamination.

6   Remove the LSP & BV bracket mounting bolts and remove the LSP & BV and bracket assembly.

7   Unbolt the LSP & BV from the bracket. Disconnect the height sensing spring from the LSP & BV.

8   Installation is the reverse of removal. Be sure to tighten the LSP & BV mounting bolts to the torque listed in this Chapter's Specifications. Set the initial length of shackle no. 2 to the value listed in this Chapter's Specifications (see illustration).

9   Bleed the brake hydraulic system (see Section 11).

10 Have the LSP & BV adjusted at a dealer service department or other qualified repair shop.

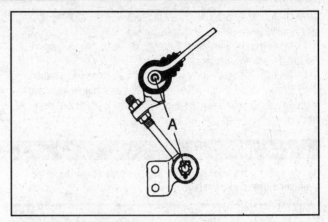

10.8 If you're installing a new LSP & BV, adjust this initial set length to the value listed in this Chapter's Specifications

11   Brake hydraulic system - bleeding

▶ **Refer to illustrations 11.8 and 11.12**

✳✳ WARNING:

Wear eye protection when bleeding the brake system. If the fluid comes in contact with your eyes, immediately rinse them with water and seek medical attention.

➡**Note: Bleeding the brake system is necessary to remove any air that's trapped in the system when it's opened during removal and installation of a hose, line, caliper, wheel cylinder or master cylinder.**

1   It will probably be necessary to bleed the system at all four brakes if air has entered the system due to low fluid level, or if the brake lines have been disconnected at the master cylinder.

2   If a brake line was disconnected only at a wheel, then only that caliper or wheel cylinder must be bled.

3   If a brake line is disconnected at a fitting located between the master cylinder and any of the brakes, that part of the system served by the disconnected line must be bled.

4   Remove any residual vacuum from the brake power booster by applying the brake several times with the engine off.

5   Remove the master cylinder reservoir cap and fill the reservoir with brake fluid. Reinstall the cap.

➡**Note: Check the fluid level often during the bleeding operation and add fluid as necessary to prevent the fluid level from falling low enough to allow air bubbles into the master cylinder.**

6   Have an assistant on hand, as well as a supply of new brake fluid, an empty clear plastic container, a length of clear tubing to fit over the bleeder valve and a wrench to open and close the bleeder valve.

7   Beginning at the right rear wheel, loosen the bleeder screw slightly, then tighten it to a point where it's snug but can still be loosened quickly and easily.

8   Place one end of the tubing over the bleeder screw fitting and submerge the other end in brake fluid in the container (see illustration).

9   Have the assistant slowly push down on the brake pedal, then hold the pedal firmly depressed.

10 While the pedal is held depressed, open the bleeder screw just enough to allow a flow of fluid to leave the valve. Watch for air bubbles to exit the submerged end of the tube. When the fluid flow slows after a couple of seconds, tighten the screw and have your assistant release the pedal slowly.

11 Repeat Steps 9 and 10 until no more air is seen leaving the tube, then tighten the bleeder screw and proceed to the left rear wheel, the right front wheel and the left front wheel, in that order, and perform the same procedure. Be sure to check the fluid in the master cylinder reservoir frequently.

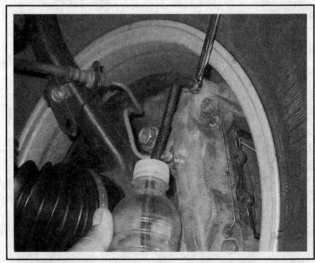

11.8 When bleeding the brakes, a hose is connected to the bleed screw at the caliper or wheel cylinder and then submerged in brake fluid - air will be seen as bubbles in the tube and container (all air must be expelled before moving to the next wheel)

12 On Tundra models, bleed the load sensing proportioning and bypass valve in the same manner (see illustration).

13 Never use old brake fluid. It contains moisture which can boil, rendering the brakes useless.

14 Refill the master cylinder with fluid at the end of the operation.

15 Check the operation of the brakes. The pedal should feel solid when depressed, with no sponginess. If necessary, repeat the entire process.

❊❊ WARNING:

Do not operate the vehicle if you are in doubt about the effectiveness of the brake system.

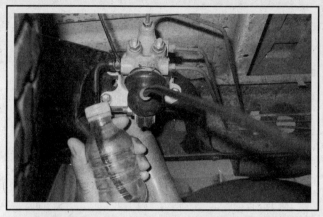

11.12 Bleeding the load sensing proportioning and bypass valve (Tundra models only)

12 Power brake booster - check, removal and installation

OPERATING CHECK

1 Depress the brake pedal several times with the engine off and make sure there's no change in the pedal reserve distance.

2 Depress the pedal and start the engine. If the pedal goes down slightly, operation is normal.

AIRTIGHTNESS CHECK

3 Start the engine and turn it off after one or two minutes. Depress the brake pedal slowly several times. If the pedal depresses less each time, the booster is airtight.

4 Depress the brake pedal while the engine is running, then stop the engine with the pedal depressed. If there's no change in the pedal reserve travel after holding the pedal for 30 seconds, the booster is airtight.

REMOVAL

▶ **Refer to illustrations 12.10 and 12.11**

5 Power brake booster units shouldn't be disassembled. They require special tools not normally found in most automotive repair sta-

tions or shops. Because of its critical relationship to brake performance, the booster should be replaced with a new or rebuilt one.

6 Disconnect the hose leading from the engine to the booster. Be careful not to damage the hose when removing it from the booster fitting.

7 Remove the brake master cylinder (see Section 8).

8 Remove the left side lower finish panel from the instrument panel (see Chapter 11).

9 Remove the pedal return spring.

10 Locate the pushrod clevis connecting the booster to the brake pedal (see illustration). Remove the clevis pin retaining clip with pliers and pull out the pin.

11 Remove the four nuts holding the brake booster to the firewall (see illustration); you may need a light to see them. Slide the booster straight out from the firewall until the studs clear the holes.

12 On Sequoia models, remove the nuts and detach the bracket from the booster.

INSTALLATION

▶ **Refer to illustrations 12.14a, 12.14b, 12.14c, 12.14d and 12.14e**

13 Installation procedures are basically the reverse of removal. Tighten the clevis locknut securely (if loosened or removed) and the

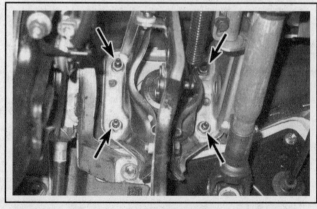

12.10 To disconnect the power brake booster pushrod from the brake pedal, remove the retaining clip and slide out the clevis pin

12.11 To detach the booster from the firewall, remove these four nuts

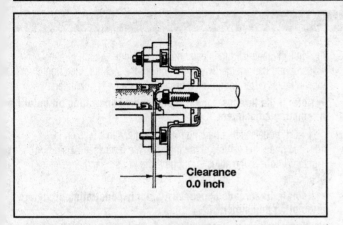

12.14a There should be no clearance between the booster pushrod and the master cylinder pushrod, but no interference either; if there is interference between the two, the brakes may drag; if there is clearance, there will be excessive brake pedal travel

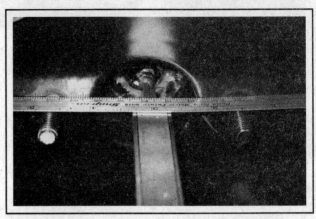

12.14b Measure the distance from the master cylinder mounting surface to the tip of the booster pushrod (dimension A)

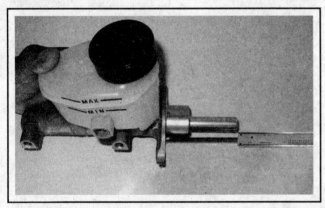

12.14c Measure the distance from the mounting flange to the end of the master cylinder (dimension B)

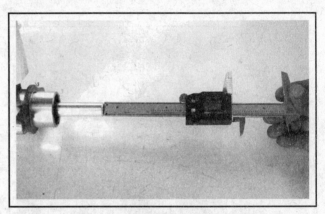

12.14d Measure the distance from the piston pocket to the end of the master cylinder (dimension C)

booster mounting nuts to the torque listed in this Chapter's Specifications.

14 If you're working on a Tundra and a new power brake booster unit is being installed, check the pushrod clearance (see illustration) as follows (this procedure doesn't apply to Sequoia models):

a) Measure the distance from the face of the booster to the tip of the pushrod. Write down this measurement (see illustration). This is "dimension A."

b) Measure the distance from the mounting flange to the end of the master cylinder piston (see illustration). Write down this measurement. This is "dimension B."

c) Measure the distance from the end of the master cylinder piston to the bottom of the pocket in the piston (see illustration). Write down this measurement. This is "dimension C."

d) Subtract measurement C from measurement B, then subtract measurement A from the difference between C and B. This the pushrod clearance.

e) Compare your calculated pushrod clearance to the pushrod clearance listed in this Chapter's Specifications. If necessary, adjust the pushrod length to achieve the correct clearance (see illustration).

15 After the final installation of the master cylinder and brake lines, the system must be bled (see Section 11) and the brake pedal height and freeplay must be checked (see Chapter 1).

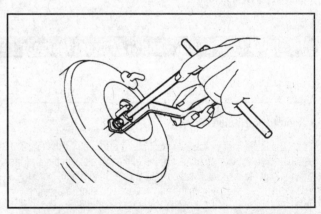

12.14e To adjust the length of the booster pushrod, hold the stationary portion of the rod with a special offset wrench and turn the adjusting screw in or out with a socket, as necessary, to achieve the desired setting

16 Carefully test the operation of the brakes before placing the vehicle in normal operation.

13 Parking brake - adjustment

◆ Refer to illustrations 13.3 and 13.4

1   The parking brake is hand-operated by a handle under the dash on manual transmission Tundra models and by a pedal under the left end of the instrument panel on all other models. All parking brake systems are self-adjusting; automatic adjusters compensate for brake shoe wear. However, additional adjustment may be needed in the event of cable stretch, which occurs as the vehicle ages. The handle or pedal should apply the parking brake within the number of clicks listed in the Chapter 1 Specifications. If the handle or pedal applies the parking brake system in less than the specified number of clicks, the cable is too tight; if it travels more than the specified number of clicks, the cable is too loose.

2   Release the parking brake handle, raise the rear of the vehicle until the wheels are off the ground and support it securely on jackstands. Block the front wheels to prevent the vehicle from rolling.

3   On models with a parking brake pedal, cable tension is adjusted at the pedal (see illustration). Remove the left-side under-dash panel,

the left kick panel and the heater duct. Using one wrench on the adjuster nut, use another wrench to loosen the locknut, then tighten or loosen the adjuster nut until the proper pedal travel is obtained.

➡Note: If the threads appear rusty, apply penetrating oil before attempting adjustment.

4   On models with a parking brake handle, cable tension is adjusted under the vehicle at the intermediate lever (see illustration). Tighten or loosen the adjuster nut until the proper handle travel is obtained.

➡Note: If the threads appear rusty, apply penetrating oil before attempting adjustment.

Verify that the bellcrank stopper screw comes into contact with the backing plate (see illustration 6.7b).

5   Tighten the locknut securely after adjustment. Make sure the rear wheels turn freely when the parking brake is released.

6   Remove the jackstands and lower the vehicle.

13.3 Parking brake cable adjuster nut - models with a pedal-operated parking brake

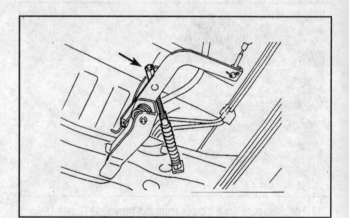

13.4 Parking brake cable adjuster nut - models with a handle-operated parking brake

14 Brake light switch - check and replacement

CHECK

◆ Refer to illustration 14.1

1   The brake light switch (see illustration) is located on the brake pedal bracket. You'll need to remove the trim panel beneath the steering column to get to the switch and connector (see Chapter 11).

2   With the brake pedal in the fully released position, the switch opens the brake light circuit. When the brake pedal is depressed, the switch closes the circuit and sends current to the brake lights.

3   If the brake lights are inoperative, check the fuse and the bulbs (see Chapter 12).

4   If the fuse and bulbs are okay, verify that voltage is available at the switch.

5   If there's no voltage to the switch, search for an open circuit condition between the fuse block and the switch. If there is voltage to the switch, close the switch (depress the brake pedal) and verify that there's voltage on the other side of the switch.

6   If there's no voltage on the other side of the switch, replace the

14.1 The brake light switch is located at the top of the brake pedal assembly; to remove it, unplug the electrical connector (upper arrow), then remove the locknut (lower arrow); the upper nut (middle arrow) is the adjustment nut

switch. If there is voltage but the brake lights still don't work, look for an open circuit condition between the switch and the brake lights.

➡**Note: There is always the remote possibility that all of the brake light bulbs are burned out, but this is not very likely.**

REPLACEMENT

7 Remove the trim panel below the steering column (see Chapter 11).
8 Unplug the electrical connector from the switch.

9   Remove the locknut (the lower nut) from the switch.
10 Remove the switch from its bracket.
11 Installation is the reverse of removal.
12 Adjust the switch when you're done (see Chapter 1, *Clutch/brake pedal height and freeplay check and adjustment*).

Specifications

General

Brake fluid type	See Chapter 1
Brake light switch-to-pedal stopper distance	0.020 to 0.094 inch
Brake pedal height and freeplay	See Chapter 1
Brake pedal reserve distance	3.74 inches minimum
LSP & BV shackle no. 2 initial set length (Tundra)	4.72 inches
Parking brake cable bellcrank dimension C (Tundra)	0.016 to 0.031 inch
Power brake booster pushrod-to-master cylinder piston clearance	0.0 inch

Disc brakes

Brake pad minimum lining thickness	See Chapter 1
Disc lateral runout limit	
Front	0.0028 inch
Rear	0.0039 inch
Disc minimum (discard) thickness	Cast into disc

Drum brakes

Brake shoe minimum lining thickness	See Chapter 1
Maximum drum diameter	Cast into drum

Torque specifications
Ft-lbs (unless otherwise indicated)

➡**Note: One foot-pound (ft-lb) of torque is equivalent to 12 inch-pounds (in-lbs) of torque. Torque values below approximately 15 ft-lbs are expressed in inch-pounds, since most foot-pound torque wrenches are not accurate at these smaller values.**

ABS wheel speed sensor bolt	71 in-lbs
Brake line-to-rear caliper banjo bolt	23
Caliper mounting bolts	
Front	90
Rear	65
Caliper mounting bracket bolts (rear)	77
Load sensing proportioning valve-to-bracket nuts	108 in-lbs
Load sensing proportioning valve bracket-to-frame nuts	108 in-lbs
Master cylinder mounting nuts	
Tundra	108 in-lbs
Sequoia	18
Power brake booster-to-bracket nuts (Sequoia)	18
Power brake booster-to-firewall nuts	108 in-lbs
Wheel cylinder mounting bolts	84 in-lbs
Wheel lug nuts	See Chapter 1

Notes

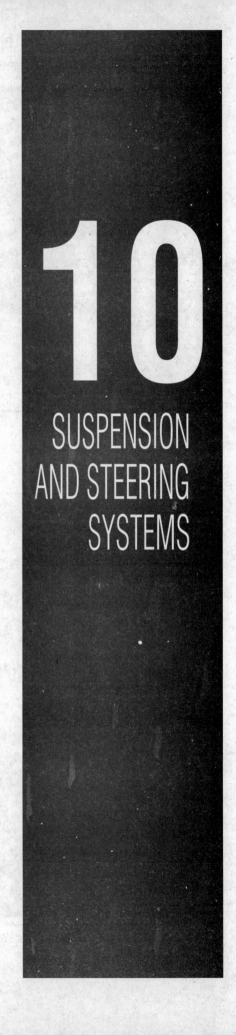

10

SUSPENSION AND STEERING SYSTEMS

Section

Reference to other Chapters

1.1 Front suspension and steering components

1   Upper control arm
2   Shock absorber/coil spring assembly
3   Steering knuckle

4   Lower balljoint
5   Tie-rod end
6   Lower control arm

7   Steering gear
8   Stabilizer bar

1.2 Rear suspension components - Tundra

1   Leaf spring shackle
2   Shock absorber

3   Leaf spring
4   Spring seat

5   Spring hanger
6   Rear axle housing

1 General information

FRONT SUSPENSION

▶ **Refer to illustration1.1**

The front suspension system is fully independent (see illustration). The steering knuckles are connected to the upper and lower control arms by balljoints. The control arms are bolted to the frame. The shock absorbers and coil springs are integral assemblies; the upper ends are bolted to brackets on the frame and the lower ends are bolted to the lower control arms. All models use a front stabilizer bar to reduce vehicle roll during cornering.

REAR SUSPENSION

▶ **Refer to illustrations 1.2 and 1.3**

The rear suspension on Tundra models consists of a pair of multi-leaf springs and two shock absorbers (see illustration). The rear axle assembly is attached to the leaf springs by U-bolts. The front ends of the springs are attached to the frame at the front hangers, through rub-ber bushings. The rear ends of the springs are attached to the frame by shackles which allow the springs to alter their length when the vehicle is in operation.

The rear suspension on Sequoia models (see illustration) consists of a pair of coil springs, two shock absorbers, four suspension arms (two lower, two upper), and a lateral control rod, which connects the right end of the axle to the left frame rail. A stabilizer bar, bolted to the axle and connected to the frame by a pair of links, reduces vehicle roll during cornering.

STEERING

All models are equipped with power-assisted rack-and-pinion steering systems. The steering gear is bolted to the back of the cross-member and is connected to the steering knuckles by a pair of tie-rods.

Frequently, when working on the suspension or steering system components, you may come across fasteners which seem impossible to loosen. These fasteners on the underside of the vehicle are continu-ally subjected to water, road grime, mud, etc., and can become rusted or "frozen," making them extremely difficult to remove. In order to

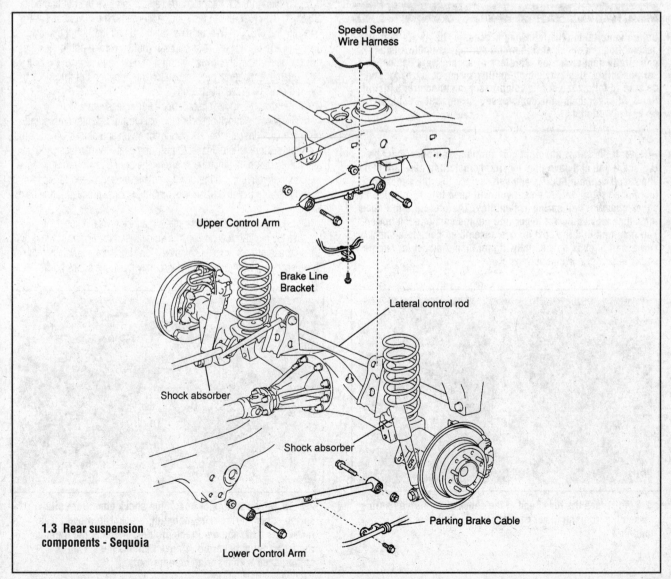

1.3 Rear suspension components - Sequoia

unscrew these stubborn fasteners without damaging them (or other components), be sure to use lots of penetrating oil and allow it to soak in for a while. Using a wire brush to clean exposed threads will also ease removal of the nut or bolt and prevent damage to the threads. Sometimes a sharp blow with a hammer and punch is effective in breaking the bond between a nut and bolt threads, but care must be taken to prevent the punch from slipping off the fastener and ruining the threads. Heating the stuck fastener and surrounding area with a torch sometimes helps too, but isn't recommended because of the obvious dangers associated with fire. Long breaker bars and extension, or "cheater," pipes will increase leverage, but never use an extension pipe on a ratchet - the ratcheting mechanism could be damaged. Sometimes, turning the nut or bolt in the tightening (clockwise) direction first will help to break it loose. Fasteners that require drastic measures to unscrew should always be replaced with new ones.

Since most of the procedures that are dealt with in this Chapter involve jacking up the vehicle and working underneath it, a good pair of jackstands will be needed. A hydraulic floor jack is the preferred type of jack to lift the vehicle, and it can also be used to support certain components during various operations.

❊❊ WARNING:

Never, under any circumstances, rely on a jack to support the vehicle while working on it. Also, whenever any of the suspension or steering fasteners are loosened or removed they must be inspected and, if necessary, replaced with new ones of the same part number or of original equipment quality and design. Torque specifications must be followed for proper reassembly and component retention. Never attempt to heat or straighten suspension or steering components. Instead, replace bent or damaged parts with new ones.

2   Shock absorber/coil spring (front) - removal, component replacement and installation

▶ Refer to illustrations 2.2, 2.3, 2.7 and 2.8

❊❊ WARNING:

Before undertaking the following procedure, be aware that disassembling the shock absorber/coil spring assemblies is a potentially dangerous job. Careless or unsafe work can cause serious injury. Use only a high-quality spring compressor and be sure to follow the spring compressor manufacturer's instructions. After removing the compressed spring, set it aside in a safe, isolated place.

➡Note: If the shock absorber/coil spring assemblies must be replaced, you can save time by simply installing new complete shock/coil assemblies. Or, you can replace just the shocks or just the springs. But, to do so, you will have to disassemble the shock absorber/coil spring assemblies. Therefore, before deciding which way you want to go, find out the cost of each option. You may find that the cost for two assembled complete shock/coil assemblies is only slightly higher than the cost for two new shock absorbers or coil springs.

1   Loosen the front wheel lug nuts. Raise the vehicle and support it securely on jackstands. Remove the front wheels.

2   Working from underneath the vehicle, remove the nut and bolt attaching the lower end of the shock absorber to the lower control arm (see illustration). On 4WD models, push the bolt through the shock absorber and bracket as far as you can (but be careful not to nick the driveaxle), then pry the lower control arm down and remove the bolt.

3   Remove the three nuts that attach the upper end of the shock to the frame bracket (see illustration).

4   Remove the shock absorber/coil spring assembly.

5   Inspect the shock absorber for leaking fluid, dents, cracks and other damage. Inspect the coil spring for chips and cracks which could cause premature failure. Inspect the spring seats for hardness and general deterioration. If either the shock or the spring is worn or damaged, replace it. If you're installing new complete units, proceed to Step 14; if you're going to install new shocks or coil springs, proceed to the next Step.

6   Secure the shock/coil assembly in a bench vise. If you're planning to reuse the old shock absorbers, line the jaws of the vise with wood or shop rags to protect the shock bodies. Don't tighten the jaws any more than necessary; overtightening the vise may crush the shock body.

2.2 To detach the lower end of the shock absorber/coil spring assembly from the lower control arm, remove this nut and bolt

2.3 To detach the upper end of the shock absorber/coil spring assembly from its mounting bracket, remove these three nuts (NOT the nut in the middle, which is the damper rod nut; it must never be removed unless the spring is compressed with a spring compressor)

2.7 Install the spring compressor(s) in accordance with the manufacturer's instructions; compress the spring until you can wiggle it before removing the damper rod nut

7 Install a spring compressor in accordance with the tool manufacturer's instructions (see illustration). (You can buy a spring compressor at most auto parts stores or rent one from most equipment yards on a daily basis.) Compress the spring far enough to relieve all pressure from the spring seat; when you can wiggle the spring, it's compressed enough to disassemble the shock/spring assembly.

✳ WARNING:

Don't compress the spring any more than necessary.

8 Hold the damper rod with a wrench and remove the damper rod nut (see illustration).

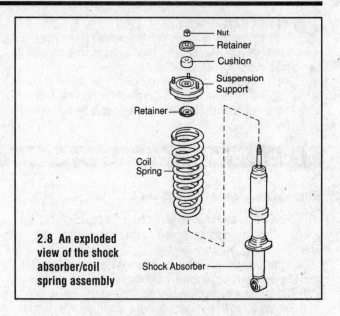

2.8 An exploded view of the shock absorber/coil spring assembly

9 Remove the upper retainer, upper cushion and suspension support.
10 Remove the compressed spring assembly and set it safely aside.
11 Remove the lower retainers and cushion.
12 Inspect the cushions for cracks and tears and general deterioration. Replace them if they're damaged. Make sure the retainers are in good condition, too. If they're distorted or otherwise damaged, replace them.
13 Reassembly is the reverse of disassembly. Make sure the lower end of the coil spring is correctly seated in the low spot in the lower spring seat. Tighten the damper rod nut to the torque listed in this Chapter's Specifications before releasing tension on the spring.
14 Installation is the reverse of removal. Be sure to tighten the upper and lower fasteners to the torque listed in this Chapter's Specifications.
15 Tighten the lug nuts to the torque listed in the Chapter 1 Specifications.

3 Stabilizer bar and bushings (front) - removal and installation

◗ **Refer to illustrations 3.2 and 3.3**

1 Raise the vehicle and support it securely on jackstands.
2 Remove the nuts from the stabilizer bar links and detach the links from the bar. Also remove the nuts attaching the links to the lower control arms, then remove the links (see illustration). Note how the bushings and retainers are arranged on the upper ends of the links.
3 Remove the stabilizer bar bushing bracket nuts and/or bolts (see illustration).

3.2 Stabilizer bar link fasteners

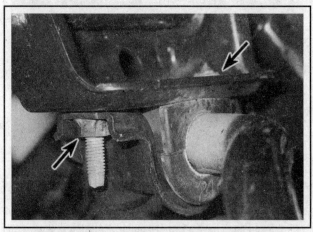

3.3 To separate the stabilizer bar from the frame, remove the bushing clamp nuts from both clamps

4   Remove the stabilizer bar.
5   Remove the rubber bushings from the stabilizer bar.
6   Inspect the rubber bushings for cracks, tears and deterioration. If they're worn or damaged, replace them.
7   Check the balljoint on the lower end of each link for looseness or other signs of excessive wear. You can check the rotational torque required to turn the balljoint by threading the nut onto the ballstud and turning it with an inch-pound torque wrench and comparing your reading to the value listed in this Chapter's Specifications.
8   When you install the rubber bushings on the stabilizer bar, be sure to position them so the slits face toward the front of the vehicle.
9   Installation is otherwise the reverse of removal. Be sure to tighten all fasteners to the torque listed in this Chapter's Specifications.

4   Upper control arm - removal and installation

▶ Refer to illustrations 4.4a, 4.4b and 4.5

1   Loosen the wheel lug nuts, raise the front of the vehicle and support it securely on jackstands. Apply the parking brake. Remove the wheel.
2   If equipped with ABS, disconnect the speed sensor wiring harness from the upper control arm and steering knuckle.
3   Remove the inner fender apron seal.
4   Remove the cotter pin, then loosen the upper balljoint nut a few turns. Using a two-jaw puller or a picklefork-type balljoint separator, separate the balljoint from the upper control arm (see illustrations).

❋ CAUTION:

Don't allow the steering knuckle to fall outward, as the brake hose may be damaged. It's a good idea to wire the steering knuckle to the coil spring so this doesn't happen.

5   Remove the nut, washer and pivot bolt and detach the upper control arm from the frame (see illustration). Remove the arm.
6   Inspect the bushings for wear and deterioration. If they're cracked or damaged, take the arm to an automotive machine shop and have new bushings installed.
7   Installation is the reverse of removal. Be sure to tighten all suspension fasteners to the torque listed in this Chapter's Specifications, and use a new cotter pin on the upper control arm balljoint nut. If necessary, tighten the balljoint nut a little more to align the hole in the ballstud with the slots in the nut - don't loosen the nut to achieve this alignment.

➡ Note: The pivot bolt/nut should be tightened with the vehicle at normal ride height. This can be done after the vehicle has been lowered to the ground (on vehicles with adequate clearance), or can be simulated by raising the lower control arm with a floor jack.

8   Tighten the lug nuts to the torque listed in the Chapter 1 Specifications.
9   It's a good idea to have the wheel alignment checked and, if necessary, adjusted.

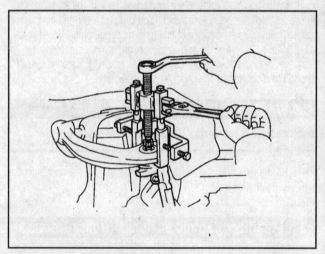

4.4a   The best way to separate the upper control arm from the balljoint is to use a two-jaw puller

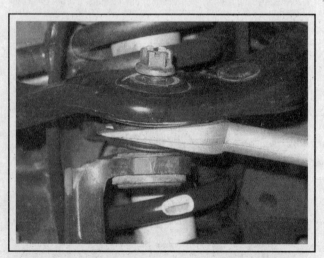

4.4b   If you don't have a two-jaw puller, a picklefork-type balljoint separator can be used, but keep in mind that this tool will probably damage the balljoint seal

4.5   The upper control arm is attached to the frame with a single, long pivot bolt; remove this nut and slide the bolt out towards the rear of the vehicle

5   Lower control arm - removal and installation

◆ **Refer to illustrations 5.6 and 5.7**

1   Loosen the wheel lug nuts, raise the front of the vehicle and support it securely on jackstands. Apply the parking brake. Remove the wheel.

2   Remove the under-vehicle splash shield (see Chapter 1, illustration 8.6).

3   Unbolt the steering gear assembly from the frame (see Section 19).

4   Unbolt the shock absorber/coil spring assembly from the lower control arm (see Section 2).

5   Remove the nuts and detach the stabilizer bar links from the lower control arms (see Section 3).

6   Remove the cotter pin, then loosen the lower balljoint nut a few turns. Using a two-jaw puller or a picklefork-type balljoint separator, separate the balljoint from the upper control arm (see illustration).

7   Make alignment marks on the front and rear adjusting cams (see illustration). Hold the adjusting cam plate nuts with a wrench and unscrew the pivot bolts. Remove the bolts and detach the control arm from the frame.

➡ **Note: When removing the rear pivot bolts, you'll have to**

reposition the steering gear up and back to make room for the bolts to slide out (see illustration 19.7).

8   Inspect the bushings for wear and deterioration. If they're cracked or damaged, take the arm to an automotive machine shop and have new bushings installed.

9   Installation is the reverse of removal. Make sure that the alignment marks you made prior to disassembly are lined up. Be sure to tighten all suspension fasteners to the torque listed in this Chapter's Specifications, and use a new cotter pin on the lower control arm balljoint nut. If necessary, tighten the balljoint nut a little more to align the hole in the ballstud with the slots in the nut - don't loosen the nut to achieve this alignment.

➡ **Note: The pivot bolts should be tightened with the vehicle at normal ride height. This can be done after the vehicle has been lowered to the ground (on vehicles with adequate clearance), or can be simulated by raising the lower control arm with a floor jack.**

10   Tighten the lug nuts to the torque listed in the Chapter 1 Specifications.

11   Have the wheel alignment checked and, if necessary, adjusted.

5.6 Remove the cotter pin, back off - but don't remove - the ballstud nut, then use a two-jaw puller to separate the ballstud from the lower control arm

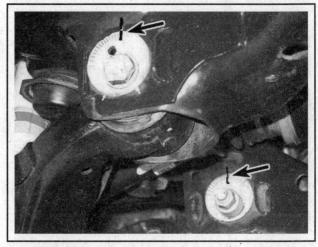

5.7 Mark the relationship of the adjusting cams to the frame

6   Steering knuckle - removal and installation

1   Loosen the wheel lug nuts. Raise the front of the vehicle and support it securely on jackstands. Apply the parking brake. Remove the wheel.

2   Detach all brake hose brackets from the steering knuckle. On vehicles with ABS, detach the wheel speed sensor from the steering knuckle.

3   Remove the brake caliper and brake disc (see Chapter 9). Hang the caliper with a length of wire - don't let it hang by the brake hose.

4   Disconnect the tie-rod end from the steering knuckle (see Section 17).

5   If you're working on a 4WD model, remove the driveaxle/hub nut

(see Chapter 8).

6   Disconnect the upper and lower control arms from the steering knuckle (see Sections 4 and 5), then remove the steering knuckle. On 4WD models, guide the driveaxle out of the hub, being careful to not overextend the inner CV joint. Support the driveaxle with a length of wire - don't let it hang by the inner CV joint.

7   Installation is the reverse of removal. Tighten all suspension fasteners to the torque values listed in this Chapter's Specifications.

8   Tighten the lug nuts to the torque listed in the Chapter 1 Specifications.

7 Hub and bearing assembly (front) - replacement

1   Remove the steering knuckle (see Section 6).
2   The front hub and bearing are pressed into the steering knuckle. Take the steering knuckle to an automotive machine shop and have the hub and old bearing pressed out and the new bearing and hub pressed in.

3   Install the steering knuckle (see Section 6).

8 Balljoints - check and replacement

CHECK

▶ **Refer to illustrations 8.4 and 8.7**

1   Inspect the upper and lower balljoints for looseness whenever the vehicle is raised for any reason. You can check the balljoints with the suspension assembled as follows.
2   Raise the front of the vehicle and support it securely on jackstands.
3   Wipe the balljoints clean and inspect the seals for cuts and tears. If a balljoint seal is damaged, replace the balljoint.

Lower balljoint

4   With the vehicle raised and supported on jackstands and the suspension hanging free, attach a dial indicator to the lower control arm, with the plunger of the dial indicator touching the steering knuckle (see illustration). Pull up and push down on the steering knuckle with a force of 65 pounds each way and check the dial indicator. If there is more than 0.020-inch of play, replace the balljoint.

Upper balljoint

5   With the vehicle raised and supported on jackstands, and a floor jack supporting the lower control arm (slightly raised), pry up and down on the upper control arm while feeling for play in the balljoint. If there is significant play, replace the balljoint.

All balljoints

6   Balljoints should also be checked whenever they're separated from the steering knuckle or the upper or lower control arm. See if you can turn the ballstud in its socket with your fingers. If the balljoint is loose, or if the ballstud can be turned, replace the balljoint.

7   Toyota also specifies a more accurate version of this bench test. Flip the balljoint stud back and forth five times (see illustration), then install the nut. Using an inch-pound torque wrench, measure the turning torque as follows: turn the nut continuously at a rate of one turn every two to four seconds. On the fifth turn, read the indicated torque and compare it to the acceptable torque range listed in this Chapter's Specifications. If the indicated turning torque is not within the specified range, replace the balljoint.

REPLACEMENT

Upper balljoint

▶ **Refer to illustrations 8.10 and 8.11**

8   Remove the steering knuckle (see Section 6).
9   Remove the dust boot wire retainer and dust boot, then remove the snap-ring with a pair of snap-ring pliers.
10 Use a two-jaw puller, with the screw of the puller bearing on a large socket that fits over the balljoint and bears on the base of the balljoint, to remove the upper balljoint from the steering knuckle (see illustration). If you don't have the right tool, have the old balljoint removed and the new one installed at an automotive machine shop.
11 If you're replacing the balljoint yourself, use a hydraulic press (or a balljoint press) and two sockets to install the new balljoint (see illustration).
12 Installation is otherwise the reverse of removal. Tighten all suspen-

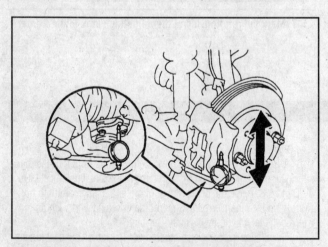

8.4 Check for play in the lower balljoint with a dial indicator mounted on the lower control arm and touching the part of the balljoint mounted on the steering knuckle; pull up and push down on the steering knuckle and read the indicator gauge

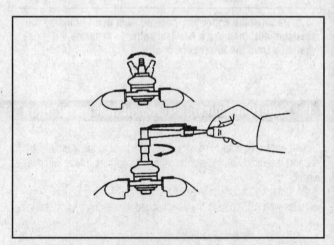

8.7 To bench test a balljoint, flip the balljoint stud back and forth five times, then measure the turning torque on the fifth turn

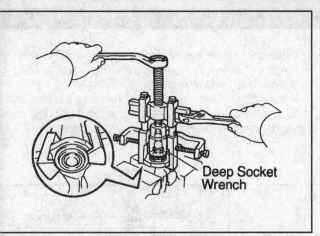

8.10 Use a two-jaw puller and a socket to remove the upper balljoint from the steering knuckle

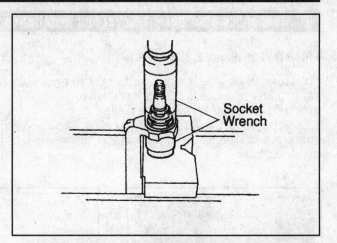

8.11 If you're replacing the balljoint yourself, use a press and two sockets of the proper size to install the new balljoint. The upper socket must fit over the balljoint and bear on the steering knuckle; the lower socket must bear on the balljoint

sion fasteners to the torque listed in this Chapter's Specifications. Tighten the wheel lug nuts to the torque listed in the Chapter 1 Specifications.

Lower balljoint

▶ **Refer to illustration 8.15**

13 Disconnect the tie-rod end from the steering arm on the lower balljoint (see Section 17).

14 Disconnect the lower control arm from the steering knuckle (see Section 5).

15 Unbolt the lower balljoint from the steering knuckle (see illustration).

16 Install the new lower balljoint and tighten the four bolts to the torque listed in this Chapter's Specifications.

17 Installation is otherwise the reverse of removal. Tighten all suspension fasteners to the torque listed in this Chapter's Specifications. Be sure to use a new cotter pin. If necessary, tighten the balljoint nut a little more to align the hole in the ballstud with the slots in the nut - don't loosen the nut to achieve this alignment. Tighten the wheel lug nuts to the torque listed in the Chapter 1 Specifications.

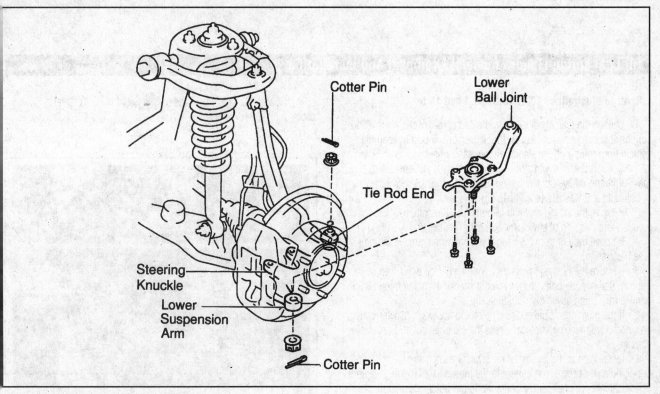

8.15 Lower balljoint installation details

9 Rear shock absorber - removal and installation

▶ **Refer to illustrations 9.2, 9.3a and 9.3b**

1 Loosen the rear wheel lug nuts. Raise the rear of the vehicle and support securely on jackstands. Block the front wheels so the vehicle doesn't roll off the stands. Remove the rear wheels.

2 Support the rear axle with a floor jack, placed near the shock absorber to be changed. Remove the shock absorber lower mounting bolt (see illustration).

3 Remove the shock absorber upper mounting nut (see illustration). Remove the shock absorber. Note the arrangement of the retainers and cushions (see illustration).

4 Installation is the reverse of removal. Tighten the mounting fasteners to the torque listed in this Chapter's Specifications. Tighten the lug nuts to the torque listed in the Chapter 1 Specifications.

9.2 Shock absorber lower mounting bolt and nut (Tundra shown; on Sequoia models the bolt threads into a boss on the rear axle)

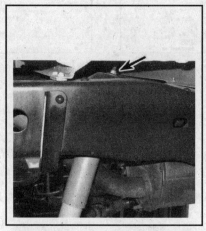

9.3a To detach the upper end of the shock absorber from the frame, remove this nut

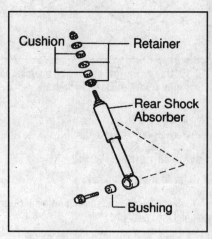

9.3b Shock absorber, cushion and retainer details

10 Leaf spring (Tundra models) - removal and installation

▶ **Refer to illustrations 10.2, 10.3, 10.4 and 10.5**

1 Loosen the rear wheel lug nuts. Raise the rear of the vehicle and support it securely on jackstands. Block the front wheels to keep the vehicle from rolling off the stands. Remove the wheel.

2 Support the axle with a floor jack and raise it just enough to take the weight and relieve the tension on the leaf springs. Remove the four U-bolt nuts and washers (see illustration).

3 Remove the spring seat and U-bolts (see illustration). Note the location and position of any caster adjusting wedges.

4 Remove the nut from the hanger pin bolt, then remove the bolt (see illustration).

5 Remove the nuts and washers from the shackle bolts (see illustration). Remove the bolts and detach the shackle from the frame, then remove the spring assembly from the vehicle.

6 If the bushings at the ends of the spring are worn or deteriorated, an automotive machine shop can press the old ones out and press new ones in.

7 Installation is the reverse of removal. Gradually tighten the U-bolt nuts in a criss-cross pattern to the torque listed in this Chapter's Specifications. The hanger pin and shackle pin bolts/nuts should also be tightened to the torque listed in this Chapter's Specifications, but this should be done with the vehicle resting at normal ride height.

8 Tighten the lug nuts to the torque listed in the Chapter 1 Specifications.

10.2 To detach the spring from the axle, remove the four U-bolt nuts and washers

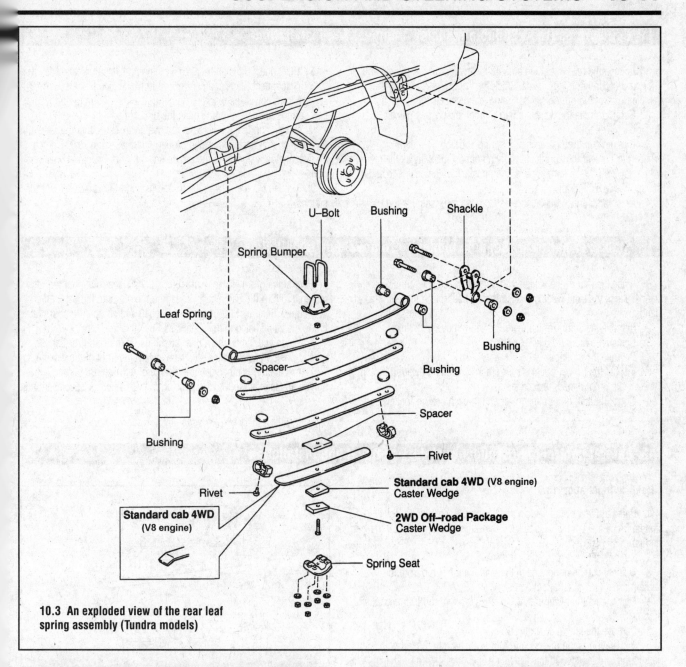

10.3 An exploded view of the rear leaf spring assembly (Tundra models)

10.4 To detach the front end of the spring from the frame, remove the nut from the hanger pin bolt, then remove the bolt

10.5 To detach the rear end of the spring, remove the nuts from the shackle bolts, pull the bolts out, then detach the spring and shackle from the vehicle

11 Coil springs (rear, Sequoia models) - removal and installation

1   Loosen the rear wheel lug nuts. Raise the rear of the vehicle and support it securely on jackstands. Block the front wheels to keep the vehicle from rolling off the stands. Remove the wheels.

2   Support the rear axle with a pair of floor jacks - one placed under each axle tube.

3   Run a chain through each coil spring and bolt their ends together to prevent the springs from flying out accidentally as the rear axle is lowered. Make sure there is enough slack in the chains to allow the springs to extend fully.

4   Disconnect the shock absorbers from the axle (see Section 9).

5   Disconnect the stabilizer bar bushing brackets from the axle.

6   Disconnect the lateral control rod from the axle (see Section 13).

7   Slowly lower the rear axle housing just far enough to remove the coil springs. Be careful not to snag the brake line.

8   Installation is the reverse of removal. Make sure that the lower end of each coil spring is seated properly on the spring seat. Tighten all suspension fasteners to the torque listed in this Chapter's Specifications.

9   Tighten the lug nuts to the torque listed in the Chapter 1 Specifications.

12 Stabilizer bar (rear, Sequoia models) - removal and installation

1   Raise the rear of the vehicle and support it securely on jackstands. Block the front wheels to keep the vehicle from rolling off the stands.

2   Remove the stabilizer bar link nuts and remove both links.

3   Detach the bushing bracket bolts.

4   Remove the stabilizer bar.

5   Inspect the rubber bushings for cracks, tears and deterioration. If they're worn or damaged, replace them.

6   Check the balljoint on the lower end of each link for looseness

or other signs of excessive wear. You can check the rotational torque required to turn the balljoint by threading the nut onto the ballstud and turning it with an inch-pound torque wrench and comparing your reading to the value listed in this Chapter's Specifications.

7   When you install the rubber bushings on the stabilizer bar, be sure to position them so the slits face toward the front of the vehicle.

8   Installation is otherwise the reverse of removal. Be sure to tighten all fasteners to the torque listed in this Chapter's Specifications.

13 Lateral control rod (Sequoia models) - removal and installation

▶ Refer to illustration 13.3

1   Raise the rear of the vehicle and support it securely on jackstands. Block the front wheels to keep the vehicle from rolling off the stands.

2   Support the rear axle with a floor jack.

3   Disconnect the upper end of the lateral control rod from the frame bracket (see illustration).

4   Disconnect the lower end of the lateral control rod from the axle bracket.

5   Remove the lateral control rod.

6   Installation is the reverse of removal. Be sure to tighten both nuts to the torque listed in this Chapter's Specifications (the vehicle should be at normal ride height when the bolts are tightened).

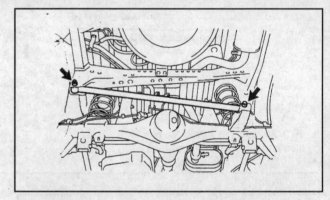

13.3 The lateral control rod is secured by a bolt, washer and nut at the frame bracket, and by a bolt at the axle housing

14 Suspension arms (rear, Sequoia models) - removal and installation

▶ Refer to illustrations 14.4 and 14.5

⁂ WARNING:

Remove and install one control arm at a time to prevent the axle from shifting.

1   Loosen the rear wheel lug nuts, raise the rear of the vehicle and support it securely on jackstands. Block the front wheels to keep the

vehicle from rolling off the stands. Remove the wheel.

2   Support the rear axle with a floor jack.

3   If you're removing an upper control arm, detach the brake hose bracket and ABS wiring harness from the upper suspension arms.

4   To remove an upper control arm, remove the nut, washer and bolt attaching the arm to the frame, and the nut, washer and bolt attaching the arm to the axle (see illustration).

5   To remove a lower control arm, remove the nut and bolt attaching the arm to the frame bracket and the nut, washer and bolt attaching the arm to the axle bracket (see illustration).

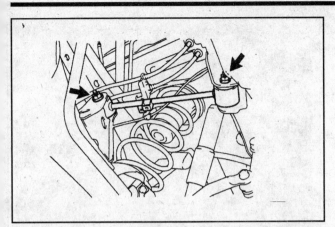

14.4 To detach the upper control arm from the frame and the axle, remove these nuts and bolts

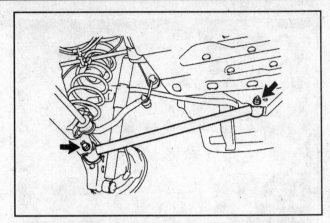

14.5 To detach the lower control arm from the frame and the axle, remove these nuts and bolts

6   Inspect the control arm bushings. If they're cracked or torn or otherwise deteriorated, replace the arm.

7   Installation is the reverse of removal. Tighten all fasteners to the

torque values listed in this Chapter's Specifications, with the vehicle resting at normal ride height.

15 Steering wheel - removal and installation

▶ **Refer to illustrations 15.3, 15.4, 15.5, 15.6, 15.7 and 15.8**

❋❋ WARNING:

The models covered by this manual are equipped with airbags. The airbag is armed and can deploy (inflate) anytime the battery is connected. To prevent accidental deployment (and possible injury), turn the ignition key to LOCK and disconnect the negative battery cable whenever working near airbag components. After the battery is disconnected, wait at least two minutes before beginning work (the system has a back-up capacitor that must fully discharge). For more information see Chapter 12.

1   Park the vehicle with the front wheels pointed straight ahead and

the steering wheel centered. Disconnect the cable from the negative terminal of the battery (see Chapter 5).

2   On airbag-equipped models, refer to Chapter 12 and disable the airbag system.

3   Pry out the two small covers, one on each side of the steering wheel, and loosen the two Torx screws that retain the airbag module (see illustration). Back out the screws until the grooves along the screw circumferences catch on the screw cases.

4   Lift the airbag module off the steering wheel and disconnect the airbag connector (see illustration).

❋❋ WARNING:

When handling the airbag module, hold it with the trim side facing away from you. Set the airbag module down in a safe location with the trim side facing up.

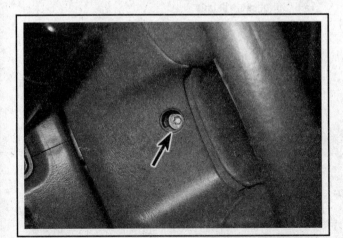

15.3 Pry off the covers on the sides of the steering wheel and, using a Torx bit of the correct size, loosen the Torx screws that retain the airbag module; back off the screws until the grooves in their circumference catch on the screw cases (when you feel the screws "catch," don't force them beyond that point)

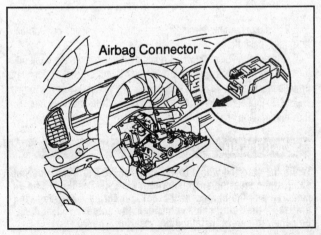

15.4 To unplug the airbag connector, slide the locking tab in the direction of the arrow, then pull the connector halves apart

15.5 Unplug any other electrical connectors that would interfere with steering wheel removal

15.6 Remove the steering wheel nut, then mark the relationship of the steering wheel to the steering shaft

15.7 Use a steering wheel puller to remove the steering wheel

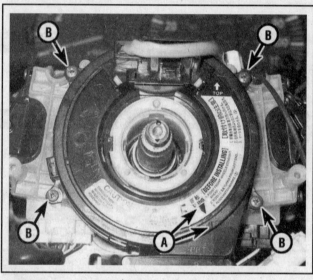

15.8 Spiral cable details

A   Alignment marks   B   Mounting screws

5   Unplug any other electrical connectors, such as the one for the horn (see illustration).

6   Remove the steering wheel retaining nut and mark the position of the steering wheel to the shaft, if marks don't already exist or don't line up (see illustration).

7   Use a puller to detach the steering wheel from the shaft (see illustration). Don't hammer on the shaft to dislodge the wheel.

✳✳ WARNING:

While the steering wheel is removed, do NOT turn the steering shaft. If the steering shaft is turned, the spiral cable will be un-centered and the harness will break, rendering the airbag inoperative. If the spiral cable is accidentally un-centered, it must be centered before installing the steering wheel.

8   Make sure the spiral cable is centered, as follows: Verify that the front wheels are pointing straight ahead. Turn the spiral cable housing counterclockwise by hand until it becomes hard to turn. Turn the spiral cable clockwise about 2-1/2 turns and align the marks (see illustration). If it's necessary to remove the spiral cable from the combination switch, remove the four screws, follow the wiring harness down the steering column and disconnect the electrical connector, then remove the spiral cable unit.

9   To install the wheel, align the mark on the steering wheel hub with the mark on the shaft and slide the wheel onto the shaft. Install the nut and tighten it to the torque listed in this Chapter's Specifications.

10   Installation is otherwise the reverse of removal.

16 Steering column - removal and installation

▶ Refer to illustrations 16.6, 16.7 and 16.8

✳✳ WARNING:

The models covered by this manual are equipped with airbags. The airbag is armed and can deploy (inflate) anytime the battery is connected. To prevent accidental deployment (and possible injury), turn the ignition key to LOCK and disconnect the negative battery cable whenever working near airbag components. After the battery is disconnected, wait at least two minutes before beginning work (the system has a back-up capacitor that must fully discharge). For more information see Chapter 12.

REMOVAL

1   Park the vehicle with the wheels pointing straight ahead. Disconnect the cable from the negative terminal of the battery.
2   Remove the steering wheel (see Section 15), then turn the ignition key to the LOCK position to prevent the steering shaft from turning.

✳✳ CAUTION:

If this is not done, the airbag spiral cable could be damaged.

3   Remove the lower finish panel under the column (see Chapter 11) and the heater duct behind it.
4   Remove the steering column covers (see Chapter 11).
5   Disconnect the electrical connectors for the steering column harness. Also, on models with an automatic transmission, detach the shift cable and shift interlock cable (see Chapter 7B).
6   Remove the column hole cover at the firewall (see illustration).
7   Mark the relationship of the intermediate shaft U-joint to the steering shaft, then remove the pinch bolt (see illustration).
8   Remove the steering column mounting fasteners (see illustration), lower the column and pull it to the rear, making sure nothing is still connected. Separate the intermediate shaft from the steering shaft and remove the column.

16.6 The steering column hole cover is retained by three bolts

INSTALLATION

9   Guide the steering column into position, then align the marks made in Step 7 and connect the intermediate shaft.
10 Install and tighten the column mounting fasteners to the torque listed in this Chapter's Specifications.
11 Install the pinch bolt, tightening it to the torque listed in this Chapter's Specifications.
12 The remainder of installation is the reverse of removal.

✳✳ WARNING:

If you're installing the steering column on a Sequoia model, the zero point calibration of the steering angle, master cylinder pressure, yaw rate and deceleration sensors for the Vehicle Skid Control (VSC) system must be performed. This will require taking the vehicle to a dealer service department or other qualified repair shop equipped with the special tool necessary to carry out this procedure.

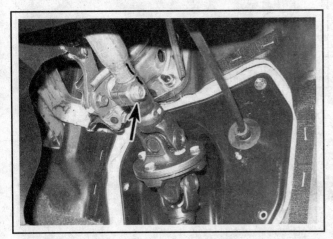

16.7 Mark the U-joint to the steering shaft, then remove the pinch bolt

16.8 Steering column mounting fasteners

17 Tie-rod ends - removal and installation

▶ **Refer to illustrations 17.2a, 17.2b, 17.3a and 17.3b**

1 Loosen the wheel lug nuts, raise the vehicle and place it securely on jackstands. Remove the wheel.

2 Loosen the tie-rod end locknut and mark the position of the tie-rod end on the threaded portion of the tie-rod (see illustrations).

3 Remove the cotter pin and loosen the castle nut from the tie-rod end balljoint stud, then install a puller and separate the tie-rod end from the steering knuckle (see illustrations). Remove the nut and

detach the tie-rod end from the steering knuckle arm.

4 Unscrew the old tie-rod end and install the new one. Make sure the new tie-rod end is aligned with the mark you made on the threads of the tie-rod.

5 Installation is the reverse of removal. Be sure to tighten the tie-rod end balljoint nut to the torque listed in this Chapter's Specifications. Tighten the locknut securely.

6 Tighten the lug nuts to the torque listed in the Chapter 1 Specifications.

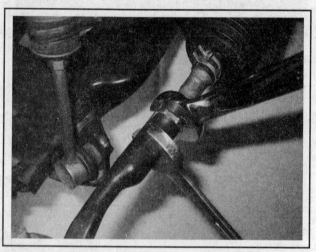

17.2a Loosen the tie-rod end locknut . . .

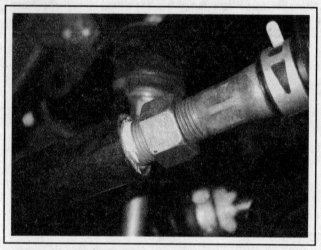

17.2b . . . and mark the position of the tie-rod end on the threaded portion of the tie-rod

17.3a Remove the cotter pin . . .

17.3b . . . loosen (don't remove) the castle nut from the tie-rod end balljoint stud, then install a puller and separate the tie-rod end from the steering knuckle

18 Steering gear boots - replacement

▶ **Refer to illustration 18.4**

1 If a steering gear boot is torn, dirt and moisture can damage the steering gear. Replace it.

2 Loosen the wheel lug nuts, raise the vehicle and place it securely on jackstands. Remove the front wheels.

3 Remove the tie-rod end and locknut (see Section 17).

4   Remove the boot clamps (see illustration) and slide the boot off the tie-rod.

5   Installation is the reverse of removal. Be sure to use new clamps on the boot. Tighten the lug nuts to the torque listed in the Chapter 1 Specifications.

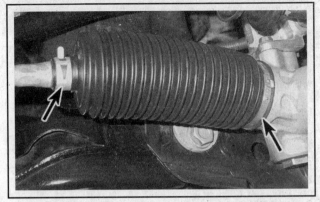

18.4 The outer clamp on the steering gear boot can be removed with a pair of pliers - the inner clamp must be cut off

19 Steering gear - removal and installation

▶ Refer to illustrations 19.5, 19.6 and 19.7

※※ WARNING 1:

The models covered by this manual are equipped with airbags. The airbag is armed and can deploy (inflate) anytime the battery is connected. To prevent accidental deployment (and possible injury), turn the ignition key to LOCK and disconnect the negative battery cable whenever working near airbag components. After the battery is disconnected, wait at least two minutes before beginning work (the system has a back-up capacitor that must fully discharge). For more information see Chapter 12.

※※ WARNING 2:

Make sure the steering column shaft is not turned while the steering gear is removed or you could damage the airbag system clockspring. To prevent the shaft from turning, turn the

ignition key to the lock position before beginning work, and run the seat belt through the steering wheel and clip it into its latch.

1   Park the vehicle with the front wheels pointing straight ahead.

2   Loosen the wheel lug nuts. Raise the front of the vehicle and support it securely on jackstands. Apply the parking brake. Remove the wheels. Remove the under-vehicle splash shield (see Chapter 1, illustration 8.6).

3   Disconnect the tie-rod ends from the steering knuckles (see Section 17).

4   Remove the stabilizer bar (see Section 3).

5   Mark the relationship of the U-joint to the intermediate shaft and the steering gear (see illustration), remove the U-joint pinch bolts, then slide the U-joint up and off the steering gear input shaft.

6   Disconnect the pressure and return lines from the steering gear (see illustration).

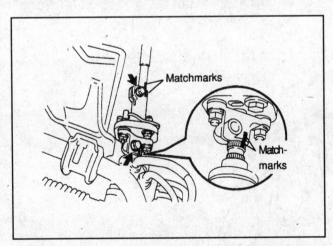

19.5 Mark the relationship of the intermediate shaft U-joint coupler to the steering gear and intermediate shaft, then remove the coupler pinch bolts and detach it from the steering gear

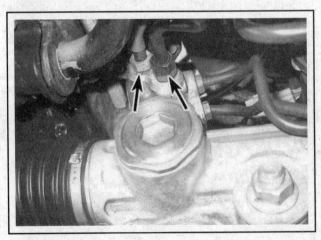

19.6 Steering gear pressure and return lines

7 Remove the steering gear mounting fasteners (see illustration).

8 Remove the steering gear assembly.

9 Installation is the reverse of removal. Be sure to align the matchmarks made in Step 5 and tighten all suspension and steering gear fasteners to the torque listed in this Chapter's Specifications. Tighten the wheel lug nuts to the torque listed in the Chapter 1 Specifications.

10 Bleed the power steering system when you're done (see Section 21).

✳✳ WARNING:

If you're installing the steering gear on a Sequoia model, the zero point calibration of the steering angle, master cylinder pressure, yaw rate and deceleration sensors for the Vehicle Skid Control (VSC) system must be performed. This will require taking the vehicle to a dealer service department or other qualified repair shop equipped with the special tool necessary to carry out this procedure.

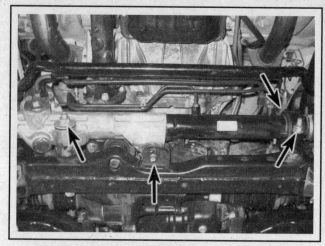

19.7 Steering gear mounting fasteners

20 Power steering pump - removal and installation

▸ **Refer to illustrations 20.4a and 20.4b**

1 Disconnect the cable from the negative terminal of the battery.

2 Remove the air filter housing (see Chapter 4).

3 Remove the drivebelt (see Chapter 1).

4 Position a drain pan under the power steering pump. Disconnect the pressure and return hoses from the backside of the pump (see illustrations). Plug the hoses to prevent contaminants from entering. Discard the sealing washers - new ones should be used during installation.

5 On V6 models, disconnect the electrical connector from the

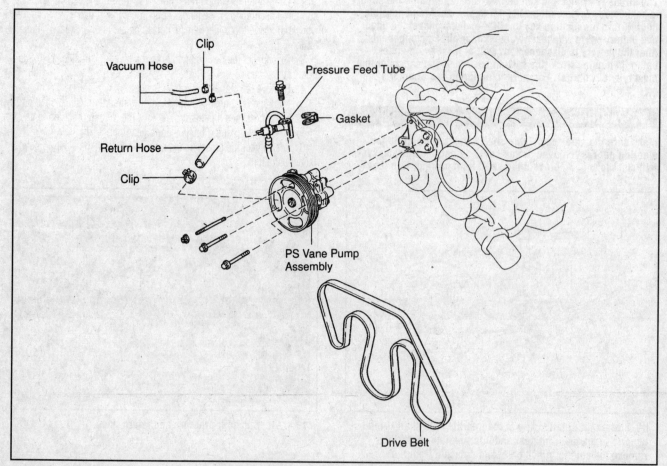

20.4a Power steering pump mounting details - V8 engine

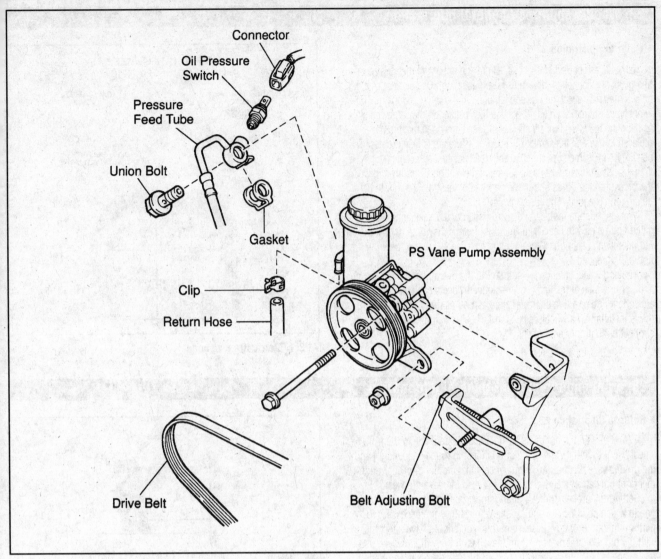

20.4b Power steering pump mounting details - V6 engine (2004 and earlier model shown)

power steering pressure switch. On V8 models, disconnect the vacuum hoses from the switch.

6 Remove the pump mounting fasteners and lift the pump from the engine compartment, taking care not to spill fluid on the painted surfaces.

7 Installation is the reverse of removal. Use new sealing washers

on either side of the pressure line fitting. Tighten the pressure line union bolt and the pump mounting bolts and nuts to the torque listed in this Chapter's Specifications.

8 Fill the power steering reservoir with the recommended fluid (see Chapter 1) and bleed the system following the procedure described in the next Section.

21 Power steering system - bleeding

1 Following any operation in which the power steering fluid lines have been disconnected, the power steering system must be bled to remove all air and obtain proper steering performance.

2 With the front wheels in the straight ahead position, check the power steering fluid level and, if low, add fluid until it reaches the Cold mark on the dipstick.

3 Start the engine and allow it to run at fast idle. Recheck the fluid level and add more if necessary to reach the Cold mark on the dipstick.

4 Bleed the system by turning the wheels from side-to-side, without hitting the stops. This will work the air out of the system. Keep the

reservoir full of fluid as this is done.

5 When the air is worked out of the system, return the wheels to the straight ahead position and leave the vehicle running for several more minutes before shutting it off.

6 Road test the vehicle to be sure the steering system is functioning normally and noise free.

7 Recheck the fluid level to be sure it's up to the Hot mark on the dipstick while the engine is at normal operating temperature. Add fluid if necessary (see Chapter 1).

22 Wheels and tires - general information

▶ **Refer to illustration 22.1**

All vehicles covered by this manual are equipped with metric-size fiberglass or steel belted radial tires (see illustration). Use of other size or type of tires may affect the ride and handling of the vehicle. Don't mix different types of tires, such as radials and bias belted, on the same vehicle as handling may be seriously affected. It's recommended that tires be replaced in pairs on the same axle, but if only one tire is being replaced, be sure it's the same size, structure and tread design as the other.

Because tire pressure has a substantial effect on handling and wear, the pressure on all tires should be checked at least once a month or before any extended trips (see Chapter 1).

Wheels must be replaced if they're bent, dented, leak air, have elongated bolt holes, are heavily rusted, out of vertical symmetry or if the lug nuts won't stay tight. Wheel repairs that use welding or peening are not recommended.

Tire and wheel balance is important to the overall handling, braking and performance of the vehicle. Unbalanced wheels can adversely affect handling and ride characteristics as well as tire life. Whenever a tire is installed on a wheel, the tire and wheel should be balanced by a shop with the proper equipment.

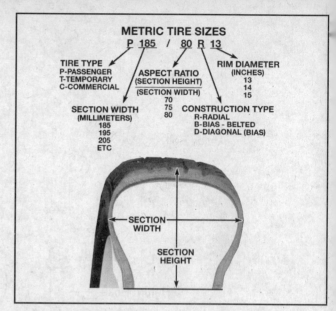

22.1 Metric tire size code

23 Front end alignment - general information

▶ **Refer to illustration 23.1**

A front end alignment refers to the adjustments made to the front wheels so they're in proper angular relationship to the suspension and the ground (see illustration). Front wheels that are out of proper alignment not only affect steering control, but also increase tire wear.

Getting the proper front wheel alignment is a very exacting process, one in which complicated and expensive machines are necessary to perform the job properly. Because of this, you should have a technician with the proper equipment perform these tasks. We will, however, use this space to give you a basic idea of what is involved with front end alignment so you can better understand the process and deal intelligently with the shop that does the work.

Toe-in is the turning in of the front wheels. The purpose of a toe specification is to ensure parallel rolling of the front wheels. In a vehicle with zero toe-in, the distance between the front edges of the wheels will be the same as the distance between the rear edges of the wheels. The actual amount of toe-in is normally only a fraction of an inch. Toe-in adjustment is controlled by the tie-rod length. Incorrect toe-in will cause the tires to wear improperly by making them scrub against the road surface.

Camber is the tilting of the front wheels from vertical when viewed from the front of the vehicle. When the wheels tilt out at the top, the camber is said to be positive (+). When the wheels tilt in at the top the camber is negative (-). The amount of tilt is measured in degrees from vertical and this measurement is called the camber angle. This angle affects the amount of tire tread which contacts the road and compensates for changes in the suspension geometry when the vehicle is cornering or traveling over an undulating surface. Camber is adjusted by rotating cam-shaped adjusters at the front and rear of the lower control arm.

Caster is the tilting of the top of the front steering axis from the vertical. A tilt toward the rear is positive caster and a tilt toward the front is negative caster. Caster is also adjusted by rotating cam-shaped adjusters at the front and rear of the lower control arm.

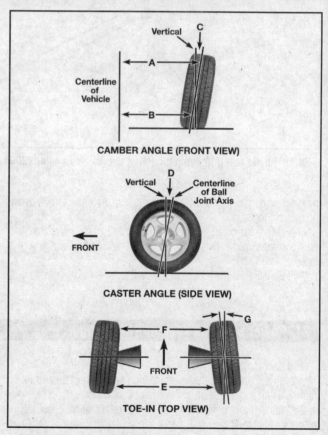

23.1 Front end alignment details

A minus B = C (degrees camber)
D = degrees caster

E minus F = toe-in (measured in inches)
G = toe-in (expressed n degrees)

Specifications

General

Power steering fluid type	See Chapter 1
Balljoint stud turning torque	
Stabilizer bar links	0.4 to 17 in-lbs
Upper control arm balljoint	6 to 39 in-lbs
Lower control arm balljoint	1 to 22 in-lbs

Torque specifications   Ft-lbs (unless otherwise indicated)

➡**Note: One foot-pound (ft-lb) of torque is equivalent to 12 inch-pounds (in-lbs) of torque. Torque values below approximately 15 ft-lbs are expressed in inch-pounds, since most foot-pound torque wrenches are not accurate at these smaller values.**

Front suspension

Shock absorber-to-lower control arm bolt/nut	100
Shock absorber upper mounting nuts	47
Shock absorber damper rod-to-suspension support nut	18
Upper control arm-to- frame bolt/nut	72
Lower control arm-to-frame bolt	96
Upper balljoint-to-upper control arm nut	77
Lower balljoint-to-lower control arm nut	103
Stabilizer bar-to-link nut	14
Stabilizer bar link-to-lower control arm nut	51
Stabilizer bar bracket-to-frame nut/bolt	
2000 Tundra	19
All others	27
Lower balljoint-to-steering knuckle bolts	59

Rear suspension

Tundra

Shock absorber-to-frame nut	15
Shock absorber-to-axle bolt/nut	64
Leaf spring	
To shackle bolt/nut	125
Shackle-to-frame bolt/nut	125
To frame bracket (front) bolt/nut	125
U-bolt-to-spring seat nuts	98

Sequoia

Rear stabilizer bar	
Bracket-to-axle bolts	27
Stabilizer bar link-to-frame nut	51
Stabilizer bar-to-stabilizer bar link nut	51
Lateral control rod	
To-frame bolt/nut	103
To-axle bolt	96
Control arms	
Upper control arm-to-frame bolt/nut	103
Upper control arm-to-axle bolt/nut	103
Lower control arm-to-frame bolt/nut	96
Lower control arm-to-axle bolt/nut	96

Torque specifications (continued)   Ft-lbs (unless otherwise indicated)

➡Note: One foot-pound (ft-lb) of torque is equivalent to 12 inch-pounds (in-lbs) of torque. Torque values below approximately 15 ft-lbs are expressed in inch-pounds, since most foot-pound torque wrenches are not accurate at these smaller values.

Steering

Airbag module mounting screws	78 in-lbs
Power steering pressure line-to-pump union bolt	34
Power steering pump mounting fasteners	33
Steering wheel nut	
Tundra	
2000	26
2001 and later	35
Sequoia	37
Steering gear	
Center mounting bolt	123
Right mounting bracket bolt/nut	123
Left mounting bolt/nut	96
Steering column mounting nuts	19
Intermediate shaft-to-steering pinch bolts (all)	26
Power steering pressure line-to-pump union bolt	34
Tie-rod end-to-steering knuckle nut	67

Section

11

BODY

1 General information

✳✳ WARNING 1:

The models covered by this manual are equipped with Supplemental Restraint systems (SRS), more commonly known as airbags. Always disable the airbag system before working in the vicinity of any airbag system component to avoid the possibility of accidental deployment of the airbag, which could cause personal injury (see Chapter 12).

✳✳ WARNING 2:

The front seat belts on these models may be equipped with pre-tensioners, which are pyrotechnic (explosive) devices designed to retract the seatbelts in the event of a collision. On models equipped with pre-tensioners, do not remove the front seatbelt retractor assemblies, and do not disconnect the electrical connectors leading to the assemblies. Problems with the pre-tensioners will turn on the SRS (airbag) warning light on the dash.

If any pre-tensioner problems are suspected, take the vehicle to a dealer service department or other qualified repair shop.

The vehicles covered by this manual are built with a body-on-frame construction. The frame is a ladder type, consisting of two box steel side rails joined by crossmembers. These crossmembers are welded or riveted to the side rails, with exception of some transmission crossmembers which are bolted into place for easy removal. The vehicle body is secured to the chassis by rubber insulated mounts and can be completely removed from the chassis.

Certain components are particularly vulnerable to accident damage and can be unbolted and repaired or replaced. Among these parts are the doors, seats, tailgate, liftgate and glass, bumpers, front fenders and door glass.

Only general body maintenance practices and body panel repair procedures within the scope of the do-it-yourselfer are included in this Chapter.

2 Body - maintenance

1 The condition of your vehicle's body is very important, because the resale value depends a great deal on it. It's much more difficult to repair a neglected or damaged body than it is to repair mechanical components. The hidden areas of the body, such as the wheel wells, the frame and the engine compartment, are equally important, although they don't require as frequent attention as the rest of the body.

2 Once a year, or every 12,000 miles, it's a good idea to have the underside of the body steam cleaned. All traces of dirt and oil will be removed and the area can then be inspected carefully for rust, damaged brake lines, frayed electrical wires, damaged cables and other problems. The steering knuckle stops should be greased after completion of this job (see Chapter 1).

3 At the same time, clean the engine and the engine compartment with a steam cleaner or water soluble degreaser.

4 The wheel wells should be given close attention, since undercoating can peel away and stones and dirt thrown up by the tires can cause the paint to chip and flake, allowing rust to set in. If rust is found, clean down to the bare metal and apply an anti-rust paint.

5 The body should be washed about once a week. Wet the vehicle thoroughly to soften the dirt, then wash it down with a soft sponge and plenty of clean soapy water. If the surplus dirt is not washed off very carefully, it can wear down the paint.

6 Spots of tar or asphalt thrown up from the road should be removed with a cloth soaked in solvent.

7 Once every six months, wax the body and chrome trim. If a chrome cleaner is used to remove rust from any of the vehicle's plated parts, remember that the cleaner also removes part of the chrome, so use it sparingly.

3 Upholstery and carpets - maintenance

1 Every three months remove the floor mats and clean the interior of the vehicle (more frequently if necessary). Use a stiff whisk broom to brush the carpeting and loosen dirt and dust, then vacuum the upholstery and carpets thoroughly, especially along seams and crevices.

2 Dirt and stains can be removed from carpeting with basic household or automotive carpet shampoos available in spray cans. Follow the directions and vacuum again, then use a stiff brush to bring back the nap of the carpet.

3 Most interiors have cloth or vinyl upholstery, either of which can be cleaned and maintained with a number of material-specific cleaners or shampoos available in auto supply stores. Follow the directions on the product for usage, and always spot-test any upholstery cleaner on an inconspicuous area (bottom edge of a back seat cushion) to ensure that it doesn't cause a color shift in the material.

4 After cleaning, vinyl upholstery should be treated with a protectant.

➡Note: Make sure the protectant container indicates the product can be used on seats - some products may make a seat too slippery.

✳✳ WARNING:

Do not use protectant on vinyl-covered steering wheels.

5 Leather upholstery requires special care. It should be cleaned regularly with saddle soap or leather cleaner. Never use alcohol, gasoline, nail polish remover or thinner to clean leather upholstery.

6 After cleaning, regularly treat leather upholstery with a leather conditioner, rubbed in with a soft cotton cloth. Never use car wax on leather upholstery.

7 In areas where the interior of the vehicle is subject to bright sunlight, cover leather seating areas of the seats with a sheet if the vehicle is to be left out for any length of time.

4 Vinyl trim - maintenance

Don't clean vinyl trim with detergents, caustic soap or petroleum-based cleaners. Plain soap and water works just fine, with a soft brush to clean dirt that may be ingrained. Wash the vinyl as frequently as the rest of the vehicle.

After cleaning, application of a high quality rubber and vinyl protectant will help prevent oxidation and cracks. The protectant can also be applied to weatherstripping, vacuum lines and rubber hoses (which often fail as a result of chemical degradation) and to the tires.

5 Hinges and locks - maintenance

Once every 3000 miles, or every three months, the hinges and latch assemblies on the doors, hood and trunk should be given a few drops of light oil or lock lubricant. The door latch strikers should also be lubricated with a thin coat of grease to reduce wear and ensure free movement. Lubricate the door and trunk locks with spray-on graphite lubricant.

6 Body repair - minor damage

REPAIR OF SCRATCHES

1   If the scratch is superficial and does not penetrate to the metal of the body, repair is very simple. Lightly rub the scratched area with a fine rubbing compound to remove loose paint and built up wax. Rinse the area with clean water.

2   Apply touch-up paint to the scratch, using a small brush. Continue to apply thin layers of paint until the surface of the paint in the scratch is level with the surrounding paint. Allow the new paint at least two weeks to harden, then blend it into the surrounding paint by rubbing with a very fine rubbing compound. Finally, apply a coat of wax to the scratch area.

3   If the scratch has penetrated the paint and exposed the metal of the body, causing the metal to rust, a different repair technique is required. Remove all loose rust from the bottom of the scratch with a pocket knife, then apply rust inhibiting paint to prevent the formation of rust in the future. Using a rubber or nylon applicator, coat the scratched area with glaze-type filler. If required, the filler can be mixed with thinner to provide a very thin paste, which is ideal for filling narrow scratches. Before the glaze filler in the scratch hardens, wrap a piece of smooth cotton cloth around the tip of a finger. Dip the cloth in thinner and then quickly wipe it along the surface of the scratch. This will ensure that the surface of the filler is slightly hollow. The scratch can now be painted over as described earlier in this Section.

REPAIR OF DENTS

▶ **See photo sequence**

4   When repairing dents, the first job is to pull the dent out until the affected area is as close as possible to its original shape. There is no point in trying to restore the original shape completely as the metal in the damaged area will have stretched on impact and cannot be restored to its original contours. It is better to bring the level of the dent up to a point which is about 1/8-inch below the level of the surrounding metal. In cases where the dent is very shallow, it is not worth trying to pull it out at all.

5   If the back side of the dent is accessible, it can be hammered out gently from behind using a soft-face hammer. While doing this, hold a block of wood firmly against the opposite side of the metal to absorb the hammer blows and prevent the metal from being stretched.

6   If the dent is in a section of the body which has double layers, or some other factor makes it inaccessible from behind, a different technique is required. Drill several small holes through the metal inside the damaged area, particularly in the deeper sections. Screw long, self-tapping screws into the holes just enough for them to get a good grip in the metal. Now the dent can be pulled out by pulling on the protruding heads of the screws with locking pliers.

7   The next stage of repair is the removal of paint from the damaged area and from an inch or so of the surrounding metal. This is easily done with a wire brush or sanding disk in a drill motor, although it can be done just as effectively by hand with sandpaper. To complete the preparation for filling, score the surface of the bare metal with a screwdriver or the tang of a file or drill small holes in the affected area. This will provide a good grip for the filler material. To complete the repair, see the Section on filling and painting.

REPAIR OF RUST HOLES OR GASHES

8   Remove all paint from the affected area and from an inch or so of the surrounding metal using a sanding disk or wire brush mounted in a drill motor. If these are not available, a few sheets of sandpaper will do the job just as effectively.

9   With the paint removed, you will be able to determine the severity of the corrosion and decide whether to replace the whole panel, if possible, or repair the affected area. New body panels are not as expensive as most people think and it is often quicker to install a new panel than to repair large areas of rust.

10 Remove all trim pieces from the affected area except those which will act as a guide to the original shape of the damaged body, such as headlight shells, etc. Using metal snips or a hacksaw blade, remove all loose metal and any other metal that is badly affected by rust. Hammer the edges of the hole on the inside to create a slight depression for the filler material.

11 Wire brush the affected area to remove the powdery rust from the surface of the metal. If the back of the rusted area is accessible, treat it with rust inhibiting paint.

12 Before filling is done, block the hole in some way. This can be done with sheet metal riveted or screwed into place, or by stuffing the hole with wire mesh.

13 Once the hole is blocked off, the affected area can be filled and painted. See the following subsection on filling and painting.

FILLING AND PAINTING

14 Many types of body fillers are available, but generally speaking, body repair kits which contain filler paste and a tube of resin hardener

These photos illustrate a method of repairing simple dents. They are intended to supplement Body repair - minor damage in this Chapter and should not be used as the sole instructions for body repair on these vehicles.

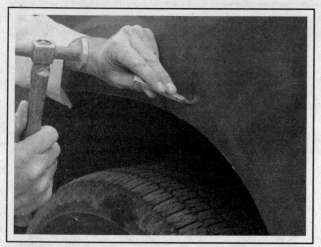

1 If you can't access the backside of the body panel to hammer out the dent, pull it out with a slide-hammer-type dent puller. In the deepest portion of the dent or along the crease line, drill or punch hole(s) at least one inch apart . . .

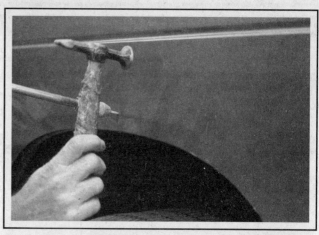

2 . . . then screw the slide-hammer into the hole and operate it. Tap with a hammer near the edge of the dent to help 'pop' the metal back to its original shape. When you're finished, the dent area should be close to its original contour and about 1/8-inch below the surface of the surrounding metal

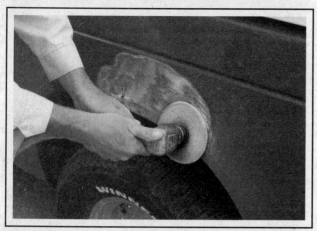

3 Using coarse-grit sandpaper, remove the paint down to the bare metal. Hand sanding works fine, but the disc sander shown here makes the job faster. Use finer (about 320-grit) sandpaper to feather-edge the paint at least one inch around the dent area

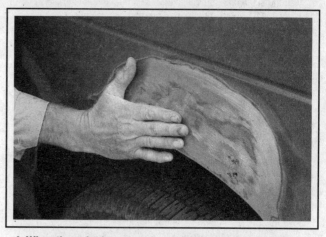

4 When the paint is removed, touch will probably be more helpful than sight for telling if the metal is straight. Hammer down the high spots or raise the low spots as necessary. Clean the repair area with wax/silicone remover

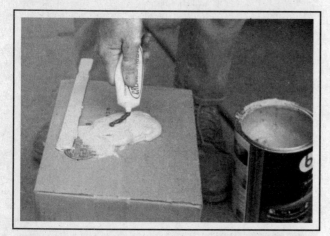

5 Following label instructions, mix up a batch of plastic filler and hardener. The ratio of filler to hardener is critical, and, if you mix it incorrectly, it will either not cure properly or cure too quickly (you won't have time to file and sand it into shape)

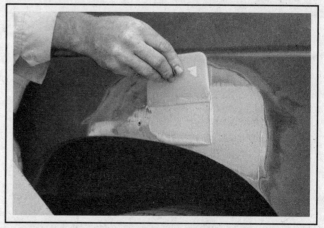

6 Working quickly so the filler doesn't harden, use a plastic applicator to press the body filler firmly into the metal, assuring it bonds completely. Work the filler until it matches the original contour and is slightly above the surrounding metal

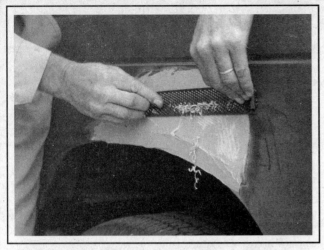

7 Let the filler harden until you can just dent it with your fingernail. Use a body file or Surform tool (shown here) to rough-shape the filler

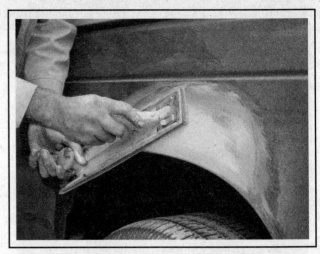

8 Use coarse-grit sandpaper and a sanding board or block to work the filler down until it's smooth and even. Work down to finer grits of sandpaper - always using a board or block - ending up with 360 or 400 grit

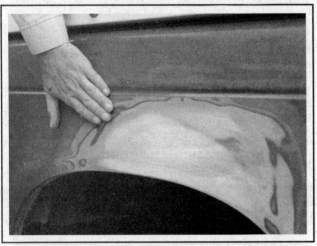

9 You shouldn't be able to feel any ridge at the transition from the filler to the bare metal or from the bare metal to the old paint. As soon as the repair is flat and uniform, remove the dust and mask off the adjacent panels or trim pieces

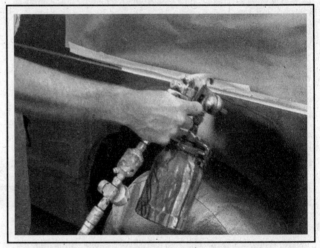

10 Apply several layers of primer to the area. Don't spray the primer on too heavy, so it sags or runs, and make sure each coat is dry before you spray on the next one. A professional-type spray gun is being used here, but aerosol spray primer is available inexpensively from auto parts stores

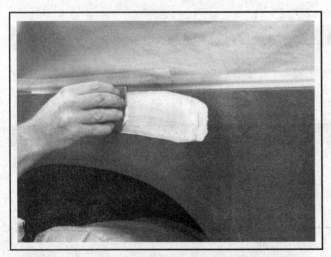

11 The primer will help reveal imperfections or scratches. Fill these with glazing compound. Follow the label instructions and sand it with 360 or 400-grit sandpaper until it's smooth. Repeat the glazing, sanding and respraying until the primer reveals a perfectly smooth surface

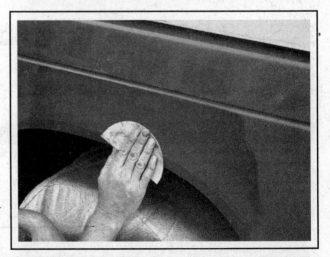

12 Finish sand the primer with very fine sandpaper (400 or 600-grit) to remove the primer overspray. Clean the area with water and allow it to dry. Use a tack rag to remove any dust, then apply the finish coat. Don't attempt to rub out or wax the repair area until the paint has dried completely (at least two weeks)

are best for this type of repair work. A wide, flexible plastic or nylon applicator will be necessary for imparting a smooth and contoured finish to the surface of the filler material. Mix up a small amount of filler on a clean piece of wood or cardboard (use the hardener sparingly). Follow the manufacturer's instructions on the package, otherwise the filler will set incorrectly.

15 Using the applicator, apply the filler paste to the prepared area. Draw the applicator across the surface of the filler to achieve the desired contour and to level the filler surface. As soon as a contour that approximates the original one is achieved, stop working the paste. If you continue, the paste will begin to stick to the applicator. Continue to add thin layers of paste at 20-minute intervals until the level of the filler is just above the surrounding metal.

16 Once the filler has hardened, the excess can be removed with a body file. From then on, progressively finer grades of sandpaper should be used, starting with a 180-grit paper and finishing with 600-grit wet-or-dry paper. Always wrap the sandpaper around a flat rubber or wooden block, otherwise the surface of the filler will not be completely flat. During the sanding of the filler surface, the wet-or-dry paper should be periodically rinsed in water. This will ensure that a very smooth finish is produced in the final stage.

17 At this point, the repair area should be surrounded by a ring of bare metal, which in turn should be encircled by the finely feathered edge of good paint. Rinse the repair area with clean water until all of the dust produced by the sanding operation is gone.

18 Spray the entire area with a light coat of primer. This will reveal any imperfections in the surface of the filler. Repair the imperfections with fresh filler paste or glaze filler and once more smooth the surface with sandpaper. Repeat this spray-and-repair procedure until you are satisfied

that the surface of the filler and the feathered edge of the paint are perfect. Rinse the area with clean water and allow it to dry completely.

19 The repair area is now ready for painting. Spray painting must be carried out in a warm, dry, windless and dust free atmosphere. These conditions can be created if you have access to a large indoor work area, but if you are forced to work in the open, you will have to pick the day very carefully. If you are working indoors, dousing the floor in the work area with water will help settle the dust which would otherwise be in the air. If the repair area is confined to one body panel, mask off the surrounding panels. This will help minimize the effects of a slight mismatch in paint color. Trim pieces such as chrome strips, door handles, etc., will also need to be masked off or removed. Use masking tape and several thickness of newspaper for the masking operations.

20 Before spraying, shake the paint can thoroughly, then spray a test area until the spray painting technique is mastered. Cover the repair area with a thick coat of primer. The thickness should be built up using several thin layers of primer rather than one thick one. Using 600-grit wet-or-dry sandpaper, rub down the surface of the primer until it is very smooth. While doing this, the work area should be thoroughly rinsed with water and the wet-or-dry sandpaper periodically rinsed as well. Allow the primer to dry before spraying additional coats.

21 Spray on the top coat, again building up the thickness by using several thin layers of paint. Begin spraying in the center of the repair area and then, using a circular motion, work out until the whole repair area and about two inches of the surrounding original paint is covered. Remove all masking material 10 to 15 minutes after spraying on the final coat of paint. Allow the new paint at least two weeks to harden, then use a very fine rubbing compound to blend the edges of the new paint into the existing paint. Finally, apply a coat of wax.

7 Body repair - major damage

1 Major damage must be repaired by an auto body/frame repair shop with the necessary welding and hydraulic straightening equipment.

2 If the damage has been serious, it is vital that the structure be checked for proper alignment or the vehicle's handling characteristics may be adversely affected. Other problems, such as excessive tire wear and wear in the driveline and steering may occur.

3 Due to the fact that all of the major body components (hood, fenders, etc.) are separate and replaceable units, any seriously damaged components should be replaced rather than repaired. Sometimes these components can be found in a wrecking yard that specializes in used vehicle components, often at considerable savings over the cost of new parts.

8 Windshield and fixed glass - replacement

Replacement of the windshield and fixed glass requires the use of special fast setting adhesive/caulk materials. These operations should

be left to a dealer or a shop specializing in glass work.

9 Hood - removal, installation and adjustment

➡**Note: The hood is somewhat awkward to remove and install - at least two people should perform this procedure.**

REMOVAL AND INSTALLATION

▸ **Refer to illustrations 9.2, 9.3 and 9.4**

1 Open the hood and place rags or covers over the windshield and fenders to protect them during the removal procedure.

2 Disconnect the windshield washer fluid lines (see illustration).

3 Mark the relationship of the hood to the hinges (see illustration).

4 Use a screwdriver to detach the hood support strut (see illustration).

5 Remove the hood retaining bolts and lift off the hood.

6 Installation is the reverse of removal.

ADJUSTMENT

▸ **Refer to illustration 9.8**

7 If necessary after installation, the entire hood latch assembly can be adjusted up-and-down as well as from side-to-side on the upper

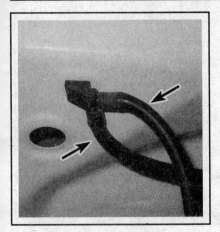

9.2 Before removing the hood, detach the windshield washer fluid lines

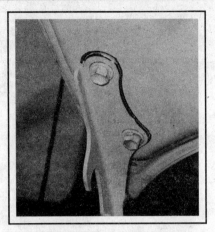

9.3 Before removing the hood hinge bolts, mark the relationship of the hood to the hood hinges

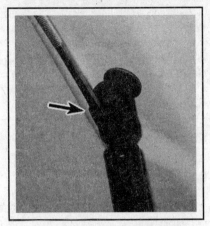

9.4 Use a screwdriver to pry the clip off and detach the hood support strut

radiator support so the hood closes securely and is flush with the fenders. To do this, scribe a line around the hood latch mounting bolts to provide a reference point. Then loosen the bolts and reposition the latch as necessary.

➡**Note: The factory bolts are "centering" type that will not allow adjustment. To adjust the hood in relation to the hinges, these bolts must be replaced with standard bolts with flat washers and lock washers.**

Following adjustment, retighten the mounting bolts.

8   Finally, adjust the hood bumpers on the hood, so when closed, it is flush with the fenders (see illustration).

9   The hood latch, as well as the hinges, should be periodically lubricated with white lithium-based grease to prevent sticking and wear.

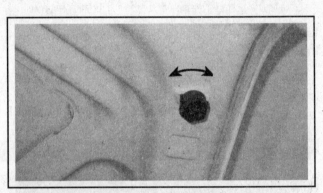

9.8 Screw the hood bumpers in or out to adjust the hood flush with the fenders

10 Hood latch and release cable - removal and installation

LATCH

▶ **Refer to illustrations 10.1a and 10.1b**

1   Remove the latch cover for access, scribe a line around the latch to aid alignment when installing, then remove the retaining bolts securing the hood latch to the radiator support (see illustrations). Remove the latch.

10.1a Remove the two screws and detach the latch cover

10.1b Remove the retaining bolts and detach the latch

2   Disconnect the hood release cable by disengaging the cable from the latch assembly.

3   Installation is the reverse of removal.

→**Note: Adjust the latch so the hood engages securely when closed and the hood bumpers are slightly compressed.**

CABLE

♦ **Refer to illustration 10.5**

4   Working in the passenger compartment, remove the driver's side kick panel.

5   Lift the hood release handle lever upward, then remove the screws and disengage the cable from the hood release lever handle (see illustration).

6   Attach a piece of thin wire or string to the end of the cable.

7   Working in the engine compartment, disconnect the hood release cable from the latch assembly as described in Steps 1 and 2. Unclip all the cable retaining clips on the radiator support and the inner fenderwell.

8   Pull the cable forward into the engine compartment until you can see the wire or string, then remove the wire or string from the old cable and fasten it to the new cable.

9   With the new cable attached to the wire or string, pull the wire or string back through the firewall until the new cable reaches the inside handle.

10   Working in the passenger compartment, reinstall the new cable

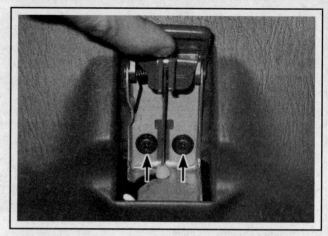

10.5 Lift the release lever for access and remove the lever retaining screws

into the hood release lever, making sure the cable housing fits snugly into the notch in the handle bracket.

→**Note: Pull on the cable with your fingers from the passenger compartment until the cable stop seats in the grommet on the firewall correctly.**

11   The remainder of the installation is the reverse of removal.

11   Radiator grille - removal and installation

❄❄ **WARNING:**

The models covered by this manual are equipped with Supplemental Restraint systems (SRS), more commonly known as airbags. Always disable the airbag system before working in the vicinity of any airbag system component to avoid the possibility of accidental deployment of the airbag, which could cause personal injury (see Chapter 12).

1   Open the hood and remove the hood-to-grille retaining nuts.

2   Detach the radiator grille from the hood.

3   Installation is the reverse of removal.

12   Bumpers - removal and installation

❄❄ **WARNING:**

The models covered by this manual are equipped with Supplemental Restraint systems (SRS), more commonly known as airbags. Always disable the airbag system before working in the vicinity of any airbag system component to avoid the possibility of accidental deployment of the airbag, which could cause personal injury (see Chapter 12).

FRONT

♦ **Refer to illustration 12.2**

1   Remove the fog light bulbs (see Chapter 12).

2   Remove the bumper arm-to-body bolts (see illustration).

3   Installation is the reverse of removal. Be sure to tighten all fasteners securely.

REAR

Tundra

♦ **Refer to illustrations 12.4 and 12.5**

4   If you're removing the bumper and brackets on a Tundra, unbolt the bumper brackets from the frame (see illustration).

5   If you're removing the Tundra bumper, unbolt it from the bumper brackets (see illustration).

6   Installation is the reverse of removal. Be sure to tighten all fasteners securely.

Sequoia

♦ **Refer to illustration 12.7**

7   If you're replacing the rear bumper or brackets on a Sequoia model, refer to the accompanying exploded view (see illustration).

8   Installation is the reverse of removal. Be sure to tighten all fasteners securely.

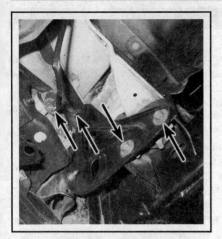

12.2 To remove the front bumper, remove these bolts from each side

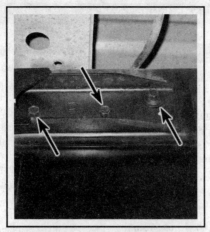

12.4 Remove the bolts from the frame rails If you're planning to remove the Tundra bumper and brackets as a unit

12.5 If you're planning to remove only the Tundra bumper, remove the bolts securing the bumper to the bumper brackets

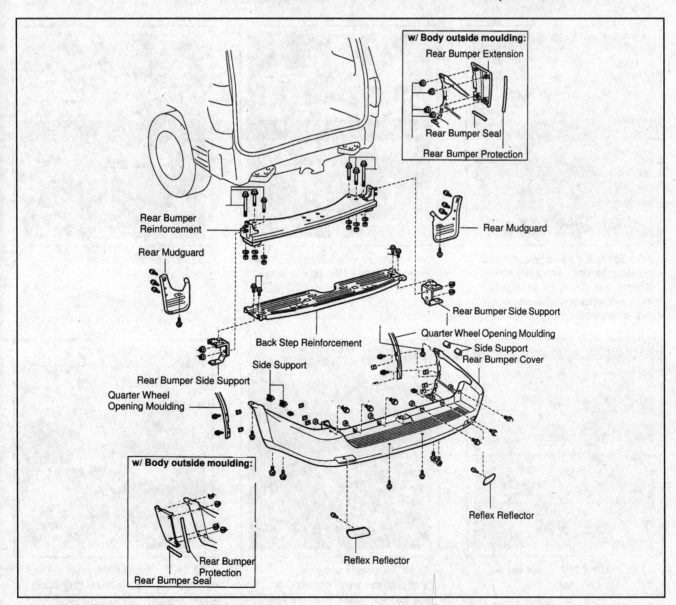

w/ Body outside moulding:

Rear Bumper Extension

Rear Bumper Seal

Rear Bumper Protection

Rear Bumper Reinforcement

Rear Mudguard

Rear Mudguard

Rear Bumper Side Support

Back Step Reinforcement

Quarter Wheel Opening Moulding

Side Support

Rear Bumper Cover

Rear Bumper Side Support

Side Support

Quarter Wheel Opening Moulding

w/ Body outside moulding:

Rear Bumper Protection

Rear Bumper Seal

Reflex Reflector

Reflex Reflector

12.7 Typical Sequoia rear bumper mounting details

13 Door trim panel - removal and installation

※ WARNING:

The models covered by this manual are equipped with Supplemental Restraint Systems (SRS), more commonly known as airbags. Always disable the airbag system before working in the vicinity of any airbag system components to avoid the possibility of accidental deployment of the airbag(s), which could cause personal injury (see Chapter 12).

REMOVAL

▶ Refer to illustrations 13.2, 13.3, 13.4a, 13.4b, 13.5a, 13.5b, 13.6, 13.7a, 13.7b, 13.8a, 13.8b and 13.8c

1   On models equipped with power windows, disconnect the negative cable at the battery.

2   To remove the window regulator handle on models without power windows, work a shop rag between the handle and the door trim panel, then pull up on both ends of the rag and pop loose the snapring that secures the handle to the regulator shaft (see illustration).

3   Remove the pull handle screws (see illustration).

4   Remove the screw from the inside door handle trim cover, detach the cover from the door (see illustrations).

5   To remove the switch control panel on power window equipped vehicles, carefully pry it upward to detach the clips at each end, then pull it out and unplug the electrical connector (see illustrations).

6   Use a small screwdriver to carefully pry out the screw covers and remove any remaining screws (see illustration).

7   To detach the door trim panel, carefully pry around the edge to detach the retaining clips, then pull it away from the door and disconnect the electrical connectors (see illustrations).

8   For access to the inner door, carefully peel back the plastic water shield (see illustrations).

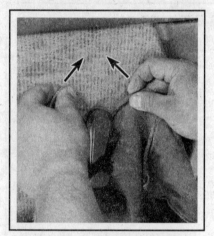

13.2 To remove the manual window regulator handle, work a clean shop rag between the handle and the door as shown, then pull up on the rag to pop off the snap-ring

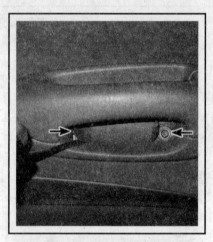

13.3 Use a screwdriver to pry up the cover and remove the screws from the pull handle well

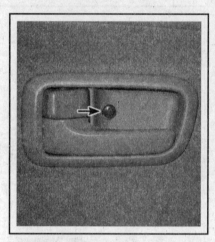

13.4a Remove the door handle screw

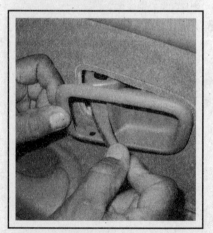

13.4b Pull the door handle out and detach the trim cover

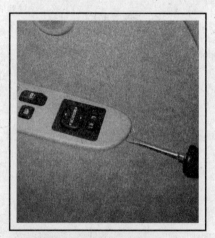

13.5a On models with power windows, use a screwdriver to pry out the switch panel from the door trim panel . . .

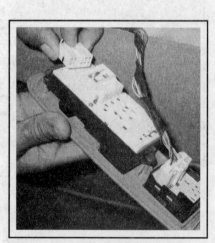

13.5b . . . then lift the panel out and disconnect the electrical connectors

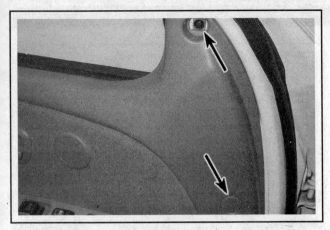

13.6 Pry out the door trim panel screw covers and remove the screws

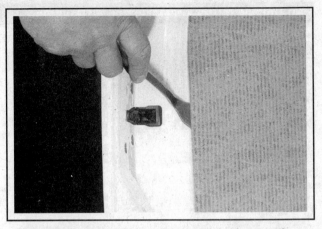

13.7a Insert a putty knife or trim removal tool between the door and the trim panel, then carefully pry the clips out

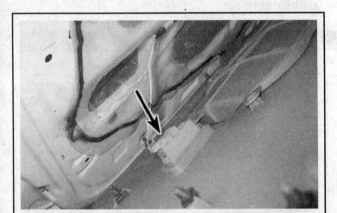

13.7b Unplug the electrical connectors from the trim panel

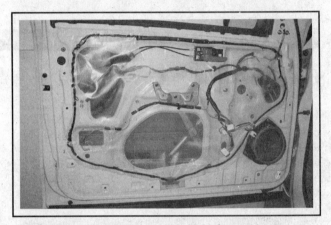

13.8a Before removing the water shield, be sure to remove any components that could damage it

INSTALLATION

9 Before installing the door trim panel, make sure the plastic water shield is correctly installed (see illustrations 13.8a and 13.8b). If the shield is torn, patch it with tape. If the edge fails to adhere to the door anywhere, seal it with a little silicone sealant. Install any grommets, clips or other fasteners which may have fallen out of the trim panel or the door when you removed the panel.

10 Place the door trim panel in position and push all the clips into their respective grommets until they're fully seated. Install the trim panel retaining screws and snap the plastic covers into place.
11 Install the inside handle trim cover and screw.
12 Install the pull handle screws.
13 Install the regulator handle or the switch control panel. Don't forget to plug in all electrical connectors.

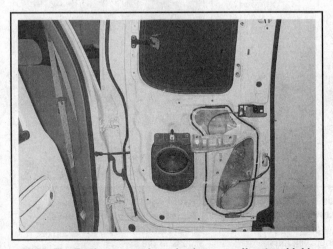

13.8b The Tundra access door also has a small water shield

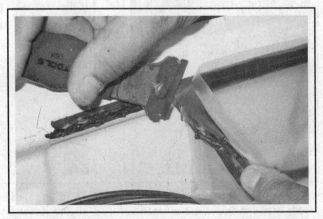

13.8c Carefully peel off the water shield - if the shield is torn during removal, repair it with tape or replace it - make sure the shield is sealed all the way around the edge

14 Outside mirror - removal and installation

▶ **Refer to illustration 14.2**

1   Remove the door trim panel (Section 13).
2   Remove the mirror-to-door retaining bolts/nuts and unplug the electrical connector (see illustration).
3   Remove the mirror assembly.
4   Installation is the reverse of removal.

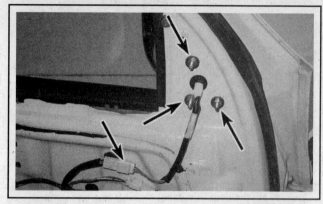

14.2 Remove the mirror retaining or nuts and disconnect the electrical connector

15 Door - removal and installation

▶ **Refer to illustrations 15.3, 15.4, 15.7a and 15.7b**

1   Disconnect the negative cable from the battery.
2   Remove the door trim panel (see Section 13). On doors with power components, unplug all electrical connections (it's a good idea to label all connections to aid the reassembly process) and remove the electrical harness from the door.
3   Open the door all the way and support it on jacks or blocks covered with cloth or pads to prevent damaging the paint, then remove the door stay bolt (see illustration).
4   Mark the relationship of the door hinges to the body by scribing or drawing a line around the hinges. With an assistant supporting the door, remove the door hinge bolts, then carefully lift off the door (see illustration).
5   When installing the door, make sure the hinges are aligned with the outlines you made, then tighten the hinge bolts securely.
6   Installation is otherwise the reverse of removal.
7   If the door does not close properly after installation, slightly loosen the door latch striker bolts (see illustrations) and carefully tap

15.3 Remove the bolt from the door stay

the striker up, down or sideways as necessary to provide positive engagement with the latch mechanism.

15.4 To detach the door from the body, scribe or draw alignment marks along the edges of the hinges, then remove the upper and lower hinge bolts

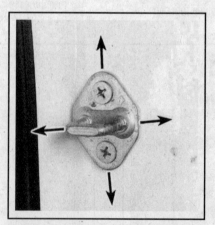

15.7a To adjust the door lock striker, slightly loosen the screws and tap the striker up, down or sideways as necessary until the door lock latch engages it correctly

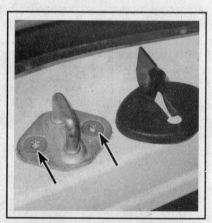

15.7b Before adjusting the lower door lock striker on Tundra access doors, remove the plastic cover for access to the screws

16 Front fender - removal and installation

▶ Refer to illustrations 16.2a, 16.2b, 16.2c, 16.3, 16.4a, 16.4b and 16.4c

❋❋ WARNING:

The models covered by this manual are equipped with Supplemental Restraint Systems (SRS), more commonly known as airbags. Always disable the airbag system before working in the vicinity of any airbag system components to avoid the possibility of accidental deployment of the airbag(s), which could cause personal injury (see Chapter 12).

1   Loosen the front wheel lug nuts. Raise the vehicle, support it securely on jackstands and remove the front wheel.

2   Detach the mud flaps, fender flares (if equipped) and wheel housing splash shields from the fender (see illustrations). The splash shields are attached between the frame and the fender with plastic screws and clips. Remove the plastic Phillips head screw in the center, then pry the clip out.

3   Remove the retaining bolts from the top edge of the fender (see illustration).

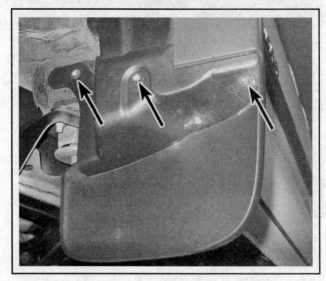

16.2a Remove the bolts retaining the mud flaps

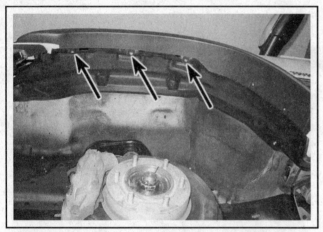

16.2b Detach the fender flares from their retaining clips after removing the bolts (not applicable for all models)

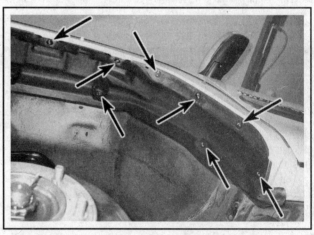

16.2c After removing the plastic clips, detach the wheel housing splash shields from the fender

16.3 To detach the upper edge of the fender, remove the bolts shown

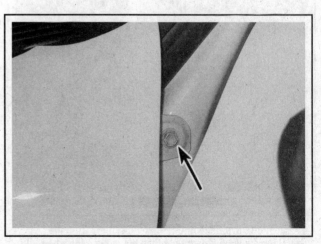

16.4a Remove the bolt securing the upper rear edge of the fender

16.4b Remove the nut retaining the front lower edge of the fender

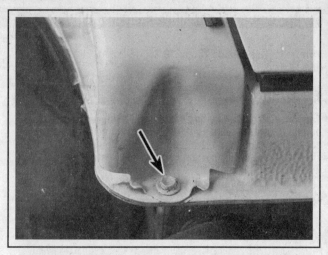

16.4c Remove the bolt retaining the rear lower edge of the fender

4   Remove the remaining bolts and detach the fender from the vehicle (see illustrations).

5   Remove the fender. It is a good idea to have an assistant support the fender while it's being moved away from the vehicle to prevent dam-age to the surrounding body panels.

6   Installation is the reverse of removal. Tighten all fasteners securely.

17 Tailgate and latch (Tundra models) - removal and installation

▶ **Refer to illustrations 17.1a, 17.1b, 17.2, 17.3a, 17.3b and 17.3c**

1   To remove the tailgate assembly, simply disengage the two tailgate cables then raise the tailgate until the slot in the right side hinge lines up with the flat on the hinge pin and lift it out (see illustrations).

2   If you want to replace the tailgate latch strikers, mark their relationship to the body, then remove the striker bolts (see illustration). If you want to replace the tailgate cables, simply remove the screw securing each end of the cable.

3   If you want to replace the tailgate handle or the latches, remove the plastic liner if equipped, then remove the service hole cover (see illustration). Disengage the latch control rods from the tailgate handle (see illustration), unbolt the handle from the tailgate or detach the latches from the tailgate (see illustration) and pull them out.

4   Installation is the reverse of removal. Tighten all fasteners securely.

17.1a To remove the entire tailgate assembly from the vehicle use a screwdriver to detach the cable ends, then raise the liftgate . . .

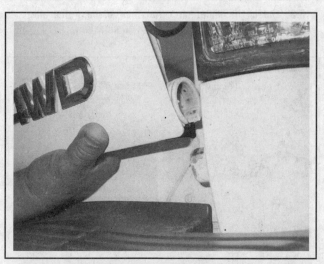

17.1b . . . until the slot in the right side hinge lines up with the flat on the hinge pin and lift the tailgate out

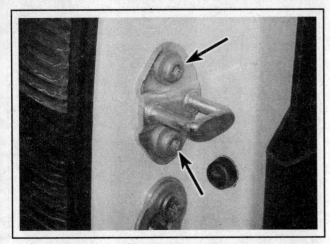

17.2 To remove the tailgate striker, mark its location and remove the bolts

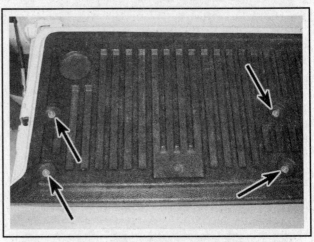

17.3a Remove the screws and detach the plastic tailgate liner (four of eight screws shown; if not equipped with a liner like this, remove the metal access panel)

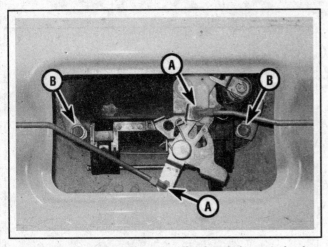

17.3b To remove the tailgate handle, detach the control rods (A) and remove the bolts (B)

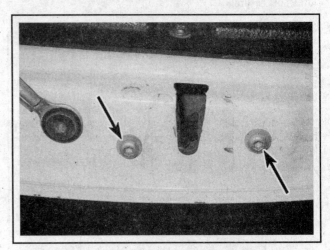

17.3c To remove a latch, unscrew the two screws and pull out the latch (the control rod must be disconnected from the back of the latch mechanism)

18 Liftgate (Sequoia models) - removal, installation and adjustment

♦ Refer to illustrations 18.4, 18.6a, 18.6b, 18.6c and 18.6d

➡Note: The liftgate is heavy and somewhat awkward to remove and install - at least two people should perform this procedure.

1   Disconnect the negative cable from the battery.
2   Open the liftgate and support it securely. Remove the liftgate struts (see illustration 9.4).
3   Disconnect the wire harness between the liftgate and the body.
4   Mark the relationship of the hinges to the liftgate, then unbolt them (see illustration).
5   Installation is the reverse of removal.
6   If the liftgate requires adjustment to close properly, there are four adjustments available. To move the liftgate forward, rearward or vertically in relation to the body, loosen the nuts and bolts that attach the hinges to the body (see illustration) and lightly tap the hinges with a small plastic hammer in the direction you want to move the liftgate, then tighten the nuts securely. To adjust the liftgate to the left or right, or up or down, in relation to the body, loosen the hinge bolts on the

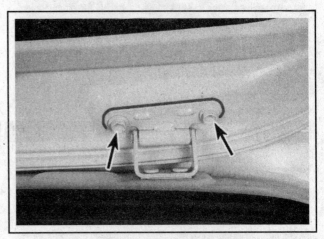

18.4 To detach the liftgate from the body, mark the relationship of both hinges to the body, then remove the bolts

18.6a To adjust the liftgate forward or rearward or vertically in relation to the body, loosen the hinge-to-body nuts and tap the hinges forward or backward as necessary, then tighten the nuts securely

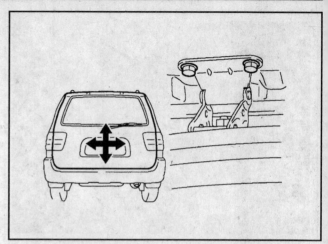

18.6b To adjust the liftgate left or right or vertically in relation to the body, loosen the hinge-to-liftgate bolts and move the liftgate as necessary, then tighten the bolts securely

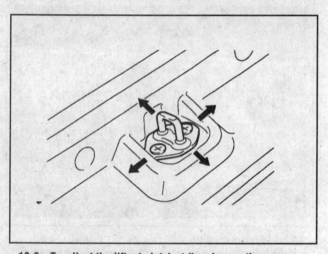

18.6c To adjust the liftgate latch striker, loosen the screws and tap the striker as necessary, then tighten the screws securely

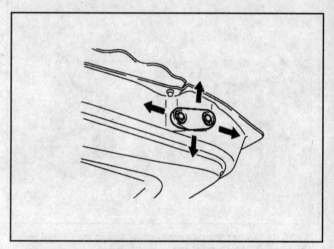

18.6d To adjust the liftgate closed height in relation to the body, loosen the stopper bolts and move the stoppers as necessary by tapping them with a small hammer, then tighten the bolts securely

liftgate (see illustration), carefully move the liftgate the direction you want to go, then tighten the bolts securely. To adjust the liftgate striker, loosen the striker screws and carefully tap the striker with a small plastic hammer until the liftgate lock and the striker are engaging properly,

then tighten the striker screws securely (see illustration). The liftgate-to-body gap is adjusted by loosening the mounting bolts and moving the stoppers in increments until the proper gap is achieved (see illustration).

19 Liftgate glass and regulator (Sequoia models) - removal and installation

▶ **Refer to illustration 19.4**

1 Disconnect the negative cable from the battery.

2 Although it's not absolutely necessary, you may want to remove the liftgate (see Section 18) before removing the liftgate glass. Servicing the liftgate components is easier with the liftgate on the floor.

3 Remove the liftgate pull strap.

4 To remove the trim panel, pry loose the trim panel clips. To

remove the service plate, unplug all electrical connectors and remove the service plate screws (see illustration).

5 Remove the liftgate glass run, the outer weather-stripping and the liftgate glass (see illustration 19.4).

6 Disconnect any remaining electrical connections and remove the window regulator (see illustration 19.4).

7 Installation is the reverse of removal.

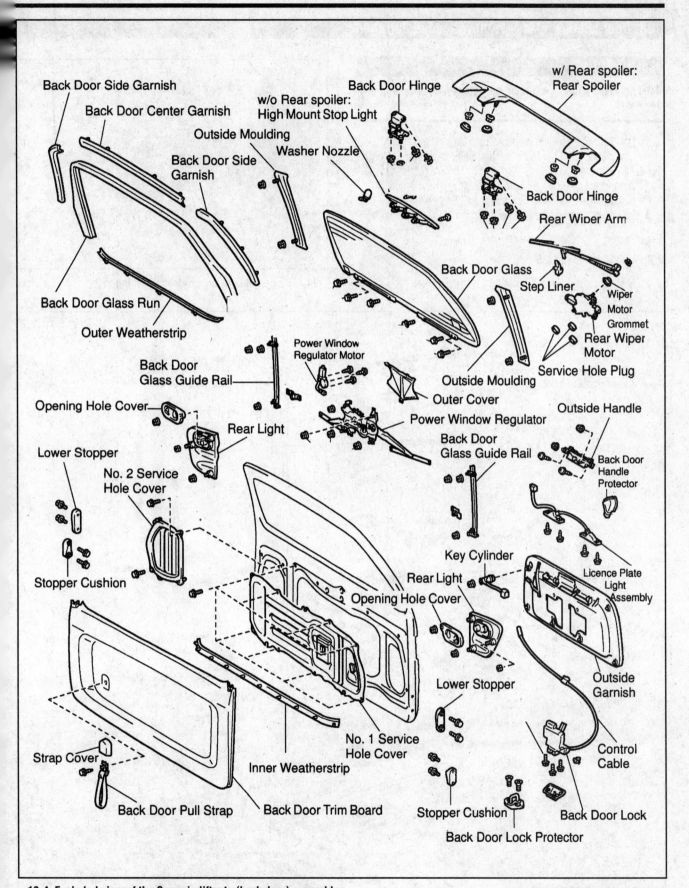

19.4 Exploded view of the Sequoia liftgate (back door) assembly

20 Door window glass - removal and installation

⁂ WARNING:

Safety glasses and gloves should be worn when performing this procedure.

1   Lower the glass fully in the door. On models equipped with power windows and/or door locks, disconnect the cable from the negative battery terminal.
2   Remove the door trim panel and water shield (see Section 13).
3   Remove the outside rear view mirror (see Section 14).
4   Pry out the window weatherstrip in the top of the door opening.

Front door

▶ **Refer to illustration 20.5**

5   Remove the door glass bolts (see illustration).

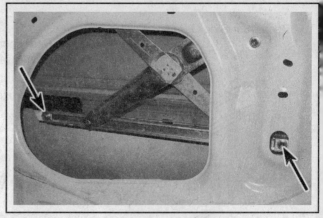

20.5 Remove the two door glass retaining bolts

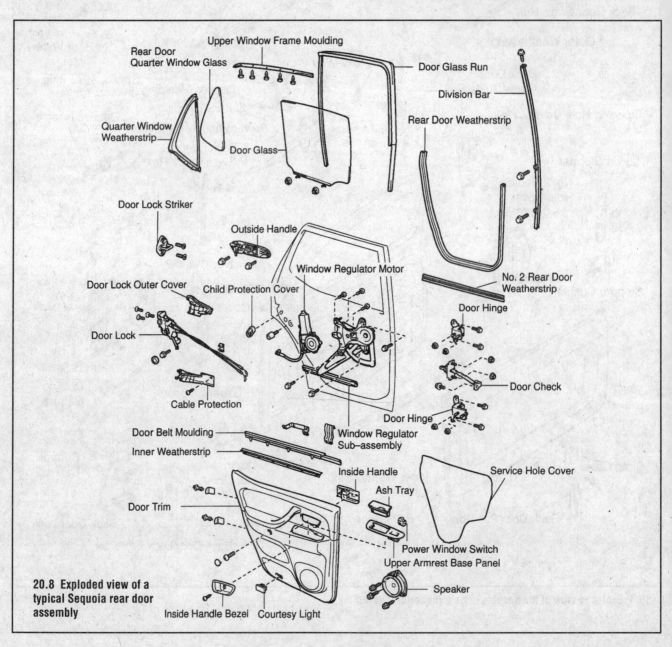

20.8 Exploded view of a typical Sequoia rear door assembly

Rear Door Quarter Window Glass
Upper Window Frame Moulding
Door Glass Run
Division Bar
Rear Door Weatherstrip
Quarter Window Weatherstrip
Door Glass
Door Lock Striker
Outside Handle
Window Regulator Motor
No. 2 Rear Door Weatherstrip
Door Lock Outer Cover
Child Protection Cover
Door Hinge
Door Lock
Door Check
Cable Protection
Door Hinge
Door Belt Moulding
Window Regulator Sub-assembly
Inner Weatherstrip
Inside Handle
Service Hole Cover
Door Trim
Ash Tray
Power Window Switch
Upper Armrest Base Panel
Speaker
Inside Handle Bezel
Courtesy Light

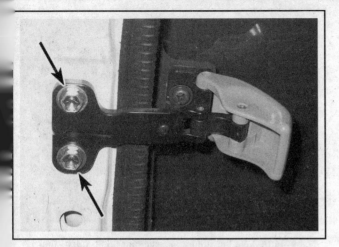

20.14 Remove the two access door window opener latch bolts (Tundra)

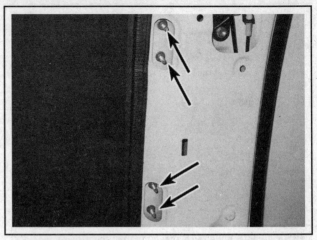

20.15 Remove the four nuts and detach the window from the access door (Tundra)

6   Remove the door glass.

7   Installation is the reverse of removal. Tighten all fasteners securely.

Rear door

▶ Refer to illustration 20.8

8   Remove the inner and outer weather-strip (see illustration).

9   Remove the division bar, the quarter window glass, the quarter window weather-strip and the glass run.

10 Remove the door latch and outside handle (see Section 22).

11 Remove the window regulator (see Section 21).

12 Remove the door glass.

13 Installation is the reverse of removal.

Tundra access door

▶ Refer to illustrations 20.14 and 20.15

14 Remove the window latch screws (see illustration).

15 Support the window, remove the retaining nuts and detach the window glass from the door (see illustration).

16 Installation is the reverse of removal.

21 Door window regulator - removal and installation

▶ Refer to illustration 21.4

1   On models equipped with power windows and door locks, disconnect the cable from the negative battery terminal.

2   Remove the door trim panel and water shield (see Section 13).

3   Remove the door glass (see Section 20).

4   To remove the window regulator, unplug the electrical connector from the motor (power windows), remove the retaining bolts (see illustration) and maneuver the regulator out through the access hole.

5   Installation is the reverse of removal. Lubricate all regulator rollers with multi-purpose grease. Tighten all fasteners securely. On rear doors, tighten the regulator bolts in a criss-cross pattern.

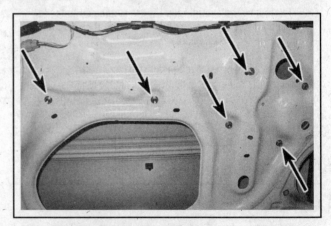

21.4 To remove the window regulator, remove the bolts, lower the regulator and remove it through the access hole (Tundra model shown, other models similar)

22 Door latch, lock cylinder and handles - removal and installation

1   On models equipped with power windows and door locks, disconnect the cable from the negative battery terminal.

2   Remove the door trim panel and the plastic water shield (see Section 13).

DOOR LATCH

▶ Refer to illustrations 22.5a and 22.5b

3   Using a flashlight, note how the control rods are connected to

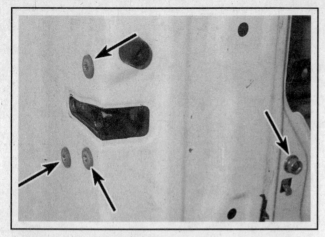

22.5a To detach the latch assembly from the driver's side door, remove three screws and one bolt

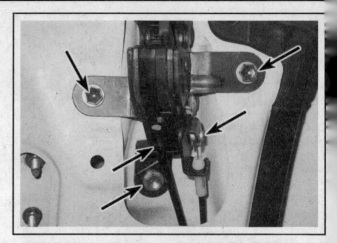

22.5b To remove the Tundra access door upper latch, first remove the bolts and disconnect the control rods and connectors

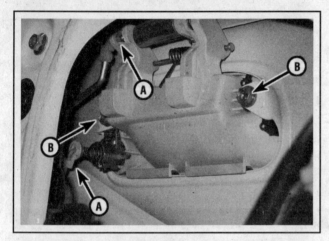

22.7 Note how the control rods (A) are connected to the key lock cylinder and to the outside door handle, then disconnect them and remove the bolts (B)

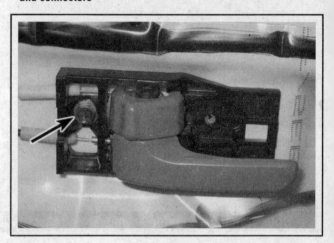

22.11 Remove the inside handle retaining bolt

the door latch. Disengage the rod(s) from the lock.

4 On models with power door locks, unplug the electrical connector from the door latch.

5 Remove the latch retaining screws (see illustrations) and remove the latch assembly.

6 Installation is the reverse of removal. Tighten the screws securely.

OUTSIDE HANDLE AND KEY LOCK CYLINDER

▶ **Refer to illustration 22.7**

7 Note how the control rods are connected to the outside door handle and to the key lock cylinder (see illustration). Disengage the link(s) from the handle and/or the lock cylinder.

8 Remove the outside handle/key lock cylinder retaining bolts and detach the handle from the door.

9 Installation is the reverse of removal. Tighten the bolts securely.

INSIDE DOOR HANDLE

▶ **Refer to illustrations 22.11 and 22.12**

10 Remove the door trim panel (see Section 13).

11 Remove the handle retaining screw and disengage the handle from the door (see illustration).

12 Disengage the handle-to-latch cables and remove the handle from the door (see illustration).

13 Installation is the reverse of removal.

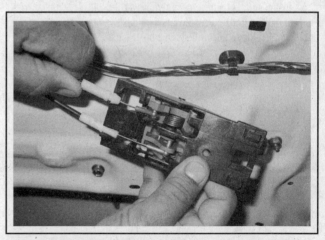

22.12 Detach the cables from the back of the inside handle

23 Steering column covers - removal and installation

▶ Refer to illustration 23.4

❊❊ WARNING:

The models covered by this manual are equipped with Supplemental Restraint Systems (SRS), more commonly known as airbags. Always disable the airbag system before working in the vicinity of any airbag system components to avoid the possibility of accidental deployment of the airbag(s), which could cause personal injury (see Chapter 12).

1 Disconnect the negative battery cable.
2 Disable the airbag system (see Chapter 12).
3 Remove the steering wheel (see Chapter 10).
4 Remove the steering column cover screws (see illustration).
5 Separate the cover halves and remove them from the column.
6 Installation is the reverse of removal.

23.4 Remove the three screws and detach the steering column cover halves

24 Console - removal and installation

▶ Refer to illustrations 24.3 and 24.4

❊❊ WARNING:

The models covered by this manual are equipped with Supplemental Restraint Systems (SRS), more commonly known as airbags. Always disable the airbag system before working in the vicinity of any airbag system components to avoid the possibility of accidental deployment of the airbag(s), which could cause personal injury (see Chapter 12).

1 Disconnect the negative battery cable.
2 Remove the manual shift lever knob (see Chapter 7).
3 Detach the cup holder from the center console (see illustration).
4 Remove the retaining screws and remove the console (see illustration).
5 Installation is the reverse of removal.

24.3 Detach the cup holder from the center console

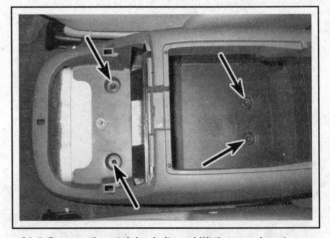

24.4 Remove the retaining bolts and lift the console out

25 Dashboard trim panels - removal and installation

❊❊ WARNING:

The models covered by this manual are equipped with Supplemental Restraint Systems (SRS), more commonly known as airbags. Always disable the airbag system before working in the vicinity of any airbag system components to avoid the possibil-

ity of accidental deployment of the airbag(s), which could cause personal injury (see Chapter 12).

1 Disconnect the negative battery cable.

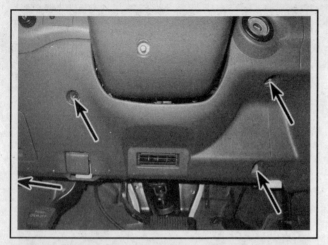

25.2 Remove the lower left finish panel screws

25.3 Remove the cluster bezel screws

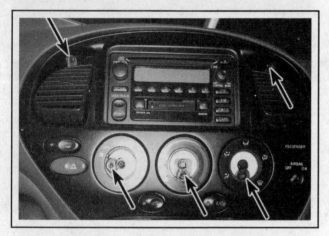

25.6a Remove the center cluster bezel screws

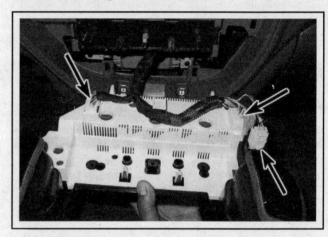

25.6b Pull the bezel out and disconnect the electrical connectors

LOWER LEFT FINISH PANEL

▶ Refer to illustration 25.2

  2 Remove the lower left finish panel retaining screws (see illustration), pull off the lower left finish panel and detach the hood release cable (see Section 10).

INSTRUMENT CLUSTER BEZEL

▶ Refer to illustration 25.3

  3 Remove the instrument cluster bezel screws (see illustration).
  4 Detach the bezel and pull it out of the dashboard.

CENTER CLUSTER FINISH PANEL

▶ Refer to illustrations 25.6a and 25.6b

  5 Remove the heater control knobs.
  6 Remove the center cluster finish panel retaining screws, pull off the center cluster finish panel and unplug the electrical connectors (see illustrations).

CENTER LOWER FINISH PANEL

▶ Refer to illustrations 25.7 and 25.8

  7 Remove the screws and detach the panel (see illustration).

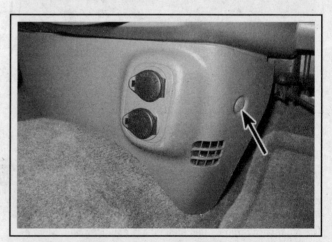

25.7 Use a small screwdriver to pry out the cover(s) and remove the center lower finish panel

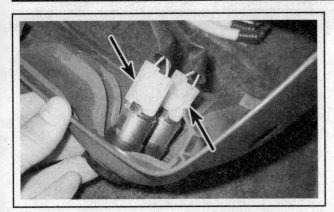

25.8 Unplug the center lower finish panel electrical connectors

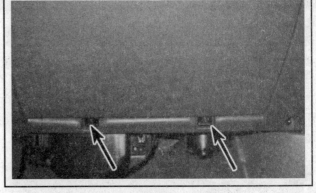

25.9 Remove the two screws and lower the glove compartment door from the dash

8   Pull the center cluster finish panel out and unplug the electrical connectors (see illustration).

GLOVE COMPARTMENT

♦ **Refer to illustrations 25.9 and 25.10**

9   Remove the screws, open the compartment door and lower it from the dashboard (see illustration).

10   Remove the screws, detach the glove compartment and lower it from the dashboard (see illustration).

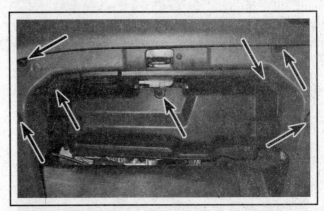

25.10 Remove the screws and detach the glove compartment

26 Cowl vent grille - removal and installation

♦ **Refer to illustrations 26.3a and 26.3b**

1   Mark the position of the windshield wiper blades on the windshield with a wax marking pen or pieces of tape.

2   Pry the wiper arm covers up, remove the nuts and detach the wipers.

3   Remove the cowl retaining clips and screws and remove the cowl (see illustrations).

4   Installation is the reverse of removal. Make sure to align the wiper blades with the marks made during removal.

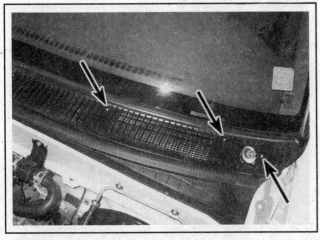

26.3a Left side cowl vent grille mounting screws

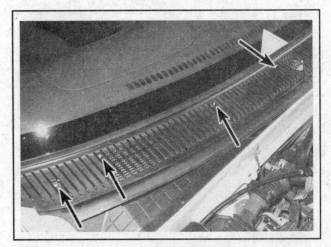

26.3b Right side cowl vent grille mounting screws

27 Seats - removal and installation

✳✳ WARNING 1:

Some models are equipped with seat belt pre-tensioners, which are pyrotechnic (explosive) devices that tighten the seat belts during an impact of sufficient force. Always disable the airbag system before working in the vicinity of any restraint system component to avoid the possibility of accidental deployment of the airbag(s) and seat belt pre-tensioners, which could cause personal injury (see Chapter 12).

✳✳ WARNING 2:

Do not use a memory saving device to preserve the ECM's memory when working on or near restraint system components.

FRONT SEAT

▶ **Refer to illustration 27.2**

1 Position the seat all the way forward and all the way to the rear to access the front seat retaining bolts.

2 Detach any bolt trim covers and remove the retaining bolts (see illustration).

3 Tilt the seat upward to access the underneath, then disconnect any electrical connectors and lift the seat from the vehicle.

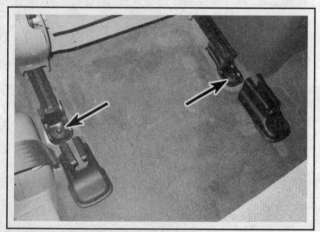

27.2 Remove the covers and the bolts, then remove the seat

4 Installation is the reverse of removal.

REAR SEAT

Tundra models

▶ **Refer to illustration 27.6**

5 Unlatch the seat cushions and lift them up.

6 Unscrew the four bolts, pull the seat up and remove it from the vehicle (see illustration).

7 Installation is the reverse of removal.

Sequoia models

Middle seat

8 Fold the seat back down, then unlatch the seat and flip the entire seat up and forward.

9 Carefully pry off the covers, then remove the mounting bolts and lift the seat from the vehicle.

10 Installation is the reverse of removal.

Back seat

11 Unlatch the seat, then flip it up and forward.

12 Remove the covers and unscrew the bolts from the seat frame. Remove the seat from the vehicle.

13 Installation is the reverse of removal.

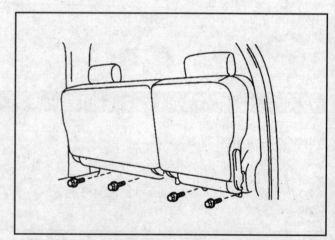

27.6 Rear seat mounting details - Tundra models

Section

12

CHASSIS ELECTRICAL SYSTEM

1   General information

The electrical system is a 12-volt, negative ground type. Power for the lights and all electrical accessories is supplied by a lead/acid-type battery which is charged by the alternator.

This Chapter covers repair and service procedures for the various electrical components not associated with the engine. Information on the battery, alternator, ignition system and starter motor can be found in Chapter 5. It should be noted that when portions of the electrical system are serviced, the negative battery cable should be disconnected from the battery to prevent electrical shorts and/or fires.

2   Electrical troubleshooting - general information

▶ **Refer to illustrations 2.5a, 2.5b, 2.6 and 2.9**

A typical electrical circuit consists of an electrical component, any switches, relays, motors, fuses, fusible links or circuit breakers related to that component and the wiring and connectors that link the component to both the battery and the chassis. To help you pinpoint an electrical circuit problem, wiring diagrams are included at the end of this Chapter.

Before tackling any troublesome electrical circuit, first study the appropriate wiring diagrams to get a complete understanding of what makes up that individual circuit. Trouble spots, for instance, can often be narrowed down by noting if other components related to the circuit are operating properly. If several components or circuits fail at one time, chances are the problem is in a fuse or ground connection, because several circuits are often routed through the same fuse and ground connections.

Electrical problems usually stem from simple causes, such as loose or corroded connections, a blown fuse, a melted fusible link or a failed relay. Visually inspect the condition of all fuses, wires and connections in a problem circuit before troubleshooting the circuit.

If test equipment and instruments are going to be utilized, use the diagrams to plan ahead of time where you will make the necessary connections in order to accurately pinpoint the trouble spot.

The basic tools needed for electrical troubleshooting include a circuit tester or voltmeter (a 12-volt bulb with a set of test leads can also be used), a continuity tester, which includes a bulb, battery and set of test leads, and a jumper wire, preferably with a circuit breaker incorporated, which can be used to bypass electrical components (see illustrations). Before attempting to locate a problem with test instruments, use the wiring diagram(s) to decide where to make the connections.

VOLTAGE CHECKS

Voltage checks should be performed if a circuit is not functioning properly. Connect one lead of a circuit tester to either the negative battery terminal or a known good ground. Connect the other lead to a connector in the circuit being tested, preferably nearest to the battery or fuse (see illustration). If the bulb of the tester lights, voltage is present, which means that the part of the circuit between the connector and the battery is problem free. Continue checking the rest of the circuit in the same fashion. When you reach a point at which no voltage is present, the problem lies between that point and the last test point with voltage. Most of the time the problem can be traced to a loose connection.

➡**Note: Keep in mind that some circuits receive voltage only when the ignition key is in the Accessory or Run position.**

FINDING A SHORT

One method of finding shorts in a live circuit is to remove the fuse and connect a test light in place of the fuse terminals (fabricate two jumper wires with small spade terminals, plug the jumper wires into the fuse box and connect the test light). There should be voltage present in the circuit. Move the suspected wiring harness from side-to-side while watching the test light. If the bulb goes off, there is a short to ground somewhere in that area, probably where the insulation has rubbed through.

GROUND CHECK

Perform a ground test to check whether a component is properly

2.5a The most useful tool for electrical troubleshooting is a digital multimeter that can check volts, amps, and test continuity

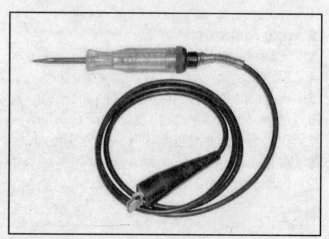

2.5b A simple test light is a very handy tool for testing voltage

2.6 In use, a basic test light's lead is clipped to a known good ground, then the pointed probe can test connectors, wires or electrical sockets - if the bulb lights, the circuit being tested has battery voltage

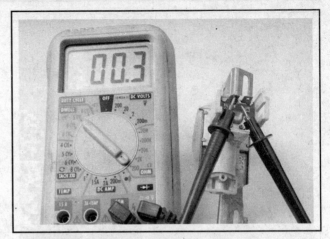

2.9 With a multimeter set to the ohms scale, resistance can be checked across two terminals - when checking for continuity, a low reading indicates continuity, a high reading or infinity indicates lack of continuity

grounded. Disconnect the battery and connect one lead of a continuity tester or multimeter (set to the ohms scale), to a known good ground. Connect the other lead to the wire or ground connection being tested. If the resistance is low (less than 5 ohms), the ground is good. If the bulb on a self-powered test light does not go on, the ground is not good.

CONTINUITY CHECK

A continuity check is done to determine if there are any breaks in a circuit - if it is passing electricity properly. With the circuit off (no power in the circuit), a self-powered continuity tester or multimeter can be used to check the circuit. Connect the test leads to both ends of the circuit (or to the "power" end and a good ground), and if the test light comes on the circuit is passing current properly (see illustration). If the resistance is low (less than 5 ohms), there is continuity; if the reading is 10,000 ohms or higher, there is a break somewhere in the circuit. The same procedure can be used to test a switch, by connecting the continuity tester to the switch terminals. With the switch turned On, the test light should come on (or low resistance should be indicated on a meter).

FINDING AN OPEN CIRCUIT

When diagnosing for possible open circuits, it is often difficult to locate them by sight because the connectors hide oxidation or terminal misalignment. Merely wiggling a connector on a sensor or in the wiring harness may correct the open circuit condition. Remember this when an open circuit is indicated when troubleshooting a circuit. Inter

mittent problems may also be caused by oxidized or loose connections.

Electrical troubleshooting is simple if you keep in mind that all electrical circuits are basically electricity running from the battery, through the wires, switches, relays, fuses and fusible links to each electrical component (light bulb, motor, etc.) and to ground, from which it is passed back to the battery. Any electrical problem is an interruption in the flow of electricity to and from the battery.

CONNECTORS

Most electrical connections on these vehicles are made with multi-wire plastic connectors. The mating halves of many connectors are secured with locking clips molded into the plastic connector shells. The mating halves of large connectors, such as some of those under the instrument panel, are held together by a bolt through the center of the connector.

To separate a connector with locking clips, use a small screwdriver to pry the clips apart carefully, then separate the connector halves. Pull only on the shell, never pull on the wiring harness as you may damage the individual wires and terminals inside the connectors. Look at the connector closely before trying to separate the halves. Often the locking clips are engaged in a way that is not immediately clear. Additionally, many connectors have more than one set of clips.

Each pair of connector terminals has a male half and a female half. When you look at the end view of a connector in a diagram, be sure to understand whether the view shows the harness side or the component side of the connector. Connector halves are mirror images of each other, and a terminal shown on the right side end-view of one half will be on the left side end view of the other half.

3   Fuses - general information

▶ **Refer to illustrations 3.1a, 3.1b and 3.3**

1   The electrical circuits of the vehicle are protected by a combination of fuses, circuit breakers and fusible links. The passenger compartment fuse block is located in the left end of the instrument panel (see illustration). The engine compartment fuse block (see illustration) is located in the left side of the engine compartment.

2   Each of the fuses is designed to protect a specific circuit, and the various circuits are identified on the fuse panel itself.

3   Miniaturized fuses are employed in the fuse block. These compact fuses, with blade terminal design, allow fingertip removal and replacement. If an electrical component fails, always check the fuse first. The best way to check the fuses is with a test light. Check for

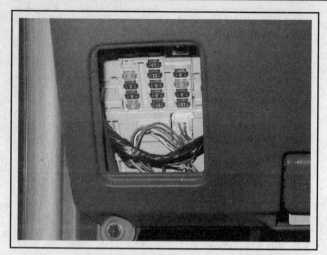

3.1a The passenger compartment fuse/relay block is located in the left end of the instrument panel behind a removable trim panel

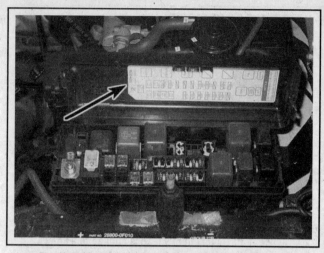

3.1b The engine compartment fuse/relay block is located on the driver's side of the engine compartment - the various circuits are identified on the fuse panel label on the inside of the cover

power at the exposed terminal tips of each fuse. If power is present on one side of the fuse but not the other, the fuse is blown. A blown fuse can also be confirmed by visually inspecting it (see illustration).

4   Be sure to replace blown fuses with the correct type. Fuses of different ratings are physically interchangeable, but only fuses of the proper rating should be used. Replacing a fuse with one of a higher or lower value than specified is not recommended. Each electrical circuit needs a specific amount of protection. The amperage value of each fuse is molded into the fuse body.

5   If the replacement fuse fails immediately, don't replace it again until the cause of the problem is isolated and corrected. In most cases, the cause will be a short circuit in the wiring caused by a broken or deteriorated wire.

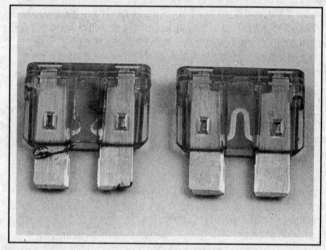

3.3 When a fuse blows, the element between the terminals burns - the fuse on the left is blown, the fuse on the right is good

4   Fusible links - general information

Some circuits are protected by fusible links. The links are used in circuits which are not ordinarily fused, such as the ignition circuit.

Cartridge type fusible links are located in the engine compartment fuse block and are similar to a large fuse (see illustration 3.3), and, after disconnecting the negative battery cable, are simply unplugged and replaced by a unit of the same amperage.

5   Circuit breakers - general information

Circuit breakers protect components such as power windows, power door locks and headlights.

On some models the circuit breaker resets itself automatically, so an electrical overload in a circuit breaker protected system will cause the circuit to fail momentarily, then come back on. If the circuit doesn't come back on, check it immediately. Once the condition is corrected, the circuit breaker will resume its normal function. Some circuit breakers must be reset manually.

6   Relays - general information and testing

GENERAL INFORMATION

1   Several electrical accessories in the vehicle, such as the fuel injection system, horns, starter, and fog lamps use relays to transmit the electrical signal to the component. Relays use a low-current circuit (the control circuit) to open and close a high-current circuit (the power circuit). If the relay is defective, that component will not operate properly. Most relays are mounted in the engine compartment fuse/relay boxes, with some specialized relays located above the interior fuse box under the dash. If a faulty relay is suspected, it can be removed and tested using the procedure below or by a dealer service department or a repair shop. Defective relays must be replaced as a unit.

TESTING

▸ **Refer to illustrations 6.3a, 6.3b and 6.6**

2   Refer to the wiring diagrams for the circuit to determine the proper connections for the relay you're testing. If you can't determine the correct connection from the wiring diagrams, however, you may be able to determine the test connections from the information that follows.

3   There are four basic types of relays used on these models (see illustrations). Some are normally open type and some normally closed, while others include a circuit of each type.

4   On most relays, two of the terminals are the relay control circuit (they connect to the relay coil which, when energized, closes the large contacts to complete the circuit). The other terminals are the power circuit (they are connected together within the relay when the control-circuit coil is energized).

5   Some relays may be marked as an aid to help you determine which terminals are the control circuit and which are the power circuit. If the relay is not marked, refer to the wiring diagrams at the end of this Chapter to determine the proper hook-ups for the relay you're testing.

6   To test a relay, connect an ohmmeter across the two terminals of the power circuit, continuity should not be indicated (see illustration). Now connect a fused jumper wire between one of the two control circuit terminals and the positive battery terminal. Connect another jumper wire between the other control circuit terminal and ground. When the connections are made, the relay should click and continuity should be indicated on the meter. On some relays, polarity may be critical, so, if the relay doesn't click, try swapping the jumper wires on the control circuit terminals.

7   If the relay fails the above test, replace it.

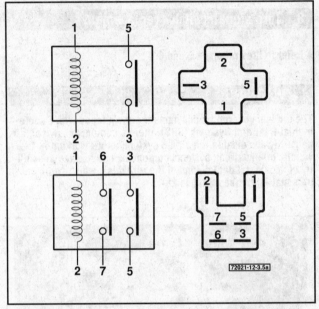

6.3a These two relays are typical normally open types; the one above completes a single circuit (terminal 5 to terminal 3) when energized - the lower relay type completes two circuits (6 and 7, and 3 and 5) when energized

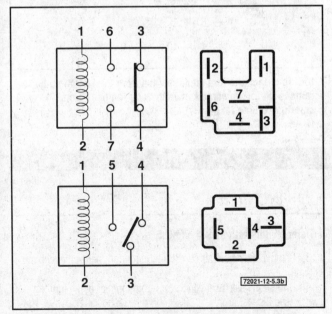

6.3b These relays are normally closed types, where current flows though one circuit until the relay is energized, which interrupts that circuit and completes the second circuit

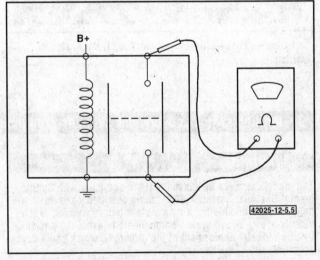

6.6 To test a typical four-terminal normally open relay, connect an ohmmeter to the two terminals of the power circuit - the meter should indicate continuity with the relay energized and no continuity with the relay not energized

7   Turn signal/hazard flasher - replacement

✳✳ WARNING:

The models covered by this manual are equipped with Supplemental Restraint Systems (SRS), more commonly known as airbags. Always disable the airbag system before working in the vicinity of any airbag system components to avoid the possibility of accidental deployment of the airbag(s), which could cause personal injury (see Section 28).

1   To replace the flasher relay, unplug it from the passenger compartment fuse block.

2   Make sure that the replacement unit is identical to the original. Compare the old one to the new one before installing it.

3   Installation is the reverse of removal.

8   Turn signal switch - replacement

◆ **Refer to illustrations 8.5 and 8.6**

✳✳ WARNING:

The models covered by this manual are equipped with Supplemental Restraint Systems (SRS), more commonly known as airbags. Always disable the airbag system before working in the vicinity of any airbag system components to avoid the possibility of accidental deployment of the airbag(s), which could cause personal injury (see Section 28).

1   Disconnect the cable from the negative terminal of the battery.

2   Disable the airbag system (see Section 28).

3   Remove the steering wheel (see Chapter 10).

4   Remove the steering column covers (see Chapter 11).

5   Unplug the electrical connector from the turn signal switch/headlight switch (see illustration).

6   Detach the turn signal switch/headlight switch from the switch body (see illustration).

7   Installation is the reverse of removal. Verify that the spiral cable is correctly centered before installing the steering wheel (see Chapter 10).

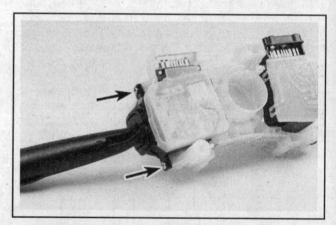

8.5 Unplug the electrical connector from the turn signal switch

8.6 To detach the turn signal switch from the switch body, remove these two screws (switch body removed from steering column for clarity)

9   Ignition switch and key lock cylinder - replacement

✳✳ WARNING:

The models covered by this manual are equipped with Supplemental Restraint Systems (SRS), more commonly known as airbags. Always disable the airbag system before working in the vicinity of any airbag system components to avoid the possibility of accidental deployment of the airbag(s), which could cause personal injury (see Section 28).

1   Disconnect the cable from the negative terminal of the battery.

2   Remove the left-side under-dash panel (see Chapter 11).

IGNITION SWITCH

◆ **Refer to illustrations 9.4 and 9.5**

3   Remove the heater/air conditioner duct from below the steering column.

4   Unplug the electrical connectors from the rear of the ignition switch (see illustration).

5   Remove the screws and detach the switch from the lock cylinder housing (see illustration).

6   Installation is the reverse of removal.

9.4 Unplug these electrical connectors from the rear of the ignition switch . . .

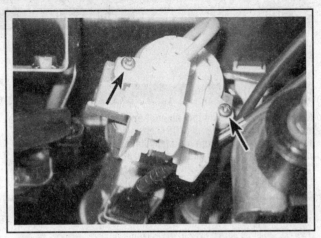

9.5 . . . then remove the ignition switch retaining screws

9.7 The ignition lock cylinder illumination ring is held on by this plastic pin

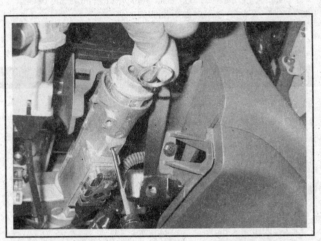

9.9 Insert a small screwdriver into the hole in the bottom of the ignition switch casting and press the release button while pulling the lock cylinder straight out

LOCK CYLINDER

♦ **Refer to illustrations 9.7 and 9.9**

7   Pull out the plastic pin and detach the lock cylinder illumination ring (see illustration).

8   Insert the ignition key and turn it to the ACC position.
9   Insert a small screwdriver into the hole in the bottom of the ignition switch casting and press the release button while pulling the lock cylinder straight out (see illustration).
10 Installation is the reverse of removal.

10 Headlight switch - replacement

✳✳ WARNING:

The models covered by this manual are equipped with Supplemental Restraint Systems (SRS), more commonly known as airbags. Always disable the airbag system before working in the vicinity of any airbag system components to avoid the possibility of accidental deployment of the airbag(s), which could cause personal injury (see Section 28).

1   Disconnect the cable from the negative terminal of the battery.
2   Disable the airbag system (see Section 28).

3   Remove the steering wheel (see Chapter 10).
4   Remove the steering column covers (see Chapter 11).
5   Unplug the electrical connector from the turn signal switch/headlight switch.
6   Detach the turn signal switch/headlight switch from the switch body (see illustration 8.6).
7   Installation is the reverse of removal. Verify that the spiral cable is correctly centered before installing the steering wheel (see Chapter 10).

11 Windshield wiper switch - replacement

▶ Refer to illustration 11.5

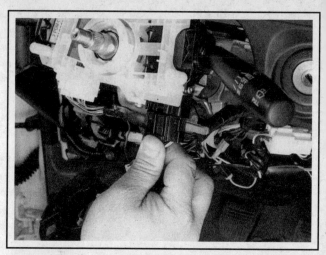

✳✳ WARNING:

The models covered by this manual are equipped with Supplemental Restraint Systems (SRS), more commonly known as airbags. Always disable the airbag system before working in the vicinity of any airbag system components to avoid the possibility of accidental deployment of the airbag(s), which could cause personal injury (see Section 28).

1 Disconnect the cable from the negative terminal of the battery.
2 Disable the airbag system (see Section 28).
3 Remove the steering wheel (see Chapter 10).
4 Remove the steering column covers (see Chapter 11).
5 Unplug the electrical connector, remove the screws and detach the windshield wiper switch (see illustration).
6 Installation is the reverse of removal.

11.5 Unplug the electrical connector from the windshield wiper switch

12 Headlight bulb - replacement

▶ Refer to illustrations 12.1, 12.2, 12.3 and 12.4

1 Unplug the electrical connector from the headlight bulb holder (see illustration).
2 Pull the rubber water shield off the bulb assembly (see illustration).
3 Detach the spring retainer (see illustration).
4 Withdraw the bulb from the housing (see illustration).
5 Installation is the reverse of removal.

✳✳ CAUTION:

Don't touch the bulb with your fingers. If you do, clean it with rubbing alcohol (the oil from your skin can cause the bulb to overheat and fail).

12.1 Unplug the electrical connector from the headlight bulb holder

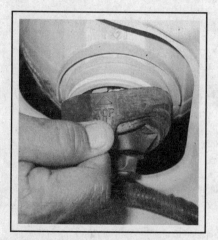

12.2 Pull off the water shield

12.3 Detach the spring retainer from the bulb holder

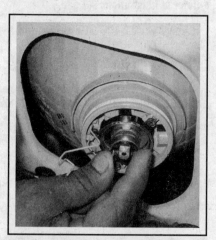

12.4 Remove the bulb holder and bulb from the headlight housing

13 Headlights - adjustment

♦ **Refer to illustrations 13.1 and 13.3**

➡**Note: The headlights must be aimed correctly. If adjusted incorrectly they could blind the driver of an oncoming vehicle and cause a serious accident or seriously reduce your ability to see the road. The headlights should be checked for proper aim every 12 months and any time a new headlight is installed or front end body work is performed. It should be emphasized that the following procedure is only an interim step that will provide temporary adjustment until a properly equipped shop can adjust the headlights.**

1   The headlights have an adjusting screw on the headlight housing to control up-and-down (vertical) movement (see illustration).

2   There are several methods of adjusting the headlights. The simplest method requires a blank wall 25-feet in front of the vehicle and a level floor.

3   Position masking tape vertically on the wall in reference to the vehicle centerline and the centerlines of both headlights (see illustration).

4   Position a horizontal tape line in reference to the centerline of all the headlights.

➡**Note: It may be easier to position the tape on the wall with the vehicle parked only a few inches away.**

5   Adjustment should be made with the vehicle sitting level, the gas tank half-full and no unusually heavy load in the vehicle.

6   Starting with the low beam adjustment, position the high intensity zone so it's two inches below the horizontal line and two inches to the side of the headlight vertical line away from oncoming traffic. Adjustment is made by turning the top adjusting screw clockwise to raise the beam and counterclockwise to lower the beam.

7   With the high beams on, the high intensity zone should be vertically centered with the exact center just below the horizontal line.

➡**Note: It may not be possible to position the headlight aim exactly for both high and low beams. If a compromise must be made, keep in mind that the low beams are the most used and have the greatest effect on driver safety.**

8   Have the headlights adjusted by a dealer service department or service station at the earliest opportunity.

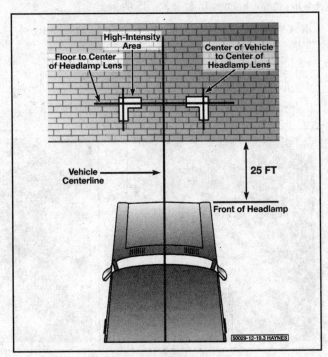

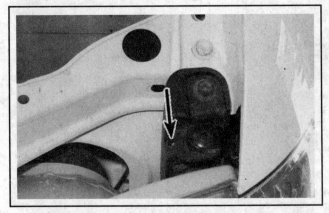

13.1 Insert a Phillips head screwdriver into the hole to make vertical headlight adjustments

13.3 Headlight adjustment details

14 Headlight housing - removal and installation

♦ **Refer to illustration 14.3**

1   Unplug the electrical connector from the headlight bulb (see illustration 12.1).

2   Remove the side marker light (see Section 15).

3   Remove the headlight housing retaining nuts and bolts (see illustration) and pull out the assembly.

4   Installation is the reverse of removal. Check and adjust headlights (see Section 13).

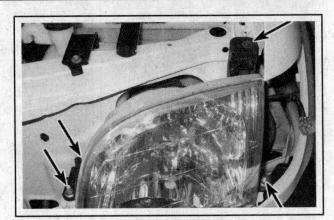

14.3 Remove the bolt and nuts and detach the headlight housing

15 Bulb replacement

FOG LIGHTS

▶ **Refer to illustrations 15.1 and 15.2**

1   Disconnect the electrical connector from the fog light bulb holder (see illustration).

2   Turn the bulb holder counterclockwise and remove it from the housing (see illustration).

3   To detach the bulb from the bulb holder, pull it straight out.

4   Installation is the reverse of removal.

FRONT SIDE MARKER/TURN SIGNAL LIGHT

▶ **Refer to illustrations 15.5, 15.6a, 15.6b and 15.7**

5   Remove the side marker light housing retaining screw, then pull out the plastic fastener (see illustration).

6   Strike the rear part of the housing in a forward direction with your hand to free the housing, then rotate the side marker light housing out of the fender and unplug the electrical connector (see illustrations).

7   Turn the bulb holder counterclockwise and remove it from the side marker light housing (see illustration).

8   To detach a bulb from the bulb holder, pull it straight out.

9   Installation is the reverse of removal. When installing the retaining screw, place it in position and press straight down until it clicks in place.

TAIL LIGHT/BRAKE LIGHT/TURN SIGNAL

▶ **Refer to illustrations 15.10 and 15.13**

10   On Tundra models, open the tailgate, remove the two bolts and rotate the tail light housing out for access to the bulbs (see illustration).

11   On Sequoia models, open the liftgate and remove the screws along the edge of the stop/tail and backup light housing, then detach the light housing from the vehicle.

12   On the liftgate-mounted tail light housings, remove the two retaining nuts, then use a screwdriver with a taped tip to carefully pry the housing off.

13   On all models, remove the bulb holder by rotating it counterclockwise, then remove the bulb by pulling it straight out (see illustration).

14   Installation is the reverse of removal.

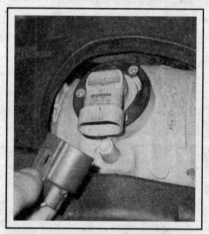

15.1 Disconnect the fog light electrical connector

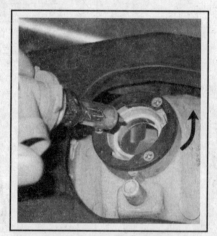

15.2 Turn the bulb holder counterclockwise and remove it from the fog light housing

15.5 Remove the side marker housing retaining screw, then pull out the plastic fastener

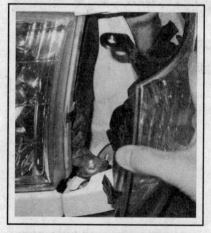

15.6a Rotate the side marker housing out of the fender

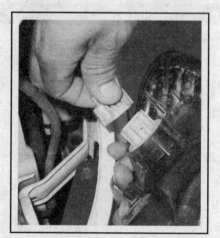

15.6b Unplug the electrical connector from the bulb holder

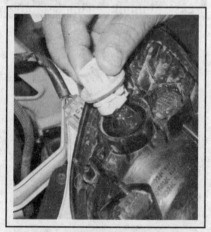

15.7 Turn the bulb holder counterclockwise and pull it out

15.10 To detach the Tundra tail light housing, remove the screws and rotate it out

15.13 To replace a bulb, simply turn the holder counterclockwise and pull it out of the tail light housing - pull the bulb straight out of the holder

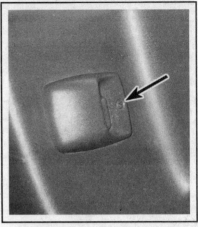

15.15 To remove the license plate bulb holder on side-mounted license plate lights, remove the screw and detach the housing from the vehicle (Tundra)

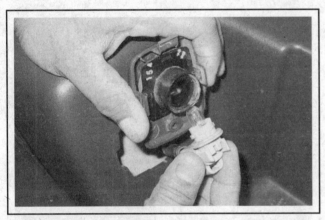

15.16 Rotate the bulb holder counterclockwise and withdraw it from the housing (Tundra)

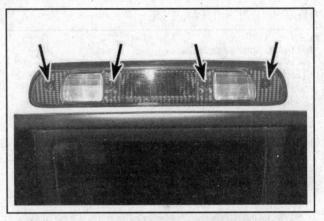

15.20 To get to the cargo light bulb (or high-mounted brake light bulbs), remove the screws and detach the lens

LICENSE PLATE LIGHT

▶ Refer to illustrations 15.15 and 15.16

15 On Tundra models with side-mounted lights, remove the screw and detach the housing from the bumper (see illustration). On center mounted lights, squeeze the tabs together and lower the bulb housing.

16 To remove the bulb holder, turn it counterclockwise and pull it out (see illustration).

17 To remove the bulb from the holder, pull it straight out.

18 On Sequoia models, remove the retaining screws and detach the lens. Remove the bulb from the holder by pulling it straight out of the housing.

19 Installation is the reverse of removal.

HIGH-MOUNTED BRAKE AND CARGO LIGHT

▶ Refer to illustrations 15.20 and 15.21

20 Remove the screws and detach the lens for bulb access (see illustration).

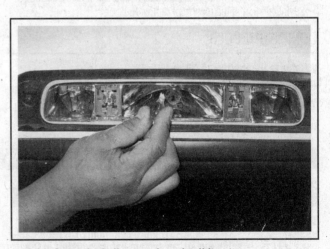

15.21 Grasp the bulb securely and pull it out

21 Pull out the defective bulb (see illustration).

22 Replace the bulb and install the lens and screws.

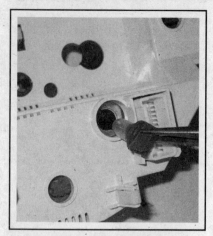

15.24 To replace the heater control panel bulbs, first remove the center dashboard trim bezel, then rotate the bulbs and pull them straight out

15.27 To remove a bulb from the instrument cluster, rotate the bulb holder counterclockwise and pull it out

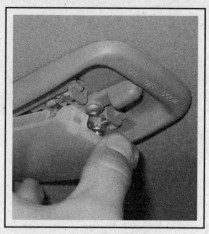

15.30 Detach the dome light lens

HEATER CONTROL PANEL LIGHT

▶ Refer to illustration 15.24

23 Remove the dashboard trim panel surrounding the heater and air conditioning control panel (see Chapter 11).
24 Pull out the defective bulb (see illustration).
25 Installation is the reverse of removal.

INSTRUMENT CLUSTER LIGHT BULBS

▶ Refer to illustration 15.27

26 Remove the instrument cluster (see Section 21).
27 Rotate the bulb holder counterclockwise and pull it out of the cluster (see illustration).
28 To remove the bulb from the holder, simply pull it straight out.
29 Installation is the reverse of removal.

DOME LIGHT BULB

▶ Refer to illustration 15.30 and 15.31

30 Detach the dome light lens (see illustration).

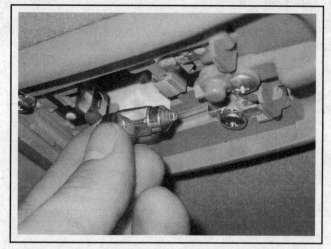

15.31 Grasp the bulb securely and detach it from the spring clip at each end

31 Grasp the bulb securely and detach it from the spring clips at each end (see illustration).
32 Installation is the reverse of removal.

16 Radio and speakers - removal and installation

✳✳ WARNING:

The models covered by this manual are equipped with Supplemental Restraint Systems (SRS), more commonly known as airbags. Always disable the airbag system before working in the vicinity of any airbag system components to avoid the possibility of accidental deployment of the airbag(s), which could cause personal injury (see Section 28).

RADIO

▶ Refer to illustration 16.3

1   Detach the cable from the negative terminal of the battery.
2   Remove the center cluster finish bezel (see Chapter 11).
3   Remove the radio retaining bolts (see illustration) and pull out the radio, then disconnect the antenna lead and the electrical connector.

16.3 Remove the mounting bolts, detach the radio from the instrument panel and disconnect the negative electrical connector

16.7a To remove a door-mounted speaker, remove the retaining screws, pull off the speaker and unplug the electrical connector

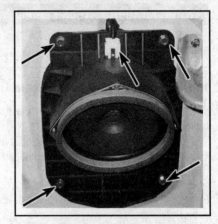

16.7b Tundra access door-mounted speaker installation details - unplug the electrical connector and remove the speaker retaining screws

4 Remove the radio from the instrument panel.
5 Installation is the reverse of removal.

SPEAKERS

▶ **Refer to illustrations 16.7a and 16.7b**

6 All models have at least one speaker in each door. Extra cab Tun-dra models are also equipped with a pair of rear speakers. To remove any door-mounted speaker, remove the door trim panel (see Chapter 11).

7 Remove the speaker retaining screws and unplug the electrical connector (see illustrations).
8 Installation is the reverse of removal.

17 Antenna - removal and installation

▶ **Refer to illustrations 17.2 and 17.3**

➡**Note: The following procedure applies to conventional external type antennas. It does not apply to the glass-printed antennas used on some Sequoia models. Glass-printed antennas should be serviced by a dealer service department or other qualified repair shop.**

1 Remove the radio and unplug the antenna cable (see Section 16). Trace the routing of the cable and detach all cable clamps and/or clips. Attach about four or five feet of string or wire to the antenna lead to help aid the installation of the new antenna lead.

2 Use a small open-end wrench to unscrew the antenna mast (see illustration).
3 Remove the antenna base retaining nut (see illustration), then detach the antenna base and pull out the antenna lead until the string or wire is exposed.
4 Attach the string or wire to the new antenna lead and pull the antenna lead back through.
5 Installation is the reverse of removal.

17.2 Use a small open-end wrench to unscrew the antenna mast

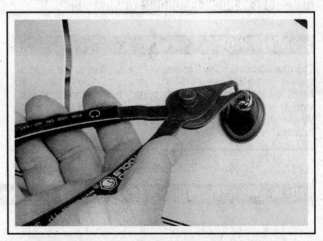

17.3 Use a pair of snap-ring pliers or similar tool to loosen and remove the antenna base retaining nut

18 Wiper motor - replacement

FRONT WIPER MOTOR

♦ **Refer to illustrations 18.2, 18.3 and 18.4**

1   Remove the cowl (see Chapter 11) and disconnect the electrical connector from the wiper motor.

2   Remove the wiper motor mounting bolts (see illustration).

3   Pull out the motor and detach the lever arm from the wiper link (see illustration).

4   To remove the wiper linkage, detach the bolts from the cowl (see illustration) and remove it through the vent hole.

5   Installation is the reverse of removal.

REAR WIPER MOTOR (SEQUOIA MODELS)

6   To detach the wiper arm, flip up the cover and remove the nut.

7   Open the liftgate and remove the trim panel and service hole cover (see Chapter 11).

8   Remove the retaining bolts, unplug the electrical connector, then lift the wiper motor from the vehicle.

9   Installation is the reverse of removal.

18.2 To remove the windshield wiper motor, remove the mounting bolts

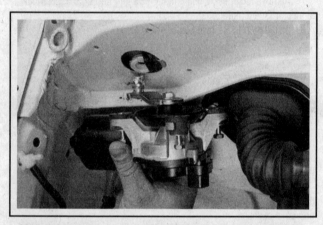

18.3 Pull out the motor and detach the lever arm from the wiper linkage

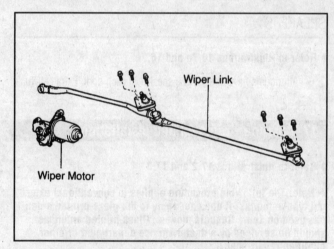

Wiper Link

Wiper Motor

18.4 A typical wiper linkage assembly

19 Rear window defogger switch (Sequoia models) - replacement

❋❋ WARNING:

The models covered by this manual are equipped with Supplemental Restraint Systems (SRS), more commonly known as airbags. Always disable the airbag system before working in the vicinity of any airbag system components to avoid the possibility of accidental deployment of the airbag(s), which could cause personal injury (see Section 28).

1   Detach the cable from the negative terminal of the battery.

2   Remove the center dash panel (see Chapter 11).

3   Remove the switch from dash panel.

4   Installation is the reverse of removal.

20 Rear window defogger (Sequoia models) - check and repair

1   The rear window defogger consists of a number of horizontal elements baked onto the glass surface.

2   Small breaks in the element can be repaired without removing the rear window.

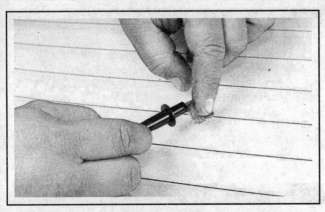

20.4 When measuring the voltage at the rear window defogger grid, wrap a piece of aluminum foil around the positive probe of the voltmeter and press the foil against the wire with your finger

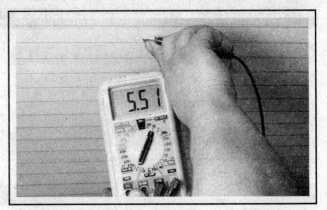

20.5 To determine if a heating element has broken, check the voltage at the center of each element - if the voltage is 6-volts, the element is unbroken - if the voltage is 12-volts, the element is broken between the center and the ground side - if there's no voltage, the wire is broken between the center of the wire and the power side

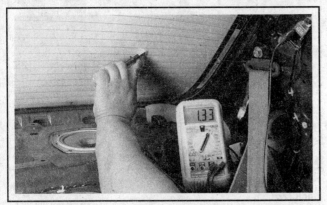

20.7 To find the break, place the voltmeter negative lead against the defogger ground terminal, place the voltmeter positive lead with the foil strip against the heating element at the positive terminal end and slide it toward the negative terminal end - the point at which the voltmeter deflects from 12-volts to zero volts is the point at which the element is broken

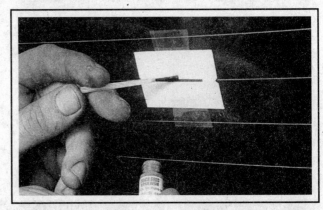

20.13 To use a defogger repair kit, apply masking tape to the inside of the window at the damaged area, then brush on the special conductive coating

CHECK

▶ **Refer to illustrations 20.4, 20.5 and 20.7**

3. Turn the ignition switch and defogger system switches to the ON position. Using a voltmeter, place the positive probe against the defogger grid positive terminal and the negative lead against the ground terminal. If battery voltage is not indicated, check the fuse, defogger switch and related wiring.

4   When measuring voltage during the next two tests, wrap a piece of aluminum foil around the tip of the voltmeter positive probe and press the foil against the heating element with your finger (see illustration).

5   Check the voltage at the center of each heating element (see illustration). If the voltage is 6-volts, the element is okay (there is no break). If the voltage is 12-volts, the element is broken between the center of the element and the ground side. If the voltage is 0-volts the element is broken between the center of the element and positive side.

6   If none of the elements are broken, connect the negative lead to a good body ground. The voltage reading should stay the same, if it doesn't the ground connection is bad.

7   To find the break, place the voltmeter negative lead against the defogger ground terminal. Place the voltmeter positive lead with the foil

strip against the heating element at the positive terminal end and slide it toward the negative terminal end. The point at which the voltmeter deflects from several volts to zero is the point at which the heating element is broken (see illustration).

REPAIR

▶ **Refer to illustration 20.13**

8   Repair the break in the element using a repair kit specifically recommended for this purpose, such as Dupont paste No. 4817 (or equivalent). Included in this kit is plastic conductive epoxy.

9   Prior to repairing a break, turn off the system and allow it to cool off for a few minutes.

10 Lightly buff the element area with fine steel wool, then clean it thoroughly with rubbing alcohol.

11 Use masking tape to mask off the area being repaired.

12 Thoroughly mix the epoxy, following the instructions provided with the repair kit.

13 Apply the epoxy material to the slit in the masking tape, overlapping the undamaged area about 3/4-inch on either end (see illustration).

14 Allow the repair to cure for 24 hours before removing the tape and using the system.

21 Instrument cluster - removal and installation

1 Detach the cable from the negative terminal of the battery.
2 Remove the instrument cluster trim bezel (see Chapter 11).
3 Remove the instrument cluster retaining screws, pull out the cluster and unplug the electrical connectors from the cluster.
4 Installation is the reverse of removal.

22 Horn - replacement

▸ **Refer to illustration 22.2**

1 The horns are located on the radiator support.
2 To replace a horn, unplug the electrical connector and remove the bracket bolt (see illustration).
3 Unbolt the bracket from the old horn and bolt it onto the new unit.
4 Installation is the reverse of removal.

22.2 To detach the horn from the body, remove this bracket bolt (arrow)

23 Power mirror control system - description and check

1 Electric rear view mirrors use two motors to move the glass; one for up and down adjustments and one for left-right adjustments.
2 The control switch has a selector portion which sends voltage to the left or right side mirror. With the ignition ON but the engine OFF, roll down the windows and operate the mirror control switch through all functions (left-right and up-down) for both the left and right side mirrors.
3 Listen carefully for the sound of the electric motors running in the mirrors.
4 If the motors can be heard but the mirror glass doesn't move, there's probably a problem with the drive mechanism inside the mirror.
5 If the mirrors do not operate and no sound comes from the mirrors, check the fuse (see Section 3).
6 If the fuse is OK, remove the mirror control switch from its mounting without disconnecting the wires attached to it. Turn the ignition ON and check for voltage at the switch. There should be voltage at one terminal. If there's no voltage at the switch, check for an open or short in the circuit between the fuse panel and the switch.
7 If the mirror motor fails to operate as described, replace the mirror assembly (see Chapter 11).

24 Cruise control system - description and check

1 The cruise control system maintains vehicle speed with an electrically operated motor located in the engine compartment, which is connected to the throttle lever by a cable. The system consists of the cruise control unit, brake switch, control switches and vehicle speed sensor. Some features of the system require special testers and diagnostic procedures which are beyond the scope of this manual. Listed below are some general procedures that may be used to locate common problems.
2 Locate and check the fuse (see Section 3).
3 Have an assistant operate the brake lights while you check their operation (voltage from the brake light switch deactivates the cruise control).
4 If the brake lights don't come on, or if they stay on all the time, correct the problem and re-test the cruise control.
5 Visually inspect the control cable between the cruise control motor and the throttle linkage for free movement; replace it if necessary.
6 The cruise control system uses a speed sensing device. The speed sensor is located in the speedometer. Check the electrical connectors at the instrument cluster (see Section 21).
7 Test drive the vehicle to determine if the cruise control is now working. If it isn't, take it to a dealer service department or an automotive electrical specialist for further diagnosis.

25 Power window system - description and check

1   The power window system operates electric motors, mounted in the doors, which lower and raise the windows. The system consists of the control switches, relays, the motors, regulators, glass mechanisms and associated wiring.

2   The power windows can be lowered and raised from the master control switch by the driver or by remote switches located at the individual windows. Each window has a separate motor which is reversible. The position of the control switch determines the polarity and therefore the direction of operation.

3   The circuit is protected by a fuse and a circuit breaker. Each motor is also equipped with an internal circuit breaker; this prevents one stuck window from disabling the whole system.

4   The power window system will only operate when the ignition switch is ON. In addition, many models have a window lockout switch at the master control switch which, when activated, disables the switches at the rear windows and, sometimes, the switch at the passenger's window also. Always check these items before troubleshooting a window problem.

5   These procedures are general in nature, so if you can't find the problem using them, take the vehicle to a dealer service department or other properly equipped repair facility.

6   If the power windows won't operate, always check the fuse and circuit breaker first.

7   If only the rear windows are inoperative, or if the windows only operate from the master control switch, check the rear window lockout switch for continuity in the unlocked position. Replace it if it doesn't have continuity.

8   Check the wiring between the switches and fuse panel for continuity. Repair the wiring, if necessary.

9   If only one window is inoperative from the master control switch, try the other control switch at the window.

➡Note: This doesn't apply to the drivers door window.

10 If voltage is reaching the motor, disconnect the glass from the regulator (see Chapter 11). Move the window up and down by hand while checking for binding and damage. Also check for binding and damage to the regulator. If the regulator is not damaged and the window moves up and down smoothly, replace the motor. If there's binding or damage, lubricate, repair or replace parts, as necessary.

11 If voltage isn't reaching the motor, check the wiring in the circuit for continuity between the switches and motors. You'll need to consult the wiring diagram for the vehicle. If the circuit is equipped with a relay, check that the relay is grounded properly and receiving voltage.

12 Test the windows after you are done to confirm proper repairs.

26 Power door lock system - description and check

1   A power door lock system operates the door lock actuators mounted in each door. The system consists of the switches, actuators, a control unit and associated wiring. Diagnosis can usually be limited to simple checks of the wiring connections and actuators for minor faults that can be easily repaired. Since this system uses an electronic control unit in-depth diagnosis should be left to a dealership service department.

2   Power door lock systems are operated by bi-directional solenoids located in the doors. The lock switches have two operating positions: Lock and Unlock. When activated, the switch sends a ground signal to the door lock control unit to lock or unlock the doors. Depending on which way the switch is activated, the control unit reverses polarity to the solenoids, allowing the two sides of the circuit to be used alternately as the feed (positive) and ground side.

3   Some vehicles may have an anti-theft system incorporated into the power locks. If you are unable to locate the trouble using the following general Steps, consult a dealer service department or other qualified repair shop.

4   Always check the circuit protection first. Some vehicles use a combination of circuit breakers and fuses.

5   Operate the door lock switches in both directions (Lock and Unlock) with the engine off. Listen for the click of the solenoids operating.

6   Test the switches for continuity. Remove the switches and have them checked by a dealer service department or other qualified automobile repair facility.

7   Check the wiring between the switches, control unit and solenoids for continuity. Repair the wiring if there's no continuity.

8   Check for a bad ground at the switches or the control unit.

9   If all but one lock solenoids operate, remove the trim panel from the affected door (see Chapter 11) and check for voltage at the solenoid while the lock switch is operated. One of the wires should have voltage in the Lock position; the other should have voltage in the Unlock position.

10 If the inoperative solenoid is receiving voltage, replace the solenoid.

11 If the inoperative solenoid isn't receiving voltage, check the circuit between the lock solenoid and the control unit.

➡Note: It's common for wires to break in the portion of the harness between the body and door (opening and closing the door fatigues and eventually breaks the wires).

27 Daytime Running Lights (DRL) - general information

The Daytime Running Lights (DRL) system used on some models illuminates the headlights whenever the engine is running. The only exception is with the engine running and the parking brake engaged. Once the parking brake is released, the lights will remain on as long as the ignition switch is on, even if the parking brake is later applied.

The DRL system supplies reduced power to the headlights so they won't be too bright for daytime use, while prolonging headlight life.

28 Airbags - general information

These models are equipped with a Supplemental Restraint System (SRS), more commonly known as an airbag. This system is designed to protect the driver and the front seat passenger from serious injury in the event of a head-on or frontal collision. It consists of an airbag module in the center of the steering wheel and another airbag module on the right side of the instrument panel plus, on some and later models, side airbags and curtain shield airbags designed to protect the occupants in a side impact and a sensing/diagnostic module which is mounted in the center of the vehicle below the instrument panel. These models are also equipped with a pair of impact sensors that are located at the front of the vehicle.

Some later models are equipped with seatbelt pre-tensioners, also part of the airbag system. The pre-tensioners are pyrotechnic (explosive) devices designed to retract the seat belts in the event of a collision.

On models equipped with pre-tensioners, do not remove the front seat belt retractor assemblies. Problems with the pre-tensioners will turn on the SRS (airbag) warning light on the dash. If any pre-tensioner problems are suspected, take the vehicle to a dealer service department.

AIRBAG MODULE

Steering wheel-mounted

The airbag inflator module contains a housing incorporating the cushion (airbag) and inflator unit, mounted in the center of the steering wheel. The inflator assembly is mounted on the back of the housing over a hole through which gas is expelled, inflating the bag almost instantaneously when an electrical signal is sent from the system. A spiral cable assembly on the steering column under the module carries this signal to the module. This spiral cable assembly can transmit an electrical signal regardless of steering wheel position.

Instrument panel-mounted

The passenger side airbag is mounted above the glove compartment and designated by the letters SRS (Supplemental Restraint System). It consists of an inflator containing an igniter, a bag assembly, a reaction housing and a trim cover.

The passenger airbag is considerably larger than the steering wheel-mounted unit and is supported by the steel reaction housing. The trim cover has a molded seam which splits when the bag inflates.

SENSING AND DIAGNOSTIC MODULE

The sensing and diagnostic module supplies the current to the airbag system in the event of the collision, even if battery power is cut off. It checks this system every time the vehicle is started, causing the "AIR BAG" light to go on then off, if the system is operating properly. If there is a fault in the system, the light will go on and stay on, flash, or the

dash will make a beeping sound. If this happens, the vehicle should be taken to your dealer immediately for service.

SIDE AND CURTAIN AIRBAGS

The passenger side airbag and inflator modules are mounted on the sides of the front seats contain an inflator containing an igniter and bag assembly. The curtain shield airbag assemblies run along the interior of the roof from the front A-pillar to the rear of the passenger compartment. In the event of a side impact both airbag assemblies are activated by the sensors mounted at the base of the center pillar behind the seats.

PRECAUTIONS

Disabling the SRS system

✷✷ WARNING 1:

Failure to follow these precautions could result in accidental deployment of the airbag and personal injury.

✷✷ WARNING 2:

Never install a "memory-saver" device, used to preserve PCM memory and radio station presets, when working on or around any of the airbag system components.

Whenever working in the vicinity of the steering wheel, instrument panel or any of the other SRS system components, the system must be disarmed. To disarm the system:

a) *Point the wheels straight ahead and turn the ignition key to the LOCK position.*
b) *Disconnect the cable from the negative terminal of the battery.*
c) *Wait at least two minutes for the back-up power supply capacitor to be depleted.*

Whenever handling an airbag module, always keep the airbag opening (trim side) pointed away from your body. Never place the airbag module on a bench or other surface with the airbag opening facing the surface. Always place the airbag module in a safe location with the airbag opening (trim side) facing up.

Never measure the resistance of any SRS component. An ohmmeter has a built-in battery supply that could accidentally deploy the airbag.

Never use electrical welding equipment on a vehicle equipped with an airbag without first disconnecting the negative battery cable.

Never dispose of a live airbag module. Return it to your dealer for safe deployment, using special equipment, and disposal.

29 Wiring diagrams - general information

Since it isn't possible to include all wiring diagrams for every year covered by this manual, the following diagrams are those that are typical and most commonly needed.

Prior to troubleshooting any circuits, check the fuse and circuit breakers (if equipped) to make sure they're in good condition. Make

sure the battery is properly charged and check the cable connections (see Chapter 1).

When checking a circuit, make sure that all connectors are clean, with no broken or loose terminals. When unplugging a connector, do not pull on the wires. Pull only on the connector housings themselves.

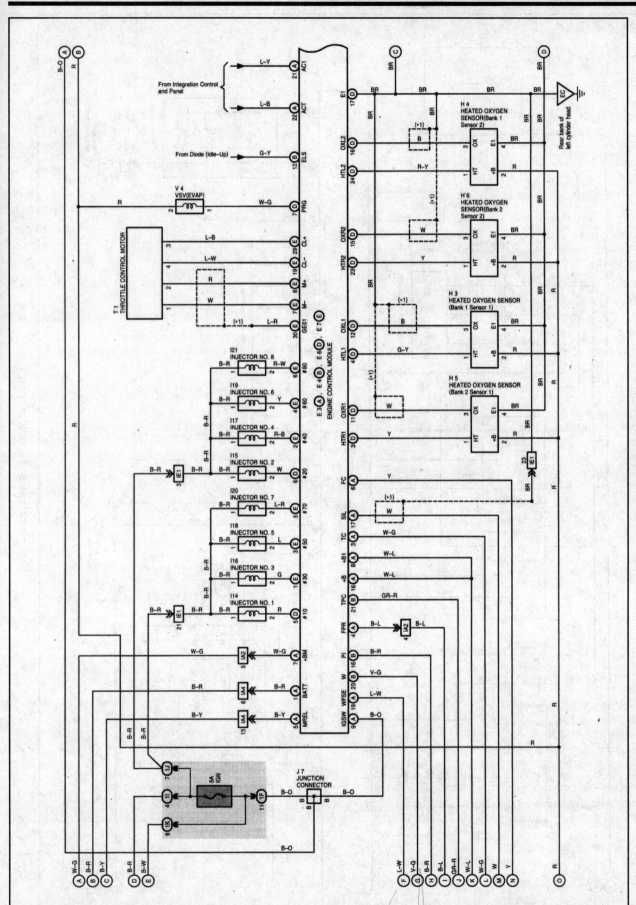

Engine control system - Tundra V8 models (1 of 2)

See Stop Light System

S 4
STOP LIGHT SW

*1 : Shielded

B–O

L–B 19 (A) ST–

L–W 16 (B) TACH

To Combination Meter

G–O 15 (B) SPD

From Park/Neutral Position SW(A/T), Clutch Start SW or Clutch Start Cancel SW(M/T) B–W 17 (B) STA

G–W 22 (IE1) G–W 11 (IA4) G–W 1

G–B 18 (IE1) G–B 18 (IA4) G–B 3

G–B

R–G 25 (IA4) R–G 2

R–G

V 1
VAPOR PRESSURE SENSOR

22 (B) PTNK

A 8
ACCEL POSITION SENSOR

G–W E2 VC G–B G–B 8 (D) VC

VPA2 L 9 (D) VPA2 7 (E)

VPA G–R 21 (D) VPA

T 2
THROTTLE POSITION SENSOR

G–W E2 VC G–B

VTA2 P–L 3 (D) VTA2 20 (D)

VTA1 B–Y 13 (D) VTA

E 2
ENGINE COOLANT TEMP. SENSOR

G–W 1 2 G 14 (D) THW

G–W G–W

M 1
MASS AIR FLOW METER

G–W 1 G–W E2 18 (D) E2

R–W 5 VG 10 (D) VG

B–W 4 EVG 19 (D) EVG

Y–G 3 2 THA 22 (D) THA

From "STA" Fuse P 20 (B) NSW

To Detection SW (Transfer L4 Position) L–R 22 (A) L4

CCS 23 (B) G

CMS 24 (B) G–B

RES/ACC SET/COAST CANCEL

CRUISE

CCS 4
CMS 5
EP 3 BR 1 (IE2) BR EC

C16 CRUISE CONTROL SW [COMB. SW]

Rear bank of left cylinder head

STP 6 (B) G–W From Stop Light SW

IGT8 6 (E) LG
IGT7 26 (E) P
IGT6 16 (E) P–L
IGT5 15 (E) G–W
IGT4 14 (E) R–W
IGT3 13 (E) B–Y
IGT1 11 (E) B–L
IGT2 12 (E) LG–B

To Igniter and Ignition Coil

IGF2 28 (E) B–W
IGF1 27 (E) B–R

From Igniter and Ignition Coil

E01 21 (E) BR BR
E02 1 (D) BR BR
E03 31 (E) BR BR
ME01 9 (E) BR

NE 23 (E) G

C 2
CRANKSHAFT POSITION SENSOR

R 1 2

(*1) BR

(*1)

NE– 22 (E) R L 1

C 1
CAMSHAFT POSITION SENSOR

G2 10 (E) Y 2

KNKL 18 (E) GR EC1 (4WD) EC2 (2WD) GR

K 1
KNOCK SENSOR 1

KNKR 17 (E) B EC1 EC2 B

K 2
KNOCK SENSOR 2

(*1) BR EC1 EC2 BR

(*1)

E3 (A) . E4 (B) . E6 (D) . E7 (E)
ENGINE CONTROL MODULE

EB Rear bank of right cylinder head

(A) B–O
(B) R
(C) BR
(D) BR

Engine control system - Tundra V8 models (2 of 2)

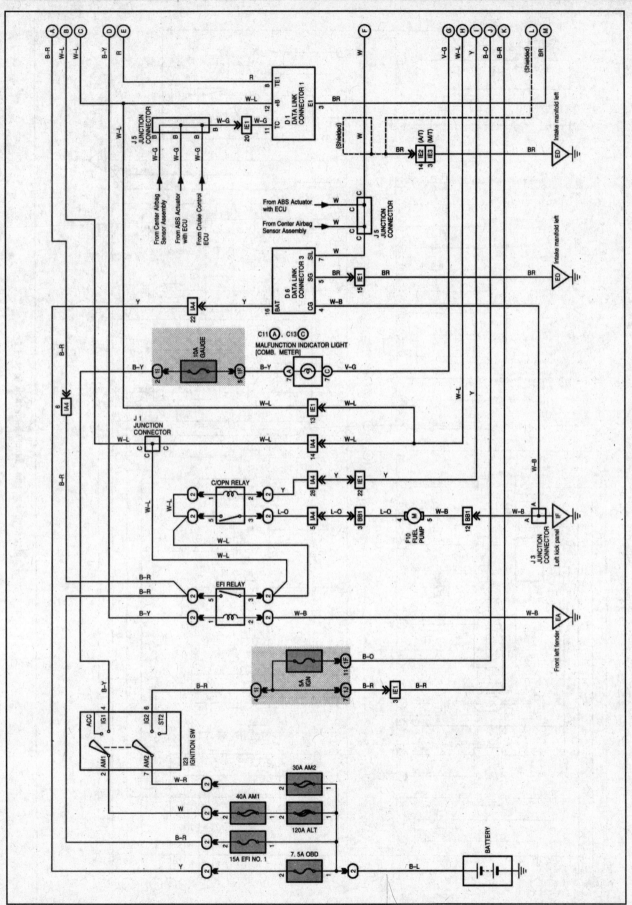

Engine control system - Tundra V6 models (1 of 2)

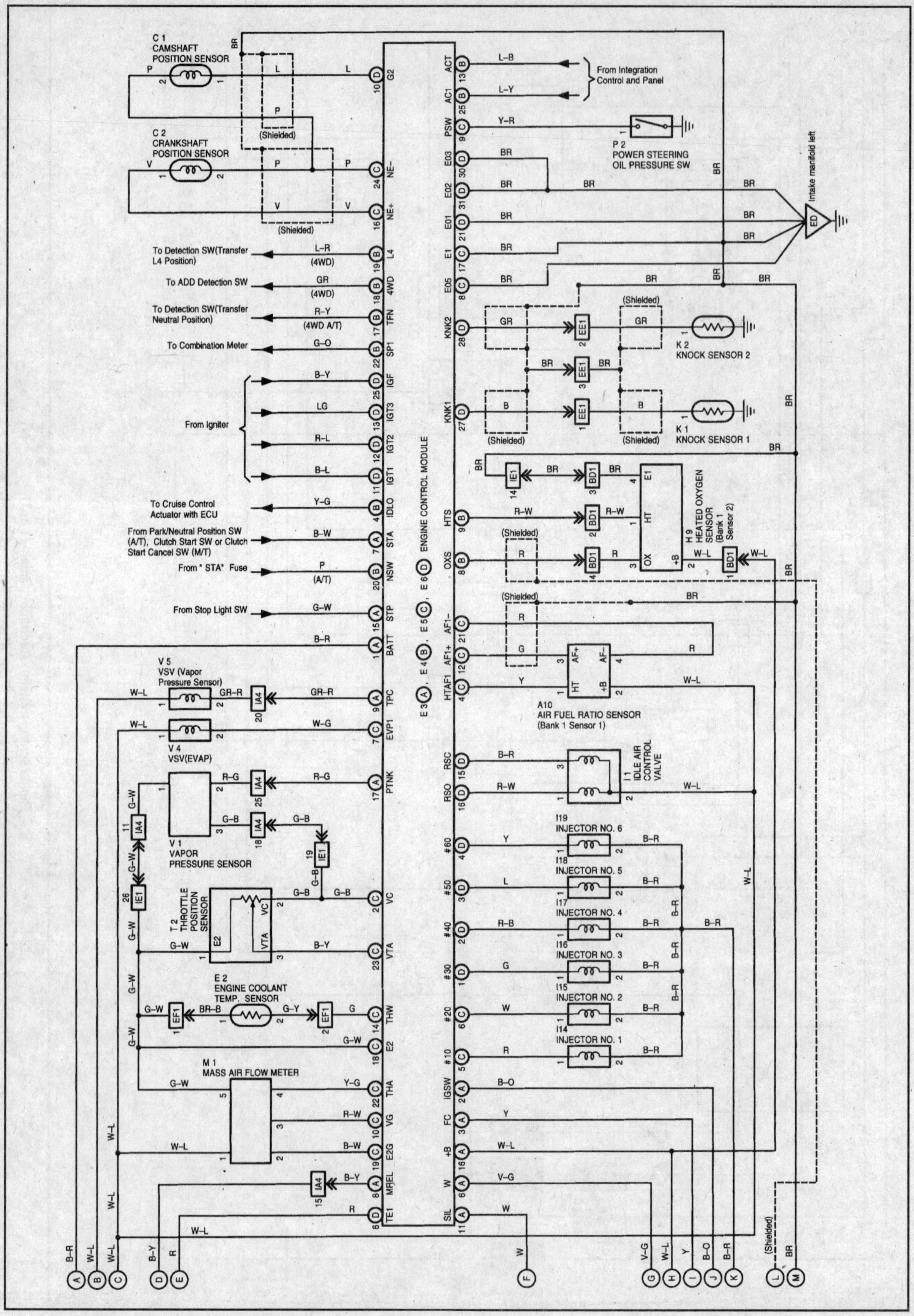

C 1
CAMSHAFT
POSITION SENSOR

C 2
CRANKSHAFT
POSITION SENSOR

(Shielded)

(Shielded)

To Detection SW(Transfer L4 Position) (4WD)
To ADD Detection SW (4WD)
To Detection SW(Transfer Neutral Position) (4WD A/T)
To Combination Meter

From Igniter

To Cruise Control Actuator with ECU
From Park/Neutral Position SW (A/T), Clutch Start SW or Clutch Start Cancel SW (M/T)
From " STA " Fuse (A/T)
From Stop Light SW

V 5
VSV (Vapor Pressure Sensor)

V 4
VSV(EVAP)

V 1
VAPOR PRESSURE SENSOR

T 2
THROTTLE POSITION SENSOR

E 2
ENGINE COOLANT TEMP. SENSOR

M 1
MASS AIR FLOW METER

ENGINE CONTROL MODULE

E 3 (A), E 4 (B), E 5 (C), E 6 (D)

From Integration Control and Panel

P 2
POWER STEERING OIL PRESSURE SW

Intake manifold left

ED

K 2
KNOCK SENSOR 2

K 1
KNOCK SENSOR 1

(Shielded)

H 9
HEATED OXYGEN SENSOR
(Bank 1 Sensor 2)

A10
AIR FUEL RATIO SENSOR
(Bank 1 Sensor 1)

I 1
IDLE AIR CONTROL VALVE

I19 INJECTOR NO. 6
I18 INJECTOR NO. 5
I17 INJECTOR NO. 4
I16 INJECTOR NO. 3
I15 INJECTOR NO. 2
I14 INJECTOR NO. 1

Engine control system - Tundra V6 models (2 of 2)

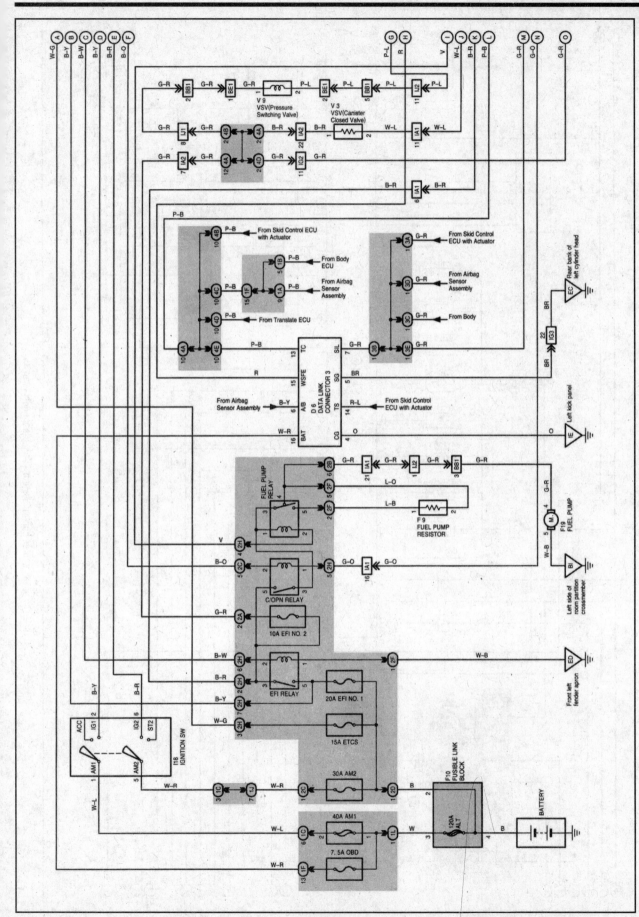

Engine control system - Sequoia models (1 of 3)

From Integration Control and Panel

P-L
W-R
LG-B

From * TAIL * Fuse G-Y

V 4 VSV(EVAP)
G-R
W-G

T 2 THROTTLE CONTROL MOTOR
L-B
L-W
V
P
(+1) L-R

I16 INJECTOR NO. 8 B-R R-W
I14 INJECTOR NO. 6 B-R Y
I12 INJECTOR NO. 4 B-R R-B
I10 INJECTOR NO. 2 B-R W
I15 INJECTOR NO. 7 B-R L-R
I13 INJECTOR NO. 5 B-R L
I11 INJECTOR NO. 3 B-R G
I9 INJECTOR NO. 1 B-R R

IG2 B-R
IG2 B-R

To Combination Meter R

W-G IA1 W-G
B-Y IA1 B-Y
B-W IA1 B-W

ENGINE CONTROL MODULE
E 4 (A) E 5 (B) E 6 (C)
E 7 (D) E 8 (E)

AC1 21 (A)
THWO 20 (A)
ACT 22 (A)
ELS 12 (B)
PRG 7 (D)
CL+ 29 (E)
CL- 19 (E)
M+ 8 (E)
M- 7 (E)
GE01 30 (E)
#80 6 (E)
#60 4 (E)
#40 2 (E)
#20 6 (D)
#70 5 (E)
#50 3 (E)
#30 1 (E)
#10 5 (D)
+BM 5 (E)
BATT 1 (A)
MREL 23 (A)

E1 17 (D)
OXL2 16 (D)
HTL2 24 (D)
OXR2 15 (D)
HTR2 23 (D)
OXL1 12 (D)
HTL 4 (D)
OXR1 11 (D)
HTR 3 (D)
FC 6 (A)
SIL 17 (A)
TC 3 (A)
+B1 8 (A)
+B 16 (A)
CCV 20 (B)
FPR 4 (A)
PI 16 (B)
W 23 (B)
WFSE 19 (A)
TBP 21 (B)
IGSW 9 (A)

BR BR BR
E1 EC

H 2 HEATED OXYGEN SENSOR (Bank 1 Sensor 2)
(+1) G
H 4 HEATED OXYGEN SENSOR (Bank 2 Sensor 2)
(+1) R
H 1 HEATED OXYGEN SENSOR (Bank 1 Sensor 1)
(+1) B
H 3 HEATED OXYGEN SENSOR (Bank 2 Sensor 1)
(+1) W

Rear bank of left cylinder head

J16 JUNCTION CONNECTOR

G-O
G-R
P-B
B-R
B-R
W-L
V IA1 V
LG-R
V-W
R
P-L

20A IGN2
10A IGN1
10A GAUGE

C 5 (A) C 6 (B)
COMBINATION METER
MALFUNCTION INDICATOR LAMP
CRUISE
V-W
R-L LG-R

Engine control system - Sequoia models (2 of 3)

A B C D E
B-O G-R BR B-R

A B C D E F G H I J K L M N O
W-G B-Y B-W D-Y B-R B-O P-L R V W-L B-R P-B G-R G-O G-R

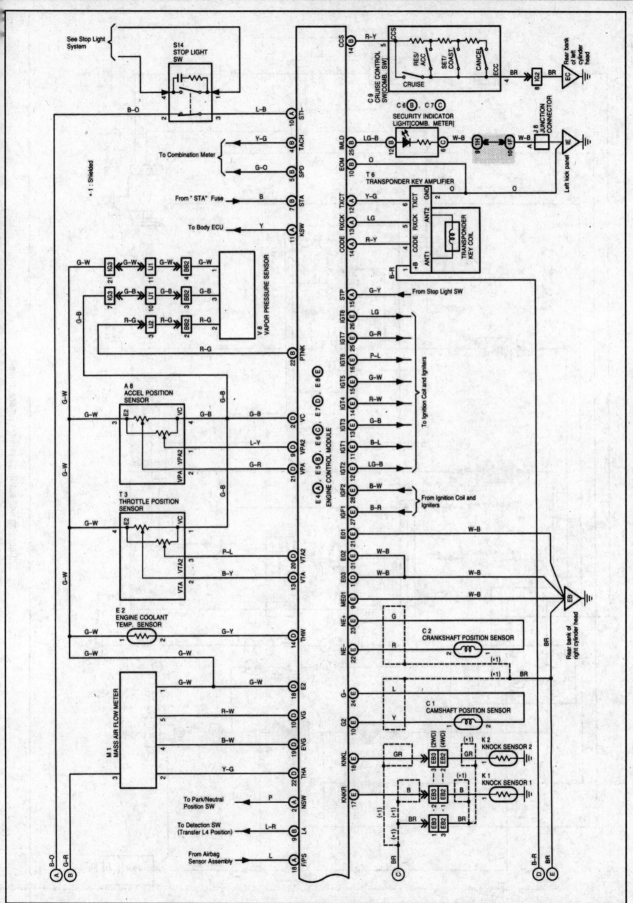

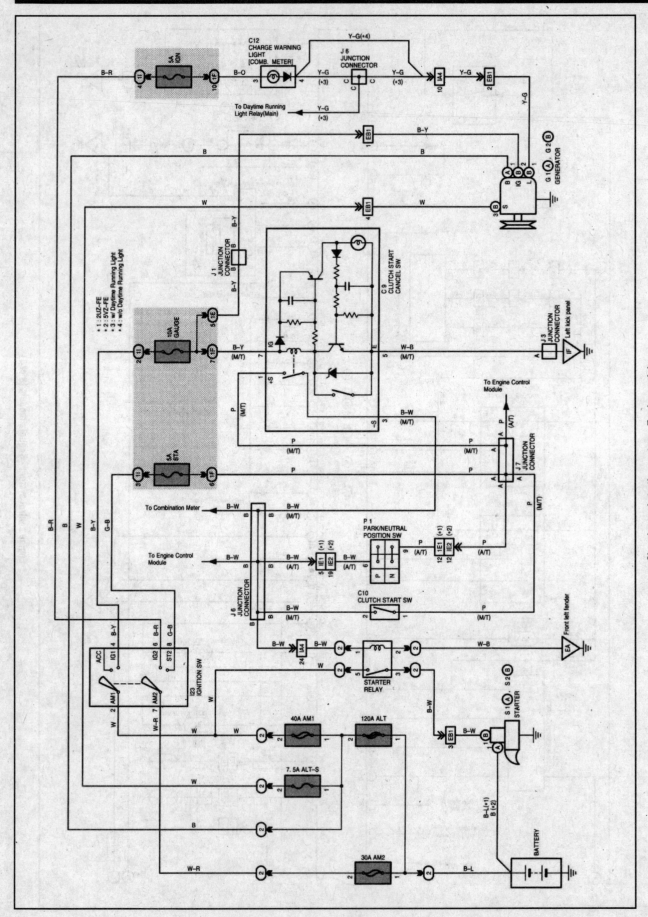

Charging and starting system - Tundra models

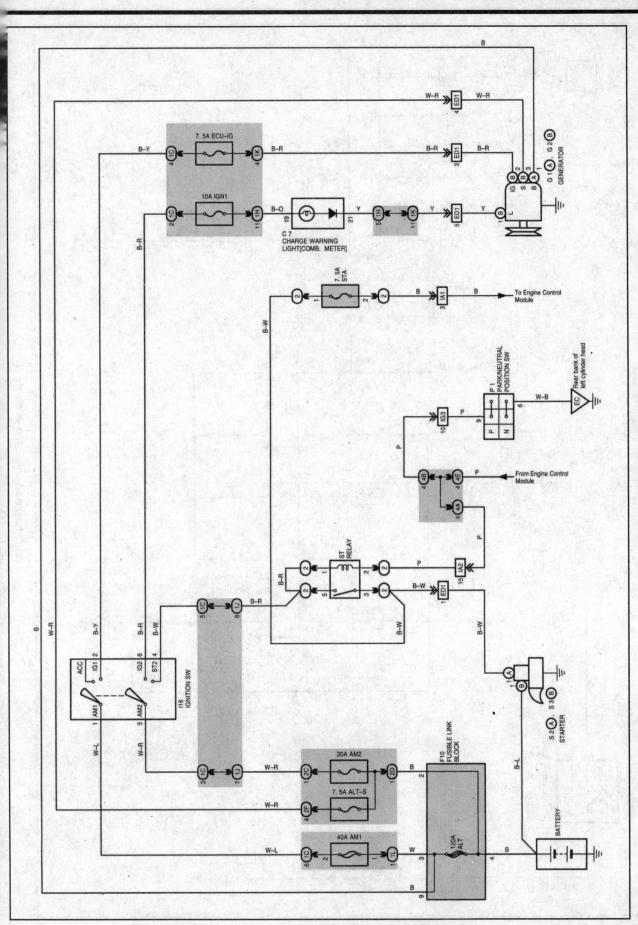

Charging and starting system - Sequoia models

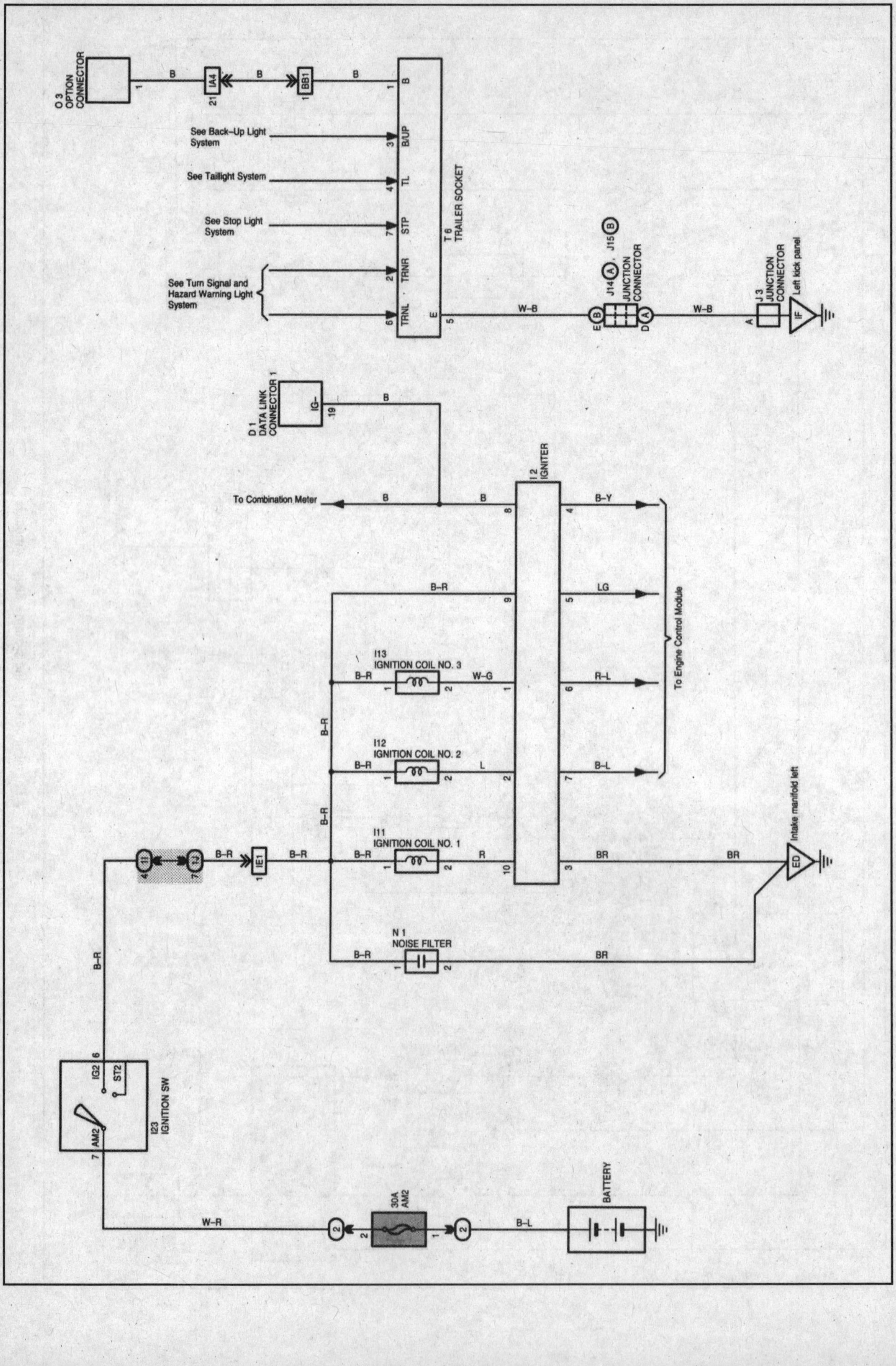

O 3
OPTION CONNECTOR

IA4
21

BB1
1

T 6
TRAILER SOCKET

See Back-Up Light System
See Taillight System
See Stop Light System
See Turn Signal and Hazard Warning Light System

J14 (A) . J15 (B)
JUNCTION CONNECTOR

J 3
JUNCTION CONNECTOR

Left kick panel

D 1
DATA LINK CONNECTOR 1

To Combination Meter

I 2
IGNITER

To Engine Control Module

I13
IGNITION COIL NO. 3

I12
IGNITION COIL NO. 2

I11
IGNITION COIL NO. 1

N 1
NOISE FILTER

ED
Intake manifold left

IE1

IG2
ST2
AM2
I23
IGNITION SW

30A
AM2

BATTERY

Engine ignition system - Tundra V6 models

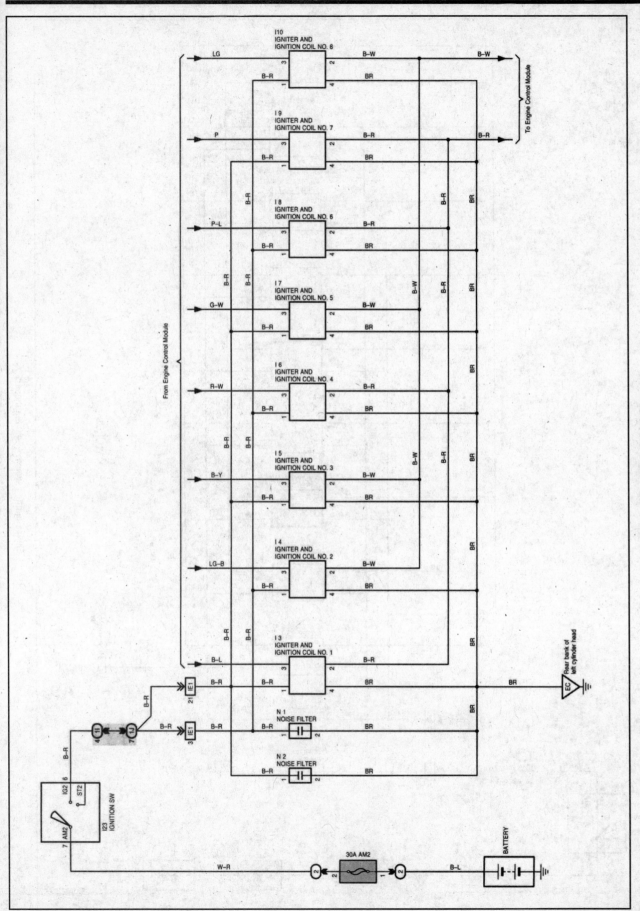

Engine ignition system - Tundra V8 models

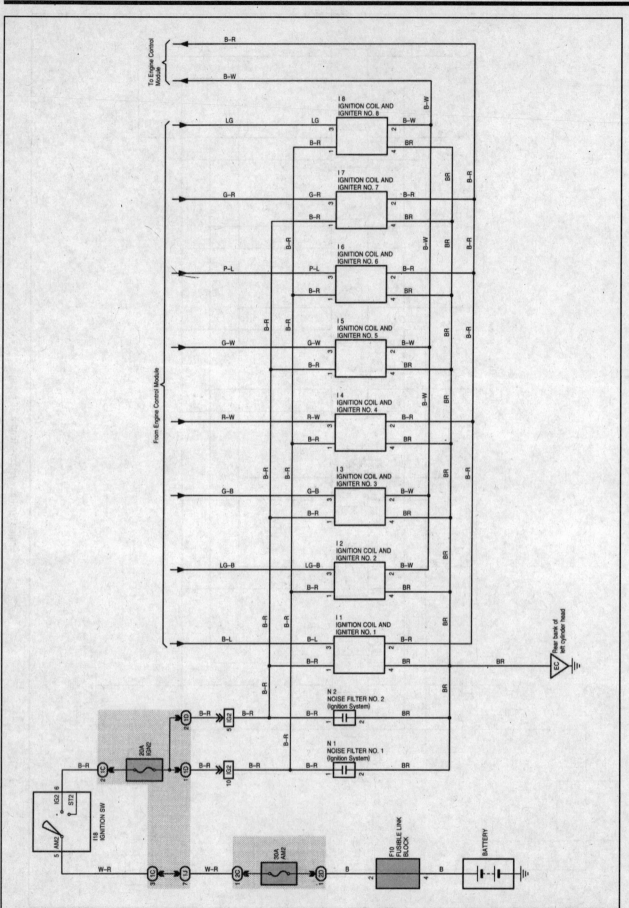

Engine ignition system - Sequoia models

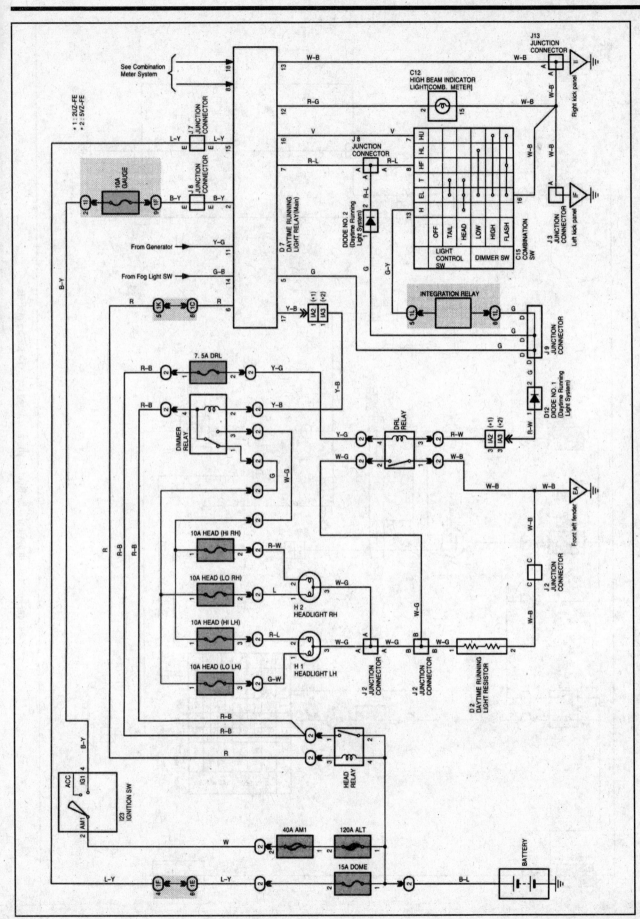

Headlight system (with Daytime Running Lights) - Tundra models

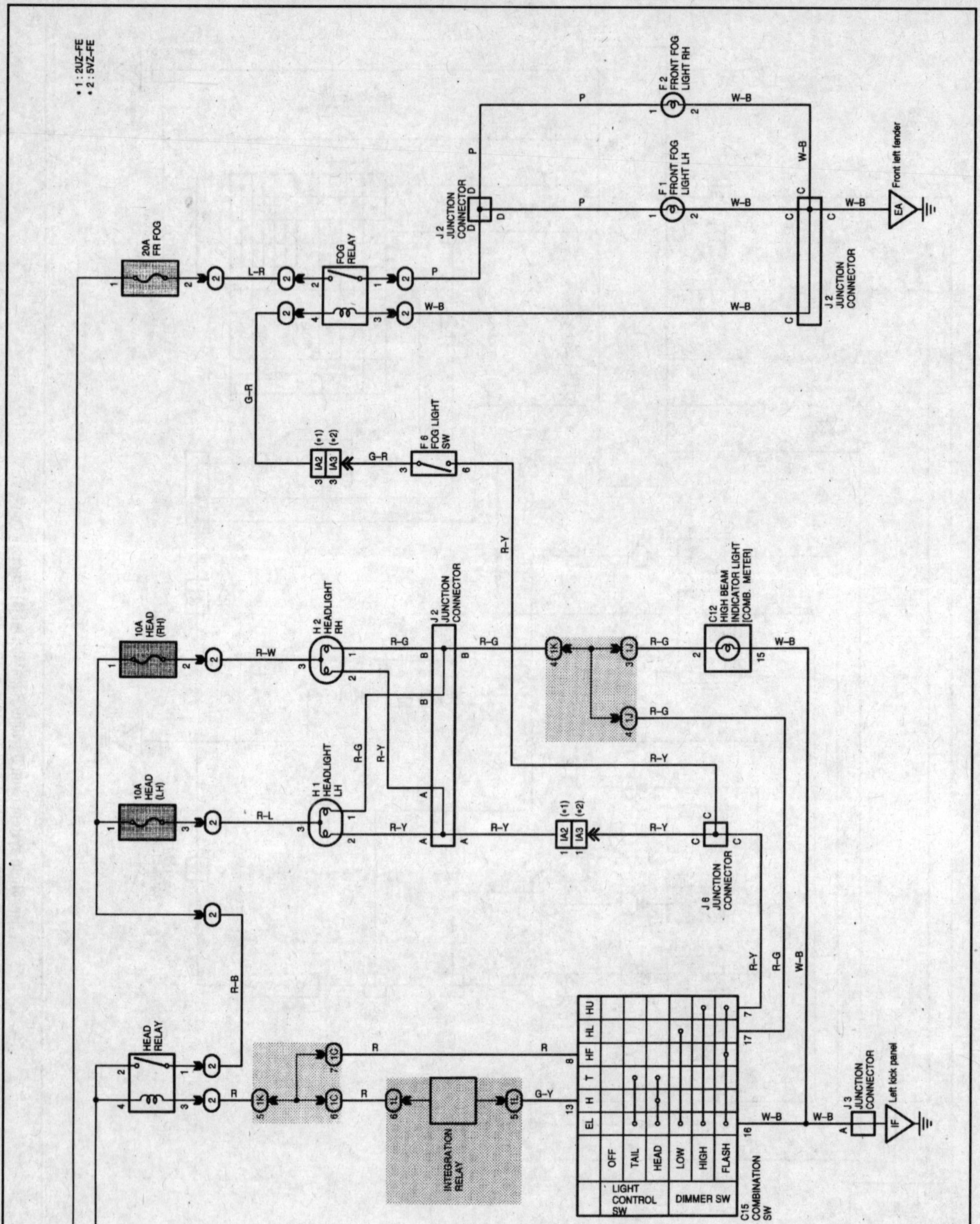

*1 : 2UZ-FE
*2 : 5VZ-FE

20A FR FOG

FOG RELAY

F 2 FRONT FOG LIGHT RH

FRONT FOG LIGHT LH

F 1 FRONT FOG LIGHT LH

J 2 JUNCTION CONNECTOR

J 2 JUNCTION CONNECTOR

EA Front left fender

L-R

P

W-B

G-R

IA2 (*1)
IA3 (*2)

F 6 FOG LIGHT SW

G-R

R-Y

10A HEAD (RH)

H 2 HEADLIGHT RH

R-W

R-G

J 2 JUNCTION CONNECTOR

C12 HIGH BEAM INDICATOR LIGHT (COMB. METER)

10A HEAD (LH)

H 1 HEADLIGHT LH

R-L

R-G

R-Y

IA2 (*1)
IA3 (*2)

J 8 JUNCTION CONNECTOR

HEAD RELAY

R-B

R

INTEGRATION RELAY

R

G-Y

C15 COMBINATION SW

	HU					
	HL					
	HF					
H	T					
EL						
	OFF	TAIL	HEAD	LOW	HIGH	FLASH
	LIGHT CONTROL SW			DIMMER SW		

J 3 JUNCTION CONNECTOR

IF Left kick panel

W-B

120A ALT

B-L

BATTERY

B-L

Headlight and foglight system - (without Daytime Running Lights) - Tundra models

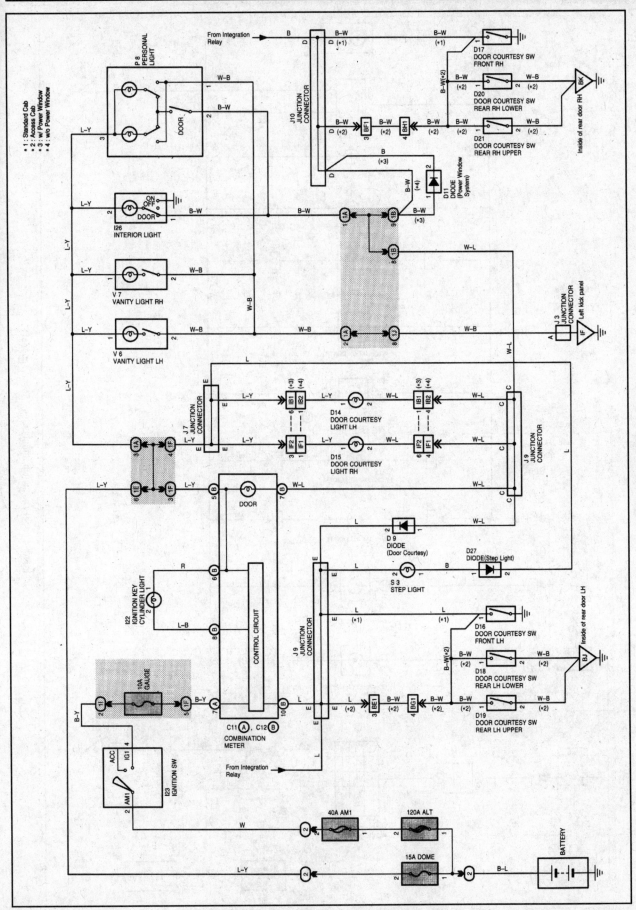

Interior light system - Tundra models

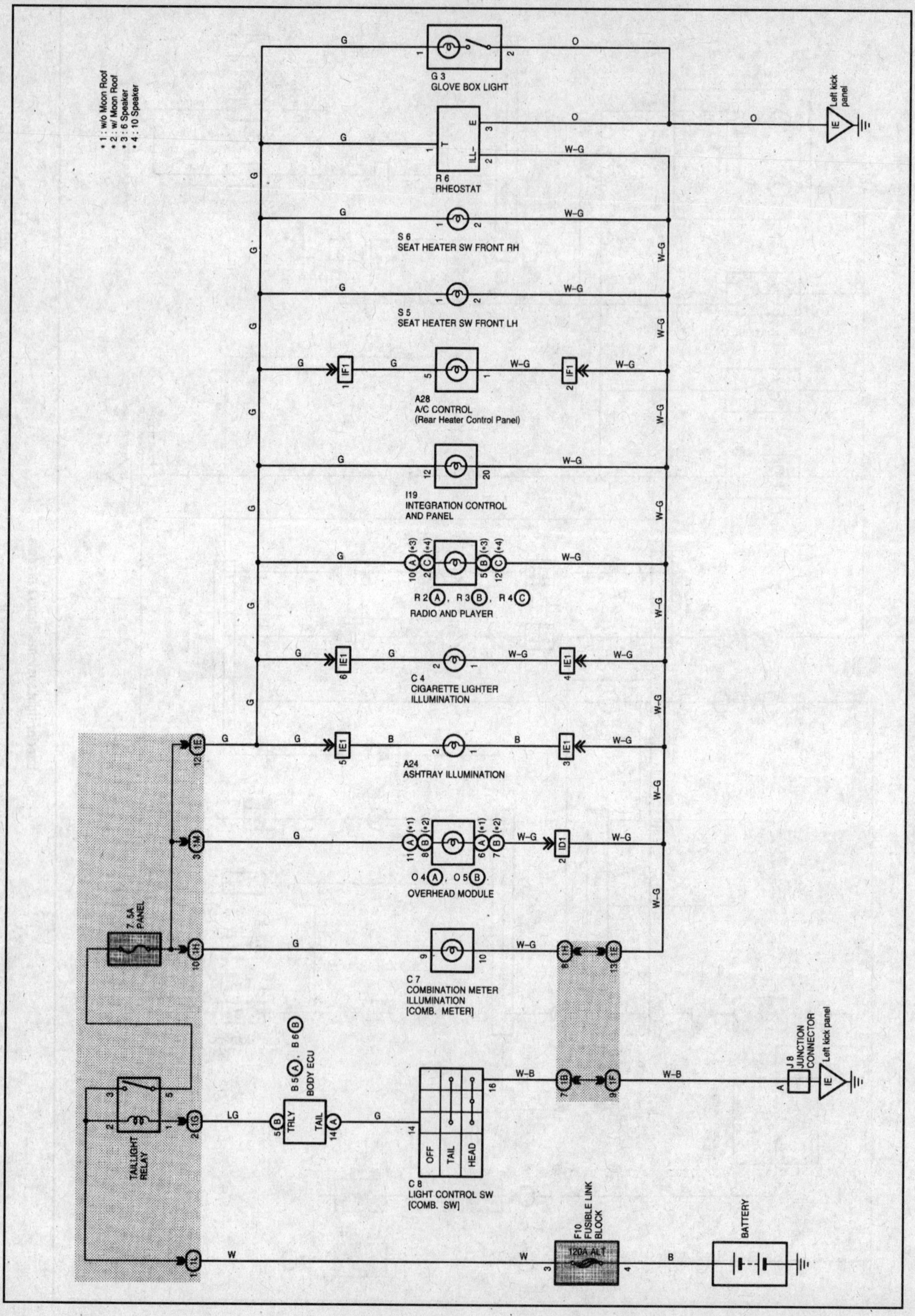

Instrument panel and switch illumination - Sequoia models

* 1 : w/o Moon Roof
* 2 : w/ Moon Roof
* 3 : 6 Speaker
* 4 : 10 Speaker

G 3
GLOVE BOX LIGHT

R 6
RHEOSTAT

S 6
SEAT HEATER SW FRONT RH

S 5
SEAT HEATER SW FRONT LH

A28
A/C CONTROL
(Rear Heater Control Panel)

I19
INTEGRATION CONTROL
AND PANEL

R 2 (A) , R 3 (B) , R 4 (C)
RADIO AND PLAYER

C 4
CIGARETTE LIGHTER
ILLUMINATION

A24
ASHTRAY ILLUMINATION

O 4 (A) , O 5 (B)
OVERHEAD MODULE

C 7
COMBINATION METER
ILLUMINATION
[COMB. METER]

B 5 (A) , B 6 (B)
BODY ECU

C 8
LIGHT CONTROL SW
[COMB. SW]

OFF
TAIL
HEAD

TAILLIGHT
RELAY

7.5A
PANEL

F10
FUSIBLE LINK
BLOCK

120A ALT

BATTERY

J 8
JUNCTION
CONNECTOR
Left kick panel

IE Left kick panel

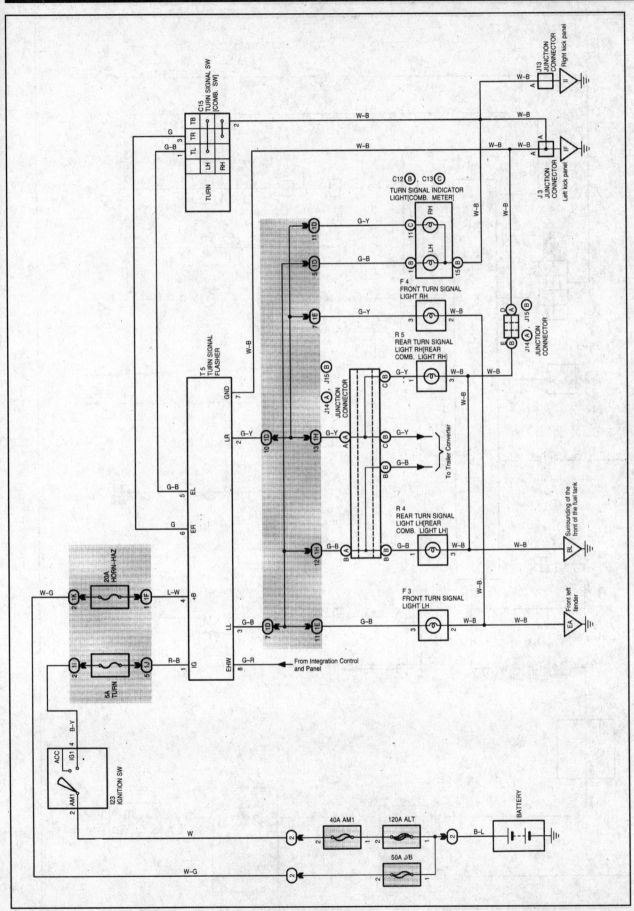

Turn signal and hazard light system - Tundra models

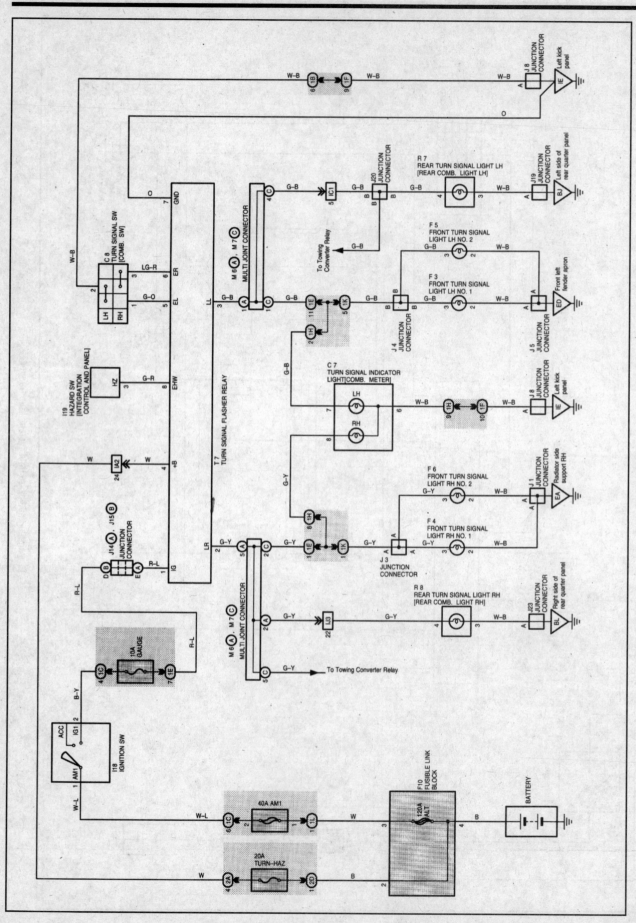

Turn signal and hazard light system - Sequoia models

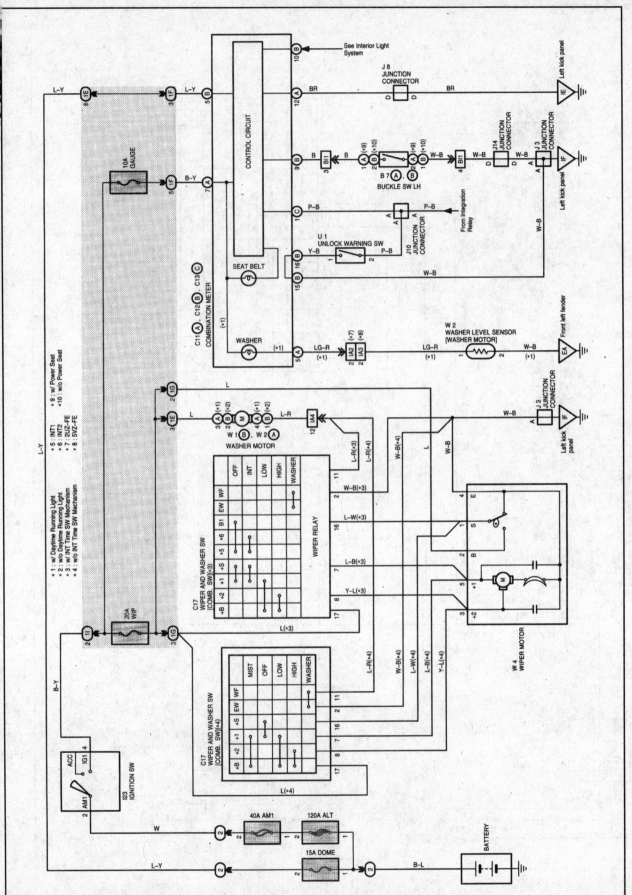

Front wiper/washer and key reminder systems - Tundra models

Front wiper/washer system - Sequoia models

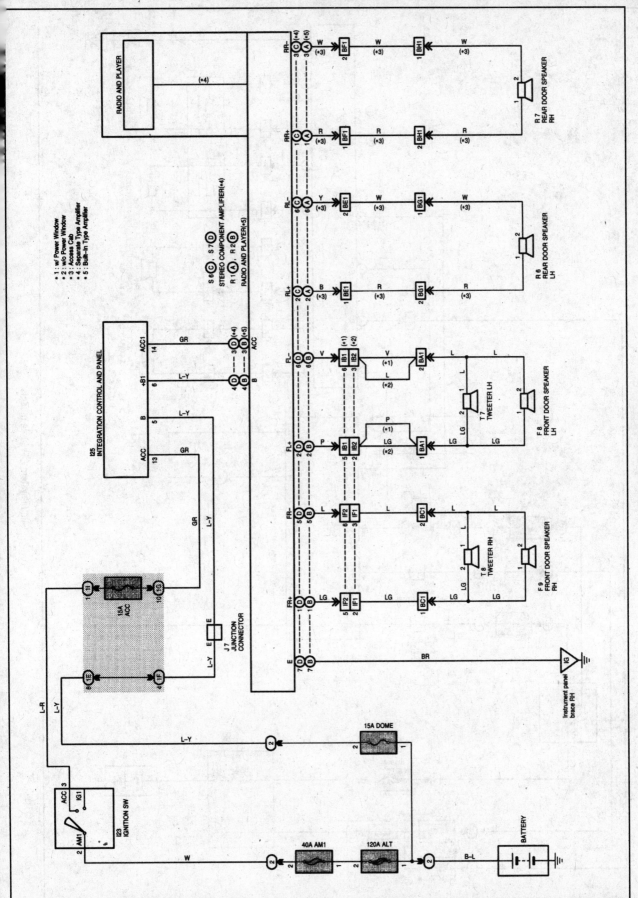

Audio system - Tundra models

Audio system - Sequoia models

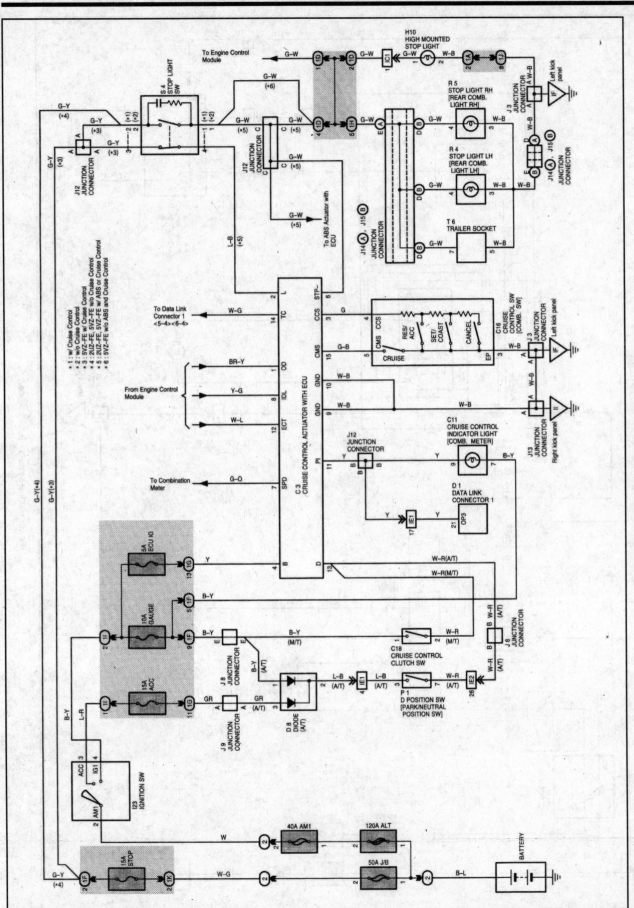

Cruise control and brake light systems - Tundra models

Power window system - Tundra models

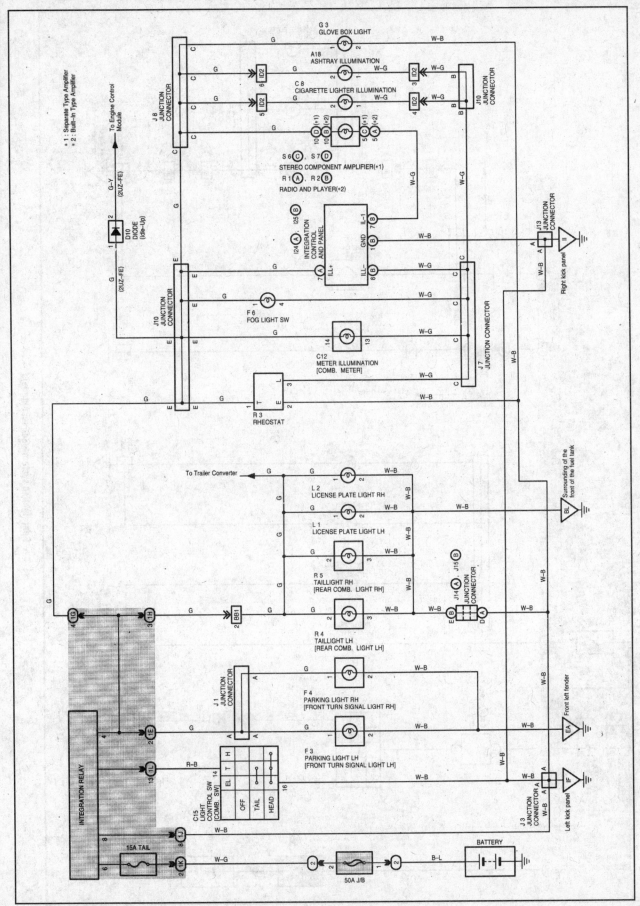

Rear lighting system - Tundra models

Rear lighting system - Sequoia models

GLOSSARY

AIR/FUEL RATIO: The ratio of air-to-gasoline by weight in the fuel mixture drawn into the engine.

AIR INJECTION: One method of reducing harmful exhaust emissions by injecting air into each of the exhaust ports of an engine. The fresh air entering the hot exhaust manifold causes any remaining fuel to be burned before it can exit the tailpipe.

ALTERNATOR: A device used for converting mechanical energy into electrical energy.

AMMETER: An instrument, calibrated in amperes, used to measure the flow of an electrical current in a circuit. Ammeters are always connected in series with the circuit being tested.

AMPERE: The rate of flow of electrical current present when one volt of electrical pressure is applied against one ohm of electrical resistance.

ANALOG COMPUTER: Any microprocessor that uses similar (analogous) electrical signals to make its calculations.

ARMATURE: A laminated, soft iron core wrapped by a wire that converts electrical energy to mechanical energy as in a motor or relay. When rotated in a magnetic field, it changes mechanical energy into electrical energy as in a generator.

ATMOSPHERIC PRESSURE: The pressure on the Earth's surface caused by the weight of the air in the atmosphere. At sea level, this pressure is 14.7 psi at 32°F (101 kPa at 0°C).

ATOMIZATION: The breaking down of a liquid into a fine mist that can be suspended in air.

AXIAL PLAY: Movement parallel to a shaft or bearing bore.

BACKFIRE: The sudden combustion of gases in the intake or exhaust system that results in a loud explosion.

BACKLASH: The clearance or play between two parts, such as meshed gears.

BACKPRESSURE: Restrictions in the exhaust system that slow the exit of exhaust gases from the combustion chamber.

BAKELITE: A heat resistant, plastic insulator material commonly used in printed circuit boards and transistorized components.

BALL BEARING: A bearing made up of hardened inner and outer races between which hardened steel balls roll.

BALLAST RESISTOR: A resistor in the primary ignition circuit that lowers voltage after the engine is started to reduce wear on ignition components.

BEARING: A friction reducing, supportive device usually located between a stationary part and a moving part.

BIMETAL TEMPERATURE SENSOR: Any sensor or switch made of two dissimilar types of metal that bend when heated or cooled due to the different expansion rates of the alloys. These types of sensors usually function as an on/off switch.

BLOWBY: Combustion gases, composed of water vapor and unburned fuel, that leak past the piston rings into the crankcase during normal engine operation. These gases are removed by the PCV system to prevent the buildup of harmful acids in the crankcase.

BRAKE PAD: A brake shoe and lining assembly used with disc brakes.

BRAKE SHOE: The backing for the brake lining. The term is, however, usually applied to the assembly of the brake backing and lining.

BUSHING: A liner, usually removable, for a bearing; an anti-friction liner used in place of a bearing.

CALIPER: A hydraulically activated device in a disc brake system, which is mounted straddling the brake rotor (disc). The caliper contains at least one piston and two brake pads. Hydraulic pressure on the piston(s) forces the pads against the rotor.

CAMSHAFT: A shaft in the engine on which are the lobes (cams) which operate the valves. The camshaft is driven by the crankshaft, via a belt, chain or gears, at one half the crankshaft speed.

CAPACITOR: A device which stores an electrical charge.

CARBON MONOXIDE (CO): A colorless, odorless gas given off as a normal byproduct of combustion. It is poisonous and extremely dangerous in confined areas, building up slowly to toxic levels without warning if adequate ventilation is not available.

CARBURETOR: A device, usually mounted on the intake manifold of an engine, which mixes the air and fuel in the proper proportion to allow even combustion.

CATALYTIC CONVERTER: A device installed in the exhaust system, like a muffler, that converts harmful byproducts of combustion into carbon dioxide and water vapor by means of a heat-producing chemical reaction.

CENTRIFUGAL ADVANCE: A mechanical method of advancing the spark timing by using flyweights in the distributor that react to centrifugal force generated by the distributor shaft rotation.

CHECK VALVE: Any one-way valve installed to permit the flow of air, fuel or vacuum in one direction only.

CHOKE: A device, usually a moveable valve, placed in the intake path of a carburetor to restrict the flow of air.

CIRCUIT: Any unbroken path through which an electrical current can flow. Also used to describe fuel flow in some instances.

CIRCUIT BREAKER: A switch which protects an electrical circuit from overload by opening the circuit when the current flow exceeds a predetermined level. Some circuit breakers must be reset manually, while most reset automatically.

COIL (IGNITION): A transformer in the ignition circuit which steps up the voltage provided to the spark plugs.

COMBINATION MANIFOLD: An assembly which includes both the intake and exhaust manifolds in one casting.

COMBINATION VALVE: A device used in some fuel systems that routes fuel vapors to a charcoal storage canister instead of venting them into the atmosphere. The valve relieves fuel tank pressure and allows fresh air into the tank as the fuel level drops to prevent a vapor lock situation.

COMPRESSION RATIO: The comparison of the total volume of the cylinder and combustion chamber with the piston at BDC and the piston at TDC.

CONDENSER: 1. An electrical device which acts to store an electrical charge, preventing voltage surges. 2. A radiator-like device in the air conditioning system in which refrigerant gas condenses into a liquid, giving off heat.

CONDUCTOR: Any material through which an electrical current can be transmitted easily.

CONTINUITY: Continuous or complete circuit. Can be checked with an ohmmeter.

COUNTERSHAFT: An intermediate shaft which is rotated by a mainshaft and transmits, in turn, that rotation to a working part.

CRANKCASE: The lower part of an engine in which the crankshaft and related parts operate.

CRANKSHAFT: The main driving shaft of an engine which receives reciprocating motion from the pistons and converts it to rotary motion.

CYLINDER: In an engine, the round hole in the engine block in which the piston(s) ride.

CYLINDER BLOCK: The main structural member of an engine in which is found the cylinders, crankshaft and other principal parts.

CYLINDER HEAD: The detachable portion of the engine, usually fastened to the top of the cylinder block and containing all or most of the combustion chambers. On overhead valve engines, it contains the valves and their operating parts. On overhead cam engines, it contains the camshaft as well.

DEAD CENTER: The extreme top or bottom of the piston stroke.

DETONATION: An unwanted explosion of the air/fuel mixture in the combustion chamber caused by excess heat and compression, advanced timing, or an overly lean mixture. Also referred to as "ping".

DIAPHRAGM: A thin, flexible wall separating two cavities, such as in a vacuum advance unit.

DIESELING: A condition in which hot spots in the combustion chamber cause the engine to run on after the key is turned off.

DIFFERENTIAL: A geared assembly which allows the transmission of motion between drive axles, giving one axle the ability to turn faster than the other.

DIODE: An electrical device that will allow current to flow in one direction only.

DISC BRAKE: A hydraulic braking assembly consisting of a brake disc, or rotor, mounted on an axle, and a caliper assembly containing, usually two brake pads which are activated by hydraulic pressure. The pads are forced against the sides of the disc, creating friction which slows the vehicle.

DISTRIBUTOR: A mechanically driven device on an engine which is responsible for electrically firing the spark plug at a predetermined point of the piston stroke.

DOWEL PIN: A pin, inserted in mating holes in two different parts allowing those parts to maintain a fixed relationship.

DRUM BRAKE: A braking system which consists of two brake shoes and one or two wheel cylinders, mounted on a fixed backing plate, and a brake drum, mounted on an axle, which revolves around the assembly.

DWELL: The rate, measured in degrees of shaft rotation, at which an electrical circuit cycles on and off.

ELECTRONIC CONTROL UNIT (ECU): Ignition module, module, amplifier or igniter. See Module for definition.

ELECTRONIC IGNITION: A system in which the timing and firing of the spark plugs is controlled by an electronic control unit, usually called a module. These systems have no points or condenser.

END-PLAY: The measured amount of axial movement in a shaft.

ENGINE: A device that converts heat into mechanical energy.

EXHAUST MANIFOLD: A set of cast passages or pipes which conduct exhaust gases from the engine.

FEELER GAUGE: A blade, usually metal, or precisely predetermined thickness, used to measure the clearance between two parts.

FIRING ORDER: The order in which combustion occurs in the cylinders of an engine. Also the order in which spark is distributed to the plugs by the distributor.

FLOODING: The presence of too much fuel in the intake manifold and combustion chamber which prevents the air/fuel mixture from firing, thereby causing a no-start situation.

FLYWHEEL: A disc shaped part bolted to the rear end of the crankshaft. Around the outer perimeter is affixed the ring gear. The starter drive engages the ring gear, turning the flywheel, which rotates the crankshaft, imparting the initial starting motion to the engine.

FOOT POUND (ft. lbs. or sometimes, ft.lb.): The amount of energy or work needed to raise an item weighing one pound, a distance of one foot.

FUSE: A protective device in a circuit which prevents circuit overload by breaking the circuit when a specific amperage is present. The device is constructed around a strip or wire of a lower amperage rating than the circuit it is designed to protect. When an amperage higher than that stamped on the fuse is present in the circuit, the strip or wire melts, opening the circuit.

GEAR RATIO: The ratio between the number of teeth on meshing gears.

GENERATOR: A device which converts mechanical energy into electrical energy.

HEAT RANGE: The measure of a spark plug's ability to dissipate heat from its firing end. The higher the heat range, the hotter the plug fires.

HUB: The center part of a wheel or gear.

HYDROCARBON (HC): Any chemical compound made up of hydrogen and carbon. A major pollutant formed by the engine as a byproduct of combustion.

HYDROMETER: An instrument used to measure the specific gravity of a solution.

INCH POUND (inch lbs.; sometimes in.lb. or in. lbs.): One twelfth of a foot pound.

INDUCTION: A means of transferring electrical energy in the form of a magnetic field. Principle used in the ignition coil to increase voltage.

INJECTOR: A device which receives metered fuel under relatively low pressure and is activated to inject the fuel into the engine under relatively high pressure at a predetermined time.

INPUT SHAFT: The shaft to which torque is applied, usually carrying the driving gear or gears.

INTAKE MANIFOLD: A casting of passages or pipes used to conduct air or a fuel/air mixture to the cylinders.

JOURNAL: The bearing surface within which a shaft operates.

KEY: A small block usually fitted in a notch between a shaft and a hub to prevent slippage of the two parts.

MANIFOLD: A casting of passages or set of pipes which connect the cylinders to an inlet or outlet source.

MANIFOLD VACUUM: Low pressure in an engine intake manifold formed just below the throttle plates. Manifold vacuum is highest at idle and drops under acceleration.

MASTER CYLINDER: The primary fluid pressurizing device in a hydraulic system. In automotive use, it is found in brake and hydraulic clutch systems and is pedal activated, either directly or, in a power brake system, through the power booster.

MODULE: Electronic control unit, amplifier or igniter of solid state or integrated design which controls the current flow in the ignition primary circuit based on input from the pick-up coil. When the module opens the primary circuit, high secondary voltage is induced in the coil.

NEEDLE BEARING: A bearing which consists of a number (usually a large number) of long, thin rollers.

OHM: (Ω) The unit used to measure the resistance of conductor-to-electrical flow. One ohm is the amount of resistance that limits current flow to one ampere in a circuit with one volt of pressure.

OHMMETER: An instrument used for measuring the resistance, in ohms, in an electrical circuit.

OUTPUT SHAFT: The shaft which transmits torque from a device, such as a transmission.

OVERDRIVE: A gear assembly which produces more shaft revolutions than that transmitted to it.

OVERHEAD CAMSHAFT (OHC): An engine configuration in which the camshaft is mounted on top of the cylinder head and operates the valve either directly or by means of rocker arms.

OVERHEAD VALVE (OHV): An engine configuration in which all of the valves are located in the cylinder head and the camshaft is located in the cylinder block. The camshaft operates the valves via lifters and pushrods.

OXIDES OF NITROGEN (NOx): Chemical compounds of nitrogen produced as a byproduct of combustion. They combine with hydrocarbons to produce smog.

OXYGEN SENSOR: Use with the feedback system to sense the presence of oxygen in the exhaust gas and signal the computer which can reference the voltage signal to an air/fuel ratio.

PINION: The smaller of two meshing gears.

PISTON RING: An open-ended ring with fits into a groove on the outer diameter of the piston. Its chief function is to form a seal between the piston and cylinder wall. Most automotive pistons have three rings: two for compression sealing; one for oil sealing.

PRELOAD: A predetermined load placed on a bearing during assembly or by adjustment.

PRIMARY CIRCUIT: the low voltage side of the ignition system which consists of the ignition switch, ballast resistor or resistance wire, bypass, coil, electronic control unit and pick-up coil as well as the connecting wires and harnesses.

PRESS FIT: The mating of two parts under pressure, due to the inner diameter of one being smaller than the outer diameter of the other, or vice versa; an interference fit.

RACE: The surface on the inner or outer ring of a bearing on which the balls, needles or rollers move.

REGULATOR: A device which maintains the amperage and/or voltage levels of a circuit at predetermined values.

RELAY: A switch which automatically opens and/or closes a circuit.

RESISTANCE: The opposition to the flow of current through a circuit or electrical device, and is measured in ohms. Resistance is equal to the voltage divided by the amperage.

RESISTOR: A device, usually made of wire, which offers a preset amount of resistance in an electrical circuit.

RING GEAR: The name given to a ring-shaped gear attached to a differential case, or affixed to a flywheel or as part of a planetary gear set.

ROLLER BEARING: A bearing made up of hardened inner and outer races between which hardened steel rollers move.

ROTOR: 1. The disc-shaped part of a disc brake assembly, upon which the brake pads bear; also called, brake disc. 2. The device mounted atop the distributor shaft, which passes current to the distributor cap tower contacts.

SECONDARY CIRCUIT: The high voltage side of the ignition system, usually above 20,000 volts. The secondary includes the ignition coil, coil wire, distributor cap and rotor, spark plug wires and spark plugs.

SENDING UNIT: A mechanical, electrical, hydraulic or electromagnetic device which transmits information to a gauge.

SENSOR: Any device designed to measure engine operating conditions or ambient pressures and temperatures. Usually electronic in nature and designed to send a voltage signal to an on-board computer, some sensors may operate as a simple on/off switch or they may provide a variable voltage signal (like a potentiometer) as conditions or measured parameters change.

SHIM: Spacers of precise, predetermined thickness used between parts to establish a proper working relationship.

SLAVE CYLINDER: In automotive use, a device in the hydraulic clutch system which is activated by hydraulic force, disengaging the clutch.

SOLENOID: A coil used to produce a magnetic field, the effect of which is to produce work.

SPARK PLUG: A device screwed into the combustion chamber of a spark ignition engine. The basic construction is a conductive core inside of a ceramic insulator, mounted in an outer conductive base. An electrical charge from the spark plug wire travels along the conductive core and jumps a preset air gap to a grounding point or points at the end of the conductive base. The resultant spark ignites the fuel/air mixture in the combustion chamber.

SPLINES: Ridges machined or cast onto the outer diameter of a shaft or inner diameter of a bore to enable parts to mate without rotation.

TACHOMETER: A device used to measure the rotary speed of an engine, shaft, gear, etc., usually in rotations per minute.

THERMOSTAT: A valve, located in the cooling system of an engine, which is closed when cold and opens gradually in response to engine heating, controlling the temperature of the coolant and rate of coolant flow.

TOP DEAD CENTER (TDC): The point at which the piston reaches the top of its travel on the compression stroke.

TORQUE: The twisting force applied to an object.

TORQUE CONVERTER: A turbine used to transmit power from a driving member to a driven member via hydraulic action, providing changes in drive ratio and torque. In automotive use, it links the driveplate at the rear of the engine to the automatic transmission.

TRANSDUCER: A device used to change a force into an electrical signal.

TRANSISTOR: A semi-conductor component which can be actuated by a small voltage to perform an electrical switching function.

TUNE-UP: A regular maintenance function, usually associated with the replacement and adjustment of parts and components in the electrical and fuel systems of a vehicle for the purpose of attaining optimum performance.

TURBOCHARGER: An exhaust driven pump which compresses intake air and forces it into the combustion chambers at higher than atmospheric pressures. The increased air pressure allows more fuel to be burned and results in increased horsepower being produced.

VACUUM ADVANCE: A device which advances the ignition timing in response to increased engine vacuum.

VACUUM GAUGE: An instrument used to measure the presence of vacuum in a chamber.

VALVE: A device which control the pressure, direction of flow or rate of flow of a liquid or gas.

VALVE CLEARANCE: The measured gap between the end of the valve stem and the rocker arm, cam lobe or follower that activates the valve.

VISCOSITY: The rating of a liquid's internal resistance to flow.

VOLTMETER: An instrument used for measuring electrical force in units called volts. Voltmeters are always connected parallel with the circuit being tested.

WHEEL CYLINDER: Found in the automotive drum brake assembly, it is a device, actuated by hydraulic pressure, which, through internal pistons, pushes the brake shoes outward against the drums.

MASTER INDEX

A

Notes